FUNDAMENTALS OF BUSINESS STATISTICS

INTERNATIONAL EDITION

FUNDAMENTALS OF BUSINESS STATISTICS

JEROME D. BRAVERMAN

Rider College

INTERNATIONAL EDITION

ACADEMIC PRESS

New York
San Francisco
London
A Subsidiary of Harcourt Brace Jovanovich, Publishers

INTERNATIONAL EDITION

ACADEMIC PRESS, INC.
111 Fifth Avenue,
New York, New York 10003

United Kingdom Edition published by
ACADEMIC PRESS, INC. (LONDON) LTD.
24/28 Oval Road, London NW1

ISBN: 0-12-128030-0
Library of Congress Catalog Card Number: 78-3328

Printed in the United States of America

Cover art by Guy Billout

Chapter opening art by
The Idea Factory

To My Father

CONTENTS

Chapter 1 **1**
Introduction to Business Statistics

1.1 Descriptive Statistics **3**
1.2 Statistical Inference **4**
1.3 Origins of Statistical Data **8**
1.4 Types of Statistical Data **9**
1.5 Statistics and the Computer **10**

Chapter 2 **13**
A Review of Prerequisites

2.1 Terminology and Notation **14**
2.2 Significant Digits and Rounding **19**
2.3 Logarithms **24**
2.4 Sets and Operations with Sets **27**
2.5 Permutations and Combinations **34**
2.6 Functions **39**
2.7 Looking Ahead **42**

Chapter 3 **51**
Data Organization

3.1 Frequency Distributions **53**
3.2 Special Frequency Distributions **61**
3.3 Categorical Distributions **64**
3.4 Histograms **65**
3.5 Practical Uses of Frequency Distributions and Histograms **67**

3.6 Other Statistical Graphs **69**
3.7 General Purpose Tables **71**

Chapter 4 **79**
Descriptive Measures—Measures of Location
4.1 Measures of Location **80**
4.2 The Mode **80**
4.3 The Median **82**
4.4 The Arithmetic Mean **86**
4.5 Comparing the Mean, the Median, and the Mode **94**
4.6 The Geometric Mean **96**
4.7 Other Measures of Location **97**

Chapter 5 **107**
Descriptive Measures—Variation and Skewness
5.1 The Need for Describing Variation **108**
5.2 The Range **108**
5.3 The Variance and the Standard Deviation **109**
5.4 Relative Variation **116**
5.5 A Measure of Skewness **118**

Chapter 6 **127**
Index Numbers
6.1 Construction of a Simple Relative **128**
6.2 Some Common Business Index Numbers **130**
6.3 Problems of Index Number Construction **131**
6.4 Price Index Numbers **133**
6.5 Quantity Index Numbers **143**
6.6 Other Business Applications of Index Numbers **145**

Chapter 7 **153**
Probability
7.1 The Nature of Probability **154**
7.2 Some Important Definitions **155**
7.3 Basic Rules of Probability **160**
7.4 Conditional Probability **163**
7.5 Statistical Independence **168**
7.6 The Rule of Elimination **170**
7.7 Bayes' Rule **172**

Chapter 8 181
Probability Distributions
8.1 Random Variables 182
8.2 The Probability Function of a Random Variable 184
8.3 Expected Value of a Random Variable 187
8.4 Variance of a Probability Distribution 189
8.5 Discrete Probability Distributions 191
8.6 Continuous Probability Distributions 201

Chapter 9 217
Sampling and Sampling Distributions
9.1 The Rationale for Sampling 218
9.2 Sample Statistics 219
9.3 Random Sampling 222
9.4 Sampling Distributions 227
9.5 The Sampling Distribution of the Mean 229
9.6 Sampling Distributions and Probabilities 233
9.7 Student's *t* Distribution 236
9.8 The Sampling Distribution of the Proportion and the Sample Number 238

Chapter 10 247
Estimation
10.1 The Estimation Process 248
10.2 Estimators and Their Properties 250
10.3 Point and Interval Estimates 253
10.4 One- and Two-Sided Confidence Intervals 256
10.5 Estimation of Means 260
10.6 Estimation of Proportions 265
10.7 Estimating Variation 270
10.8 Estimating Sample Size 272

Chapter 11 285
Hypothesis Testing—One Sample Tests
11.1 The Rationale for Hypothesis Testing 287
11.2 A General Procedure for Hypothesis Testing 289
11.3 One-Tailed and Two-Tailed Tests 292
11.4 Errors in Hypothesis Testing 294
11.5 Tests about Means—Large Samples 296
11.6 Tests about Means—Small Samples 299
11.7 Tests about Proportions 302
11.8 Calculating the Probability of the Type II Error 306
11.9 Testing Hypotheses with Critical Values of *z* and *t* 312

Chapter 12 321
Hypothesis Testing—Two Sample Tests
12.1 The Test Statistic for Tests of Differences of Means and Proportions **323**
12.2 Test of Difference between Means—Large Samples **324**
12.3 Test of Difference between Means—Small Samples **326**
12.4 Test of Difference between Proportions **328**
12.5 Tests Using Paired Observations **330**
12.6 The *F* Distribution **333**
12.7 Test of the Equality of Variances **334**

Chapter 13 343
Analysis of Variance; Multiple-Sample Tests of Means
13.1 Notation for Analysis of Variance **345**
13.2 The Rationale for Analysis of Variance **347**
13.3 Sums of Squares and the ANOVA Table **350**
13.4 One-Factor Analysis of Variance **354**
13.5 Least Significant Difference **359**
13.6 Two-Factor Analysis of Variance **361**

Chapter 14 377
Chi-Square Tests
14.1 The Chi-Square Distribution **378**
14.2 The Chi-Square Statistic **379**
14.3 Tests of Goodness-of-Fit **381**
14.4 Tests of Independence—Contingency Tables **389**

Chapter 15 401
Nonparametric Statistical Methods
15.1 The Runs Test for Randomness **402**
15.2 The Sign Test **407**
15.3 The Rank Sum Test for Two Samples **409**
15.4 The Rank Sum Test for Three or More Samples: The Kruskal–Wallace Test **413**
15.5 Rank Correlation **416**

Chapter 16 427
Regression Analysis
16.1 Correlation **429**
16.2 Inferences about the Correlation Coefficient **435**
16.3 Simple Linear Regression **436**
16.4 Confidence Intervals for Predictions **447**
16.5 Multiple Regression **452**

Chapter 17 467
Time Series
17.1 The Time Series Model **469**
17.2 Trend **472**
17.3 The Cycle **476**
17.4 The Seasonal Index **481**
17.5 Forecasting **487**

Chapter 18 493
Introduction to Statistical Decision Theory
18.1 Decision Making under Uncertainty **495**
18.2 States of Nature **496**
18.3 Notation Used in Statistical Decision Theory **497**
18.4 Payoff Tables and Functions **497**
18.5 Prior Distributions and Bayes' Criterion **501**
18.6 Opportunity Loss **504**
18.7 Decision Trees **507**
18.8 The Expected Value of Perfect Information **509**
18.9 Sampling and Revising the Prior Distribution **511**
18.10 Preposterior Analysis **516**
18.11 The Expected Value of Sample Information **517**

Glossary 527

Answers to Even-Numbered Problems 535

Appendixes 553

Index 589

PREFACE

Business programs in both two- and four-year colleges have been growing rapidly in recent years. Since most of these programs require one and sometimes two semesters of statistics, there appears to be a need for a textbook suitable for both one- and two-semester courses that covers all of the basic statistical techniques and their business applications in a concise, straightforward, and simple manner. My goal is to provide such a book. A second objective is to make the study of statistics interesting by including a large number of examples that illustrate how statistical techniques can actually be used in a wide variety of business situations. The sooner the student realizes that statistics can really be useful in business, the more likely he or she will become motivated to master the subject matter.

Because of its intended audience, the mathematical prerequisites for this book are minimal. Essentially, they are limited to those topics covered in a course in high school algebra. Chapter 2 provides a brief review of these prerequisites. For students with the necessary background, most of Chapter 2 can be omitted. I have, however, deviated slightly from the usual presentation of topics by covering permutations and combinations in this review chapter rather than as part of the introductory probability chapter. My rationale is that permutations and combinations are often included in algebra courses, and many students may already be conversant with the topic. For students without this background the instructor should review this portion of Chapter 2 before proceeding with probability.

This text has sufficient material for a two-semester sequence in statistics. For a one-semester course there are, of course, more options regarding the inclusion and exclusion of topics. Descriptive statistics and the classical approach to statistical inference are covered thoroughly with an introduction to Bayesian inference in the last chapter. I chose to concentrate on the classical approach for two reasons. First, I believe that an understanding of classical statistics is necessary before the student can fully appreciate the Bayesian technique. Second, it has been my experience that practically all of the statistical methods actually employed in business today are based on the classical methodology. Consequently, it is this approach that should prove to be of the greatest benefit to today's business student.

Each chapter begins with a brief introduction to the topic covered and the rationale for its inclusion and closes with a summary of the important terms and all of the formulas introduced in the chapter. A glossary of terms is also included as part of the appendix. All of the appendix tables were selected or constructed with a view toward easy usage. The use of all of the statistical tables is explained in the text as well as by example in the appendix.

Now a final word about notation. Unfortunately, a standard set of statistical symbols does not exist. As a result, notation varies among books written by different authors. In this book I have tried to use symbols that are common and generally accepted. It is inevitable, however, that my notation will differ from that used in some other books. Regarding such differences, my choice was to keep the notation as simple as possible and the number of different symbols to a minimum.

I would like to express my appreciation to the staff of Academic Press for their valuable assistance in the preparation of this manuscript. Also, for their constructive advice and professional comments, I am grateful as well to Professors Marvin J. Karson of the University of Alabama and Otto W. Taylor of the New York Agricultural and Technical College, State University of Farmingdale, New York.

FUNDAMENTALS OF BUSINESS STATISTICS

INTERNATIONAL EDITION

CHAPTER 1 INTRODUCTION TO BUSINESS STATISTICS

1.1
Descriptive statistics

1.2
Statistical inference

1.3
Origins of statistical data

1.4
Types of statistical data

1.5
Statistics and the computer

Why should all business students be required to study statistics? This is a logical question, frequently asked by students enrolled in programs in business administration. After all, statistics is a branch of applied mathematics, and business people do not have to be mathematicians, do they? Of course they don't. However, business people do have to make decisions, and the more tools they have at their disposal to help them make good decisions, the more likely they are to succeed in business. Statistics, as you will soon discover, is concerned with the problem of obtaining useful information from numerical data, that is, from large sets of numbers. In almost any type of business, people have always had to work with numbers, because numbers result naturally and logically from all sorts of business transactions.

Numbers represent prices and quantities of products produced and sold, hours on time cards, and accrued sick leave. Numbers stand for taxes collected or paid, dollars of sales and returns, and delinquent accounts. Numbers also represent inventories, customers, raw materials, costs and profits, assets and liabilities, and income and expenses. In short, information about every possible business transaction can be represented numerically. Whenever numbers are collected and compiled, regardless of what they represent, they become statistics.

The word STATISTICS has more than one meaning, as do many other words in the English language. One definition states that *a group of statistics is simply a set of numbers or numerical data.* Unfortunately, numbers, particularly when there are a lot of them, can be difficult to comprehend or interpret. Statistics, by another definition, refers to *the procedures employed for extracting or deriving information from sets of numerical data.*

Business people, particularly managers, are frequently confronted with large quantities of numerical data or statistics. These data contain information that may reflect the current status of some operational unit, such as a retail store, a brokerage office, or a department in a manufacturing firm. They may also reflect the status of some critical operating factor within the unit, such as sales, debt, absenteeism, or costs of one kind or another. The manager may be required to make a decision based on this information that will affect the future status of the organization. The information might be in the form of reports or computer printouts, or it might simply consist of records that are kept in ledgers or other record books, or in file folders in a filing cabinet. The ability of the manager to extract pertinent information from the data and use it in making decisions can have a significant effect on his or her own future as well as that of the organization.

1.1 Descriptive Statistics

Numerical data at the disposal of a manager can be used to describe the status or condition of the organization or some part of the organization for which he or she is responsible. For example, consider the case of the credit manager of a large department store. The data might consist of a list of all delinquent accounts and the balance due on each. The list might also contain information regarding the number of days, weeks, or months that each account is past due. A list such as this one might have been generated by a computer, or it might have been prepared by a clerk who simply examined each folder in a file of accounts receivable and noted on a tablet the name and number of each delinquent account, its dollar balance, and the number of days past due. The important information about each account is the number of dollars of the balance and the number of days past due. However, whether generated by a computer or prepared by hand, the accounts might be listed according to the name of the account holder or by the identifying number of the account. In either case, the order in which the accounts are listed may have absolutely no relation to the important information about the size of the balance or the period of the delinquency. Faced with a list of, say, a thousand such accounts, the credit manager would have a difficult time obtaining useful information regarding the general status of the entire set of delinquent accounts. However, there are methods for rearranging, grouping, and summarizing sets of data to obtain information about the entire set that can be used by the manager and easily communicated to others. These methods belong to a general category of statistical techniques called DESCRIPTIVE STATISTICS.

The methods of statistical description are used for a definite purpose. Descriptive measures provide the manager with information for making decisions. In the case of the delinquent accounts, the manager may decide to turn over all accounts more than four months past due to a collection agency. He may decide that collection letters should be sent to past due accounts with balances in excess of \$100. He might also decide that the total amount of debt represented by accounts with balances less than \$10 is so small that the collection costs would be greater than the amount recovered. In that case the credit manager might decide to write those accounts off as bad debts. However, before he can make any of these decisions, he must first know the current status of these accounts. The techniques of descriptive statistics help us determine exactly where we are by providing information in intelligible and useable form.

Everyone who reads newspapers and magazines has seen descriptive statistics. It is almost impossible to pick up a newspaper or news magazine without being confronted with some statistical summary, whether it be a chart, a graph, a statistical table, or simply a descriptive measure such as an average. Stock market averages, such as the Dow-Jones average, appear in the newspaper every day. Charts comparing increases in the cost of living among various cities or countries are common. Changes in the rate of inflation are frequently illustrated using different types of charts and graphs. These devices summarize large quantities of numerical data for easy comprehension by the reader. Various types of averages, for instance, can reduce a large mass of data to a single, descriptive number. However, statistical summaries must be used and interpreted with care because they can be, and often are, misused and abused. For example, the use of the wrong kind of average can be extremely misleading. Also, the use of an average value alone to describe a set of statistical data can often lead to erroneous conclusions. The complacent sales manager who believes that he has a very effective sales force because the average weekly sales per individual is $2000 may be mistaken. It might be that half of the staff is working hard and producing sales of $3500 per week each, while the other half is producing weekly sales of only $500 each. Average sales figures tell us nothing whatsoever about the range of individual performances. Think of the improvement that could be achieved if all of the $500 producers were replaced by $3500 producers!

1.2 Statistical Inference

Methods for summarizing and describing numerical data correctly and accurately are very important to the business person. The difference between success and failure in business often depends upon the ability to obtain and use accurate information. However, there are many situations in which the existing amount of numerical information is too great to be collected and analyzed.

For example, suppose the vice-president of a nationwide chain of shoe stores believes that information about the characteristics, tastes, and preferences of the stores' customers would be extremely useful in planning merchandise, styles, and advertising. However, in order to obtain this information about all of their customers, they would have to contact every person who had ever made a purchase at any of their stores. Since the number of customers could run into the millions, an attempt to identify and contact every customer would be prohibitively expensive and time consuming. Nevertheless, decisions regarding styles and sizes to be stocked, prices, and advertising appeals and programs should be based on information about the customer.

The market research department of the shoe store chain, having been presented with the problem, might decide that the required information can be obtained from a representative sample of customers chosen from the universe or population of all customers. Information from this sample can then be used to make inferences about the entire population. When conclusions about some population are based on information obtained from some portion or sample of that population, the techniques used for doing this fall into the category of STATISTICAL INFERENCE. The object of statistical inference is to use information obtained from some small set that constitutes the SAMPLE to draw conclusions about the larger set that makes up the population.

The physical units that are observed to obtain information are referred to as ELEMENTARY UNITS. Elementary units can be people, houses, accounts, parts produced on an assembly line, or any other observable physical entity. The set of all elementary units of interest in a statistical study is generally called the STATISTICAL UNIVERSE. The term STATISTICAL POPULATION, or simply POPULATION, is frequently used synonymously with universe. When a distinction is made between the two terms, universe generally refers to the set of elementary units while population refers to the set of observations made on the elementary units. In this book, the term population is used to refer to both the elementary units and the observations.

In discussing descriptive statistics, we will assume that the data represent a population, since our interest will focus on the problem of describing the set of data at hand. It is only when we use the data at hand to draw conclusions about some larger set or population that the data is considered to represent a sample.

The uses of the techniques of statistical inference in business are widespread and growing. Several typical situations in business that require the use of statistical inference are described below.

A firm of certified public accountants is called in to audit the books of a large company engaged in retail trade. One of the assets listed on the company's balance sheet is accounts receivable. In order to check the balance of that account as stated, the C.P.A. in charge of the audit plans to contact charge account customers and ask them to verify their individual balances. Since there are thousands of customers with accounts it would cost too much to contact every one of them. Instead, the C.P.A. takes a random sample of customers with outstanding balances and contacts the members of the sample. On the basis of the results of the sampling, he is able to verify the stated balance.

The marketing vice-president of a nationwide chain of shoe stores needs information about the preferences and characteristics of the stores' customers in order to make merchandising and advertising plans for the coming year. Since it is impossible to identify and contact all of the customers of the various stores in the chain, the vice-president has the market research department devise a sampling plan. The plan has two stages. First a random sample of stores in the chain is taken. Then a random sample of customers is selected at each of the sampled stores. The selected customers are asked several questions by an interviewer who writes their answers on a questionnaire. In this way, a sample that is representative of the population of customers is obtained and, on the basis of their responses, inferences are made about the entire population. The merchandising and advertising plans are then based on this information.

The purchasing manager of a company that manufactures office equipment buys large quantities of a particular part from a supplier. The purchased part is used in one of the company's products. One of the physical dimensions of the part is very critical, for if it is too large or too small it cannot be assembled into the finished product. The part is ordered and shipped in large quantities, and the quality of the shipments varies substantially. The production manager of the company constantly complains that the number of defective parts that he receives is much too high. He can never be sure if a given shipment is good or bad. He wants the purchasing manager to begin inspecting each part in every incoming shipment. The purchasing manager responds that the cost of such a procedure is prohibitive. At this point, the quality control manager intervenes and claims that he can establish a sampling plan that effectively detects shipments in which the number of defectives is excessively large. The quality control manager also claims that he can save the company money in the bargain. The sampling plan that is established requires that a few parts be randomly selected from each shipment and inspected. The plan provides a decision rule that specifies the number of parts to be inspected from each shipment and the allowable number of defectives in the sample. If any sample contains more defectives than that allowed number, the entire shipment is rejected and returned to the supplier. Although there is a cost associated with the sampling plan, the quality of the parts that enter the production process improves so greatly that the cost of manufacturing the product is actually reduced.

Though generally unadmitted, it is common practice among motel and hotel managers to overbook rooms. This is done under the assumption that a certain proportion of people with reservations will either cancel at the last minute or simply fail to show up. Overbooking, however, creates a problem for the managers. If too many reservations are made and not enough cancellations occur, the manager is faced with a number of irate travelers demanding that their reservations be honored for rooms already occupied. On the other hand, if too few reservations are made, it is very likely that some rooms will remain unoccupied and the hotel or motel will suffer lost revenues.

The manager of one hotel decides to apply statistics to the problem of overbooking. He selects a sample of days from the past year and uses the sample data to estimate the proportion of confirmed reservations that result in cancellations or no-shows. Based on this estimate, he establishes a policy for overbooking that reduces the number of vacancies and at the same time guarantees that the probability of having to turn away people with confirmed reservations is very small.

The marketing manager of a company that manufactures and markets ladies' cologne and toilet water has three different point-of-sale displays that she wants to evaluate. These are displays that are used at the cosmetics departments of the various department stores that sell the products. The marketing manager is not sure which of the displays is best in terms of stimulating sales, so she decides to conduct a test. Since the products are sold at hundreds of department stores throughout the country, it is impossible to test the displays at all of them. Instead, she selects thirty stores at random and randomly assigns ten stores to each of the three displays. The displays are used in the test stores for a period of one month, and the effect on sales in each store is noted. On the basis of these data, the marketing manager is able to determine that one display is indeed superior to the others and that display is then distributed to all of the stores that handle the product.

Each year, a general housing contractor is faced with the problem of predicting the demand for new houses in his building area. This is necessary because he is forced to make advance commitments to various subcontractors and suppliers in order to be assured that labor and materials are available when needed. In the past, his predictions were more on the order of guesses based on the demand in the year just past. Since this procedure results in either overcommitments or shortages, he is determined to find a better way. On consulting a statistician, the contractor is advised to try to determine if the demand for new housing appears to be related to some other measurable phenomenon. Research indicates that a definite relationship appears to exist between the number of housing permits issued in a given year and the number of marriage licenses issued during the previous year. The builder goes to the county courthouse and obtains data on building permits and marriage licenses for a period of twenty years. The statistician then takes the data and performs a statistical procedure, called a correlation and regression analysis, to determine the relationships, if any, that exist between these factors. The analysis indicates that there is a relationship strong enough to be used in making predictions. The statistician is also able to determine the nature of the relationship and express it in the form of an equation. On the basis of this analysis, the builder is now able to predict the demand for new housing with much greater accuracy. This, in turn, leads to much greater profits for the builder.

All of the previous illustrations are examples of how statistical inference can be used in business. All of the problems discussed here are real ones and the procedures leading to their solutions are covered in this book.

1.3 Origins of Statistical Data

Statistical data, as we have seen, result from a process of observation. These observations are made of some characteristic of the elementary units that make up the population of interest. In keeping records of transactions, every business organization is constantly creating or generating statistical data. These data are the raw material of statistical analysis. In fact, such data, the result of original observations, are generally referred to as RAW DATA. One of the primary purposes of the various statistical techniques that we will cover is to process this raw data to make it more intelligible and useable in making business decisions.

An organization's transactions and the records of those transactions provide the most important source of statistical data for use by the firm in its decision making. Sometimes, however, it is necessary to go outside the organization to collect statistical information from customers, suppliers, or other parties. Any planned process for collecting statistical data, whether within or outside of the organization, is generally referred to as a SURVEY. An example of a survey conducted to obtain data external to the organization is a market survey. In a market survey, customers or potential customers might be interviewed and their answers to questions noted on a questionnaire. The set of completed questionnaires constitutes the raw data for statistical analysis.

In a statistical survey, whenever observations are made on every elementary unit in the population, the survey is called a CENSUS. If the observations are made on less than the complete population, the survey is referred to as a SAMPLE SURVEY.

Large quantities of useful statistical data are available from various organizations and agencies. The United States government, for example, is an excellent source of statistical data. Various trade associations also publish statistics that relate to their own areas of interest. In referring to published statistical reports, the terms *primary* and *secondary sources* are frequently used. A primary source is the agency that actually collects the data and first publishes it. A secondary source refers to a republication of the data by some other organization or agency. Since this book is concerned with methods for compiling and analyzing statistical data, we will make few references to published data. We will instead devote our attention to the techniques used for collecting, describing, and drawing conclusions from the data itself.

1.4 Types of Statistical Data

When observations are made on some characteristic of an elementary unit, the result is not necessarily a number. Characteristics that are not numerical are called qualitative characteristics or ATTRIBUTES. An attribute cannot be measured or counted; only its presence or absence can be observed. Attributes of customers that might interest the vice-president of a chain of shoe stores might include sex, race, and home ownership. Although a customer can be classified into one of two or more categories corresponding to each attribute, there is no numerical value that is inherently associated with such a classification. When dealing with qualitative or attributes data, numerical values result from counting the number of elementary units that fall into each category. The credit manager who requests a list of accounts receivable with each account classified as either current or delinquent is dealing with attributes data.

On the other hand, when a number can be naturally associated with some characteristic of an elementary unit by a process of measuring or counting, that characteristic is said to be a quantitative characteristic. Height, weight, shoe size, income, number of members in the family, hours worked per week, are all quantitative characteristics of an elementary unit. Such quantitative characteristics are generally called VARIABLES, and the data resulting from such observations are referred to as variables data.

We must recognize that there are two types of variables. A variable that can assume any value within some range of values is said to be a CONTINUOUS VARIABLE. Height, weight, and shoe size are examples of continuous variables and result from a process of measurement. Even though all measurements are reported in terms of units such as pounds, inches, or tenths of an inch, the value of the characteristic measured is not restricted to any arbitrary unit of measurement. No matter how fine the gradations on a measuring instrument, the actual value of the characteristic being measured can always fall between any two of them.

On the other hand, a variable that is restricted to certain values only, usually the whole numbers, is said to be a DISCRETE VARIABLE. Values of a discrete variable result from a process of counting. Family size is a discrete variable, since its values can only be whole numbers. A family may consist of one, two, or three members or more. It cannot contain 3.5 or 4.7 members. The variable representing income or the balance of an account in dollars and cents is also discrete, because the value is always restricted to some whole number of cents.

1.5 Statistics and the Computer

It seems to be a fairly common impression that the study of statistics must necessarily involve the study of computers. This is not true. Computer science and statistics are two entirely different disciplines. The computer is, however, a very useful computational tool which can be applied to the solution of statistical and many other types of problems. The value of the large-scale digital computer lies in its ability to perform mathematical computations at enormous speeds. Since many statistical investigations and analyses involve the compilation and analysis of large masses of data, the computer makes it possible to undertake statistical studies that would have been impossible before its advent. Furthermore, a great many computations which are required in order to find solutions to many practical statistical problems that occur in business are lengthy and tedious. These computations can be performed quickly and easily by the computer, particularly when standard or packaged programs are available.

This textbook does not cover the use of the computer for several reasons. First, a basic course in statistics should be devoted entirely to statistics. Second, the problems in this book, which should be solved by the student, are neither too long nor too complicated to be easily solved with a pencil and paper. Third, solving problems is a part of the learning process. Statistical methods cannot be learned nor will the logic of the techniques be assimilated if the computer does all or most of the work. Finally, in the real world, where computers are available, technicians and programmers are also available for assistance in using the computer.

It may be comforting to know, however, that in practical business situations, when the statistical problems become more lengthy and more complicated than typical examples in a textbook, computers can be used to do the computation. However, remember that no computer now in existence is capable of formulating the problem. Only a human being can do that.

Definitions

Many terms used in this chapter will reappear throughout the rest of this book. These terms have special meanings with respect to the study of statistics and their definitions should be memorized. To make this easier, a list of these terms and their definitions follows. Additional terms are defined as they are introduced.

Attribute: A characteristic of an elementary unit that can only be observed as to its presence or absence.

Census: A survey in which observations are made on the entire population of elementary units.

Continuous variable: A variable that can assume any value within some range of values. Observations on a continuous variable are the result of measurements.

Descriptive statistics: The branch of statistics which is concerned only with describing and summarizing sets of numerical data. In general, we assume that descriptive statistics deals with statistical populations.

Discrete variable: A variable that is restricted to certain values within some range of values. Generally these values are integers. Observations of a discrete variable are counts.

Elementary unit: The physical entity on which an observation is made.

Population or **statistical population:** The entire set of possible observations that may be made on the statistical universe. Although the term universe typically refers to the elementary units themselves while the term population refers to the observations made on the units, the two terms are frequently used interchangeably. In this book we use the term population to refer to both the elementary units and the observations.

Raw data: The statistical data in its original form, before any statistical techniques are used to refine, process, or summarize.

Sample: Any portion of the statistical population. Generally, a sample consists of fewer elementary units or observations than are contained in the population.

Sample survey: A survey in which observations are made on a sample of elementary units drawn from the population of interest.

Statistical inference: The branch of statistics which is concerned with drawing conclusions about populations on the basis of incomplete data, that is, on the basis of data obtained from a sample.

Statistics: The branch of applied mathematics which is concerned with the collection and analysis of numerical data. Also a body or set of numerical data.

Survey or **statistical survey:** A planned process of collecting statistical data by making observations on elementary units.

Universe or **statistical universe:** The entire body of elementary units that are of interest and are the subject of a statistical investigation.

Variable: An observable quantitative characteristic of an elementary unit that may vary from unit to unit.

2.x y
=
z-xy

CHAPTER 2
A REVIEW OF PREREQUISITES

2.1
Terminology and notation

2.2
Significant digits and rounding

2.3
Logarithms

2.4
Sets and operations with sets

2.5
Permutations and combinations

2.6
Functions

2.7
Looking ahead

Statistics is a branch of mathematics. However, a first course in statistics does not require a great deal of mathematical preparation. The most important requirement is simply the ability to think logically and to relate symbols to their real-world counterparts. Ideally, preparation for a noncalculus based course in statistics should include college algebra. This ideal is seldom achieved, and a year or so of high school algebra is probably the most common preparation among business students. Even with this background, many students need some review. For that reason, this chapter is devoted to a brief review of a few topics that are not really statistics but are essential for easy comprehension and use of the material that follows.

2.1 Terminology and Notation

The language of statistics consists of many terms and symbols having special meanings that are peculiar to this field. Some of these terms, such as population, sample, and variable have already been defined. For example, a *variable* is an observable quantitative characteristic of an elementary unit that may vary from unit to unit. In statistics, it is necessary to have some method for designating a variable. By convention, a letter from the latter part of the alphabet, such as x, y, or z, is usually used for this purpose.

Suppose the equation $y = ax + b$ represents the relationship between income and years of experience for a particular class or group of skilled workers. If y stands for income and x for experience, then x and y are variables because their values will vary from worker to worker.

Certain quantities always remain the same within the context of some specified problem. Such quantities are called *constants* and are usually designated by letters in the first part of the alphabet such as a, b, and c. If the letter a represents a constant value in some problem, then whenever a appears in that problem it will have the same value. However, in an entirely different problem the same letter can be used to represent some other constant value.

In the equation $y = ax + b$ given above, a and b are constants. That is, for the particular group of workers referred to, a and b will always have the same value. However, for some other group, those values might be different.

There are certain constants, however, that have values that never change. One such constant is represented by the Greek letter π, pi. This constant stands for the ratio of the area of a circle to the square of its radius and always has the approximate value 3.1416. Whenever π appears in an equation, regardless of what the equation represents, π always has the same value. A constant whose value never changes is called an *absolute constant*.

In referring to the absolute constant π, we use a letter of the

Greek alphabet. Greek letters are very common in mathematics and statistics and it is useful to be able to recognize the most commonly used Greek letters. There are actually two Greek alphabets, roughly corresponding to our alphabets of capital and lowercase letters. Some of the Greek letters are similar to the Roman letters, while others are very different. For example, the capital Greek letter alpha is A, while the lowercase alpha looks like α. These letters are similar to A and a. On the other hand, the capital and lowercase Greek letter phi look like Φ and ϕ respectively. There is no similar letter in our alphabet. The Greek alphabets are reproduced here as Table 2.1 for your reference.

Greek letters are most frequently used in statistics to represent population parameters. A PARAMETER is some constant characteristic of a population. For example, a decision about opening a store in a new shopping center might be based on the average income of the population of families living within some specified distance from the center. The average income is a parameter of that population. In the context of this particular problem, the value of the parameter is considered to be constant even though a year later or in the context of a different problem that parameter may have a different value. The lowercase Greek letter μ, mu, is generally used to represent population averages or means. (Other Greek letters and the concepts they represent are introduced at appropriate points in the text.)

It is unfortunate that statistical notation and terminology are not completely standardized. Different authors may use different symbols to represent the same quantity or concept. For this reason, the reader is cautioned to make sure of the meaning of every symbol and every term encountered when referring to more than one text or reference book on statistics.

Summation and Subscripting. If the letter x represents some particular variable, and if several observations of the variable x are

TABLE 2.1
GREEK ALPHABET

A	α	alpha	N	ν	nu
B	β	beta	Ξ	ξ	xi
Γ	γ	gamma	O	o	omicron
Δ	δ	delta	Π	π	pi
E	ϵ	epsilon	P	ρ	rho
Z	ζ	zeta	Σ	σ	sigma
H	η	eta	T	τ	tau
Θ	θ	theta	Υ	υ	upsilon
I	ι	iota	Φ	ϕ	phi
K	κ	kappa	X	χ	chi
Λ	λ	lambda	Ψ	ψ	psi
M	μ	mu	Ω	ω	omega

made on each of several elementary units, then we have several different values of x. For instance, the first observation results in a value of x that we designate x_1. The second observation can be designated x_2 and so on. The numbers appended to the right and below the x symbol are called *subscripts*. They are used to identify one particular observation of x. If ten such observations are made they can be designated x_1 through x_{10}. Sometimes we may want to refer to a single observation of x without being specific. In that case we use a letter such as *i* or *j* as the subscript instead of a number. For example, x_i or x_j, refer to the *i*th or *j*th value of x. The substitution of a number for *i* or *j* would then make that designation specific.

One of the most common operations performed on a series of observations is addition or *summation*. If several observations of the variable x are made, it may be desirable to add or sum the values obtained. Rather than write out an instruction such as "add up all of the x values," a convenient notational device which means the same thing is Σx. The Σ is the capital Greek letter sigma. When Σ appears, it simply means: add up everything that follows. A symbol such as Σ is called an *operator*, since it indicates that a particular operation is to be performed.

A more precise way of specifying which x values are to be summed is

$$\sum_{i=1}^{10} x_i$$

This is read, "sum all the values of x_i where *i* goes from one to ten." The *i* below and to the right of the x is the subscript. It identifies a specific value of x. For example, if $i = 3$, we are designating the third x value. If $i = 5$, the fifth value of x is specified. The 1 and the 10 which appear below and above the summation sign are called the *limits* of the summation. By designating different limits we can indicate partial sums from a series of values. For example, if a series of values of x are provided, then

$$\sum_{i=1}^{5} x_i$$

means: Find the sum of the first five values of x. However,

$$\sum_{i=2}^{4} x_i$$

means that only the second through the fourth values should be included in the sum. (See Example 2.1 below for an illustration.) When all of the values in a series are to be included in the sum, the limits and even the subscript can be dropped and the operation can be specified by Σx_i or simply Σx.

Example 2.1

Given a series of five observations of the variable x,

$$x_1 = 1,\ x_2 = 3,\ x_3 = 4,\ x_4 = 6,\ x_5 = 9$$

a) Find $\sum_{i=1}^{5} x_i$.

$$\sum_{i=1}^{5} x_i = x_1 + x_2 + x_3 + x_4 + x_5 = 1 + 3 + 4 + 6 + 9 = 23$$

b) Find $\sum_{i=2}^{4} x_i$.

$$\sum_{i=2}^{4} x_i = x_2 + x_3 + x_4 = 3 + 4 + 6 = 13$$

There are some special rules that govern the use of the summation operator. These rules are summarized as follows.

Rule 1 $\sum_{i=1}^{n} ax_i = a \sum_{i=1}^{n} x_i$. The sum of a constant times a variable is equal to the constant times the sum of the variable.

Example 2.2

If each value of x in the previous example is multiplied by the constant $a = 2$,

find $\sum_{i=1}^{5} ax_i$

$$\sum_{i=1}^{5} ax_i = ax_1 + ax_2 + ax_3 + ax_4 + ax_5$$
$$= (2)(1) + (2)(3) + (2)(4) + (2)(6) + (2)(9)$$
$$= 2 + 6 + 8 + 12 + 18 = 46$$

However, by rule 1,

$$\sum_{i=1}^{5} ax_i = a \sum_{i=1}^{5} x_i = (2)(23) = 46$$

Rule 2 $\sum_{i=1}^{n} a_i = na$. The sum of a series of n constant values is equal to n times the value of the constant. The reader should verify rule 2.

Rule 3 $\sum_{i=1}^{n} (x_i + b) = \sum_{i=1}^{n} x_i + nb$. The sum of a variable plus a constant is equal to the sum of the variable plus n times the value of the constant.

Example 2.3 If the constant $b = 3$ is added to each value of x in Example 2.1, find $\sum_{i=1}^{5} (x_i + b)$.

$$\sum_{i=1}^{5} (x_i + b) = (x_1 + 3) + (x_2 + 3) + (x_3 + 3) + (x_4 + 3) + (x_5 + 3)$$
$$= (1 + 3) + (3 + 3) + (4 + 3) + (6 + 3) + (9 + 3)$$
$$= 4 + 6 + 7 + 9 + 12 = 38$$

However, by rule 3,

$$\sum_{i=1}^{5} (x_i + b) = \sum_{i=1}^{5} x_i + nb = 23 + (5)(3)$$
$$= 23 + 15 = 38$$

Rule 4 $\sum_{i=1}^{n} (ax_i + b) = a \sum_{i=1}^{n} x_i + nb$. The sum of a constant times a variable plus another constant is equal to the first constant times the sum of the variable plus n times the value of the second constant.

Example 2.4 If each value of x in Example 2.1 is multiplied by the constant $a = 2$, and the constant $b = 3$ is then added to each product,

find $$\sum_{i=1}^{5} (ax_i + b).$$

$$\sum_{i=1}^{5} (ax_i + b) = (ax_1 + b) + (ax_2 + b) + \cdots + (ax_5 + b)$$
$$= [(2)(1) + 3] + [(2)(3) + 3] + [(2)(4) + 3] + [(2)(6) + 3] + [(2)(9) + 3]$$
$$= [2 + 3] + [6 + 3] + [8 + 3] + [12 + 3] + [18 + 3]$$
$$= 5 + 9 + 11 + 15 + 21 = 61$$

However, by rule 4,

$$\sum_{i=1}^{5} (ax_i + b) = a \sum_{i=1}^{5} x_i + nb = (2)(23) + (5)(3)$$
$$= 46 + 15 = 61$$

Rule 5 $\sum_{i=1}^{n} (x_i + y_i) = \sum_{i=1}^{n} x_i + \sum_{i=1}^{n} y_i$. The sum of a series consisting of one variable plus a second variable is equal to the sum of the first variable plus the sum of the second variable.

Example 2.5

Given a series of five observations of the variables x and y,

$$x_1 = 1,\ x_2 = 3,\ x_3 = 4,\ x_4 = 6,\ x_5 = 9$$
$$y_1 = 5,\ y_2 = 8,\ y_3 = 11,\ y_4 = 14,\ y_5 = 15$$

find $$\sum_{i=1}^{5} (x_i + y_i).$$

$$\sum_{i=1}^{5} (x_i + y_i) = (x_1 + y_1) + (x_2 + y_2) + (x_3 + y_3) + (x_4 + y_4) + (x_5 + y_5)$$
$$= (1 + 5) + (3 + 8) + (4 + 11) + (6 + 14) + (9 + 15)$$
$$= 6 + 11 + 15 + 20 + 24 = 76$$

However, by rule 5,

$$\sum_{i=1}^{5} (x_i + y_i) = \sum_{i=1}^{5} x_i + \sum_{i=1}^{5} y_i = 23 + 53 = 76$$

2.2 Significant Digits and Rounding

In recording and then performing computations with statistical data, it is important not to claim greater accuracy for the results than is justified by the original observations. The data result from observations that are either *counts* or *measurements*. Counts, if made accurately, are exact numbers but are frequently rounded for reporting purposes. For example, monetary values are discrete numbers and the number of dollars and cents involved in some transaction can be counted and reported exactly. However, when the amounts involved are large, such data are frequently rounded to the nearest dollar and sometimes even to the nearest thousand dollars.

On the other hand, data resulting from measurements are never exact, but are always rounded off. Weights are typically recorded to the nearest pound or ounce. Dimensions may be reported to the nearest inch or tenth of an inch, or in some cases to the nearest hundredth or thousandth of an inch. No matter how precise a measuring instrument is, the number resulting from a measurement is always an approximation. Therefore, whenever a statistic is reported, it is always reasonable to inquire as to how many of the digits in the number are significant; that is, how many of the digits convey information about the accuracy of the observation and how many are there to locate the decimal point. Before considering how significant digits are determined, we will first specify a set of rules for rounding numbers.

Rounding Conventions. The following rules are commonly used for rounding numbers.

1. If the leftmost of the digits *to be dropped* is less than five, the preceding digit is not changed. For instance, rounding 67.263 to four digits results in 67.26.

2. If the leftmost of the digits *to be dropped* is greater than five, or is five followed by a digit greater than zero, the preceding digit is increased by one. Therefore, 14.78 rounded to three digits becomes 14.8, and 9.45006 rounded to two digits is 9.5.

3. If the leftmost of the digits *to be dropped* is five or five followed only by zeros, then the digit preceding the five determines how the number is rounded. If that digit is an even number, it remains unchanged. If that digit is an odd number, it is increased by one. For example, 42.75 rounded to three digits becomes 42.8, but 42.65 rounded to three digits is 42.6.

Example 2.6 Round the following numbers to three digits.

	Original Number	*Rounded Number*	*Rule*
a)	78.634	78.6	1
b)	7.8684	7.87	2
c)	100.50	100	3
d)	100.505	101	2
e)	10.15	10.2	3
f)	64.25	64.2	3
g)	112.495	112	1

Significant Digits. After rounding, every digit except zero is always significant. In order to determine if a zero is significant, we use the following rules.

1. A zero that falls before the first significant digit is *not* significant. In the number 0.08, the zeros are not significant.

2. A zero that falls after a significant digit is always significant if the number contains a decimal point. The number 4.60 has three significant digits since, in this case, the last zero is significant.

3. A zero that falls after the last significant digit of an integer may or may not be significant. In the number 150, the zero is significant if this number results from rounding to the nearest digit, that is, the original observation was a number from 149.5 to 150.4. The zero is not significant if this number results from a process of rounding to the nearest ten units. That is, if the original observation was a number from 145 to 154.

4. A zero that falls between two significant digits is significant. In the number 405, the zero is significant.

Example 2.7

How many significant digits are there in the following numbers?

a) 0.086 **d)** 204,086 **g)** 63.003
b) 3.02 **e)** 0.004 **h)** 6,400,000
c) 3.020 **f)** 14,200

a) The number 0.086 contains two significant digits, by rule 1. We can explain this by realizing that the choice of the unit of measurement is arbitrary. Note that 0.086 kilometers is the same as 86 meters or 8600 centimeters. The first two of these numbers have two significant digits. The third may have two, three, or four significant digits, by rule 3. However, since the number 8600 appears in the same context as the other two numbers, we might assume that only two of the digits are significant and the zeros only indicate that the measurement is accurate to the nearest hundred centimeters.

b) The number 3.02 has three significant digits, by rule 4.

c) The number 3.020 has four significant digits, by rule 2.

d) The number 204,086 has six significant digits, by rule 4.

e) The number 0.004 has one significant digit, by rule 1.

f) The number 14,200 may have anywhere from three to five significant digits. The context in which the number appears may indicate how many of the digits are significant. If this number is the balance of an account reported to the nearest dollar, then all five of the digits are significant. If it is an approximate average income for some population of wage earners, then as few as three of the digits may be significant. In that case, the ambiguity might be resolved by reporting the number as 14.2 thousands.

g) The number 63.003 has five significant digits, by rule 4.

h) The number 6,400,000 may have from two to seven significant digits. The ambiguity may be eliminated by reporting the number as follows:

1) 6.4 millions or 6.4×10^6 to indicate that the two digits are significant.

2) 640 ten thousands or 640×10^4 to indicate that three digits are significant.

3) 6400 thousands or 6400×10^3 to indicate that four digits are significant.

Significant Digits Resulting from Calculations with Rounded Numbers. The following rules determine the number of significant digits that may be carried in an answer that results from calculations involving rounded numbers.

Rule 1 When numbers are added or subtracted, they are usually arranged in a column with the decimal points aligned. Digits to the right of the column in which the last significant digit occurs in *any* of the numbers are not significant and should not appear in the answers.

Example 2.8 Add and round to the appropriate number of significant digits 320.8 + 4.687 + 27.59.

$$\begin{array}{r} 320.8 \\ 4.687 \\ 27.59 \\ \hline 353.077 \end{array}$$

The answer should be recorded as 353.1.

Example 2.9 Subtract and round to the appropriate number of significant digits 320.8 − 27.59.

$$\begin{array}{r} 320.8 \\ -27.59 \\ \hline 293.21 \end{array}$$

The answer should be recorded as 293.2.

Rule 2 When numbers are multiplied or divided, the answer should contain no more significant digits than the number with the fewest significant digits in the calculation.

Example 2.10 Multiply and round to the appropriate number of significant digits 27.1 × 3.523.

$$\begin{array}{r} 3.523 \\ 27.1 \\ \hline 3523 \\ 24661 \\ 7046 \\ \hline 95.4733 \end{array}$$

The answer should be recorded as 95.5.

Example 2.11

Divide and round to the appropriate number of significant digits $180.574 \div 4.7$.

```
        38.42
   47)1805.74
      141
       395
       376
        197
        188
          94
          94
```

The answer should be recorded as 38.

Rule 3 When raising a number to a power or extracting a root, the answer should not have more significant digits than the original number.

Example 2.12

Extract the root and round to the appropriate number of significant digits $\sqrt{87.4} = 9.3487967$.
The answer should be recorded as 9.35.

Example 2.13

Square and round to the appropriate number of significant digits:

$$(4.73)^2 = 22.3729$$

The answer should be recorded as 22.4.

Rule 4 If one of the numbers in a computation is an EXACT NUMBER, that is, one that has not been rounded, it is considered to have an infinite number of significant digits. Therefore, if in computing 6×123.7, 6 is an exact number, then the number of significant digits in the answer should be four.

Example 2.14

Divide the following number and round it to the appropriate number of significant digits. Assume that 6 is an exact number.

$$385.4 \div 6 = 64.233333$$

The answer should be recorded as 64.23.

Although final answers should not report digits which are not significant, extra digits may be carried during intermediate stages of the calculations to reduce rounding error.

If a number based on a count is exact, then the number of significant digits in the result of a computation using this number is not limited by the significant digits of the exact number. For example, if six different people are to be given three books each, then the total number of books provided would be $3 \times 6 = 18$. In this case, both of the digits in the number 18 are significant.

2.3 Logarithms

The ability to calculate using logarithms is extremely useful in situations when lengthy calculations are required and a pocket calculator is not available. Since this is a review chapter, we will not discuss the theory of logarithms. We will simply review the procedures for finding the logarithms of numbers and using them to simplify calculations. Although there are two types of commonly used logarithms, common logs and natural logs, we will limit our discussion to common logs.

The logarithm of a number is itself a number consisting of two parts, a whole number called the *characteristic*, and a decimal fraction called the *mantissa*. For example:

$$\log 40.0 = 1.60206$$
$$\log 400.0 = 2.60206$$
$$\log 4000.0 = 3.60206$$

In each of these logarithms, the mantissa is .60206. The characteristics are 1, 2, and 3, respectively. Notice that for the same sequence of digits in the original number, the mantissa is the same regardless of the position of the decimal point. Only the characteristic changes.

Finding the Logarithm of a Number. The mantissa of the log of a number can always be obtained from a table such as the one provided in Appendix A.7. If the number for which the log is desired contains more than three significant digits, *interpolation* is necessary. For example, suppose we want to find the mantissa for the logarithm of 54.28. Referring to the table in Appendix A.7, a portion of which is reproduced in Table 2.2, we find that the mantissas for the logarithms of 54.2 and 54.3 are

Number	*Mantissa*
54.2	73400
54.3	73480

The number for which we want the mantissa, 54.28, lies 0.8 of the way from 54.2 to 54.3. Therefore, its mantissa should lie 0.8 of the way between 73400 and 73480. We find that mantissa by taking 0.8 of the difference between these two mantissas and adding that value to the lower of the two as follows.

TABLE 2.2
PARTIAL TABLE OF COMMON LOGARITHMS

N	0	1	2	3	4	5	6	7	8	9
50	69897	69981	70070	70157	70243	70329	70415	70501	70586	70672
51	70757	70812	70927	71012	71096	71181	71265	71349	71433	71517
52	71600	71681	71767	71850	71933	72016	72099	72181	72263	72346
53	72428	72509	72591	72673	72754	72835	72916	72997	73078	73159
54	73239	73320	73400	73480	73560	73640	73719	73799	73878	73957

Step 1. Find the difference.

$$73480 - 73400 = 80$$

Step 2. Multiply this difference by 0.8.

$$0.8 \times 80 = 64$$

Step 3. Add 64 to 73400.

$$64 + 73400 = 73464$$

The mantissa of the logarithm of 54.28 is 73464 (see Table 2.2).
The characteristic of the logarithm of a number is determined by applying one of the following rules.

1. For any number with a value of 1 or greater, the characteristic of its log is one less than the number of digits to the *left* of the decimal point. For example, since the number 40.0 has two digits to the left of the decimal, the characteristic of its log is 1. The characteristic of the log of 4.0 is 0.
2. For any number with a value less than 1, the characteristic is negative and has a value equal to 1 plus the number of zeros between the decimal point and the first nonzero digit to the right of the decimal point. For example, the characteristics of the logs of 0.1763, 0.0176, and 0.0017 are, respectively, −1, −2, and −3.

The table of common logarithms in Appendix A.7 contains only mantissas. It is assumed that there is a decimal point to the left of each set of digits in the body of the table. Since the mantissa is exactly the same for the same sequence of digits regardless of the position of the decimal point in the original number, it is possible to find the log of a wide range of numbers from this single table.

Example 2.15

Find the logs of:

a) 0.2 **b)** 2.0 **c)** 20.0 **d)** 200.0

a) Since 0.2 is less than 1, the characteristic is negative. Also, because there are no zeros between the decimal point and the first nonzero digit, the characteristic is -1. Referring to the table, the number 2 is found at the top and the corresponding mantissa is .30103. Therefore, $\log 0.2 = -1.30103$.

b) The characteristic of the log of 2.0 is one less than the number of digits to the left of the decimal point, that is, $1 - 1 = 0$. The mantissa corresponding to the number 2 is, from the table, .30103. Therefore, $\log 2.0 = 0.30103$.

c) The characteristic of the logarithm of 20.0 is 1. The mantissa corresponding to the number 20 in the table is .30103. Notice that the mantissa for 20 is the same as the mantissa for 2. Therefore, $\log 20.0 = 1.30103$.

d) The characteristic of the log of 200.0 is 2. The mantissa is the same as for 0.2, 2.0, and 20.0. Therefore, $\log 200.0 = 2.30103$.

The Antilog—Finding the Number Corresponding to a Logarithm. When the log of a number is known, it is necessary to find the corresponding number, or *antilog*. The procedure is to first find the mantissa in the table. The digits in the column on the extreme left of the table are the first two digits in the number. The digit at the top of the column is the third digit of the number. The position of the decimal point is located according to the value of the characteristic.

Example 2.16 Find the antilog of the following.

a) -1.39094 **b)** 0.39094 **c)** 2.39094

The mantissa is located in the table and the digits of the corresponding number are 2, 4, and 6. The position of the decimal point is obtained from the value of the characteristic as follows.

a) A characteristic of -1 means that the number is less than 1 and that there are no zeros between the decimal point and the first nonzero digit. Therefore, antilog $-1.39094 = 0.246$.

b) A characteristic of 0 means that the number is 1 or greater and that there is one digit to the left of the decimal point. Therefore, antilog $0.39094 = 2.46$.

c) A characteristic of 2 means that there are three digits to the left of the decimal. Therefore, antilog $2.39094 = 246.0$.

Calculations with Logarithms. The following four rules for calculating with logarithms simplify a great many calculations that may have to be performed in solving statistical problems.

For any two positive numbers x and y,

1. $\log xy = \log x + \log y$. The log of the product of two numbers is equal to the sum of their logs.

2. $\log x/y = \log x - \log y$. The log of the quotient of two numbers is equal to the difference of their logs.
3. $\log x^c = c(\log x)$. The log of x raised to the power c is equal to the product of c and log x.
4. $\log \sqrt[c]{x} = (\log x)/c$. The log of the cth root of x is equal to log x divided by c.

Example 2.17

Perform the following calculations using logarithms.

a) $(24.7)(4.1)$ **b)** $24.7 \div 4.1$ **c)** $(4.1)^3$ **d)** $\sqrt[3]{24.7}$

a) $(24.7)(4.1)$

$$\begin{aligned} \log 24.7 &= 1.39270 \\ \log 4.1 &= \underline{0.61278} \\ \text{sum} &= 2.00548 \end{aligned}$$

antilog 2.00548 = 101.27, therefore (24.7)(4.1) = 101.27.

b) $24.7 \div 4.1$

$$\begin{aligned} \log 24.7 &= 1.39270 \\ \log 4.1 &= \underline{0.61278} \\ \text{difference} &= 0.77992 \end{aligned}$$

antilog 0.77992 = 6.024, therefore $24.7 \div 4.1 = 6.024$.

c) $(4.1)^3$

$$\begin{aligned} \log 4.1 &= 0.61278 \\ & \underline{\qquad\quad 3} \\ \text{product} &= 1.83834 \end{aligned}$$

antilog 1.83834 = 68.92, therefore $(4.1)^3 = 68.92$.

d) $\sqrt[3]{24.7}$

$$\log 24.7 = 1.39270$$

$$3\overline{)1.39270} = 0.46423$$

antilog 0.46423 = 2.912, therefore $\sqrt[3]{24.7} = 2.912$.

2.4 Sets and Operations with Sets

A SET is defined as any collection of objects having some describable similarity. The similarity among members of a given set allows us to distinguish between those objects that belong to the set and those that do not. There is a set that consists of all human beings in the world. All nonhuman objects are excluded from that set. There is a set consisting of all presidents of United States corporations. Anyone who is not the president of a corporation chartered in the United States is not a member of that set. There is a set of all 1979 Ford automobiles that have less than 25,000 miles on their odometers. A 1979 Chevrolet is not a member of this set. Neither is a 1979 Ford with 32,000 miles on its odometer.

A set is designated by establishing a rule that identifies members of the set. If they are small enough, sets are also designated by listing the objects that are members. When the latter method is used, the objects included in the set are usually separated by commas and enclosed in braces. Frequently, a capital letter is used to identify a set. This capital letter, followed by an equals sign, normally precedes the listing of the members in the set. For example, if we let the letter S represent the set of integers greater than zero and less than ten, this is written $S = \{1,2,3,4,5,6,7,8,9\}$. The rule that defines this same set is written: $S = \{x | x \text{ is an integer greater than zero and less than ten}\}$. This rule simply says that S is the set of x's where x is an integer from 1 to 9 inclusive.

A set that consists of some of the members of another set is a SUBSET of the original set. For instance, consider a set consisting of all employees of the Advance Machine Company and a second set consisting of the production employees of the Advance Machine Company. The set of production employees is a subset of the set of all employees. If we use capital letters to identify sets, then a set A is a subset of another set B if every member of A is also a member of B.

There is a special set that contains *no* elements. This special set is called the EMPTY SET and is designated by the symbol $\varnothing$. The empty set is a subset of every other set.

Whenever, in the course of an inquiry or investigation, interest is focused on some specific set and all other sets are subsets of the original set, the original set is called the UNIVERSAL SET. To be more specific, the universal set is defined as that set of elementary units which is of interest in the context of some particular investigation. For example, if an investigation has as its goal the determination of the average family income within some standard metropolitan area, the universal set for this investigation is the set of all families living in that metropolitan area. If the investigator is interested in determining the characteristics of charge account customers of a particular department store, the universal set is the set of all charge account customers of that department store.

In studying the relationships among sets and subsets, a very useful device is the Venn diagram. A *Venn diagram* is a pictorial representation that uses a large rectangle to represent the universal set and one or more figures within the rectangle to represent subsets of this universal set. A simple Venn diagram is illustrated in Figure 2.1. We designate the universal set S. In this diagram, the set A, which includes everything contained in the circle, is a subset of S. Notice that we have also named the set that is inside the rectangle but is not part of the circle. This subset has been designated A'. Any part of the universal set not contained in some subset, such as A, is called the *complement* of A. The "prime" above and to the

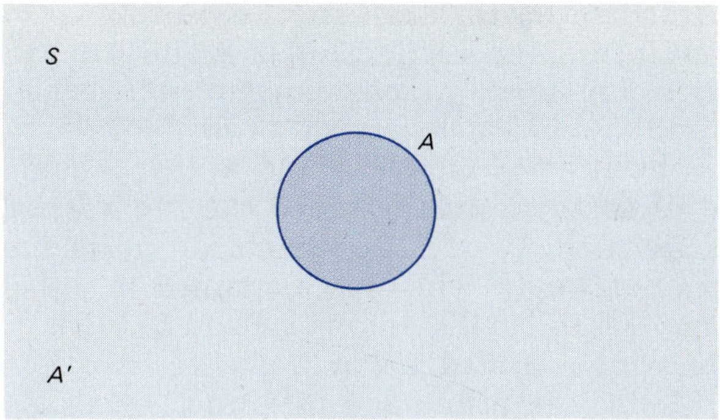

Figure 2.1
Venn diagram. Set A *is a subset of the universal Set* S. *Set* A′, *or what is not in Set* A, *is also a subset of the universal Set* S.

right of the letter designating a set refers to the complement of that set. In the case of A, the complement of A, A', can simply be read as "not A."

Figure 2.2 shows Venn diagrams with two intersecting subsets

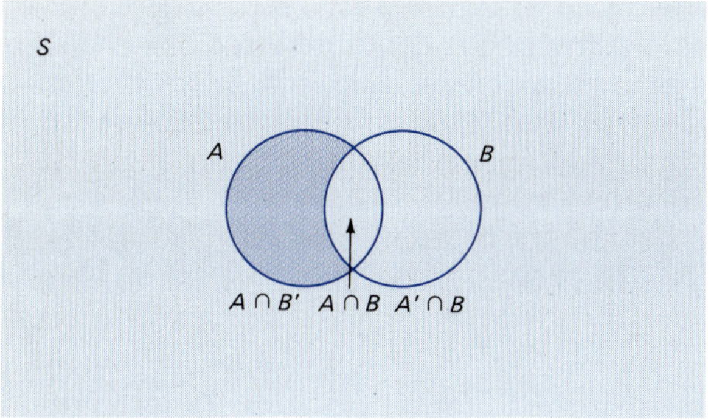

Figure 2.2a
Intersecting subsets.

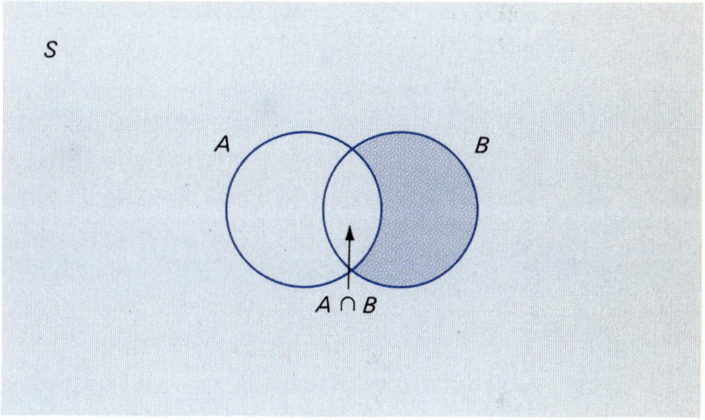

Figure 2.2b
Intersection of Set A *and Set* B.

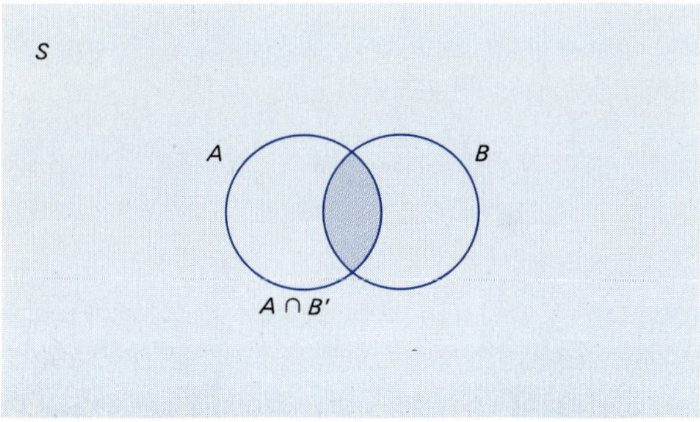

Figure 2.2c
Intersection of Set A *and the complement of Set* B.

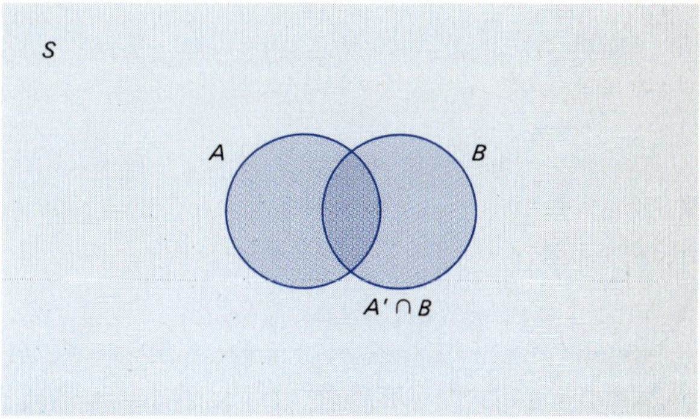

Figure 2.2d
Intersection of the complement of Set A *and Set* B.

A and B. It also illustrates and labels several other subsets that result from the relationship of A and B to each other and to S. The area contained in each complete circle belongs to the subset designated by the single letter A or B. Since the two circles intersect, it is obvious that there is a portion that is common to both A and B. This common area is called the *intersection* of A and B, and is designated $A \cap B$. The symbol $\cap$ means intersection and can be read as "and." The area contained in either or both of the two circles is called the *union* of A and B, and is designated $A \cup B$. The $\cup$ symbol is read as "or."

Using the prime and the "and" symbols, the two intersecting subsets, A and B, can be subdivided into three MUTUALLY EXCLUSIVE subsets. The term mutually exclusive implies that these three subsets do not intersect, that is, they have no common area. The intersection of A and B is designated $A \cap B$, as indicated previously. That portion of subset A that is not in B is designated $A \cap B'$. Finally, that portion of subset B that is not in A is $A' \cap B$. The subset $A \cup B$ is illustrated in Figure 2.3a. The complement of $A \cup B$, $(A \cup B)'$, is that area which is contained in S but is not in $A \cup B$. $(A \cup B)'$ is illustrated in Figure 2.3b.

Of course, if the subsets A and B are mutually exclusive, they have no points in common and appear as illustrated in Figure 2.4. Since A and B have no intersection, $A \cap B = \emptyset$.

Other relationships that should be apparent from Figure 2.1 are

$$A \cup A' = S \tag{2.1}$$

$$A \cap A' = \emptyset \tag{2.2}$$

Equation 2.1 indicates that the union of any set with its complement is the universal set. Equation 2.2 specifies that the intersection of any set with its complement is the empty set.

Figure 2.3a
The union of Set A *and Set* B.

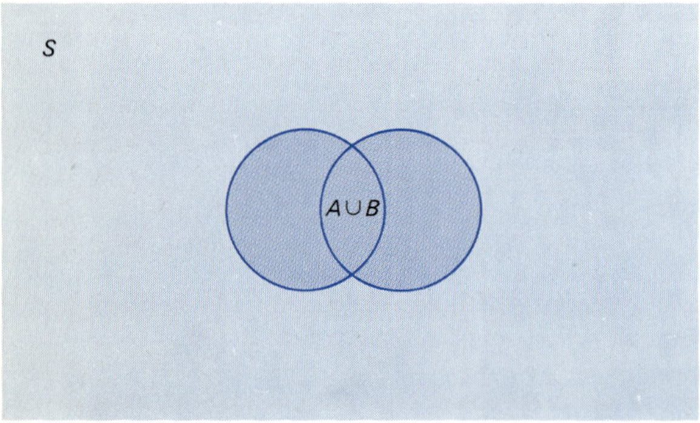

Figure 2.3b
The complement of the union of Set A *and Set* B.

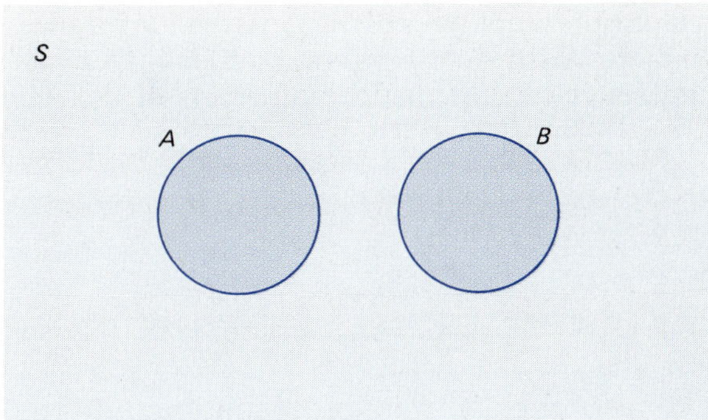

Figure 2.4
Mutually exclusive subsets.

Example 2.18

The Guardian Insurance Agency employs a staff of ten persons. The marketing or sales force consists of Allen, Baker, Collins, Davis, and Egan. We will designate this set by the letter R. That is, $R = \{$Allen, Baker, Collins, Davis, Egan$\}$. Other personnel employed by Guardian constitute the office staff. We will designate this set with the letter P so that $P = \{$Fox, Gordon, Hall, Johnson$\}$. Mr. King is the president and manager of the agency. If we let S represent the universal set of all Guardian employees, these relationships can be shown on the Venn diagram, Figure 2.5.

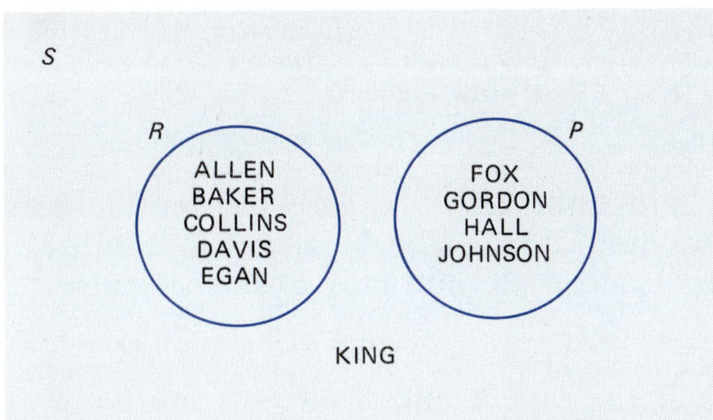

Figure 2.5
Venn diagram illustrating Example 2.18, the Guardian Life Insurance Company.

Example 2.19

The sales force, designated set R in the previous example, sells both life and casualty insurance. Allen and Baker sell only life insurance; Collins and Davis sell casualty insurance; and Egan sells both kinds. If C represents the set of casualty insurance salespersons and L represents the set of life insurance salespersons, list the members of the following sets.

a) $C \cup L = \{$Allen, Baker, Collins, Davis, Egan$\}$

b) $C \cap L = \{$Egan$\}$

c) $C \cap L' = \{$Collins, Davis$\}$

d) $C' \cap L = \{$Allen, Baker$\}$

e) $(C \cup L)' = \{$Fox, Gordon, Hall, Johnson, King$\}$

Example 2.20 Among the employees of the Guardian Insurance Agency are the following women: Allen, Fox, Gordon, and Hall. Designate the set of women employees as W. Now list the members of the following sets.

a) $W' = \{\text{Baker, Collins, Davis, Egan, Johnson, King}\}$

b) $P \cap W' = \{\text{Johnson}\}$

c) $L \cap W = \{\text{Allen}\}$

d) $C \cap W = \{\varnothing\}$

e) $P \cup W = \{\text{Allen, Fox, Gordon, Hall, Johnson}\}$

One of the common problems encountered in dealing with sets is that of counting the number of elements that belong to some specified set. When all elements of a set are listed, it is easy to count the members. However, when the set of interest results from an operation performed on other sets, some simple counting rules can be very helpful. We will use the symbol $n(A)$ to represent the number of elements in set A. For example, in (a) above, $n(W') = 6$. From Equation 2.1, we see that since $A \cup A' = S$, then

$$n(A \cup A') = n(S)$$

and

$$n(A) + n(A') = n(S) \tag{2.3}$$

Then, by simple algebra,

$$n(A') = n(S) - n(A) \tag{2.4}$$

Since the sets A and A' are mutually exclusive by definition, Equation 2.3 is applicable to any mutually exclusive sets. Therefore, if A and B are mutually exclusive, then

$$n(A \cup B) = n(A) + n(B) \tag{2.5}$$

If the sets A and B are *not* mutually exclusive, Equation 2.5 cannot be used to find the number of elements in their union. By referring once more to Figures 2.2a, 2.2b, 2.2c, and 2.2d, you can see that the union of two sets, A and B, can be divided into three mutually exclusive subsets labeled $A \cap B$, $A \cap B'$, and $A' \cap B$. Therefore:

$$n(A \cup B) = n(A \cap B) + n(A \cap B') + n(A' \cap B)$$

If we simply add $n(A) + n(B)$ as indicated in Equation 2.5, we will be counting the points in the intersection of the two sets twice. That is, since

$$n(A) = n(A \cap B') + n(A \cap B)$$

and

$$n(B) = n(A' \cap B) + n(A \cap B)$$

then $n(A) + n(B) = n(A \cap B') + n(A \cap B) + n(A' \cap B) + n(A \cap B)$. Note that $n(A \cap B)$ appears in the total twice. In order to compensate for this double counting, it is necessary to subtract the number of elements in the intersection. Therefore, the number of elements in the union of two sets that are *not* mutually exclusive is given by

$$n(A \cup B) = n(A) + n(B) - n(A \cap B) \tag{2.6}$$

If Equation 2.6 is used to find the number of elements in the union of two sets which happen to be mutually exclusive, then $(A \cap B) = \varnothing$ and $n(A \cap B) = 0$. In this case, Equation 2.6 simply reduces to Equation 2.5.

Example 2.21

Of the ten people working for the Guardian Insurance Agency, six are men, five are salespersons, and four are office staff.

a) How many are either salespersons or office staff?

$$\begin{aligned} n(R \cup P) &= n(R) + n(P) \\ &= 5 + 4 = 9 \end{aligned}$$

by Equation 2.5. Since these sets are mutually exclusive, there is no intersection.

b) How many are either men or salespersons? Let M represent the set of men.

$$\begin{aligned} n(M \cup R) &= n(M) + n(R) - n(M \cap R) \\ &= 6 + 5 - 4 = 7 \end{aligned}$$

by Equation 2.6.

c) How many are neither office staff nor salespersons?

$$\begin{aligned} n(P \cap R)' &= n(S) - n(P \cap R) \\ &= 10 - 9 = 1 \end{aligned}$$

by Equations 2.3 and 2.4.

Example 2.22

A survey made by a manufacturer of men's shaving products produced the following information about 100 men.

64 men use an electric razor.

56 men use an after-shave lotion.

31 men use a pre-shave lotion.

5 men use all three of the preparations.

35 men use an electric razor and an after-shave lotion.

6 men use both an after-shave lotion and a pre-shave lotion.

20 men use a pre-shave lotion and an electric razor, but not an after-shave lotion.

Let E represent the set of men using an electric razor, A the set of men using an after-shave lotion, and P the set using a pre-shave lotion.

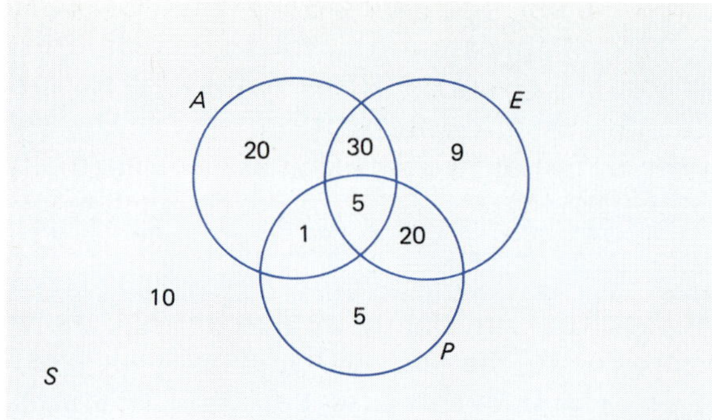

Figure 2.6
Venn diagram showing results of survey regarding the use of shaving products (Example 2.22).

With this information, draw a Venn diagram and show the number of elements that belong to each of the mutually exclusive subsets of the universal set of the 100 men surveyed. This Venn diagram is illustrated in Figure 2.6.

2.5 Permutations and Combinations

In dealing with sets of discrete objects, it is necessary at times to count the number of elements that belong to some specified subset. When these sets or subsets of interest are large, it is inconvenient and sometimes impossible to list all the members and make an actual count. The methods of permutations and combinations are used to provide counts under these circumstances.

A PERMUTATION is an arrangement of elements in some specified order. For example, identify three elements of a set by the letters **a, b,** and **c.** Each arrangement of these elements represents a different permutation. The possible permutations of these three elements are

abc, acb, bac, bca, cab, cba

There are no other possible arrangements of three elements. By listing the possible arrangements and then counting them, we have determined that there are six permutations of three elements when all three elements are taken together.

On the other hand, a COMBINATION is simply a grouping of elements without regard to the order in which the elements appear. If order is not important, then the arrangement **acb** and the arrangement **cba** constitute the same grouping of elements. Therefore, even though these two arrangements represent two different permutations, they are the same combination of elements. All six of the permutations of the letters **a, b,** and **c** listed above represent a single combination of those letters.

The number of possible permutations and the number of possible combinations of some set of elements depend on two factors, the total number of elements in the set, which we will designate n, and the number of elements to be selected at the same time, which we will designate x.

There is a simple rule that provides the basis for determining the number of permutations or combinations of a set of elements. This rule is called the *multiplication rule* and states:

> If an operation consists of k different stages and if the first stage can be performed in n_1 different ways, the second stage in n_2 different ways, and so on for each stage, then the total number of ways in which the entire operation can be performed is $n_1 \cdot n_2 \cdot n_3 \cdot \cdots \cdot n_k$.

Example 2.23

The Futuristic Car Company sells customized automobiles. Each car can be ordered in one of two body styles, any of five exterior colors, and in one of three interior color combinations. In how many different ways can the car be made to order?

The preparation of the car for delivery to the customer is an operation consisting of three stages, that is, $k = 3$. The first stage is the selection and construction of the proper body style, the second is the painting of the car in the desired color, and stage three is the installation of the interior. Since there are two body styles, $n_1 = 2$. There are five possible exteriors, therefore, $n_2 = 5$. The number of interiors is three, so $n_3 = 3$. The multiplication rule states that the total number of ways in which this operation can be performed is $n_1 \cdot n_2 \cdot n_3 = 2 \times 5 \times 3 = 30$.

Permutations. The personnel manager of a small company has three positions to fill. One is for a mail clerk, the second is for a typist, and the third is for a receptionist. Three equally qualified candidates, Abbot, Barnes, and Collins, have applied for a job. In how many different ways can the personnel manager fill the three positions with the three applicants? We will refer to each applicant by the first letter of his or her last name.

The assignment of **a** to the mail clerk's position, **b** to the typing job, and **c** to the job of receptionist is a different way of filling the positions than if **a** is given the receptionist's job, **b** the job as mail clerk, and **c** the typing job.

Think of this assignment of the three people to the three jobs as the total operation. Each stage of the operation consists of filling one position. We now have a three-stage operation and we assign a name to each stage as follows.

Stage 1. Filling the position of mail clerk

Stage 2. Filling the position of typist

Stage 3. Filling the position of receptionist

The first stage of the operation can be performed in three ways, that is, the personnel manager can fill the mail clerk position by assigning either **a, b,** or **c** to that job. However, having made his selection for that job, he has two choices left for the position of typist. Suppose applicant **a** is assigned to the job of mail clerk. Then only **b** and **c** remain for assignment to the next position. Therefore, stage two can be performed in two ways. Once an assignment is made to the typist position, say applicant **b,** then there is only one way to fill the position of receptionist. Since the first position may be filled in three ways, the second in two ways, and the third in only one way, the total number of ways in which the total operation can be performed is $3 \times 2 \times 1 = 6$. We have just determined the number of permutations of three things taken three at a time, which is abbreviated ${}_3P_3$.

It seems logical that if ${}_3P_3 = 3 \times 2 \times 1$, then ${}_4P_4 = 4 \times 3 \times 2 \times 1$ and ${}_nP_n = n(n-1)(n-2) \cdots (4)(3)(2)(1)$. A simple way of specifying this sequence of multiplications is $n!$, which is pronounced "n factorial." The factorial of a number is the product of all the decreasing integers beginning with the number and ending with one.

By convention, $0! = 1$. The number of permutations of n things taken n at a time is

$$ {}_nP_n = n! \tag{2.7} $$

It is also possible to determine the number of permutations of a set of elements when the number of objects selected and arranged at one time is less than the total number of elements in the set. If the personnel manager mentioned above has five applicants for the three positions that are available, add Davis and Egan to the other three, then the problem is to find the number of permutations of five things taken three at a time. This is written ${}_5P_3$.

The reasoning in this problem is the same as in the previous one. We still have a three-stage operation. However, the number of choices for the first stage is now five instead of three, that is, either **a** or **b** or **c** or **d** or **e.** Once a selection is made for the first position, say applicant **a,** then there are only four choices left for the second position. When the second position is filled, there are three choices left for the third. However, when the third position is filled, there are no more positions and the operation is completed. Since there are five ways to fill the first position, four ways to fill the second, and three ways to fill the third, the total number of ways in which this operation can be performed, by the multiplication rule, is $5 \times 4 \times 3 = 60$. This is an illustration of the number of permutations of n things taken x at a time, where x is less than n. The abbreviation commonly used is ${}_nP_x$.

In order to develop a general rule for ${}_nP_x$, we must determine the product of the sequence of decreasing integers beginning with n and ending with the number $(n - x + 1)$. This is the equivalent of beginning with $n!$ and cancelling the sequence of multiplications that begins with $(n - x)$ and ends with 1.

Symbolically, this is written:

$$ {}_nP_x = \frac{n!}{(n - x)!} \tag{2.8} $$

Example 2.24

The art director of the Xavier, Young, and Zale Advertising Agency has created five different displays for a client of the agency. She is required to make a presentation to the client in which she shows and discusses the best displays. Since she is allocated only 45 minutes for her presentation and she feels that she should devote at least 15 minutes to each display, she will have time to present only three of the five displays. The art director believes that the order in which she presents the displays is important. If each order of presentation of three displays is considered to be a different presentation, how many different presentations are possible?

Since it is indicated that the order of presentation is important, this is a problem in permutations. The number of permutations of five things taken three at a time, or ${}_5P_3$, is

$$ {}_5P_3 = \frac{5!}{(5 - 3)!} = \frac{5!}{2!} = 60 \tag{2.8} $$

Example 2.25

The Riggsville Comets, a professional basketball team in the Mid-Continent League, is limited by the league rules to a squad of twelve players. How many different five-member teams are available to the coach if we consider that two lineups consisting of the same five members but playing in different positions constitute two different teams?

Since the order of assignment of the members to positions is important in distinguishing among the possible teams, this is also a problem in permutations. Specifically, it is the number of permutations of twelve things taken five at a time.

$$ {}_{12}P_5 = \frac{12!}{(12 - 5)!} = \frac{12!}{7!} = 95{,}040 \tag{2.8} $$

Combinations. The problem in permutations involves selecting a subset of elements from some larger set and then arranging these elements in some order. Each of these arrangements is a separate permutation. Sometimes interest is centered only on the number of different subsets that can be selected from the larger set without regard to the order of selection or arrangement. Each of these distinct subsets is called a COMBINATION. As illustrated previously, although the groupings **abc, bac,** and **cba** represent different permutations of these three elements, they constitute only a single subset or combination of the elements.

Consider the personnel manager of a small company that has three identical jobs to fill. Five equally qualified candidates, Abbot,

Barnes, Collins, Davis, and Egan, have applied. In how many different ways can the three identical jobs be filled using the five applicants?

Remember that if **a, b,** and **c** are hired to fill the jobs, the order of assignment is completely immaterial, since all the jobs are the same. Therefore, we are interested only in determining the number of different subsets of size three that can be selected from a set of size five. In this case, the subset **a, b, c** is different from the subset **a, b, d.** However, **a, b, c** is the same subset as **c, a, b.** We call this a problem in combinations and state the problem as one of determining the number of combinations of five things taken three at a time. This is abbreviated ${}_5C_3$, a notational device similar to that used for permutations, or $\binom{5}{3}$. For the general case of n things taken x at a time, the notation is either ${}_nC_x$ or $\binom{n}{x}$. We will show how the formula for combinations is derived from the formula for permutations, Equation 2.8.

If the three jobs to be filled are different jobs, that is, distinguishable as mail clerk, typist, and receptionist, then the order of assignment of people to jobs is important. In this case, the permutation formula is used to count the number of ways in which the assignment of people to jobs can be accomplished. That is,

$$(2.8) \qquad {}_5P_3 = \frac{5!}{(5-3)!} = \frac{5!}{2!} = 60$$

These 60 permutations include all possible arrangements of each combination of three applicants. For any combination of three applicants—say **a, b, c**—there are 3! or 6 ways in which they can be arranged. Since the jobs are all identical and the arrangement or order is immaterial, these 3! arrangements must be eliminated from the total number provided by the permutation formula in order to obtain the proper number of combinations. This is accomplished by dividing the total number of permutations by the number we want to eliminate; in this case, we divide by 3!. Therefore,

$${}_5C_3 = \frac{{}_5P_3}{3!} = \frac{5!}{3!(5-3)!} = \frac{5!}{3!2!} = 10$$

To generalize for any n and any x, the number of combinations of n things taken x at a time is

$$(2.9) \qquad {}_nC_x = \binom{n}{x} = \frac{n!}{x!(n-x)!}$$

Example 2.26 If the coach of the Riggsville Comets in Example 2.25 wants to know the number of different five-member teams that can be fielded from a twelve-member squad, and if the same five people constitute a single team regardless of how they are assigned to positions, then the problem is one of combinations. Specifically,

$$\binom{12}{5} = \frac{12!}{5!(12-5)!} = \frac{12!}{5!7!} = 792 \tag{2.9}$$

This result is the same as that obtained by dividing the number of permutations obtained in Example 2.25 by 5!, since 5! equals 120 and 95,040 ÷ 120 = 792.

Example 2.27

If the art director of the Xavier, Young, and Zale Advertising Agency in Example 2.24 is unconcerned with the order in which she presents the displays, we can then use the formula for combinations. In this case, each set of three displays is a single presentation regardless of the order in which the displays are shown. Therefore, the art director wants to know how many distinct sets of three displays can be obtained from the set of five.

Since the order of selection is not to be considered here, the number of combinations can be obtained simply by dividing the number of permutations determined in Example 2.24 by $x!$. Since x, in this case, is 3, and 3! is equal to 6, the number of combinations is 60 ÷ 6 = 10. Using the combinatorial formula, Equation 2.9, the result is the same. That is,

$$\binom{5}{3} = {}_5C_3 = \frac{5!}{3!(5-3)!} = \frac{5!}{3!2!} = 10 \tag{2.9}$$

2.6 Functions

In Chapter 1, the term "variable" was defined as some observable characteristic of an elementary unit which may vary from unit to unit. The letter or symbol, such as x or y, that is used to represent the observation on any element in a specified set, is called a variable over that set. This set of elements over which the variable is defined is called the *domain*, or the *domain set*, of the variable. It is frequently convenient or necessary to establish a relationship or correspondence between the domain set of a variable and the members of a different set. Such a correspondence is called a *function*.

Quite simply, a function assigns to each member of the domain of a variable a single, unique element in another set. This set of elements which is uniquely assigned to the elements in the domain is called the *range*, or the *range set*, of the function. Therefore, a function consists of a domain, a range, and a *rule of association*.

Suppose x represents any member of the domain of a function and y represents any member of the range. Then the functional relationship between x and y can be written $y = f(x)$, where the letter f represents the rule of association.

In mathematics and statistics, functional relationships are usually written as mathematical expressions. However, such relationships can sometimes be specified in tabular form. For example, suppose that the Purple Cab Company operates a fleet of four taxis in Gibbsville. Four drivers work for the company and each week it

TABLE 2.3
PURPLE CAB COMPANY
ASSIGNMENT SHEET,
WEEK OF APRIL 22

Drivers	*Cabs*
Evans	#1
Fisher	#2
Grant	#3
Higgins	#4

is necessary to assign the drivers to the cabs. Therefore, an assignment sheet such as the one illustrated in Table 2.3 is posted on a bulletin board in the company's office each week. If X represents the set of drivers and Y represents the set of cabs, then this assignment sheet represents a function. We have followed a common convention of using capital letters to represent sets and small letters to represent the members of the sets. If x represents any driver in the set of drivers, and y represents any cab in the set of cabs, then the set X is the domain and the set Y is the range of the function. The rule of association that establishes a correspondence between cabs and drivers is represented by the assignment sheet itself.

Suppose that because of the additional revenue that can be generated, the cab company decides to operate an evening shift. Therefore, they hire two new drivers, Jordan and Knight. Since they have only four cabs and do not intend to add any new cabs to the fleet, it is necessary in some cases for two drivers to use the same cab. A new assignment sheet reflecting this change is shown in Table 2.4. Again, the set of drivers is the domain of the function, the set of cabs is the range, and the rule of association is provided by the assignment sheet. This table constitutes a function, since a single unique element of the range is associated with each element in the domain.

If the rule of association assigns a single driver to more than one cab, the relationship is not a function. In order for a relationship to

TABLE 2.4
PURPLE CAB COMPANY
ASSIGNMENT SHEET,
WEEK OF MAY 6

Drivers	*Cabs*
Evans	#1
Fisher	#2
Grant	#3
Higgins	#4
Jordan	#1
Knight	#4

TABLE 2.5
NUMERICAL FUNCTIONS

$f(x) = y = 4x + 5$		$f(x) = y = 2x^2 + 5x + 3$	
x	y	x	y
1	9	1	10
2	13	2	21
3	17	3	36
4	21	4	55
5	25	5	78
6	29	6	105
7	33	7	136

be a function, each element in the domain must be associated with one and only one element in the range.

When the elements in both the range and domain of a function are numbers, the function is called a numerical function. We will be dealing primarily with numerical functions in this book. Functions of this type are usually expressed in the form of a mathematical equation. The general form of this expression is $y = f(x)$, which is read, "y equals a function of x." For any value of x in the domain of the function, the rule of association designates a specific value of y in the range.

Two simple functions that you probably recognize from previous mathematics courses are of the form $y = mx + b$, a linear function, and $y = ax^2 + bx + c$, a quadratic function. In these functions, the letter x is a variable which represents any element in the domain, y is a variable which represents the associated element in the range, and the letters a, b, c, and m are constants. Sometimes the variable x is called the *independent variable* and y the *dependent variable*. If $a = 2$, $b = 5$, $c = 3$, and $m = 4$, then these two functions are

$$y = 4x + 5$$

and

$$y = 2x^2 + 5x + 3$$

If the domain of x is the set of positive integers, then by substituting a value for x in the function and performing the indicated operations we get the corresponding value for y. Table 2.5 shows the value of y which corresponds to several values of x for each of these two functions.

Example 2.28

The president of the Advance Machine Company is interested in determining how the cost of production of a particular product varies with the quantity produced. After examining data provided by the accounting department, he determines that the cost function for this product is

$$y = \$0.75x + \$2000$$

where x represents the quantity produced and y represents the total cost. The numbers 0.75 and 2000 are constants which represent the cost of each individual item produced and the fixed costs of production. The total costs for $x = 100$ units and $x = 1000$ units can be determined from the cost function:

$$y = \$0.75(100) + \$2000 = \$2075$$
$$y = \$0.75(1000) + \$2000 = \$2750$$

Example 2.29 Economic theory indicates that the quantity of a product that will be demanded at some specific time is a function of the price charged. Generally, the higher the price the lower the demand and vice versa. A function that describes this correspondence is called a *demand function*.

The marketing department of the Advance Machine Company determines that the demand function for one of their products could be described by

$$x = f(p) = 1200 - 50p^2$$

where x is now the dependent variable representing the quantity demanded and p is the independent variable representing the price in dollars. Table 2.6 shows the demand for the product for several values of p, the price.

TABLE 2.6
DEMAND AS A FUNCTION OF PRICE. $f(p) = x = 1200 - 50p^2$

p	x
0	1200
1	1150
2	1000
3	750
4	400
5	−50

This demand function shows that even if the price is \$0, there is a demand for only 1200 units of the product. This is not an unreasonable result if the product is extremely specialized and has a very limited application. As the price increases, the demand decreases. At some price between \$4 and \$5 the demand is zero. A negative value for demand in this product means that if the price rises above a certain point, potential buyers might find it less costly to manufacture the product themselves.

2.7 Looking Ahead

The preceding material was not intended as a course in mathematics, but was included as review material for those readers who require it. If you understand all of these topics you should be well prepared for the study of statistics. Chapters 3 through 6 are devoted to topics in descriptive statistics. Because these topics require that you deal with sets of numerical data, the material relating to number accuracy, significant digits, and rounding conventions will be useful to you.

Chapters 7 through 9 discuss probability and sampling concepts. These topics provide the basis for the study of statistical inference. In this part of the book you will begin to apply the ideas and techniques illustrated in the sections on sets, permutations and combinations, and functions.

Chapters 10 through 18 discuss statistical inference and related topics. Since statistical techniques are developed logically from the foundations previously established, you will find that all the topics covered to this point will be used throughout these last chapters. Therefore, you should be sure that you understand each topic thoroughly before going on to the next one.

Definitions

Function: A relationship between two sets, called the domain and the range, which assigns a single element of the range to each element in the domain.

Parameter: Some constant characteristic of a population.

Set: A collection of elements having some describable similarity.

Formulas

Rules pertaining to sets:

(2.1) $$A \cup A' = S$$

(2.2) $$A \cap A' = \phi$$

(2.3) $$n(A) + n(A') = n(S)$$

(2.4) $$n(A') = n(S) - n(A)$$

(2.5) $n(A \cup B) = n(A) + n(B)$ if A and B are mutually exclusive

(2.6) $$n(A \cup B) = n(A) + n(B) - n(A \cap B)$$

Permutations:

(2.7) $$_nP_n = n!$$

(2.8) $$_nP_x = \frac{n!}{(n-x)!}$$

Combinations:

(2.9) $$_nC_x = \binom{n}{x} = \frac{n!}{x!(n-x)!}$$

Problems

Summation and subscripting.

2.1 Ten observations on the variable x result in the following data.

$x_1 = 17$ $x_2 = 45$ $x_3 = 4$ $x_4 = 27$ $x_5 = 30$

$x_6 = 11$ $x_7 = 9$ $x_8 = 15$ $x_9 = 44$ $x_{10} = 10$

Find **a)** $\sum x$ **b)** $\sum_{i=1}^{10} x_i$ **c)** $\sum_{i=1}^{5} x_i$ **d)** $\sum_{i=4}^{8} x_i$

2.2 Ten observations on the variable y result in the following data.

$y_1 = 67$ $y_2 = 95$ $y_3 = 54$ $y_4 = 77$ $y_5 = 80$
$y_6 = 61$ $y_7 = 59$ $y_8 = 65$ $y_9 = 94$ $y_{10} = 60$

Find **a)** $\sum y$ **b)** $\sum_{i=4}^{10} y_i$ **c)** $\sum_{i=1}^{8} y_i$ **d)** $\sum_{i=3}^{6} y_i$

2.3 Find the following sums from the data in problem 2.1 if $c = 5$ and $k = 2$.

a) $\sum cx$ **b)** $\sum kx$ **c)** $\sum (cx + k)$ **d)** $\sum (kx + c)$

2.4 Find the following sums for the data in problem 2.2 if $a = 3$ and $b = 10$.

a) $\sum ay$ **b)** $\sum by$ **c)** $\sum (ay + b)$ **d)** $\sum (by + a)$

2.5 From the data in problems 2.1 and 2.2, find the following sums if $a = 4$, $b = 6$, and $c = 2$.

a) $\sum (x + y)$ **b)** $\sum (ax + by)$ **c)** $\sum (bx + ay)$
d) $\sum (x + by + c)$

Significant digits and rounding.

2.6 How many significant digits are there in the following numbers?

a) 7.640 **b)** 0.002 **c)** 540 **d)** 5.042
e) 88 **f)** 2.30

2.7 How many significant digits are there in the following numbers?

a) 947 **b)** 941.02 **c)** 4.10 **d)** 28,000
e) 6002 **f)** 0.10

2.8 Perform the following operations and round the answer to the proper number of significant digits.

a) $64.32 + 121 + 24.3 + 19.08$
b) $500.10 - 25.2$
c) 24.01×9.34
d) 4.2×231
e) $(6.39)^2$
f) $\sqrt{87}$

2.9 Perform the following operations and round the answers to the proper number of significant digits.

a) $872.48 \div 12$
b) $104.7 \div 38.235$
c) $82.2 - 59.689$
d) $(12.7)^2$
e) $86.5 + 1.0103 + 104.7 + 507$
f) $\sqrt{422.2}$

Logarithms.

2.10 Perform the following operations using logarithms:

a) (62.4) (7.8) **b)** 15.7 ÷ 3.2 **c)** $(17.6)^2$ **d)** $\sqrt[2]{323}$

2.11 Perform the following operations using logarithms:

a) (35.2) (15) **b)** 78.4 ÷ 18.1 **c)** $(4.33)^3$ **d)** $\sqrt[2]{856}$

2.12 Perform the following operations using logarithms:

a) (17.7) (24) **b)** 432.0 ÷ 23.8 **c)** $(5.8)^4$ **d)** $\sqrt[3]{769}$

Sets and set operations.

2.13 Let $A = \{1,2,3,4,5,6\}$, $B = \{2,4,6,8,10\}$, and $S = \{1,2,3,4,5,6,7,8,9,10\}$. List the elements of the following sets.

a) $A \cap B$ **b)** $A \cup B$ **c)** $A' \cap B$ **d)** B' **e)** $(A' \cap B')$
f) $(A \cap B)'$ **g)** $(A \cup B)'$

2.14 Let $A = \{1,2,3\}$, $B = \{3,4,5,6\}$, $C = \{4,5,6,7\}$, and $S = \{1,2,3,4,5,6,7,8,9,10\}$. List the elements of the following sets.

a) $A \cap C$ **b)** $B \cup C$ **c)** $B \cap C'$ **d)** $(A \cup B) \cap C$
e) $(A \cup B)' \cap C$ **f)** $(A \cap B \cap C)'$ **g)** $(A \cup C)'$

2.15 For the sets specified in problem 2.13 and using Equations 2.3 through 2.6 in the text, find the number of elements in the following sets.

a) $A \cup B$ **b)** B' **c)** $(A \cup B)'$ **d)** $A' \cup B'$ **e)** A'

2.16 For the sets specified in problem 2.14 and using Equations 2.3 through 2.6 in the text, find the number of elements in the following sets.

a) $A \cap B \cap C$ **b)** $A \cup C$ **c)** $B \cup C$ **d)** $A \cup C'$
e) $(A \cup C)'$ **f)** $A' \cup B$ **g)** $A \cup B \cup C$

2.17 Draw a Venn diagram illustrating the sets designated in problem 2.13 and place all the elements in their proper mutually exclusive subsets.

2.18 Draw a Venn diagram illustrating the sets designated in problem 2.14 and place all the elements in their proper mutually exclusive subsets.

2.19 Draw a Venn diagram illustrating sets A, B, and C so that all of the following requirements are met.

1. $A \cap B = \{x\}$
2. $B \cap C = \{\varnothing\}$
3. $A \cup B = \{w,x,y,z\}$
4. $A \cap C = \{y\}$
5. $A \cup C = \{v,x,y,z\}$

2.20 For the previous problem:

a) List the elements in set A.
b) List the elements in set B.
c) List the elements in set C.

2.21 Using only the "and," "or," and "prime" symbols, describe the set designated by the shaded portion of each of the following Venn diagrams.

a)

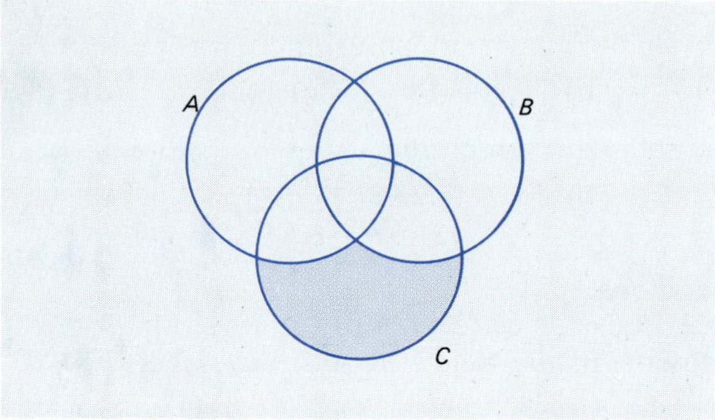

b)

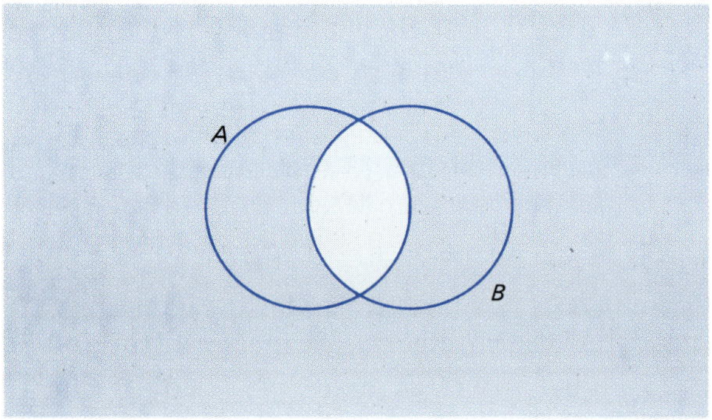

c)

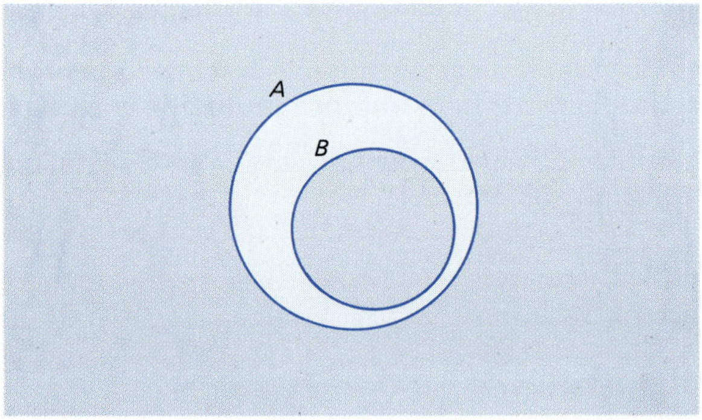

2.22 Using only the "and," "or," and "prime" symbols, describe the set designated by the shaded portion of each of the following Venn diagrams.

a)

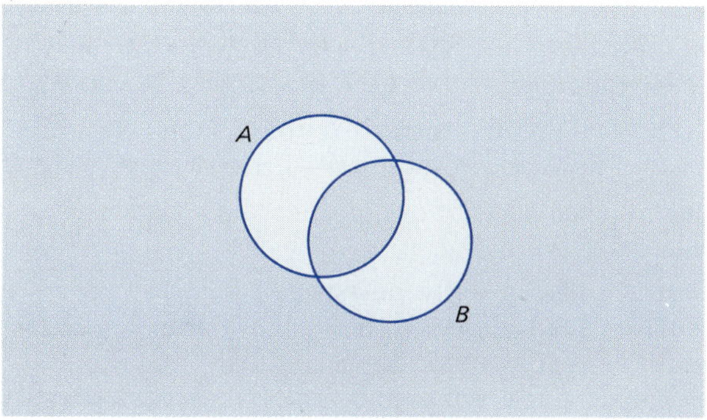

b)

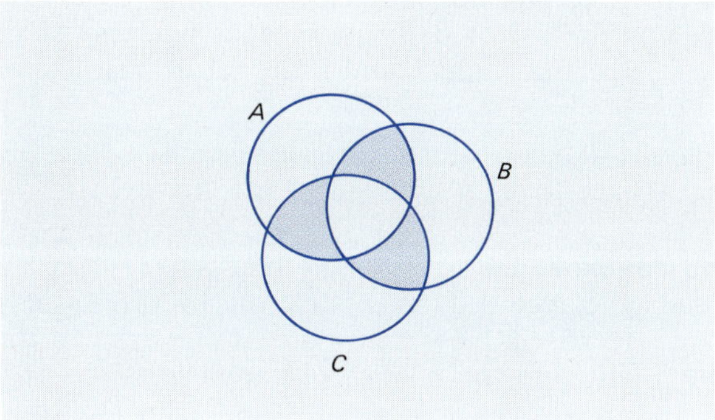

c)

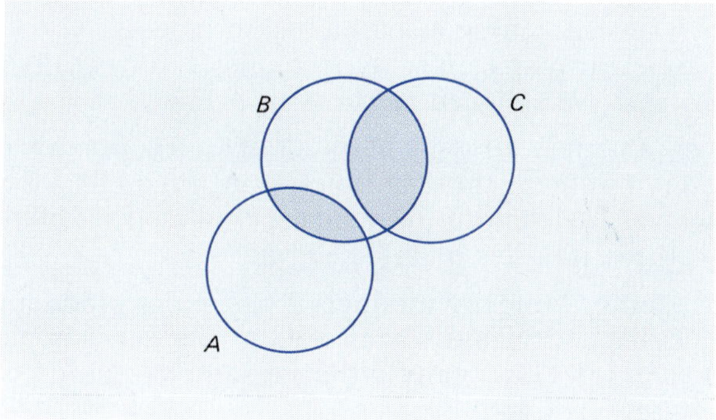

2.23 A survey of employees of Metric Systems, Inc. indicated that 550 owned their own homes, 610 owned cars, 700 owned television sets, 450 owned both cars and homes, 500 owned homes and television sets, and 400 owned all three. If the set of car owners is designated C, the set of home owners is designated H, and the set of television owners is designated T:

a) Draw a Venn diagram illustrating these facts.
b) How many employees responded to the survey?

2.24 A survey of top-level executives conducted by a management association revealed the following.

350 executives read a daily newspaper.
180 executives read a weekly news magazine.
230 executives read a business periodical.
100 read both a daily newspaper and a business periodical.
85 read both a weekly news magazine and a business periodical.
90 read both a daily newspaper and a weekly news magazine.
15 read all three types of publications.

Let N represent the set of daily newspaper readers, M the set of weekly news magazine readers, and P the set of business periodical readers:

a) Draw a Venn diagram illustrating these facts.
b) How many executives responded to the survey?

Permutations and combinations.

2.25 National Foods, Inc. received ten applications for three sales positions which it had advertised. How many different ways can these three positions be filled from the pool of ten applicants?

2.26 National Foods, Inc. constructed five new executive offices facing Main Street on the tenth floor of their headquarters building.

a) If five managers are to be assigned to these offices, how many different ways can these assignments be made?
b) If there are eight managers who need offices, how many different ways can they be assigned to these five offices?

2.27 In conducting a sample survey, a market research company intends to conduct interviews with three families living on a city block which contains twelve single-family residences. How many different ways could these three families be selected?

2.28 Advance Machine Company produces a particular product in lots of 20 pieces. Their quality control department inspects samples of five units from each lot. How many different samples could possibly be selected from a single lot?

2.29 A housing developer has seven different house plans. He intends to build five sample houses on one block in a new development. How many different arrangements of these houses are possible?

2.30 Metric Systems, Inc. has twelve directors on its board. At each annual meeting a chairman, a vice-chairman, a secretary, and a treasurer are elected. How many different slates of officers are possible?

Functions.

2.31 National Foods has determined that profit is a function of sales and variable costs, that is, $P = f(x, y)$, where x represents sales in units and y represents the variable costs in dollars. If $f(x, y) = \$35x - 15y - c$, and c is a constant \$15,000, find the profit when $x = 2000$ units and $y = \$3000$.

2.32 If income in dollars is a function of the number of units of a particular product that is sold, this relationship can be expressed as $R = f(x) = 15x$. In this equation, R stands for income and x represents sales in units of the product. Find the value of income for sales of 100, 200, 300, 400, and 500 units of the product.

2.33 The personnel manager of Metric Systems, Inc. devised a formula for allocating overtime work to production employees. This formula makes the maximum amount of overtime that might be worked by an employee in any one month a function of the employee's seniority. Seniority is defined as the number of full years of service that the employee has been with the company. The maximum number of years that might count toward determining overtime is ten. If x is years of service and y is overtime hours per month, then

$$y = f(x) = 10 - x + x^2 \quad \text{for} \quad x \leq 10$$

a) How many hours of overtime could an employee with one year of service work?

b) How many hours of overtime could an employee with four years of service work?

c) How many hours of overtime could an employee with eight years of service work?

d) What is the maximum number of overtime hours that anyone is allowed to work?

2.34 In studying the profit function, National Foods determined that variable costs, y, are also a function of x, sales in units. The relationship between y and x is, $y = f(x) = \$20x$. If P represents profit and $P = \$35x - y - c$, find the profit when $x = 3500$ units and $c = \$15,000$.

2.35 A door-to-door salesman determined that he closed one sale for every five calls he made. Express this relationship between sales and calls in the form of a function. Let x represent sales and z represent calls.

CHAPTER 3 DATA ORGANIZATION

3.1
Frequency distributions

3.2
Special frequency distributions

3.3
Categorical distributions

3.4
Histograms

3.5
Practical uses of frequency distributions and histograms

3.6
Other statistical graphs

3.7
General purpose tables

We have seen in Chapter 1 how statistical data result naturally from the process of collecting and recording all types of transactions in business organizations. We have also discussed how these data are used by managers to help them make decisions that can affect their organizations. In dealing with statistical data, the first problem managers face is to make sense of a large and confusing set of numbers that are presented in no logical order and appear to make little if any sense. This is done by using the methods of descriptive statistics.

Remember that in descriptive statistics we assume that the raw data represent a complete population. Raw data are obtained by observing some characteristic of each elementary unit in the set. Whenever observations are made on an elementary unit and recorded, information is obtained about that unit as well as the population to which that unit belongs. Each individual observation contains a bit of information. An entire population of observations contains all of the information regarding the observed characteristic.

There is no statistical procedure or technique that can extract more information from a set of data than was in the data to begin with. In order to create something from nothing, one would have to be a magician, and statisticians are not magicians. Statistical procedures, however, are designed to obtain information from data in a usable form. Some of these procedures concentrate on picturing the entire set of data, while others concentrate on a single characteristic of the set. The most important characteristics of a population of numerical values are variation, location, and symmetry.

The amount or degree to which members of a population differ from one another is probably the most important statistical characteristic of the population. Differences among individuals, called variation or variability in statistics, appear to be inherent in nature. No two objects are ever exactly alike. Even in a manufacturing process, where parts or products are turned out on a production line from a single set of plans or specifications, variation occurs. Every unit produced is always slightly different from every other one. (Methods of describing variation are covered in Chapter 5.)

Location refers to the position of the data with reference to some scale of numerical values. For example, wages earned by various types of production workers employed by a large manufacturing company range from $2.75 per hour for unskilled workers to $8.75 per hour for highly skilled technicians. The scale of numerical values which describes these wages runs from 2.75 to 8.75. A lathe operator's wages may be located on this scale between the values 5.75 and 6.25. A value which separates the highest paid half of the company's employees from the lowest paid half might be located on the scale at the value 6.15. There are several measures of location that can be obtained from a set of data. These are covered in Chapter 4.

Measures of symmetry describe the degree to which the data deviate from being symmetrical about some middle point in the set. Symmetry refers to the correspondence of equal numbers of observations on each side of the midpoint. If the data are not symmetrical, they are said to be *skewed*. A measure of skewness can also be computed from a set of data. Symmetry and skewness are discussed in Chapter 5.

If measures of variation, location, and symmetry are all computed from a set of data, they provide a reasonably good description of the data set. However, one can have values for all of these descriptive measures and still not be able to visualize the population they describe. This situation is similar to that of having several descriptive measures of a person, such as height, weight, and age, and still not being able to picture that person. Statisticians use a variety of tables and graphs to visually present characteristics of a set of statistical data. The most useful type of statistical table is called a FREQUENCY TABLE or FREQUENCY DISTRIBUTION.

3.1 Frequency Distributions

Suppose that variables data have been collected on 5000 production employees of a very large manufacturing company. To be more specific, assume that for each employee an observation is made of the number of hours worked during the previous week. Each employee is listed by name or by social security number and you, as production manager, are now confronted with 5000 numbers that represent production hours worked. These data contain useful information. However, the arrangement of the numbers is completely unrelated to their magnitudes or any other characteristic of the data. Your problem is to find a way to extract the useful information from the data.

Without any knowledge of the methods of descriptive statistics, you might find that it is extremely difficult to get any useful information from the data. You might even throw your hands in the air, throw the data in the wastebasket, and make whatever decisions that are required without any reference to the data. Frustration resulting from inadequate procedures for handling large masses of data has often led to similar reactions. In these situations despair is not necessary but statistical methods are.

A set of data does not need 5000 elements to be incomprehensible. A set containing as few as 50 numbers can lead to the same reaction. Suppose a small manufacturing company has 50 production workers of a particular type. These 50 workers produce a single component part which is sold to a large office machine manufacturer for incorporation into one of its products. Every week, the production manager of our company receives two reports. One lists the number of hours worked and the second lists the number of units produced by each of the workers. Tables 3.1 and 3.2 illustrate these weekly reports.

If these tables appear confusing, imagine what they would look like if each one contained 5000 numbers instead of 50. Before any useful information can be obtained from these tables, we must rearrange the data. The first step is to arrange the numbers in the table according to their size. That is, arrange them in order from smallest to largest or largest to smallest. This arrangement of data is called a STATISTICAL ARRAY or simply an ARRAY. Tables 3.3 and 3.4 show the data from Tables 3.1 and 3.2 in an array.

When the raw data are rearranged into an array, some of their characteristics immediately become apparent. For example, the range of data values is now easy to determine. The range of hours worked by these production employees varies from 36.1 to 46.0 hours. The range of output per employee varies from 67 to 83 units. This in itself is useful information for a manager.

The array also makes it easy to locate an approximate midpoint

TABLE 3.1

TABULATION OF PRODUCTION HOURS WORKED, WEEK OF APRIL 12

42.7	45.1	41.3	36.4	40.8
39.0	37.7	40.5	39.8	46.0
41.3	40.8	40.3	41.4	38.9
40.4	41.0	40.8	40.6	41.2
39.8	45.3	39.2	41.7	40.4
40.7	38.4	40.0	42.5	38.4
36.2	39.6	39.9	39.9	40.3
36.8	40.5	36.1	40.0	38.6
45.8	42.1	43.8	37.6	44.2
41.7	40.0	43.0	39.4	40.8

TABLE 3.2

TABULATION OF PARTS PRODUCED, WEEK OF APRIL 12

74	75	77	72	75
71	79	69	74	70
83	75	76	77	76
75	79	73	73	74
76	75	74	77	79
81	71	67	77	81
70	71	74	78	80
68	77	75	74	75
74	70	76	72	78
77	75	74	77	75

TABLE 3.3
TABULATION OF PRODUCTION HOURS WORKED ARRANGED IN ORDER OF MAGNITUDE, WEEK OF APRIL 12

36.1	39.0	40.0	40.8	42.1
36.2	39.2	40.3	40.8	42.5
36.4	39.4	40.3	40.8	42.7
36.8	39.6	40.4	41.0	43.0
37.6	39.8	40.4	41.2	43.8
37.7	39.8	40.5	41.3	44.2
38.4	39.9	40.5	41.3	45.1
38.4	39.9	40.6	41.4	45.3
38.6	40.0	40.7	41.7	45.8
38.9	40.0	40.8	41.7	46.0

TABLE 3.4
TABULATION OF PARTS PRODUCED ARRANGED IN ORDER OF MAGNITUDE, WEEK OF APRIL 12

67	72	74	76	77
68	73	75	76	78
69	73	75	76	78
70	74	75	76	79
70	74	75	77	79
70	74	75	77	79
71	74	75	77	80
71	74	75	77	81
71	74	75	77	81
72	74	75	77	83

for each set of data. Start at one end of the array and find a value that is approximately half the distance from each end. For instance, you can determine from Table 3.4 that the number 75 is located in the center of the array and might be considered to be fairly representative of the output of all 50 employees.

However, even an array is confusing when there are a great many values. Therefore, it is usually desirable to condense the data into a more compact form that tends to highlight the more important characteristics. This reorganization of the array to highlight its essential characteristics leads to a FREQUENCY DISTRIBUTION.

A frequency distribution divides the entire range of values covered by the data into a few classes and records the number of observations that fall into each class. The number of observations in each class is referred to as the FREQUENCY.

Two major problems arise in constructing frequency distributions. The first involves the number of classes into which the data are grouped. The second is that of defining the individual classes.

Determining the number of classes in the distribution is somewhat arbitrary and depends largely on the judgment of the analyst, although there are certain guidelines that can be used. As a general rule, a frequency distribution should contain at least five classes and not more than twenty. If there are too few classes, the distribution of the data is not apparent. On the other hand, too many classes also tend to obscure the characteristics of the data. In the extreme case, there can be a separate class for every data element in the set. This, of course, represents no improvement over the array.

There is a simple rule often used to approximate the number of classes in a frequency distribution. This rule is based on the number of observations in the set of data, and is stated as follows.

> If k represents the number of classes and N the number of observations, let the value of k be the smallest exponent of the number 2, so that $2^k \geq N$.

Applying this rule to the data on hours worked and units produced, where, in both cases $N = 50$, we get:

$$2^5 = 32, \text{ which is smaller than } 50,$$
$$2^6 = 64, \text{ which is greater than } 50,$$

and, therefore, we choose $k = 6$ as the number of classes.

For other values of N, we get the following results:

$N = 100$, $2^6 = 64 < N$, $2^7 = 128 > N$, therefore use 7 classes.
$N = 500$, $2^8 = 256 < N$, $2^9 = 512 > N$, therefore use 9 classes.
$N = 1000$, $2^9 = 512 < N$, $2^{10} = 1024 > N$, therefore use 10 classes.
$N = 10000$, $2^{13} = 8192 < N$, $2^{14} = 16384 > N$, therefore use 14 classes.

It should be emphasized that this rule provides only a rough guide to the number of classes and should not be applied uncritically.

The final choice of the number of classes to be used in any situation is a matter of judgment, which should be based on several other considerations, including these:

1. Since each class covers a finite range of values, called the class interval, the number of classes and the class intervals should be chosen so that the individual observations are distributed fairly evenly over each interval. If any clustering of observations does occur, it should occur near the middle of the intervals.

2. The number of classes and the class intervals should be chosen to avoid the occurrence of empty classes, that is, classes with zero frequency.

3. The classes must be all inclusive, that is they must be chosen so that they cover the entire range of values included in the data.

4. The classes must be mutually exclusive. Every observation must fall into one and only one class.

5. It is desirable that the class intervals are the same for all classes.

The data on hours worked and units produced by 50 employees of the Advance Machine Company, displayed in Tables 3.1 through 3.4, are organized into frequency distributions in Tables 3.5 and 3.6. Note that each frequency distribution is constructed with six classes, which is consistent with the guidelines suggested above.

The second major problem in constructing a frequency distribution is to define the limits of the classes. Each class interval must have an upper and lower CLASS LIMIT, which specifies the highest and the lowest recorded value included in that class. For example, in Table 3.6, in the first class the upper limit is 68 and the lower limit is 66. Class limits must be stated so that each data element belongs to one and only one class. The manner in which the class limits are specified depends on the nature of the data being grouped and the way the data are recorded and rounded.

TABLE 3.5
FREQUENCY DISTRIBUTION OF PRODUCTION HOURS WORKED

Hours Worked	Number of Employees
36.0 and under 37.8	6
37.8 and under 39.6	7
39.6 and under 41.4	24
41.4 and under 43.2	7
43.2 and under 45.0	2
45.0 and under 46.8	4
Total	50

TABLE 3.6
FREQUENCY DISTRIBUTION OF PARTS PRODUCED

Parts Produced	Number of Employees
66 to 68	2
69 to 71	7
72 to 74	12
75 to 77	20
78 to 80	6
81 to 83	3
Total	50

Consider a set of continuous variables data, such as the number of hours worked by 50 production employees of the Advance Machine Company. These employees punch a time clock when they arrive at work and again when they leave. The time clock measures time and records it on the employee's time card, which is his record for the week. These time cards are processed to produce a number corresponding to the time worked during the week for each employee. However, the time clock measures time only to the *last* tenth of an hour. If an employee works eight hours and five minutes on some specific day, credit is given for 8.0 hours, since five minutes is less than one-tenth of an hour. However, if the employee works eight hours and six minutes, credit is given for 8.1 hours, since six minutes is one-tenth of an hour.

Table 3.5 shows an unambiguous method for specifying the class limits for continuous data, which is rounded to the last recorded unit, in this case the last tenth of an hour. In this table, all time observations from 36 hours up to, but not including, 37.8 hours belong to the first class of the distribution. An observation of 37.8, however, goes into the second class, as do all values up to but not including 39.6. In this table the stated class limits and the actual class limits are identical. This is not always the case. Suppose, for example, that Advance recorded time to the *nearest* full hour for pay purposes. Therefore, although the time clock measures the time to the tenth of an hour, when the time card data are summarized for the week, the hours worked are rounded to the nearest hour. In the total for the week, fractions of an hour less than 0.5 are rounded to the previous whole hour. Fractions 0.5 or greater are rounded to the next greater number of hours. Using the data from Table 3.3, an acceptable method for stating the class limits is

Hours Worked		*Number of Employees*
35 to 36		3
37 to 38		5
39 to 40		17
41 to 42		16
43 to 44		5
45 to 46		4
	Total	50

Notice how this distribution differs from the one in Table 3.5 as a result of the different method used for rounding the data. In determining the frequencies, which the reader should verify by referring to Table 3.3, all values recorded, after rounding, as 35 and 36 go into the first class. A recorded value of 37 goes into the second class. Notice, also, that because of the rounding process, the stated class limits are not the same as the actual class limits. The first class includes all observations that have values from 34.5 up to, but not including, 36.5. The second class includes all observations with values from 36.5 up to, but not including, 38.5, and so on. These actual class limits are called class BOUNDARIES. The fact that the class boundaries of adjacent classes overlap creates no problems, since none of the observations are recorded to the tenth of an hour and therefore cannot possibly fall on a class boundary. Class boundaries are always defined to a degree of precision that is greater than that of the recorded data. It is important to remember that the boundary values are numbers that cannot possibly occur among the data that are to be classified or grouped.

The CLASS INTERVAL, which is a measure of the range of values covered by each class, can be obtained in several ways. The lower limit of a class can be subtracted from the lower limit of the next higher class. The same result can be obtained by subtracting the upper limit of a class from the upper limit of the next higher class. Finally, the class interval can be obtained by subtracting the lower class boundary from the upper class boundary of the same class. For the frequency distribution of hours worked in Table 3.6, the class interval is 3. This can be obtained in several ways:

1. The difference between the lower class limits of the first two classes is $69 - 66 = 3$.

2. The difference between the upper class limits of the first two classes is $71 - 68 = 3$.

3. The lower and upper class boundaries of the second class are 68.5 and 71.5, the difference between which is $71.5 - 68.5 = 3$.

The frequency distribution of parts produced in Table 3.6 involves discrete data that do not require any rounding. Here the class limits are stated unambiguously in the same units in which the original observations are made. In this case, the stated class limits and the actual class limits are identical. However, in order to maintain consistency in our terminology, the class boundaries are still defined as falling halfway between the upper class limit of one class and the lower class limit of the next higher class.

Whenever data are grouped into a frequency distribution, the values of the individual observations are lost. This does not mean that they have been misplaced, but simply that these individual values do not appear in the frequency distribution, which only indicates class limits and frequencies. Therefore, when calculations of various summary measures of the distribution have to be made, some value in each class must be selected to represent all the observations that fall into that class. The representative value in each class is the midpoint and is called the CLASS MARK. The class mark is obtained most easily by averaging the two class boundaries or the two class limits of the same class. For example, for the frequency distribution of Table 3.5, where the limits and boundaries are identical, the class mark of the first class is obtained by adding the upper and lower class boundaries and dividing by two,

$$\frac{36.0 + 37.8}{2} = 36.9$$

The class marks of every class in the distribution can be obtained in the same way. However, it is much simpler to find the class mark of the first class in the distribution and then add the class interval to this value to obtain the class mark of the next succeeding class. Since the class interval for this distribution is 1.8, we add 1.8 to the first class mark of 36.9 to get 38.7, which is the class mark of the second class. By repeating this procedure, we can find the other class marks.

Referring to the frequency distribution of Table 3.6, where the class limits and class boundaries are not the same, the class mark of the first class is obtained by adding the upper and lower class limits and dividing by two,

$$\frac{66 + 68}{2} = 67$$

or add the upper and lower class boundaries and divide by two,

$$\frac{65.5 + 68.5}{2} = 67$$

Since the class interval for this distribution is 3, the class marks of the other five classes are most easily obtained by adding three to each class mark to get the class mark of the next class. The class marks of the six classes of this distribution are 67, 70, 73, 76, 79, and 82. This distribution with the class marks added is shown as Table 3.7.

TABLE 3.7
FREQUENCY DISTRIBUTION OF PARTS PRODUCED

Parts Produced (Classes)	*Class Mark*	*Number of Employees (Frequency)*
66 to 68	67	2
69 to 71	70	7
72 to 74	73	12
75 to 77	76	20
78 to 80	79	6
81 to 83	82	3
	Total	50

One question that is frequently asked about the construction of frequency distributions is: Must the class intervals always be the same for all the classes? The answer is that it is not absolutely necessary but it is most certainly desirable. Unequal class intervals introduce complications and make it more difficult to compute various descriptive measures from the distribution. It also makes it more difficult to compare and interpret class frequencies.

Unequal class intervals usually occur when open-ended classes are defined. An open-ended class interval always occurs at one end of the distribution and has only one defined class limit. That is, if it is at the upper end of the distribution, it has only a lower limit; if it is at the lower end, it has only an upper limit. For example, if the frequency distribution of parts produced in Table 3.6 has its highest class defined as 81 and over, this is an open-ended interval. The lower class limit is 81, but no upper class limit is specified. At the other end of the distribution, a class defined as 68 and under is also open-ended, with no specified lower limit. The difficulty with open-ended classes is that it is impossible to find the class mark or midpoint of the class. Since the class mark is a representative value for that class and is necessary for various computations, you can see that complications arise if the class mark can't be determined.

Unfortunately open-ended classes are sometimes difficult to avoid. This can occur when the data contain a few extreme values. An open-ended class to accommodate one or two extreme values avoids the necessity for having a great many empty classes in the distribution, that is, classes with zero frequency. It also avoids the need for having more classes than the guidelines recommend. When open-ended classes appear to be unavoidable, it is advisable to note the values of the observations that fall into these classes in a footnote to the table.

3.2 Special Frequency Distributions

Frequency distributions, such as those just discussed, are sometimes called *absolute frequency distributions* because they provide the absolute number of values that fall into each class. There may be circumstances, however, where characteristics of the data can be emphasized by some other arrangement of the frequencies. Suppose that in considering the distribution of hours worked, the production manager is primarily interested in knowing how many production employees worked less than a certain specified number of hours. This type of information is emphasized in a CUMULATIVE FREQUENCY DISTRIBUTION. Such a distribution is constructed from the absolute distribution by starting at the lowest class and successively adding the frequencies as shown in Table 3.8, which is a cumulative distribution of hours worked. A cumulative distribution that starts at the lowest class and accumulates upward is called a "less than" cumulative frequency distribution.

TABLE 3.8
CUMULATIVE FREQUENCY DISTRIBUTION OF HOURS WORKED

Hours Worked	*Cumulative Frequency*
Less than 36.0	0
Less than 37.8	6
Less than 39.6	13
Less than 41.4	37
Less than 43.2	44
Less than 45.0	46
Less than 46.8	50

Note that the class of 13 people who worked less than 39.6 hours includes the six people who worked less than 37.8 hours. Note also that everyone worked at least 36 hours and no one worked 46.8 hours or more.

In converting an absolute frequency distribution to a cumulative frequency distribution, not only are the frequencies accumulated, but the method of defining the classes is changed. The cumulative frequency that corresponds to each class designation represents the total number of observations that fall in that class and all the classes that lie below it. Therefore, for a cumulative distribution, each class is defined by a single limit and a description as to how the frequencies relate to that limit. For the sake of completeness, in Table 3.8, we indicate that no observations fall below the lowest class of the absolute distribution.

In some circumstances, interest might focus on the number of observed values that are greater than certain specified levels. For example, the production manager might want an arrangement of the data that emphasize production output greater than various levels. Such a distribution is referred to as a "more than" cumula-

TABLE 3.9
CUMULATIVE FREQUENCY DISTRIBUTION OF PARTS PRODUCED

Parts Produced	*Cumulative Frequency*
66 or more	50
69 or more	48
72 or more	41
75 or more	29
78 or more	9
81 or more	3
84 or more	0

tive distribution, and is illustrated in Table 3.9. In constructing this distribution, we start with the highest class and accumulate the frequencies downward. The highest cumulative frequency, whether we are accumulating upward or downward, is always equal to the sum of all the frequencies in the absolute frequency distribution.

An absolute frequency distribution can also be revised to show the relative frequency or the percent of observations that fall into each class. When the data is presented in this manner, the distribution is called a RELATIVE FREQUENCY DISTRIBUTION or a PERCENTAGE DISTRIBUTION.

A relative frequency is the same as a PROPORTION. Absolute frequencies are converted to relative frequencies by dividing each absolute frequency by the total number of frequencies in the distribution. For example, since the total of all frequencies in the distribution of parts produced in Table 3.6 is 50, and since three of these observations fall in the class 81–83, the relative frequency of this class is $3 \div 50 = 0.06$. If we convert the absolute frequency distribution of Table 3.6 to a relative frequency distribution, we get the distribution of Table 3.10.

Relative frequencies such as 0.06 are converted to percents simply by multiplying by 100. The percentages corresponding to the relative frequencies are also provided in Table 3.10. Notice that the sum of the relative frequencies is equal to 1.00 and that the sum of the percents is equal to 100.

There is one other type of distribution that can be constructed from a set of data. This is obtained by combining the cumulative distribution with the relative distribution resulting in a CUMULATIVE RELATIVE FREQUENCY DISTRIBUTION. In other words, relative frequencies are accumulated in exactly the same way as absolute frequencies whenever it is desirable to show the proportion of observations that fall below some specified value. This is illustrated for the data on parts produced in Table 3.11.

TABLE 3.10
RELATIVE FREQUENCY DISTRIBUTION OF PARTS PRODUCED

Parts Produced		*Relative Frequency*	*Percent*
66 to 68		0.04	4
69 to 71		0.14	14
72 to 74		0.24	24
75 to 77		0.40	40
78 to 80		0.12	12
81 to 83		0.06	6
	Total	1.00	100

TABLE 3.11
CUMULATIVE RELATIVE FREQUENCY DISTRIBUTION OF PARTS PRODUCED

Parts Produced	*Cumulative Relative Frequency*
Less than 66	0
Less than 69	0.04
Less than 72	0.18
Less than 75	0.42
Less than 78	0.82
Less than 81	0.94
Less than 84	1.00

3.3 Categorical Distributions

In Chapter 1 we discussed the difference between attributes, or qualitative data, and variables, or quantitative data. When some observable characteristic of an elementary unit is not inherently numerical, we call that characteristic an attribute. Such a characteristic can only be observed as to its presence or absence. However, numbers can be associated with an attribute by counting the elementary units that possess or lack the specific attribute. When classes in a frequency distribution are specified in terms of attributes rather than numerical values, the distribution is called a *categorical frequency distribution* or simply a *categorical distribution*. The number of observations that fall into each category provide the class frequencies.

Categorical distributions are very common. Classification of registered voters by political party results in a categorical distribution. The classification of licensed drivers by sex, or the classification of accounts receivable as either current or delinquent, also results in categorical distributions. Suppose, as another illustration, that the Advance Machine Company is bidding for a government contract. In order to qualify, they are required to submit data on the racial distribution of their employees. The personnel manager

TABLE 3.12
DISTRIBUTION OF EMPLOYEES BY RACE

Race	*Number of Employees*
White	174
Black	22
Oriental	3
Other	1
Total	200

TABLE 3.13
DISTRIBUTION OF EMPLOYEES BY RACE

Race	Number of Employees
White	174
Nonwhite	26
Total	200

of the company conducts a survey of the 200 employees of the company, which results in the categorical distribution shown in Table 3.12.

Note that in constructing a categorical distribution there are no real problems regarding class limits, class boundaries, or class marks. These characteristics of a distribution are essentially quantitative in nature and simply do not apply to categorical distributions. Care must be taken, however, to be sure that the categories are all inclusive and mutually exclusive. As with distributions of quantitative observations, "all inclusive" means that every observation must fit into one of the classes or categories. "Mutually exclusive" means that the categories must not overlap. Categories such as "black" and "nonwhite" are not mutually exclusive and would create a problem in classifying the data.

The number of classes or categories in a categorical distribution cannot be governed by general rules. It is foolish to require at least five classes in the distribution when five pertinent categories of the attribute simply do not exist. The number of classes used must be related to the type of information required and the purpose for which the data will be used. It is possible for the Advance Machine Company to classify its employees into only two categories, "white" and "nonwhite," as shown in Table 3.13. Whether this classification satisfies the government requirements for information is questionable, however, and those requirements are the determining factor.

3.4 Histograms

Though not *every* picture is worth a thousand words, most people agree that a picture is generally easier to comprehend than a set of numbers in a table. And a picture in conjunction with a table is even more useful. The most common and useful picture or graph of a frequency distribution is the HISTOGRAM. A histogram is a graph in which the horizontal scale represents the values of the observed variable, and the vertical scale represents the frequencies or relative frequencies. Each class in the distribution is represented by a vertical rectangle or bar whose base corresponds to the class interval and whose height is proportional to the frequency or relative

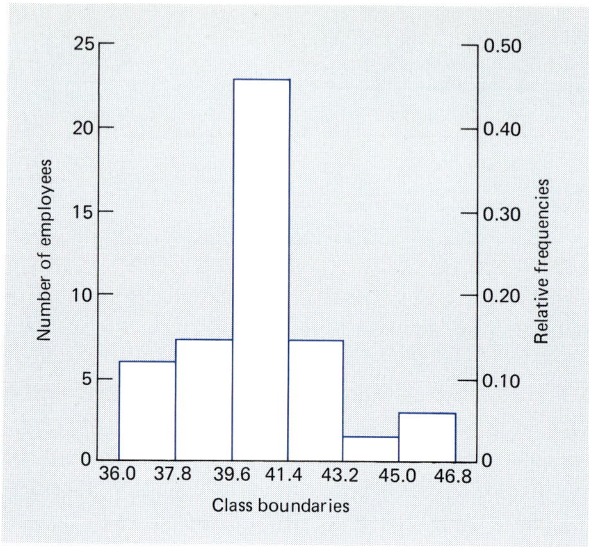

Figure 3.1
Histogram of production hours worked (from Table 3.5).

frequency of that class. The vertical edges of each bar correspond to class boundaries and consequently there are no gaps between the bars. The numerical values on the horizontal scale can be placed to correspond with either the class boundaries or the class marks.

Histograms corresponding to the frequency distributions of Tables 3.5 and 3.7 appear as Figures 3.1 and 3.2. The values on the horizontal scale in Figure 3.1 correspond to class boundaries, while those in Figure 3.2 correspond to class marks. Notice that as long as

Figure 3.2
Histogram of parts produced (from Table 3.7).

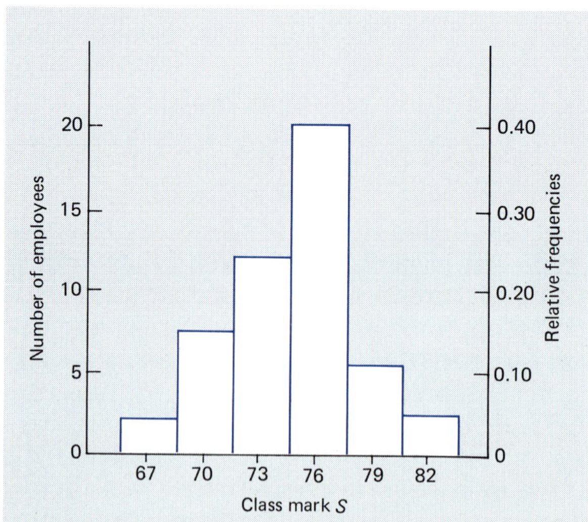

the class intervals in the frequency distributions are the same, the bars in the histogram will be of equal width, and therefore the area of each bar, as well as its height, will be proportional to the frequency of the class. As you may have suspected, a histogram cannot be constructed for a frequency distribution with open-ended classes.

Two vertical scales are provided for each of the histograms in these figures, although one scale is sufficient. The scale on the left corresponds to absolute frequencies, while the scale on the right shows relative frequencies. Typically, histograms are presented with one or the other of these vertical scales.

3.5 Practical Uses of Frequency Distributions and Histograms

The primary purpose of constructing frequency distributions and histograms is to condense and summarize large quantities of numerical data. Presentations using these forms make it much easier for a company's management to comprehend the significant data characteristics and obtain useful information. The following examples illustrate how frequency distributions and histograms are used.

Example 3.1

The typical work week for the Advance Machine Company is 40 hours. The personnel manager of the company is interested in determining exactly what proportion of the production employees actually work 40 hours (plus or minus a few minutes) and what proportion work more than and less than that amount. Reference to Tables 3.5 and 3.8 make it easy for the manager to determine that:

24 out of 50 production workers, 48%, work between 39.6 and 41.4 hours

13 out of 50, 26%, work less than 39.6 hours

13 out of 50, 26%, work 41.4 hours or more

The actual number of hours worked by all production employees varies from 36.0 to 46.8, a range of 10.8 hours. The histogram in Figure 3.1 indicates that the frequencies of hours worked tend to cluster around 40 or 41, and drop off rapidly on either side of these values.

Example 3.2

The production manager of the Advance Machine Company focuses his attention on the production figures for the week. He wants to find the most common level of production output and the general characteristics of production for the week. A glance at the histogram in Figure 3.2 and at the frequency distributions in Tables 3.6, 3.9, and 3.11 provide this information. The most frequent production output lies in the class 75–77, which accounts for 40% of the workers. Only nine production workers, 18% of the total, produced more than 77 parts, while 21 (42%) produced less than 75. In other words, more than twice as many workers were below the most common class in output as were above it. Production output for the week varied from 66 to 83 units, a range of 17. Two employees (4%)

produced less than 69 units for the week while three employees (6%) produced more than 80 units.

There are many ways in which this type of information can be used. Of course, even more valuable information can be obtained by using frequency distributions and histograms in conjunction with other descriptive measures (covered in Chapter 4). However, with only this information, the personnel manager might want to investigate why six employees worked less than 36 hours for the week. The production manager might want to see if there is an explanation for the difference in production performance between the two lowest and the three highest producers. If such a difference can be explained, it might be possible to improve the output of the least productive workers.

The above data, which represent only a single week, may not seem overwhelmingly important. However, when similar data are collected and analyzed over a period of several weeks, they can be extremely valuable. Managers can use data of this type to look for changes in patterns and to identify trends. Such changes can be either good or bad from the standpoint of the manager, but before anything can be done about them they must be identified. Information regarding production and hours worked might be useful in determining the costs of overtime and absenteeism, in analyzing production efficiency, and in developing production incentive plans and training programs.

When decisions are based on statistical data, it is usually necessary to assume that the data are taken from a homogeneous population. A homogeneous population is simply a population of similar units. Mixing different kinds of observations, or observations on different kinds of elementary units, is often misleading. A histogram of the data is useful in detecting a mixing of unlike populations. Consider a histogram of production output that looks like the one in Figure 3.3. No scale of values appears on this histogram, since its purpose is simply to illustrate a particular point and the actual numerical values are not important. This histogram shows quite clearly that the frequencies cluster around two different values. This distribution is said to be *bimodal*. One interpretation of these data might be that there has been a mixing of populations. Investigation by the production manager, for example, might reveal that these data include the production figures for two different shifts. The values on the left of the diagram might represent the production of the night shift, while those on the right might represent the day shift. If this type of pattern persists, an attempt might be made to determine the reasons why production output is lower at night than during the day. If the causes of this difference can be determined, it might be possible to improve production on

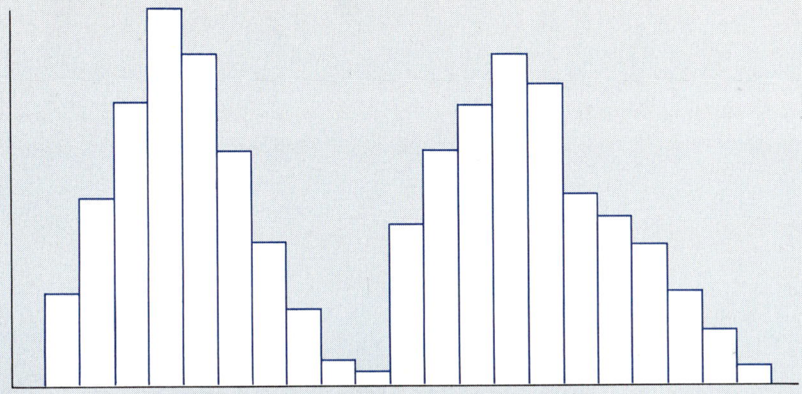

Figure 3.3
Histogram showing a possible mixing of two populations.

the night shift, resulting in better production figures overall.

The investigation and correction of the causes of differences such as those discussed above might reasonably be undertaken by specialists in fields such as engineering or psychology rather than by statisticians or businessmen. However, it is the study of statistical data and the use of statistical techniques that provide the information that a problem exists—and, of course, you can't begin to solve a problem and take corrective action until you become aware that there is a problem.

3.6 Other Statistical Graphs

Statistical data can be displayed in a wide variety of charts and graphs. The most useful of these is the histogram. However, an alternate method for graphing a frequency distribution is the FREQUENCY POLYGON. In a frequency polygon, the frequencies or relative frequencies are plotted at the class marks and these points are connected with straight lines. Classes with zero frequencies are added at both ends of the distribution so that the graph begins and ends on the horizontal axis. A frequency polygon of the data from Table 3.7 is shown in Figure 3.4.

When a frequency polygon is plotted for a cumulative distribution, the graph is called an OGIVE. It is important to note that the cumulative frequencies are not plotted at the class midpoint. Depending on whether it is a "less than" or "more than" cumulative distribution, the cumulative frequencies are plotted at the upper or lower class boundaries of each class. For the cumulative distribution of Table 3.8, we get the ogive shown in Figure 3.5.

In the case of categorical distributions, the bar chart is probably

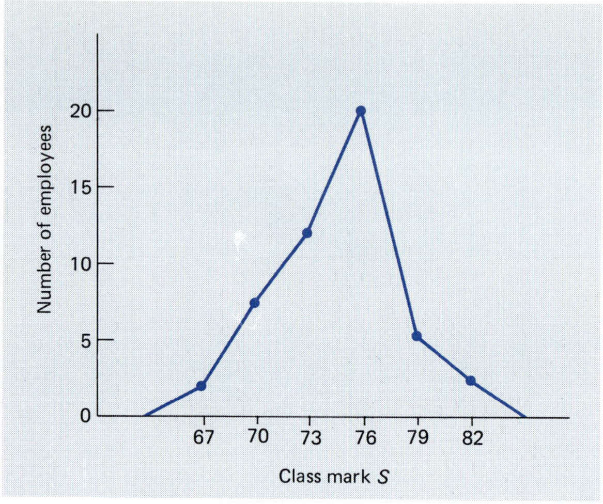

Figure 3.4
Frequency polygon of parts produced (from Table 3.7).

the most commonly used graphical representation. The bar chart is similar to the histogram in that bars are used to represent the various classes. The height of each bar is proportional to the frequency or relative frequency of each class. However, since the categories represent attributes, the horizontal axis is nonnumerical and therefore has no numerical scale. The fact that one category is displayed on the bar chart to the right or to the left of another does

Figure 3.5
Ogive of production hours worked (from Table 3.8).

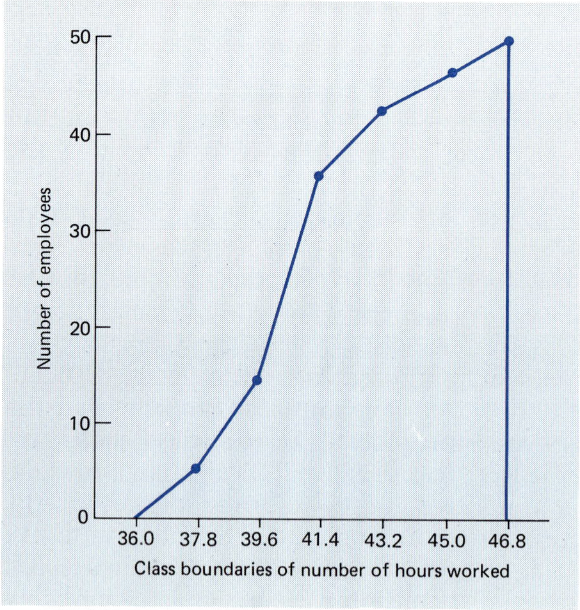

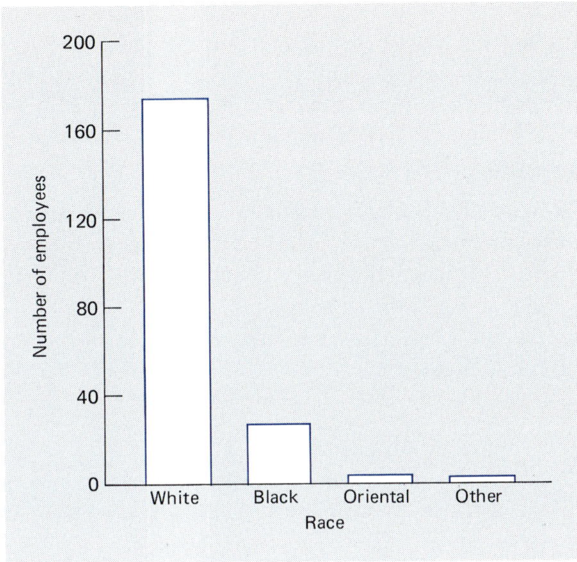

Figure 3.6
Bar chart showing distribution of employees by race (from Table 3.12).

not imply that one is greater than or less than another in any type of valuation. A bar chart for the categorical distribution of Table 3.12 is shown in Figure 3.6.

3.7 General Purpose Tables

A frequency distribution or frequency table which is used for analytical purposes need not follow any particular style other than the general rules that govern its construction. However, tables are frequently published solely for the purpose of conveying certain types of statistical information to a reader. Such tables, constructed for displaying statistical facts without having any special analytical purpose, are usually called *general purpose tables* or *reference tables.* Although tabular representations vary, every general purpose table should contain the following.

1. *Title and number.* Every table should have a title which indicates briefly what the data are, where the data originated, the time period covered by the data, and how the data are classified. If several tables appear in a publication, each table should also have a number for identification.

2. *Columns and stubs.* The body of the table consists of a stub and one or more columns. The stub shows the classifications into which the data are grouped and the stubhead identifies what is in the stubs. The column or columns contain the actual data and the captions at the head of each column describe the contents of the column.

3. *Source.* The source or origin of the data should always be indicated. If the table is a reproduction from some book or periodical, information regarding the author, agency, book, or other type of publication should be included. If the table is the first publication based on original data, the origin of the data should be specified.

4. *Footnotes.* Any explanatory information about the data in the table should be provided in footnotes. The footnotes appear immediately below the body of the table and above the source.

Tables 3.14 and 3.15 are examples of the form used in general purpose tables.

TABLE 3.14
NUMBER OF BUILDING PERMITS ISSUED FOR ROCKHARD COUNTY, 1970 THROUGH 1977, BY TYPE OF CONSTRUCTION

	Number of Permits		
Year	*Residential*	*Commercial*	*Total*
1970	801	409	1210
1971	964	510	1474
1972	1011	505	1516
1973	1045	483	1528
1974	950	473	1423
1975	816	487	1303
1976	963	499	1462
1977	965	501	1466

Source: Business Month Magazine, February 1978, p. 38.

TABLE 3.15
PRICE-TO-EARNINGS RATIO OF STOCKS LISTED ON THE NATIONAL STOCK EXCHANGE, WEEK OF MAY 14, 1977[1]

Price to Earnings[2] Ratio		*Number of Stocks*
0– 9		641
10–19		948
20–29		322
30–39		56
40–49		37
50–59		3
60–69		2
70–79		1
	Total	2010

[1]As of closing, May 14, 1977.
[2]Rounded to the nearest whole number.
Source: Bank Street Journal, May 20, 1977.

Definitions

Array: A set of numerical data arranged in order of magnitude.

Class boundaries: The highest and lowest actual values of observations that belong to a specific class.

Class interval: The range of actual values that belong to a given class.

Class limits: The highest and lowest recorded values that can go into a specified class.

Class mark: The class midpoint. A representative value for the class.

Cumulative frequency distribution: A frequency distribution that indicates the number of observations that are "less than" or "more than" certain specified values.

Cumulative relative frequency distribution: A frequency distribution that indicates the proportion of observations that are "less than" or "more than" certain specified values.

Frequency: The number of observations that fall into a specified class.

Frequency distribution: A tabular arrangement or grouping of numerical data into classes and a count of the number of observations that fall into each class.

Frequency polygon: A graphical representation of a frequency distribution in which frequencies are plotted by placing a dot above the class mark and connecting these dots with straight lines.

Ogive: A frequency polygon of a cumulative frequency distribution.

Relative frequency distribution: A frequency distribution which indicates the proportion of observations that fall into each class.

Problems

3.1 The following data were recorded by the chief actuary of the Reliable Insurance Company and represent the age at death of 100 policy holders. The ages are rounded to the last birthday.

a) Construct a frequency distribution with seven equal classes and make the lower class limit of the lowest class equal to 20.

b) Specify the class mark of each class.

c) What is the class interval?

75	37	47	59	67	46	59	44	63	59
41	84	51	63	58	63	38	57	74	79
81	63	74	61	74	58	54	68	68	48
44	67	42	69	36	27	60	78	71	76
70	75	68	72	75	69	40	62	70	56
78	55	68	78	74	66	62	59	49	67
62	61	86	54	61	55	77	86	65	84
73	73	72	48	71	70	89	45	35	44
69	78	61	74	62	80	44	63	54	75
71	73	77	77	79	53	46	71	78	69

3.2 The following data were recorded by the credit manager of The Emporium, a small department store. They represent the outstanding balances on sixty delinquent accounts rounded to the nearest dollar.

a) Group the data into a frequency distribution and determine the number of classes to use by the rule given in Section 3.1. Make the class limit of the lowest class equal $5.

b) Specify the class mark of each class.
c) Specify the class boundaries of each class.
d) What is the class interval?

6	27	85	72	8	84	25	37	9	43
16	91	121	75	45	122	24	44	41	8
28	7	29	68	51	61	38	71	37	67
19	44	36	11	12	46	41	14	83	54
50	47	101	65	27	44	18	6	12	11
66	36	87	59	22	39	69	49	26	15

3.3 The industrial engineering department of the Merit Manufacturing Company conducted a time study on one of its manufacturing operations. Forty production workers in one department were timed with a stop watch as they performed one of the required assembly operations. These times were recorded to the nearest tenth of a minute.

11.2	8.4	7.7	11.6	8.1	11.2	6.4	7.2
7.1	12.1	6.4	9.8	6.4	7.0	10.9	8.7
9.9	13.2	8.5	8.4	9.3	11.6	9.0	10.9
8.4	4.7	10.0	11.1	9.8	7.1	7.4	6.1
10.8	8.0	10.3	8.0	8.8	9.1	6.8	7.8

a) Construct a frequency distribution of the data with five classes, a class interval of two minutes, and make the class mark of the lowest class equal 4.95.
b) Show the class limits and the class boundaries of every class.

3.4 Construct both a "less than" cumulative frequency distribution and a "more than" cumulative frequency distribution for the data in problem 3.1.

3.5 Construct a relative frequency distribution from your distribution of problem 3.2.

3.6 Construct a "less than" cumulative relative frequency distribution from your distribution of problem 3.3.

3.7 Construct a "less than" cumulative frequency distribution from your distribution of problem 3.1.

3.8 Specify the class boundaries and class marks for the frequency distribution of parts produced given in Table 3.6.

3.9 The following is a frequency distribution of wages paid to hourly employees of the Merit Manufacturing Company:

Hourly Wage (in dollars)	*Number of Employees*
\$2.30 to 2.79	50
2.80 to 3.29	70
3.30 to 3.79	102
3.80 to 4.29	110
4.30 to 4.79	96
4.80 to 5.29	41
5.30 to 5.79	14
5.80 to 6.29	9
6.30 to 6.79	8

a) What is the size of the class interval?
b) What is the highest wage paid to any employee?
c) What is the class mark of the lowest class?
d) How many employees are paid between \$3.30 and \$4.29 per hour?
e) How many employees are paid between \$4.80 and \$6.29 per hour?
f) How many employees are paid exactly \$5.25 per hour?
g) What are the boundaries of the lowest and the highest classes?

3.10 Construct a "less than" cumulative frequency distribution from the distribution provided in problem 3.9 and answer the following:
a) How many employees earn less than \$4.80 per hour?
b) How many employees earn less than \$6.30 per hour?
c) How many employees earn more than \$6.29 per hour?
d) How many employees earn less than \$4.25 per hour?

3.11 Construct a relative frequency distribution from the distribution provided in problem 3.9 and answer the following.
a) What proportion of employees earns between \$2.80 and \$3.29 per hour?
b) What proportion of employees earns between \$4.30 and \$4.79 per hour?
c) What proportion of employees earns less than \$2.80 per hour?
d) What proportion of employees earns more than \$5.79 per hour?

3.12 From the relative frequency distribution of problem 3.5, determine:
a) What proportion of delinquent accounts had balances between \$25 and \$44?
b) What proportion of delinquent accounts had balances between \$65 and \$124?
c) What proportion of delinquent accounts had balances less than \$65?

3.13 From the cumulative relative distribution of problem 3.6, determine:
a) How many employees required 9.9 minutes or less to perform the operation which was timed?
b) How many employees required 9.9 minutes or less to perform the operation?
c) What proportion of employees required between 6.0 and 7.9 minutes?
d) What proportion of employees required more than 11.9 minutes?
e) How many employees required more than 7.9 minutes?

3.14 The owner of the Axleright Tire Store kept a record of the number of tires of each brand carried that were sold during the past year. He arranged the data as follows.

Brand		*Number of Tires Sold*
Goodply		60
Maximile		220
Freeway		110
Extron		20
Solar		90
	Total	500

a) What kind of a distribution is this?
b) What are the class limits and class boundaries of each class?
c) Construct an appropriate graph of this distribution.

3.15 The personnel manager of the Advance Machine Company classified all employees into four categories: unskilled, semi-skilled, skilled, and professional. According to his latest records, the numbers in each category were, respectively, 20, 40, 95, and 45.
a) Display this information in a frequency table.
b) Construct an appropriate graph of the distribution.

3.16 **a)** Construct a "more than" cumulative frequency distribution for the data on hours worked, displayed in Table 3.5.
b) Draw the ogive for this distribution.

3.17 **a)** Construct a "less than" cumulative frequency distribution for the data on parts produced, displayed in Table 3.6.
b) Draw the ogive for this distribution.

3.18 Construct a histogram and a frequency polygon for the data in problem 3.1.

3.19 Construct a histogram and frequency polygon for the data in problem 3.2.

3.20 Construct a histogram and frequency polygon for the data in problem 3.3.

3.21 Draw ogives for the distributions of problem 3.4.

3.22 Draw an ogive for the cumulative frequency distribution of problem 3.6.

3.23 Draw an ogive for the cumulative frequency distribution of problem 3.7.

3.24 Study the histograms and frequency polygons you constructed in problems 3.18, 3.19, and 3.20 and comment on which type of graph provides more information about the frequency distribution that they represent.

CHAPTER 4 DESCRIPTIVE MEASURES—MEASURES OF LOCATION

4.1
Measures of location

4.2
The mode

4.3
The median

4.4
The arithmetic mean

4.5
Comparing the mean, the median, and the mode

4.6
The geometric mean

4.7
Other measures of location

In Chapter 3, various procedures for organizing statistical data were discussed. These included the grouping of the data into several types of frequency distributions and the construction of pictures of the data such as histograms, bar charts, and frequency polygons. We also mentioned that certain measures can be computed from the data which will summarize and/or describe some of the important characteristics of the data set. These important characteristics are LOCATION, VARIATION, and SYMMETRY. The numerical values of these measures, used together with frequency tables and graphs, provide a better picture of the population than any single measure, chart, or table.

In addition to their use in describing single populations, these numerical measures are also useful in making comparisons among two or more populations. This chapter covers measures of location while the next discusses measures of variation and symmetry.

4.1 Measures of Location

A measure of location is computed to find some important point in the population on a scale of values. For example, we might be interested in finding a middle or central point that divides the population into two equal or nearly equal parts. Central points are often representative values and are called MEASURES OF CENTRAL TENDENCY. These measures are the most commonly used location measures. Measures of central tendency are often called averages. The most common of these averages are the MODE, the MEDIAN, and the ARITHMETIC MEAN. Another less common, but useful, average is the GEOMETRIC MEAN.

4.2 The Mode

The MODE is that value in a set of data which occurs with the greatest frequency. A set of data may have a single mode, in which case it is said to be *unimodal*, it may have two modes, which makes

TABLE 4.1
TABULATION OF PRODUCTION HOURS WORKED, ARRANGED IN ORDER OF MAGNITUDE; WEEK OF APRIL 12

36.1	39.0	40.0	40.8	42.1
36.2	39.2	40.3	40.8	42.5
36.4	39.4	40.3	40.8	42.7
36.8	39.6	40.4	41.0	43.0
37.6	39.8	40.4	41.2	43.8
37.7	39.8	40.5	41.3	44.2
38.4	39.9	40.5	41.3	45.1
38.4	39.9	40.6	41.4	45.3
38.6	40.0	40.7	41.7	45.8
38.9	40.0	40.8	41.7	46.0

TABLE 4.2
TABULATION OF PARTS PRODUCED, ARRANGED IN ORDER OF MAGNITUDE; WEEK OF APRIL 12

67	72	74	76	77
68	73	75	76	78
69	73	75	76	78
70	74	75	76	79
70	74	75	77	79
70	74	75	77	79
71	74	75	77	80
71	74	75	77	81
71	74	75	77	81
72	74	75	77	83

it *bimodal*, or it may have several modes and be called *multimodal*. On the other hand, for a given set of data, there may be no mode at all. For a set which is unimodal, the mode is probably the most typical value in the set.

Look at the statistical array in Table 4.1. The mode is 40.8, since that value occurs four times and no other value occurs with that frequency. For the array in Table 4.2, the mode is 75, which occurs nine times. If the production manager of the Advance Machine Company wanted to select typical values of hours worked and parts produced to report to the company president, he might select these two modes.

When data are grouped into frequency distributions, it isn't possible to find the exact value of the mode because the values of the individual observations do not appear. There are several methods for estimating the value of the mode from grouped data. Since precise methods are seldom required, it is usually sufficient to specify the MODAL CLASS, that is, the class containing the greatest number of observations. When a specific value is required, it is common practice to select the class mark of the modal class as the approximate value. The class mark of the modal class is generally referred to as the CRUDE MODE.

For the frequency distribution of production hours worked in Table 4.3, the modal class is the class with the limits 39.6 and 41.4. The crude mode of this distribution is 40.5. The modal class of the distribution of parts produced in Table 4.4 has the limits 75 and 77, and the crude mode is 76.

The mode is useful in statistical description primarily because its use as a typical value is easily justified to nonstatisticians. If a mode exists, its approximation is a simple matter, even though the exact location may be difficult to determine from grouped data. The disadvantages result from the fact that the mode does not always exist, and when it does, it does not lend itself to algebraic treatment as do some other averages.

TABLE 4.3
FREQUENCY DISTRIBUTION OF PRODUCTION HOURS WORKED

Hours Worked	*Number of Employees*
36.0 and under 37.8	6
37.8 and under 39.6	7
39.6 and under 41.4	24
41.4 and under 43.2	7
43.2 and under 45.0	2
45.0 and under 46.8	4
Total	50

TABLE 4.4
FREQUENCY DISTRIBUTION OF PARTS PRODUCED

Parts Produced	*Number of Employees*
66 to 68	2
69 to 71	7
72 to 74	12
75 to 77	20
78 to 80	6
81 to 83	3
Total	50

4.3 The Median

A more useful and more common average is the MEDIAN. The median is the value that divides a set of data in half. Exactly 50% of the values will lie on each side of the median. The median family income in the United States in 1976 was approximately $14,000. This means that half of the families in the United States had an income greater than $14,000 and half earned less. The median is said to be a positional average, since it is located rather than computed. For ungrouped data arranged in the form of an array, the median is equal to the value of the middle element if N, the number of elements, is an odd number. If N is an even number, the median is generally taken to be a value exactly halfway between the two middle elements.

A simple formula is available for locating the position of the median for ungrouped data. If Md represents the value of the median and Md_p represents the position of the median, then

$$\text{Md}_p = \frac{N + 1}{2} \tag{4.1}$$

Example 4.1

Find the median of the following set of values of x: 8, 15, 21, 4, 3, 19, 25.

Step 1. Arrange the data into an array.

$$3, \quad 4, \quad 8, \quad 15, \quad 19, \quad 21, \quad 25$$

Step 2. Determine N, the number of elements in the set.

$$N = 7$$

Step 3. Find Md_p, the position of the median.

$$\mathrm{Md}_p = \frac{N+1}{2} = \frac{7+1}{2} = \frac{8}{2} = 4 \qquad (4.1)$$

Therefore, the median is the fourth value from either end of the array.

Step 4. Count to the median value.

$$\mathrm{Md} = 15.$$

Example 4.2

Find the median of the following set of values of x: 54, 76, 22, 47, 14, 84, 86, 61.

Step 1. Arrange the data into an array.

$$14, \quad 22, \quad 47, \quad 54, \quad 61, \quad 76, \quad 84, \quad 86$$

Step 2. Determine N, the size of the data set.

$$N = 8$$

Step 3. Find Md_p, the position of the median.

$$\mathrm{Md}_p = \frac{N+1}{2} = \frac{8+1}{2} = \frac{9}{2} = 4.5 \qquad (4.1)$$

This means that the median is that value located halfway between the fourth and fifth values from either end of the array.

Step 4. Find the values of x_4 and x_5 by counting.

$$x_4 = 54, \; x_5 = 61$$

Step 5. Find the value of the median by adding the two middle values and dividing the sum by 2.

$$\mathrm{Md} = \frac{54+61}{2} = \frac{115}{2} = 57.5.$$

For data that are grouped into a frequency distribution, the position of the median must be approximated, since the values of the individual data elements are no longer specified. Since there are N values in the set of data, by common practice we specify that the median corresponds to the $N/2$ item.

The first step in this procedure is to find the *median class*, that is, the class which contains the middle or $N/2$ item. This is easily accomplished by finding the cumulative frequency of each class until we reach the class for which the cumulative frequency is equal to or greater than $N/2$. The value of the median within that class is found by a process of interpolation. The following formula summarizes this entire procedure:

$$\text{Md} = L_{\text{Md}} + \left(\frac{N/2 - F_{\text{Md}-1}}{f_{\text{Md}}}\right) c_{\text{Md}} \tag{4.2}$$

where

L_{Md} is the lower class boundary of the median class

N is the number of data elements in the set

$F_{\text{Md}-1}$ is the cumulative frequency in the class with values just below the median class

f_{Md} is the absolute frequency in the median class

c_{Md} is the class interval of the median class

Example 4.3 In order to find the median number of parts produced by the production employees of the Advance Machine Company, we use the frequency distribution below.

Parts Produced	*Frequency (f)*	*Cumulative Frequency (F)*
66 to 68	2	2
69 to 71	7	9
72 to 74	12	21
75 to 77	20	41
78 to 80	6	
81 to 83	3	
	50	

Since we have specified that the median is the value of the $N/2$ element, we are looking for the 25th element in the set. By referring to the column of cumulative frequencies, we see that the first two elements fall in the first class, elements 3 through 9 are in the second class, 10 through 21 are in the third class, and element number 25 must fall in the fourth class, which contains elements 22 through 41. Once we locate the median class, it is not necessary to continue to accumulate the frequencies. The median class, then, is the class with the limits 75 and 77.

We can now determine the approximate value of the median by applying Equation 4.2. The lower class boundary of the median class L_{Md} is 74.5. The class interval of all the classes in the distribution, including the median class, is 3. The required frequencies are obtained from the frequency distribution. Therefore, the median is

$$\begin{aligned} \text{Md} &= 74.5 + \left(\frac{25 - 21}{20}\right)3 = 74.5 + \left(\frac{4}{20}\right)3 \\ &= 74.5 + 0.6 = 75.1 \end{aligned} \tag{4.2}$$

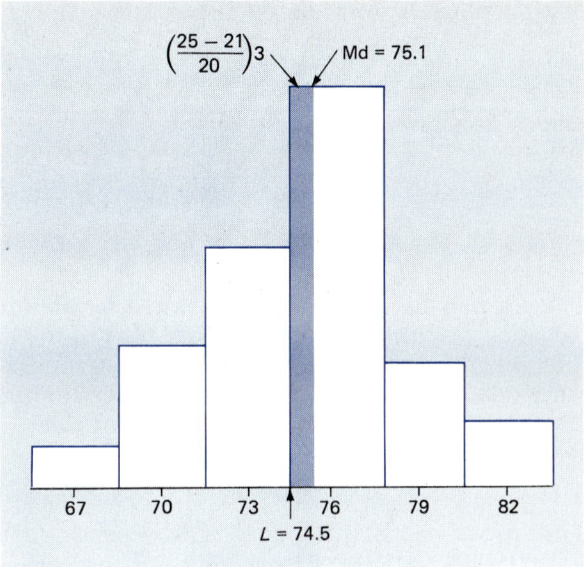

Figure 4.1
Histogram of parts produced showing median. (Illustrates Example 4.3.)

The median of this frequency distribution is illustrated in Figure 4.1.
Since the median number of parts produced is 75.1, the production manager of Advance Machine Company knows that half of the production employees produced less than 75.1 parts, while the other half produced more than 75.1 during the period covered by that frequency table. We have previously determined that the mode of this frequency distribution is 76.

Example 4.4

Find the median number of hours worked by production employees of the Advance Machine Company from the frequency table below.

Hours Worked	*Frequency* (f)	*Cumulative Frequency* (F)
36.0 and under 37.8	6	6
37.8 and under 39.6	7	13
39.6 and under 41.4	24	37
41.4 and under 43.2	7	
43.2 and under 45.0	2	
45.0 and under 46.8	4	
	50	

Step 1. Locate the median class from the cumulative frequencies. The median class is the class with the boundaries and limits, 39.6 and 41.4.

Step 2. State the formula for the median, Equation 4.2.

$$\text{Md} = L_{\text{Md}} + \left(\frac{N/2 - F_{\text{Md}-1}}{f_{\text{Md}}}\right)c_{\text{Md}} \tag{4.2}$$

Step 3. Evaluate the formula.

$$\begin{aligned}\text{Md} &= 39.6 + \left(\frac{25 - 13}{24}\right)1.8 = 39.6 + \left(\frac{12}{24}\right)1.8 \\ &= 39.6 + 0.9 \\ &= 40.5\end{aligned} \tag{4.2}$$

The median number of hours worked by production employees of the Advance Machine Company is 40.5. Half of the employees worked more than 40.5 hours while half worked less than 40.5 hours. We have previously determined that the mode of this distribution is also 40.5 so, in this case, the median and the mode are identical.

4.4 The Arithmetic Mean

The most useful and commonly used statistical average is the ARITHMETIC MEAN. Although there are other statistical measures that are special types of means, whenever we refer to a mean in this book without any qualifying adjective, we are referring to the arithmetic mean.

The mean is the sum of a set of values divided by the number of values in the set. We use the Greek letter μ, mu, to represent this mean. Expressed as an equation, the mean is defined as:

$$\mu = \frac{\sum_{i=1}^{N} x_i}{N} \tag{4.3}$$

or, when there can be no confusion about the limits of the summation, as:

$$\mu = \frac{\Sigma x}{N} \tag{4.4}$$

If a set of data consists of the six values

$$x_1 = 12,\ x_2 = 7,\ x_3 = 9,\ x_4 = 21,\ x_5 = 17,\ x_6 = 30$$

the mean of this data set is

$$\mu = \frac{\sum_{i=1}^{6} x_1}{N} = \frac{96}{6} = 16 \tag{4.3}$$

The Weighted Arithmetic Mean. There are situations in which values in a data set do not have equal importance. In these cases, computing the mean by the above methods may result in a value that is not truly representative and may, in fact, be misleading. Under these circumstances, it is usual to weight each value according to its importance and compute a WEIGHTED ARITHMETIC MEAN. The weighted mean is defined as:

$$\mu = \frac{\sum_{i=1}^{N} x_i w_i}{\sum_{i=1}^{N} w_i} \tag{4.5}$$

where w_i represents the weight applied to the ith value in the set.

Example 4.5

The Friendly Neighborhood Store sells five different products. The profit per unit on the sale of each of these products is

Product 1: $4.00
Product 2: $9.00
Product 3: $6.00
Product 4: $2.00
Product 5: $12.00

Without weighting these values in any way, we can compute an average profit per unit sold, using Equation 4.3. The result is

$$\mu = \frac{\$4.00 + \$9.00 + \$6.00 + \$2.00 + \$12.00}{5} = \frac{\$33.00}{5} = \$6.60.$$

This value, $6.60, is the mean of the five numbers, but it is not necessarily the mean profit per unit if different quantities of the various products are sold. Suppose that during the past accounting period, The Friendly Store sold the following quantities of each product:

Product 1: 150 units
Product 2: 50 units
Product 3: 250 units
Product 4: 450 units
Product 5: 100 units

In determining the average profit per unit sold, it is necessary to consider the number of units of each product sold and to weight the profit on each product by that value. The calculations are shown below in tabular form.

Product	*Profit per Unit* (x)	*Quantity Sold* (w)	xw
1	$4.00	150	$ 600
2	$9.00	50	$ 450
3	$6.00	250	$1500
4	$2.00	450	$ 900
5	$12.00	100	$1200
		1000	$4650

The weighted mean profit per unit sold is therefore

$$(4.5) \qquad \mu = \frac{\sum_{i=1}^{5} x_i w_i}{\sum_{i=1}^{5} w_i} = \frac{\$4650}{1000} = \$4.65$$

The value \$4.65 is substantially different from the value of the unweighted mean of \$6.60. The unweighted mean is incorrect in this case because it does not consider the importance of each individual product in terms of the quantity sold.

Note that the weighting factors can be converted to proportions and used in that form. In this example, 150 units of product 1 are sold out of a total of 1000 units sold. Therefore, product 1 represents 150/1000 or 0.15 of all units sold. Converting all of the quantities sold to proportions and using these proportions as weights, we obtain the weighted mean as follows:

Product	*Profit per Unit* (x)	*Quantity*	*Proportion of Total* (w)	xw
1	\$4.00	150	0.15	\$0.60
2	\$9.00	50	0.05	\$0.45
3	\$6.00	250	0.25	\$1.50
4	\$2.00	450	0.45	\$0.90
5	\$12.00	100	0.10	\$1.20
		1000	1.00	\$4.65

Notice that the sum of the xw column is the same as the weighted average we found previously. Since the sum of the weights is now equal to 1, Equation 4.5 results in the same value as follows:

$$(4.5) \qquad \mu = \frac{\sum_{i=1}^{5} x_i w_i}{\sum_{i=1}^{5} w_i} = \frac{\$4.65}{1.00} = \$4.65$$

The Mean for Grouped Data. When data are grouped into a frequency distribution, Equation 4.3 must be modified. This is because when the data are grouped the values of the individual observations are no longer available from the distribution. The frequency distribution simply specifies certain intervals, which we call classes, and indicates the number of observations that fall into each of these intervals.

To calculate the mean, or any other computational measure, from a frequency distribution, it is first necessary to designate some value in each of the classes to be representative of all the observations in that class. This value is the midpoint of the class or the class mark. Therefore, in all computations that involve data grouped into a frequency distribution, the class mark represents all the values of the observed variable that falls into that class.

Consider, once again, the frequency distribution of parts produced by production employees of the Advance Machine Company.

Parts Produced	*Class Mark* (x)	*Frequency* (*f*)	x*f*
66 to 68	67	2	134
69 to 71	70	7	490
72 to 74	73	12	876
75 to 77	76	20	1520
78 to 80	79	6	474
81 to 83	82	3	246
		50	3740

Equations 4.3 and 4.4 for the mean require that all the individual values of x be summed. However, these individual values are not available from the frequency distribution. Therefore, the class mark of each class is used to represent all the values in that class. The sum of the values in each class is obtained by multiplying the class mark by the frequency. These products are listed in the column above headed by x*f*. The sum of that column, Σxf, is 3740. Also, the sum of the column of frequencies is equal to the total number of observations, that is, $\Sigma f = N$. The mean computed from this frequency distribution is

$$\mu = \frac{3740}{50} = 74.8$$

In general, the mean of a frequency distribution is

$$\mu = \frac{\sum_{i=1}^{k} x_i f_i}{N} \tag{4.6}$$

where k represents the number of classes in the distribution. When the limits of the summation are obvious, the equation may be written as:

$$\mu = \frac{\Sigma xf}{N} \tag{4.7}$$

The mean of a frequency distribution is a form of weighted mean in which the x's are class marks and the frequencies are the weights.

Example 4.6 Find the mean number of hours worked by production employees of the Advance Machine Company from the following frequency distribution.

Hours Worked	*Class Mark* (x)	*Frequency* (f)	xf
36.0 and under 37.8	36.9	6	221.4
37.8 and under 39.6	38.7	7	270.9
39.6 and under 41.4	40.5	24	972.0
41.4 and under 43.2	42.3	7	296.1
43.2 and under 45.0	44.1	2	88.2
45.0 and under 46.8	45.9	4	183.6
		50	2032.2

Step 1. The formula for the mean of a frequency distribution is

$$\mu = \frac{\sum_{i=1}^{k} x_i f_i}{N} \tag{4.6}$$

where $k = 6$ and $N = 50$.

Step 2. Evaluating the formula results in

$$\mu = \frac{2032.2}{50} = 40.644 \doteq 40.6$$

Characteristics of the Mean. The mean is representative of a set of numerical values, in the sense that it represents a point of balance for the set. If the arithmetic mean is subtracted from every member of the set and the resulting differences added algebraically, the sum is zero. This is illustrated easily with the following set of four numbers.

$$x_1 = 1,\ x_2 = 4,\ x_3 = 7,\ x_4 = 12$$

The mean of these numbers is

$$\mu = \frac{\Sigma x}{N} = \frac{1 + 4 + 7 + 12}{4} = \frac{24}{4} = 6 \tag{4.4}$$

If we subtract 6 from every value in the original set of numbers, we obtain the following result.

$x - \mu$	=	d
$1 - 6$		-5
$4 - 6$		-2
$7 - 6$		$+1$
$12 - 6$		$+6$
		$\Sigma = 0$

The arithmetic mean has two other interesting and useful characteristics. First, if a constant is added to or subtracted from each element of a set of data, then the mean of the resulting set is equal to the mean of the original set plus or minus that constant.

Set 1 (x)	*Constant* (c)	*Set 2* (x + c)
1	3	4
4	3	7
7	3	10
12	3	15
24		36

$$\mu_1 = \frac{24}{4} = 6 \qquad \mu_2 = \frac{36}{4} = 9$$

Notice that μ_2, the mean of the second set, is simply $\mu_1 + 3$.

Second, if every element of a set of data is multiplied by a constant, the mean of the resulting set is equal to the mean of the original set multiplied by that constant.

Set 1 (x)	*Constant* (c)	*Set 2* (cx)
1	3	3
4	3	12
7	3	21
12	3	36
24		72

$$\mu_1 = \frac{24}{4} = 6 \qquad \mu_2 = \frac{72}{4} = 18$$

The mean of set 2 is simply the original mean multiplied by 3.

These characteristics make it possible to devise a short method for computing the mean from grouped data. This method is described below.

Shortcut Method for the Mean of a Frequency Distribution. Probably the most common shortcut for computing means from grouped data is the STEP DEVIATION METHOD. The only requirement for using this method is that all the class intervals in the distribution must be equal. The use of the step deviation method simplifies the arithmetic considerably, particularly if the data set is large and the recorded values involve several digits.

The formula for the mean by the step deviation method is

$$\mu = A + \left(\frac{\sum_{i=1}^{k} u_i f_i}{N}\right)c \tag{4.8}$$

where A is an assumed value for the mean, k represents the number of classes, u represents deviations from the assumed mean in units of the class interval, and c is the class interval.

When the limits on the summation are obvious, this formula can be written:

$$\mu = A + \left(\frac{\Sigma uf}{N}\right)c \tag{4.9}$$

The step deviation method involves the following.

1. Select a value of A, the assumed mean. Although the value selected is arbitrary, it should be the class mark of one of the classes near the center of the distribution.

2. Calculate u for every class in the distribution. Since u is the deviation from the assumed mean in units of the class interval, $u = (x_i - A)/c$, where x_i is the class mark of the ith class and c is the class interval.

3. Calculate the mean of the u values. This is $\Sigma uf/N$.

4. Multiply the mean value of u by the class interval c. This product $(\Sigma uf/N)c$ is called the correction factor.

5. Add the correction factor algebraically to A, the assumed mean. The result is the mean of the distribution.

Example 4.7 Compute the mean number of hours worked by production employees of the Advance Machine Company using the step deviation method.

Hours Worked	*Class Mark* (x)	*Frequency* (*f*)	*u*	*fu*
36.0 and under 37.8	36.9	6	−3	−18
37.8 and under 39.6	38.7	7	−2	−14
39.6 and under 41.4	40.5	24	−1	−24
41.4 and under 43.2	42.3	7	0	0
43.2 and under 45.0	44.1	2	+1	+2
45.0 and under 46.8	45.9	4	+2	+8
		50		−46

Step 1. Select the class mark of a class near the center of the distribution as the assumed mean. In this example we selected 42.3 as the assumed mean. It would be better to let $A = 40.5$, since it appears from the distribution of frequencies that 40.5 is closer to the true mean. However, we let $A = 42.3$ to demonstrate that the true mean can be obtained with this method regardless of the exact value chosen for A.

Step 2. If A is subtracted from each class mark in the distribution and this difference is divided by c, the values in the u column result. However, these same values for u can be obtained simply by placing a zero in the row corresponding to the assumed mean and then counting upward with consecutive negative numbers and downward with consecutive positive numbers.

Step 3. Compute the mean value of u. Multiply each value of u by the frequency in that class then add these products algebraically and divide by N. For the data in this example, we get:

$$\frac{\Sigma uf}{N} = \frac{-46}{50} = -0.92$$

Step 4. Compute the correction factor.

$$\left(\frac{\Sigma uf}{N}\right)c = (-0.92)(1.8) = -1.656$$

Step 5. Determine the true mean from Equation 4.9.

$$\mu = A + \left(\frac{\Sigma uf}{N}\right)c \qquad (4.9)$$

$$= 42.3 + (-1.656)$$
$$= 42.3 - 1.656$$
$$= 40.644 \doteq 40.6$$

This is identical to the result obtained in Example 4.6.

Example 4.8

The president of Metric Systems, Inc. asks the personnel manager to provide information regarding wages paid to hourly employees of the company. The president wants to use this information to make comparisons between Metric's pay scale and industry-wide wages as reported by an industry association.

The personnel manager collects the company data and constructs the following frequency distribution. He then computes the mean using the step deviation method. He also computes the median. The distribution and the computations follow.

Hourly Wages (in dollars)	*Class Mark (x)*	*Frequency (f)*	*u*	*uf*	*Cumulative Frequency F*
2.50 to 2.89	2.695	14	−4	−56	14
2.90 to 3.29	3.095	97	−3	−291	111
3.30 to 3.69	3.495	115	−2	−230	226
3.70 to 4.09	3.895	230	−1	−230	456
4.10 to 4.49	4.295	201	0	0	657
4.50 to 4.89	4.695	97	+1	+97	754
4.90 to 5.29	5.095	88	+2	+176	842
5.30 to 5.69	5.495	78	+3	+234	920
5.70 to 6.09	5.895	69	+4	+276	989
6.10 to 6.49	6.295	6	+5	+30	995
6.50 to 6.89	6.695	5	+6	+30	1000
		1000		+36	

The assumed mean is the class mark of the class with the limits 4.10 and 4.49, therefore $A = 4.295$. This class was selected because it is approximately at the center of the distribution and the class mark should be reasonably close to the true mean. The class interval c is 0.40.

The mean of the distribution is obtained from Equation 4.9 as follows.

$$\text{(4.9)} \qquad \begin{aligned} \mu &= A + \left(\frac{\Sigma uf}{N}\right)c \\ &= 4.295 + \left(\frac{36}{1000}\right)(0.40) \\ &= 4.295 + (.036)(.40) \\ &= 4.295 + .0144 \\ &= 4.31 \qquad \text{or} \qquad \$4.31 \end{aligned}$$

The median is also obtained by using Equation 4.2 as follows.

$$\text{(4.2)} \qquad \begin{aligned} \text{Md} &= L_{\text{Md}} + \left(\frac{N/2 - F_{\text{Md}-1}}{f_{\text{Md}}}\right)c_{Md} \\ &= 4.095 + \left(\frac{500 - 456}{201}\right)(.40) \\ &= 4.095 + \left(\frac{44}{201}\right)(.40) \\ &= 4.095 + .088 \\ &= 4.18 \qquad \text{or} \qquad \$4.18 \end{aligned}$$

By finding the largest absolute frequency in the table above, you can see that the crude mode of this distribution is 3.895 or $3.90.

The personnel manager is now able to report to the president that:

a) The modal or most commonly paid wage is $3.90 per hour.

b) The median wage is $4.18 per hour.

c) The mean wage is $4.32 per hour.

4.5 Comparing the Mean, the Median, and the Mode

We have discussed three different measures of central tendency. All three are considered representative values of a set of data. Is there any reason to prefer any one of these over the others, and if so, under what circumstances?

The mean is by far the most commonly used and most useful of these averages. It is a rigidly defined, computed measure that uses every value in the data set. It also lends itself to algebraic manipulation. For example, if we know the means of two related data sets, the mean of the combined set can be determined algebraically.

The primary disadvantage of the mean is that it is affected by the presence of extreme values in the data set, particularly if the extreme values all fall to the same side of the distribution. When this occurs, the mean may be less representative of the total set than one of the other averages.

The median is also useful as a representative value. Unlike the mean, which is a computational average, the median is a positional average. It is the true middle point of the set and divides the data into two equal parts. However, the median does not lend itself to algebraic manipulation. Also, since the median is affected by the number of elements in a data set but not by their values, the presence of extreme values does not have as great an effect on the median as on the mean. For this reason, when a data set does contain a few extreme values, the median may be the most representative measure.

The mode is probably the least useful of the three averages. Since it is the value which occurs with the greatest frequency, it is the most typical single value. Also, the mode is unaffected by extreme values. However, the mode does not always exist in a set of data. On the other hand, a given set may have several modes. Furthermore, if the data are grouped into a frequency distribution, the exact value of the mode may be difficult to determine, although the crude mode is easy to find. Like the median, the mode does not lend itself to algebraic manipulation.

There are certain relationships that exist among the mean, median, and mode of a data set. If the set is symmetrical, all three of these averages have the same value. However, if the set is not symmetrical, that is, if there is a preponderance of extreme values to one side of the distribution, the mode is unaffected while the median is moved in the direction of the extremes. The median also moves in the direction of the extremes, but not as far as the mean.

Figure 4.2 illustrates these concepts in terms of physical objects. Suppose that a set of seven blocks of unit weight are arranged on a weightless foundation, as shown in Figure 4.2a. The numbers 1 through 3 in Figure 4.2a and 1 through 6 in Figure 4.2b represent values assigned to the blocks corresponding to their positions on

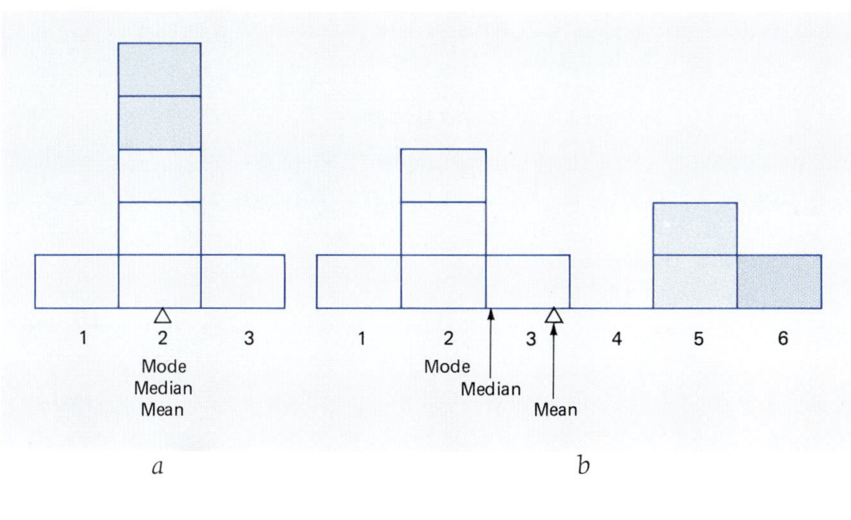

Figure 4.2
Illustration of mean, median, and mode.

the foundation. For example, in Figure 4.2a, one block has the value 1, five blocks have the value 2, and one block has the value 3. We can observe that the mode of this distribution of blocks is 2. The median is also 2, and the mean, from Equation 4.7, is

$$\mu = \frac{\Sigma xf}{N} = \frac{(1)(1) + (5)(2) + (1)(3)}{7} = 2$$

where x represents the positional value of each block and f is the number of blocks having that value.

Now suppose that the two shaded blocks with the value 2 in Figure 4.2a are moved to position 5 in Figure 4.2b, and a new shaded block is added at the extreme position 6. The mode of the distribution still remains at 2, since that is still the value with the greatest frequency. However, the median moves to the position 2.5, since that is now the point that divides the set of data blocks in half. The mean now becomes:

$$\mu = \frac{\Sigma xf}{N} = \frac{(1)(1) + (3)(2) + (1)(3) + (2)(5) + (1)(6)}{8} = \frac{26}{8} = 3.25 \qquad (4.7)$$

The small triangle in these figures represents a fulcrum, or point of balance. The point of balance in a distribution corresponds to the position of the mean. Notice in the diagrams that the mean and median are affected by extreme values, whereas the mode is not.

4.6 The Geometric Mean

The GEOMETRIC MEAN is a descriptive measure used in statistics in certain special situations. While the arithmetic mean is obtained by finding the sum of a set of values and then dividing by N, the geometric mean is the Nth root of the product of a set of N values. If we let G symbolize the geometric mean, then

$$G = \sqrt[N]{(x_1)(x_2)(x_3)\cdots(x_N)} \qquad (4.10)$$

The geometric mean is the proper measure for averaging a series of ratios or percentages. However, the geometric mean cannot be computed if any value in the set is negative, and the geometric mean is zero if any value in the set is zero.

Example 4.9 Net profits after taxes of Metric Systems, Inc. for the past five years were \$200,000, \$500,000, \$600,000, \$800,000, and \$1,200,000. The president of the company wants to determine the average annual percentage increase in profits for inclusion in a report to the stockholders at the next annual meeting. Using the first year in the series as a basis for comparison, the data is arranged in tabular form as follows.

Year	Net Profits (in thousands of dollars)	Percent of Previous Year (x)	Log x
1974	200	—	—
1975	500	250	2.39794
1976	600	120	2.07918
1977	800	133	2.12385
1978	1200	150	2.17609
		653	8.77706

The president computes the arithmetic mean of these percentages by adding them and dividing by 4. This results in

$$\mu = \frac{653}{4} = 163.25$$

The conclusion is that, on the average, net profits increased by 163.25% each year for the past five years. However, if we start with the year 1974 and increase each year's net profits by 163.25%, we arrive at a figure of $1,420,000 for the year 1978, which is considerably higher than the actual profits for that year of $1,200,000.

The proper average to use for percentages is the geometric mean. However, using Equation 4.10 to compute this value is extremely tedious. There is a much easier procedure which uses logarithms and simplifies the arithmetic considerably. This procedure is summarized in the following equation.

$$\log G = \frac{\sum_{i=1}^{N} \log x_i}{N} \tag{4.11}$$

where G, the geometric mean of the x's, is the antilog of the result.

We find the logarithms of the percentages from Table $A.7$ in the Appendix. These values and their sum are shown above. Substituting, in Equation 4.11 we get

$$\log G = \frac{8.77706}{4} = 2.19427$$

and

$$G = \text{antilog}(2.19427) = 156.42$$

Therefore, the president of Metric Systems, Inc. can report that, on the average, net profits increased by 156.4% each year.

4.7 Other Measures of Location

The measures of central tendency that we have discussed all have the purpose of locating some central value in the set of data. This value is generally considered to be representative of the entire set.

TABLE 4.5
ILLUSTRATION OF A SET OF DATA WITH QUARTILES AND DECILES INDICATED

3	11	22	29	38	45	53	63
4	12	24	31	39	46	55	65
5	15	26	33	40	47	57	67
7	17	27	34	42	49	58	69
9	18	28	36	44	51	60	70
$Q_1 = 20$		$Q_2 = 37$		$Q_3 = 52$			

3	9	17	26	31	38	44	49	57	65
4	11	18	27	33	39	45	51	58	67
5	12	22	28	34	40	46	53	60	69
7	15	24	29	36	42	47	55	63	70
$D_1 = 8$	$D_2 = 16$	$D_3 = 25$	$D_4 = 30$	$D_5 = 37$	$D_6 = 43$	$D_7 = 48$	$D_8 = 56$	$D_9 = 64$	

There are occasions, however, when some value other than a central value may be of interest. For example, a social service agency might be interested in determining that value of income below which 10% of the population falls. This particular value is called the first DECILE or the tenth PERCENTILE, since it divides the lowest 10% from the upper 90% of the population. An automobile dealer in the same community might be interested in knowing that value of income which locates the top 25% of the population. This value is called the third QUARTILE, because it divides the lower three quarters of the population from the highest one quarter.

An educational institution, in examining a set of SAT (Scholastic Aptitude Test) scores might want to determine the value above which only 5% of its applicants scored. That value is called the 95th percentile, since 95% of the scores are below this value and 5% are above it.

Measures of location that divide a data set into four equal parts are called QUARTILES. Measures that divide a data set into tenths are called DECILES, and those that divide a set into hundredths are called PERCENTILES. For any set of data there are 3 quartiles, 9 deciles, and 99 percentiles. The median, which divides a set of data into two equal parts, corresponds to the second quartile, the fifth decile, and the 50th percentile. Table 4.5 illustrates these concepts.

Each of these location measures is computed using the same method that is used for finding the median. Where $N/2$ or $0.5N$ appears in the equation for the median, however, it is necessary to substitute the number corresponding to that proportion of the population that we want to fall *below* the desired measure. Equation 4.2 is repeated below with the population proportion expressed as a decimal number instead of a fraction.

$$\text{Md} = L_{\text{Md}} + \left(\frac{0.5N - F_{\text{Md}-1}}{f_{\text{Md}}}\right) c_{\text{Md}} \tag{4.2}$$

A similar equation for the first quartile, designated Q_1, is

$$Q_1 = L_{Q_1} + \left(\frac{0.25N - F_{Q_1-1}}{f_{Q_1}}\right) c_{Q_1} \tag{4.12}$$

where

L_{Q_1} is the lower class boundary of the class containing Q_1.
N is the number of values in the set.
F_{Q_1-1} is the cumulative frequency in the class with a value just below the class containing Q_1.
f_{Q_1} is the frequency in the class containing Q_1.
c_{Q_1} is the class interval of the class containing Q_1

The equations for Q_2, the second quartile, which equals the median, and Q_3 are identical to Equation 4.12, except that the first term in the parentheses becomes $0.5N$ for Q_2 and $0.75N$ for Q_3. This equation is modified in the same way to provide formulas for deciles and percentiles.

Example 4.10

Find Q_3 for the frequency distribution in Example 4.8.

Step 1. Locate the class containing Q_3 from the cumulative frequencies. Since Q_3 is the value below which 75% of the values fall, this involves locating the class that contains the $0.75N$ observation. Since $(0.75)(1000) = 750$, we see that the 750th observation falls in the class with the limits 4.50 and 4.89.

Step 2. State the formula.

$$Q_3 = L_{Q_3} + \left(\frac{0.75N - F_{Q_3-1}}{f_{Q_3}}\right) c_{Q_3} \tag{4.13}$$

Step 3. Evaluate the formula.

$$\begin{aligned} Q_3 &= 4.495 + \left(\frac{750 - 657}{97}\right)(0.40) \\ &= 4.495 + \left(\frac{93}{97}\right)(0.40) \\ &= 4.495 + 0.3835 \\ &= 4.8785 \quad \text{or} \quad \$4.88 \end{aligned}$$

Q_3 indicates that 75% of the employees of Metric Systems, Inc. earn less than \$4.88 per hour, while 25% earn more than that amount.

A formula for deciles is illustrated by providing an equation for locating the fourth decile, D_4:

$$D_4 = L_{D_4} + \left(\frac{0.4N - F_{D_4-1}}{f_{D_4}}\right) c_{D_4} \tag{4.14}$$

Example 4.11 Find D_4 for the frequency distribution in Example 4.8.

Step 1. Locate the class containing D_4 from the cumulative frequencies. Since $(0.4)(1000) = 400$, we can see that the 400th observation falls in the class with the limits 3.70 and 4.09.

Step 2. State the formula.

$$D_4 = L_{D_4} + \left(\frac{0.4N - F_{D_4-1}}{f_{D_4}}\right) c_{D_4} \tag{4.14}$$

Step 3. Evaluate the formula.

$$\begin{aligned} D_4 &= 3.695 + \left(\frac{400 - 226}{230}\right) 0.40 \\ &= 3.695 + \left(\frac{174}{230}\right)(0.40) \\ &= 3.695 + 0.3026 \\ &= 3.998 \qquad \text{or} \qquad \$4.00 \end{aligned}$$

Therefore, we conclude that 40% of the company's employees earn less than $4.00 per hour, while 60% earn more.

Recall that the words quartile, decile, and percentile are used to refer to the value that separates the data into the desired number of parts. Sometimes these terms are used to refer to the part of the distribution into which a particular item falls. If an individual's earnings are said to be in the highest quartile, this means that his earnings are among the top 25%. However, for clarity, it is better to say that his earnings are above the third quartile, or even that they are in the upper 25% and reserve the use of the terms quartiles, deciles, and percentiles to refer to the measures themselves.

The procedure for locating percentiles is similar to the ones shown above and will not be illustrated specifically.

Definitions

Arithmetic mean: The arithmetic mean is defined best by Equations 4.3 and 4.6.

Crude mode: The class mark of the class in a frequency distribution that has the greatest number of observations.

Deciles: Those values that divide a set of data into ten equal parts.

Geometric mean: The geometric mean is defined best by Equation 4.10.

Median: The value in a data set that divides the set into two equal parts.

Mode: The value in a set of data that occurs with the greatest frequency.

Percentiles: Those values that divide a set of data into 100 equal parts.

Quartiles: Those values that divide a set of data into four equal parts.

Weighted arithmetic mean: A modified arithmetic mean that weights each value in a set according to its relative importance.

Formulas

Position of the median for ungrouped data:

(4.1) $$\mathrm{Md}_p = \frac{N + 1}{2}$$

Median from grouped data:

(4.2) $$\mathrm{Md} = L_{\mathbf{Md}} + \left(\frac{N/2 - F_{\mathbf{Md}-1}}{f_{\mathbf{Md}}}\right)(c_{\mathbf{Md}})$$

Mean for ungrouped data:

(4.3) $$\mu = \frac{\sum_{i=1}^{N} \mathrm{x}_i}{N}$$

(4.4) $$\mu = \frac{\Sigma \mathrm{x}}{N}$$

Weighted arithmetic mean:

(4.5) $$\mu = \frac{\sum_{i=1}^{N} \mathrm{x}_i \mathrm{w}_i}{\sum_{i=1}^{N} \mathrm{w}_i}$$

Mean for grouped data:

(4.6) $$\mu = \frac{\sum_{i=1}^{k} \mathrm{x}_i f_i}{N}$$

(4.7) $$\mu = \frac{\Sigma \mathrm{x} f}{N}$$

Step deviation method for the mean from grouped data:

(4.8) $$\mu = A + \left(\frac{\sum_{i=1}^{k} u_i f_i}{N}\right) c$$

(4.9) $$\mu = A + \left(\frac{\Sigma u f}{N}\right) c$$

Geometric mean:

(4.10) $$G = \sqrt[N]{(\mathrm{x}_1)(\mathrm{x}_2) \cdots (\mathrm{x}_N)}$$

Logarithm of geometric mean:

(4.11) $$\log G = \frac{\sum_{i=1}^{N} \log \mathrm{x}_i}{N}$$

The first quartile:

$$Q_1 = L_{Q_1} + \left(\frac{0.25N - F_{Q_1-1}}{f_{Q_1}}\right) c_{Q_1} \tag{4.12}$$

The third quartile:

$$Q_3 = L_{Q_3} + \left(\frac{0.75N - F_{Q_3-1}}{f_{Q_3}}\right) c_{Q_3} \tag{4.13}$$

The fourth decile:

$$D_4 = L_{D_4} + \left(\frac{0.4N - F_{D_4-1}}{f_{D_4}}\right) c_{D_4} \tag{4.14}$$

Problems

4.1 Find the mean, median, and mode of the following set of values of x: 14, 9, 14, 3, 5, 9, 21, 6, 3, 9, 18, 5, 12.

4.2 **a)** Find the mean, median, and mode of the following set of values of x: 59, 75, 27, 63, 27, 48, 56, 42.

b) Which of these averages seems to be most representative of the set?

4.3 Find the mean, median, and mode of the following set of values of x: 150, 200, 125, 310, 330, 250, 225.

4.4 If the wholesale prices of beef, pork, lamb, and veal are, respectively, \$1.35, \$1.25, \$1.55, and \$1.10 per pound, and if a restaurant manager buys 90 pounds of beef, 75 pounds of pork, 40 pounds of lamb, and 25 pounds of veal, what is the average price per pound of meat purchased?

4.5 A recent college graduate was hired by a large manufacturing corporation and placed in their management training program. As part of her training, she was assigned to five different departments of the corporation for various periods of time. At the end of the training period in each department, the supervisor graded her performance on a scale from zero to ten. At the end of the training program, the training director computed an overall mean score based on the following considerations:

The marketing and manufacturing phases of her training were assumed to be of equal importance. Both of these were considered to be three times as important as the purchasing and financial phases. The accounting training was twice as important as the latter two.

If the supervisor's ratings were as follows, compute an appropriate mean score.

Department	*Score*
Accounting	6
Finance	5
Marketing	8
Manufacturing	7
Purchasing	7

4.6 Compute the mean of the following frequency distribution using the long method (Equation 4.5).

Class	*Frequency*
0 and less than 3	4
3 and less than 6	3
6 and less than 9	5
9 and less than 12	6
12 and less than 15	7

4.7 The production division of Continental Motors planned a campaign to improve plant safety. In preparation, they compiled accident data for a period of fifty weeks. These data are grouped into the frequency distribution below. Compute the mean number of accidents per week using the step deviation method (Equation 4.6).

Number of Accidents	*Number of Weeks*
0 to 4	6
5 to 9	25
10 to 14	11
15 to 19	7
20 to 24	1

4.8 From the distribution of problem 4.7, determine the values of the median and the crude mode.

4.9 The credit manager of a department store compiled information on 200 of his accounts receivable that were delinquent. The data consisted of the number of days each account was past due. He then grouped these data into the following frequency distribution. Compute the mean number of days using the step deviation method.

Days Past Due	*Number of Accounts*
30 to 44	40
45 to 59	45
60 to 74	40
75 to 89	25
90 to 104	25
105 to 119	20
120 to 134	5

4.10 From the distribution of problem 4.9, determine the median and the crude mode.

4.11 The High Voltage Electric Company placed a pilot production run of 100 new-type light bulbs on life test. The bulbs were allowed to burn until every one failed, and the time to failure of every bulb was recorded to the nearest hour. The data were then grouped into the frequency distribution below. Find the mean life of these 100 bulbs.

Life in Hours	*Number Failed*
1 to 50	2
51 to 100	8
101 to 150	15
151 to 200	20
201 to 250	25
251 to 300	20
301 to 350	10

4.12 **a)** Find the crude mode and the median for the frequency distribution of problem 4.11.

b) Interpret each of these averages in the context of that problem.

4.13 During 1978, 250 new single-family dwellings were constructed and sold in Buckley County. A frequency distribution of prices paid for these houses is as follows.

Price (in thousands of dollars)	*Number of Houses*
25 and under 30	30
30 and under 35	40
35 and under 40	45
40 and under 45	35
45 and under 50	20
50 and under 55	30
55 and under 60	25
60 and under 65	15
65 and under 70	10

a) Compute the mean price using the step deviation method.

b) Compute the mean price using the long method and compare the value of Σxf with that of Σuf in part (a).

4.14 From the data of problem 4.13, determine the median price paid for a new house in Buckley County in 1978 and explain what this means.

4.15 The following is a frequency distribution of the diameters of 1000 parts of a particular type produced by the Advance Machine Company for a customer. Find the mean diameter of these parts.

Diameter (in centimeters)	*Number of Parts*
2.05 to 2.14	3
2.15 to 2.24	39
2.25 to 2.34	65
2.35 to 2.44	200
2.45 to 2.54	378
2.55 to 2.64	196
2.65 to 2.74	71
2.75 to 2.84	41
2.85 to 2.94	7

4.16 A survey was conducted by a large shoe store to gather information about the ages of its charge account customers. The data obtained were grouped into the following frequency distribution. Find the mean and median ages.

Ages (in years)	*Frequency*
16 to 20	22
21 to 25	94
26 to 30	101
31 to 35	49
36 to 40	18
41 to 45	11
46 to 50	3
51 to 55	1
56 to 60	1

4.17 Find Q_3, the third quartile for the distribution of problem 4.9, and explain its meaning in the context of that problem.

4.18 Find D_2, the second decile for the distribution of problem 4.13, and explain its meaning in the context of that problem.

4.19 Find the 95th percentile for the distribution of diameters in problem 4.15.

4.20 **a)** Devise a formula for finding the third quintile of a distribution. (Quintiles divide a set of data into five equal parts.) Let F_3 stand for the third quintile.

b) Find the value of F_3 for the distribution of problem 4.9.

4.21 The total number of industrial accidents occurring in the production division of Continental Motors over the past five years is given below.

Year	*Number of Accidents*	*Percent of Previous Year*
1974	300	—
1975	315	105
1976	375	119
1977	420	112
1978	460	110

Find the average yearly percent increase in accidents over this period using the appropriate average.

4.22 The number of new homes constructed in Buckley County over the past five years is given below.

Year	*Number of Homes*	*Percent of Previous Year*
1974	85	—
1975	125	147
1976	190	152
1977	210	110
1978	250	119

Find the average annual percent increase in houses constructed in Buckley County over this period using the appropriate average.

CHAPTER 5 DESCRIPTIVE MEASURES—VARIATION AND SKEWNESS

5.1
The need for describing variation

5.2
The range

5.3
The variance and the standard deviation

5.4
Relative variation

5.5
A measure of skewness

In Chapter 3, we stated that the degree to which members of a population vary is one of the most important statistical characteristics of that population. We are now ready to begin the study of some of the most common and useful measures of variation. Following our discussion of variation, we will consider another measure which indicates the extent to which a population or data set deviates from being symmetrical. This last descriptive measure is called *skewness*.

5.1 The Need for Describing Variation

In Chapter 4, several measures of central tendency were discussed. These measures are considered to be representative values to the extent that any single number is representative of an entire population. However, we noted that an attempt to describe an entire population with a single measure is extremely misleading. No single number can describe a population, just as no single number such as a height or weight can adequately describe an individual. In both cases, several values are required to provide a reasonable description.

As an illustration, consider three sets of data consisting of the following values:

$$\text{Set } A = \{10, \quad 20, \quad 30, \quad 30, \quad 40, \quad 50\}$$
$$\text{Set } B = \{5, \quad 15, \quad 30, \quad 30, \quad 45, \quad 55\}$$
$$\text{Set } C = \{25, \quad 30, \quad 30, \quad 30, \quad 30, \quad 35\}$$

If we attempt to describe each of these three sets with a single representative value, that is, a measure of central tendency, we might select the mode, the median, or the mean. A simple observation tells us that the mode of each set is 30. It is easy to determine that the median of all the sets is also 30. Finally, if we performed the calculations necessary to find the mean, we would find that the mean of each of these sets is also 30. Consequently, if we attempt to describe these sets by using any one of these three averages, we would have to conclude that the three sets are identical. However, by observing the data it is obvious that they are not identical. The principal difference among these sets is their variation. In order to provide an adequate description of these sets of data, some measure of variation is required in conjunction with a measure of central tendency.

5.2 The Range

The simplest measure of variation or dispersion is called the RANGE. The range is simply the difference between the largest value in a set and the smallest. If we let R stand for the range, then we define this measure as:

(5.1) $$R = x_l - x_s$$

where x_l is the largest value of x and x_s is the smallest.

When the data are grouped into a frequency distribution and the individual values of x are not available, the range is defined slightly differently. In this case, an approximate range for the data is determined by finding the difference between the upper class limit of the class containing the largest values and the lower class limit of the class containing the smallest values.

The primary advantage of using the range as a measure of variation is that it is so easy to determine. Its chief disadvantage is that it uses only part of the available data. The range is determined from only two values in the set of data. Consequently it ignores the way the data varies in the other $N - 2$ values in the set. Nevertheless, the range is useful as a very quick and easy measure of variation.

Example 5.1

Find the range of each set of data shown in Section 5.1. Indicate which of the sets is most variable and which is least variable.

(5.1)
$$R_A = x_l - x_s = 50 - 10 = 40$$
$$R_B = x_l - x_s = 55 - 5 = 50$$
$$R_C = x_l - x_s = 35 - 25 = 10$$

Set *B*, with a range of 50, is the most variable.
Set *C*, with a range of 10, is the least variable.

Example 5.2

Find the range of the frequency distribution of wages paid by Metric Systems, Inc., as shown in Example 4.8.

The upper class limit of the class containing the largest values is \$6.89.

The lower class limit of the class containing the smallest values is \$2.50. Therefore, the range of this frequency distribution is

(5.1) $$R = x_l - x_s = \$6.89 - \$2.50 = \$4.39$$

5.3 The Variance and the Standard Deviation

The most commonly used measures of variation in statistics are the VARIANCE and the STANDARD DEVIATION. These two measures are related in that the standard deviation is the positive square root of the variance. In illustrating computational methods for these two measures, we always compute the variance first, and the standard deviation is obtained by taking the square root of the variance. It is very difficult to define these two terms nonmathematically except to say that they measure the variation in a set of data.

The variance and standard deviation have the following desirable characteristics:

1. They utilize all the data. Every value in the set enters into the calculation of these measures.

2. Their magnitudes are related to the amount of the variation present in the data. That is, the value of the measure is large when the variation is large and small when the variation is small.

3. They measure variation about the mean of the data set.

In addition to the above characteristics, which apply to both measures, the standard deviation is always expressed in the same units of measurement as the original data. This is not true of the variance, which is expressed in terms of the original units squared.

The Greek letter σ, sigma, is used to represent the standard deviation of a population. The symbol for the variance of a population is σ^2.

The variance and standard deviation of a population are defined as:

$$\sigma^2 = \frac{\sum_{i=1}^{N} (x_i - \mu)^2}{N} \tag{5.2}$$

and

$$\sigma = \sqrt{\frac{\sum_{i=1}^{N} (x_i - \mu)^2}{N}} \tag{5.3}$$

respectively, where μ = the mean and N = the total number of elements in the set.

Equation 5.2 indicates that the variance of a population is the sum of the squared deviations of each value about the mean, divided by the total number of elements in the set or population. Although the variance and standard deviation can be computed from these formulas, there are other formulas that are algebraically equivalent and easier to use in performing calculations. The formulas that are equivalent to Equations 5.2 and 5.3 are

$$\sigma^2 = \frac{\sum_{i=1}^{N} x_i^2}{N} - \left(\frac{\sum_{i=1}^{N} x_i}{N}\right)^2 \tag{5.4}$$

and

$$\sigma = \sqrt{\frac{\sum_{i=1}^{N} x_i^2}{N} - \left(\frac{\sum_{i=1}^{N} x_i}{N}\right)^2} \tag{5.5}$$

respectively.

Notice that in Equations 5.4 and 5.5, the factor $\Sigma x/N$ is simply the mean.

Example 5.3 Find the variance and standard deviation of Set A, given in Section 5.1 using equations 5.2 and 5.3. The computations are performed easily by arranging the data into a table with the headings as follows:

	x	x − μ	(x − μ)²
	10	10 − 30 = −20	(−20)² = 400
	20	20 − 30 = −10	(−10)² = 100
	30	30 − 30 = 0	(0)² = 0
	30	30 − 30 = 0	(0)² = 0
	40	40 − 30 = 10	(10)² = 100
	50	50 − 30 = 20	(20)² = 400
Sums	180	0	1000

(4.4) $\mu = \Sigma x/N = 180/6 = 30$

(5.2) $\sigma^2 = (x - \mu)^2/N = 1000/6 = 166.7$

(5.3) $\sigma = \sqrt{166.7} = 12.9$

Example 5.4

Find the variance and standard deviation of Set *B*, given in Section 5.1 using equations 5.4 and 5.5. The data are arranged in tabular form, with the headings as follows:

	x	x²
	5	25
	15	225
	30	900
	30	900
	45	2025
	55	3025
Sums	180	7100

(4.4) $\mu = \Sigma x/N = 180/6 = 30$

(5.4) $\sigma^2 = \frac{\Sigma x^2}{N} - \left(\frac{\Sigma x}{N}\right)^2 = \frac{7100}{6} - (30)^2 = 1183.33 - 900$

$\sigma^2 = 283.33$

(5.5) $\sigma = \sqrt{283.33} = 16.8$

Variance and Standard Deviation of Grouped Data. When the data are grouped into a frequency distribution, the formulas for the variance and standard deviation are modified similarly as for the mean. Equations 5.2 and 5.3 which are the definitional formulas become:

(5.6)
$$\sigma^2 = \frac{\sum_{i=1}^{k}(x_i - \mu)^2 f_i}{N}$$

(5.7)
$$\sigma = \sqrt{\frac{\sum_{i=1}^{k}(x_i - \mu)^2 f_i}{N}}$$

where the x values are now the class marks of the distribution, k represents the number of classes, and f equals the frequency.

The variance and standard deviation of frequency distributions would probably never be computed using Formulas 5.6 and 5.7. Instead, the more convenient computational forms are used. These are

$$\sigma^2 = \frac{\sum_{i=1}^{k} x_i^2 f_i}{N} - \left(\frac{\sum_{i=1}^{k} x_i f_i}{N} \right)^2 \tag{5.8}$$

and

$$\sigma = \sqrt{\frac{\sum_{i=1}^{k} x_i^2 f_i}{N} - \left(\frac{\sum_{i=1}^{k} x_i f_i}{N} \right)^2} \tag{5.9}$$

Once again, remember that the factor $\Sigma(x_i f_i)/N$ is the mean of the distribution.

Example 5.5 Find the variance and standard deviation of the number of parts produced by production employees of the Advance Machine Company.

Parts Produced	*Class Mark* (x)	*Frequency* (f)	xf	x^2	x^2f
66 to 68	67	2	134	4489	8978
69 to 71	70	7	490	4900	34300
72 to 74	73	12	876	5329	63948
75 to 77	76	20	1520	5776	115520
78 to 80	79	6	474	6241	37446
81 to 83	82	3	246	6724	20172
		50	3740		280364

Step 1. Multiply each value of x by its frequency and sum over all classes. This is $\Sigma xf = 3740$.

Step 2. Square each value of x.

Step 3. Multiply each value of x^2 by its frequency and sum over all classes. This is $\Sigma x^2 f = 280{,}364$.

Step 4. State the formula for the variance, Equation 5.8.

$$\sigma^2 = \frac{\Sigma x^2 f}{N} - \left(\frac{\Sigma xf}{N}\right)^2 \tag{5.8}$$

Step 5. Evaluate the formula.

$$\sigma^2 = \frac{280{,}364}{50} - \left(\frac{3740}{50}\right)^2$$

$$\sigma^2 = 5607.28 - (74.8)^2$$

$$\sigma^2 = 5607.28 - 5595.04$$

$$\sigma^2 = 12.24$$

Step 6. Take the square root of σ^2.

$$\sigma = \sqrt{12.24} = 3.5 \tag{5.9}$$

The standard deviation of the distribution of parts produced is 3.5 parts. Used in conjunction with the mean or some other measure of central tendency, it is simply a device used to describe the data. However, under certain circumstances, the standard deviation is used to provide other information about the population from which it is computed. These circumstances are discussed in Chapter 8.

A Shortcut Computational Method for the Variance and Standard Deviation. In Chapter 4 we made use of two characteristics of the arithmetic mean to develop a shortcut called the step deviation method. A similar procedure can be developed for the variance and standard deviation of a frequency distribution. We showed that if a constant is added to or subtracted from every element of a set of data, the mean of the new set is equal to the original mean plus or minus the constant. In effect, the addition or subtraction of a constant, c, simply moves the entire distribution c units to the right or to the left without affecting the variation of the distribution.

Consider the two distributions shown below. Distribution 2 is simply distribution 1 with the constant, $c = 20$, added to each value of x.

Distribution 1

x	f	xf
5	1	5
10	3	30
15	1	15
	5	50

Distribution 2

x + 20	f	(x + 20)f
5 + 20 = 25	1	25
10 + 20 = 30	3	90
15 + 20 = 35	1	35
	5	150

$$\mu_1 = \frac{50}{5} = 10 \qquad \mu_2 = \frac{150}{5} = 30$$

Both of these distributions are plotted as frequency polygons in Figure 5.1. This figure shows that these distributions are identical except for their positions on the x scale. Most important is the fact that the variation of the second distribution is exactly the same as that of the first. Using ranges to illustrate this point:

$$R_1 = x_l - x_s = 15 - 5 = 10$$
$$R_2 = x_l - x_s = 35 - 25 = 10$$

A second characteristic of the variance and standard deviation has to do with the effect of multiplying every element of a distribution by a constant. Without demonstration, we will state that multiplication of every element of a set by a constant changes the standard deviation by a factor equal to the constant and changes the variance by a factor equal to the constant squared. That is, if

$$\text{SD}(x) = \sigma \quad \text{and} \quad \text{Var}(x) = \sigma^2$$

then

$$\text{SD}(cx) = c\sigma \quad \text{and} \quad \text{Var}(cx) = c^2\sigma^2$$

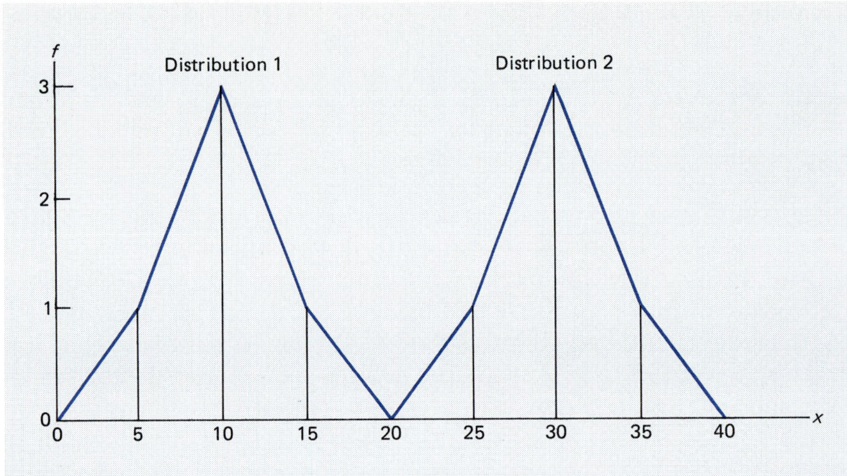

Figure 5.1
Frequency polygons of two distributions which differ by the constant c = 20.

These characteristics allow us to use the step deviation method for computing the variance and standard deviation of data grouped into a frequency distribution. The original variable, x, is converted into a new variable, u, which represents deviations from a central point of the distribution in units corresponding to the class interval. The variance and standard deviation of x are obtained from u by the following equations.

$$\sigma^2 = c^2\left[\frac{\sum_{i=1}^{k} u_i^2 f_i}{N} - \left(\frac{\sum_{i=1}^{k} u_i f_i}{N}\right)^2\right] \tag{5.10}$$

and

$$\sigma = c\sqrt{\frac{\sum_{i=1}^{k} u_i^2 f_i}{N} - \left(\frac{\sum_{i=1}^{k} u_i f_i}{N}\right)^2} \tag{5.11}$$

The steps in this procedure are

1. Select a class near the center of the distribution and let $u = 0$ for this class. If the mean of the distribution is also being computed, then let $u = 0$ correspond to the position of the assumed mean.

2. Calculate u for each of the remaining classes of the distribution in the same way as when using this method for calculating the mean. That is, add a unit for each successive row below the row containing the zero and subtract one unit for each successive row above.

3. Calculate u^2, uf, and u^2f for each value of u and sum the values of the latter two.

4. Compute the variance and standard deviation using Equations 5.10 and 5.11.

Example 5.6

Compute the variance and standard deviation of the distribution of parts produced by production employees of the Advance Machine Company using the step deviation method and compare the calculations with those in Example 5.5.

Parts Produced	*Class Mark (x)*	*Frequency (f)*	u	uf	u^2	u^2f
66 to 68	67	2	−3	−6	9	18
69 to 71	70	7	−2	−14	4	28
72 to 74	73	12	−1	−12	1	12
75 to 77	76	20	0	0	0	0
78 to 80	79	6	1	6	1	6
81 to 83	82	3	2	6	4	12
		50		−20		76

Step 1. Let $u = 0$ for the class with the limits 75 and 77.

Step 2. Calculate u for each remaining class as indicated.

Step 3. Calculate uf, u^2, and u^2f and sum the columns of uf and u^2f.

Step 4. State the formula for the variance.

$$\sigma^2 = c^2\left[\frac{\sum_{i=1}^{k} u_i^2 f_i}{N} - \left(\frac{\sum_{i=1}^{k} u_i f_i}{N}\right)^2\right] \tag{5.10}$$

Step 5. Evaluate the formula.

$$\begin{aligned}\sigma^2 &= (3)^2\left[\frac{76}{50} - \left(-\frac{20}{50}\right)^2\right] \\ &= 9[1.52 - (-.40)^2] \\ &= 9[1.52 - .16] \\ &= 9(1.36) \\ &= 12.24\end{aligned}$$

Step 6. Take the square root of σ^2.

$$\sigma = \sqrt{12.24} = 3.5 \tag{5.11}$$

Comparison of the calculations required in this example with those in the previous one, shows that these are simpler because the magnitudes of the numbers involved are much smaller. If we compare the numbers in the x^2f column in Example 5.5 with the numbers in the u^2f column above, the differences are striking.

Example 5.7

Find the variance and standard deviation of the number of hours worked by production employees of the Advance Machine Company.

Hours Worked	f	u	uf	u^2	u^2f
36.0 and under 37.8	6	−2	−12	4	24
37.8 and under 39.6	7	−1	−7	1	7
39.6 and under 41.4	24	0	0	0	0
41.4 and under 43.2	7	1	7	1	7
43.2 and under 45.0	2	2	4	4	8
45.0 and under 46.8	4	3	12	9	36
$c = 1.8$	50		+4		82

(5.10)
$$\begin{aligned}\sigma^2 &= c^2\left[\frac{\Sigma u^2 f}{N} - \left(\frac{\Sigma uf}{N}\right)^2\right]\\ &= (1.8)^2\left[\frac{82}{50} - \left(\frac{4}{50}\right)^2\right]\\ &= (3.24)[1.64 - (.08)^2]\\ &= (3.24)[1.64 - .0064]\\ &= (3.24)(1.6336)\\ &= 5.29\end{aligned}$$

(5.11)
$$\sigma = \sqrt{5.29} = 2.30$$

5.4 Relative Variation

The standard deviation is a measure of variation which is expressed in the same measurement units as the original data. We mentioned previously that this is one of the desirable features of the standard deviation. However, when it is necessary or desirable to compare two or more different populations with respect to their variation, simply comparing the standard deviations of the respective populations might be misleading. This is particularly true if the measurement units of the populations are different or if the magnitudes of the observations vary greatly from distribution to distribution.

To illustrate, consider the standard deviations of the distributions of parts produced and hours worked by production employees of the Advance Machine Company. These values were computed in Examples 5.6 and 5.7. The standard deviations are, respectively, $\sigma_1 = 3.5$ units of output and $\sigma_2 = 2.30$ hours worked. It might be of interest to know which of these two distributions exhibits the greater amount of variation.

Since the standard deviation of units is 3.5 as compared to the standard deviation of hours of 2.30, we might be inclined to conclude that the first distribution is about one and a half times as variable as the second. But can we compare hours with production units? No, not in terms of these original units.

In order to make such a comparison, it is necessary to convert both of the standard deviations into relative values that are independent of the measurement units of the original data. This is a very simple process and is accomplished by dividing each standard deviation by the mean of the same distribution. This new value is called the COEFFICIENT OF VARIATION and is designated by the letter V.

The coefficient of variation is defined as:

$$V = \frac{\sigma}{\mu} \tag{5.12}$$

In comparing the variation of the distributions of production units and hours worked, the means and standard deviations are

Units produced: $\mu_1 = 74.8$ units, $\sigma_1 = 3.5$ units
Hours worked: $\mu_2 = 40.6$ hours, $\sigma_2 = 2.30$ hours

Referring to Equation 5.12, we find the coefficient of variation for each of the two distributions to be

(5.12)

$$V_1 = \frac{\sigma_1}{\mu_1} = \frac{3.5}{74.8} = 0.047$$

$$V_2 = \frac{\sigma_2}{\mu_2} = \frac{2.30}{40.6} = 0.057$$

By comparing the coefficients of variation we see that the distribution of hours is slightly more variable than the distribution of units produced.

Erroneous conclusions can also be drawn about the variation of two distributions when both are distributions of the same variable reported in exactly the same units. This occurs if there is a large difference in the magnitudes of the observations between the two distributions.

Example 5.8

The average wage of unskilled workers in a particular industry is \$2.80 per hour, with a standard deviation of \$0.30. In the same industry, the average wage of a particular class of skilled craftsmen is \$6.50, with a standard deviation of \$0.60. Which set of wages exhibits the greater amount of variation?

If we compare the respective standard deviations, it appears as though the wages of the skilled workers are twice as variable as those of the unskilled. However, if the coefficients of variation are computed, the conclusions are quite different.

$$V_1 = \frac{\$0.30}{\$2.80} = 0.107$$

$$V_2 = \frac{\$0.60}{\$6.50} = 0.092$$

Considering the coefficients of variation, we see that the wages of the unskilled workers are slightly more variable than those of the skilled.

5.5 A Measure of Skewness

The third characteristic of a population that helps to complete its description is SKEWNESS. When a population of values is not symmetrical, it is skewed. Measures of skewness indicate the degree to which a population or data set deviates from symmetry and also indicates in which direction it is skewed. One common method of measuring skewness makes use of the relative positions of the mean and either the mode or the median of a distribution. Symmetrical and skewed distributions are illustrated in Figure 5.2.

Skewness results from a preponderance of extreme values at one side or in one tail of a distribution, thereby elongating that tail. When the extreme values occur in the upper or right tail, the distribution is positively skewed. When the extreme values occur in the lower or left tail, the distribution is negatively skewed.

We have previously demonstrated that the mean, median, and mode are affected differently by extreme values. The mean is the measure most affected by the presence of extreme values in one tail of a distribution. It is pulled substantially in the direction of the extremes. The mode is unaffected by the extreme values, while the median, which is affected by the number but not the values of the extremes, is pulled in the direction of the extremes, but not as far as the mean. In fact, the median moves about $\frac{2}{3}$ as far as the mean in the direction of the preponderance of extreme values.

The relationship between the mean and either the mode or the median is used to develop a measure of skewness called the *coefficient of skewness*. Since a mode does not always exist, it is more convenient to define this measure using the median. The coefficient of skewness is

$$\text{Sk} = \frac{3(\mu - \text{Md})}{\sigma} \tag{5.13}$$

When the distribution is symmetrical, the mean and the median are equal and Sk = 0. For positively skewed distributions, the mean

Figure 5.2
Symmetrical and skewed distributions.

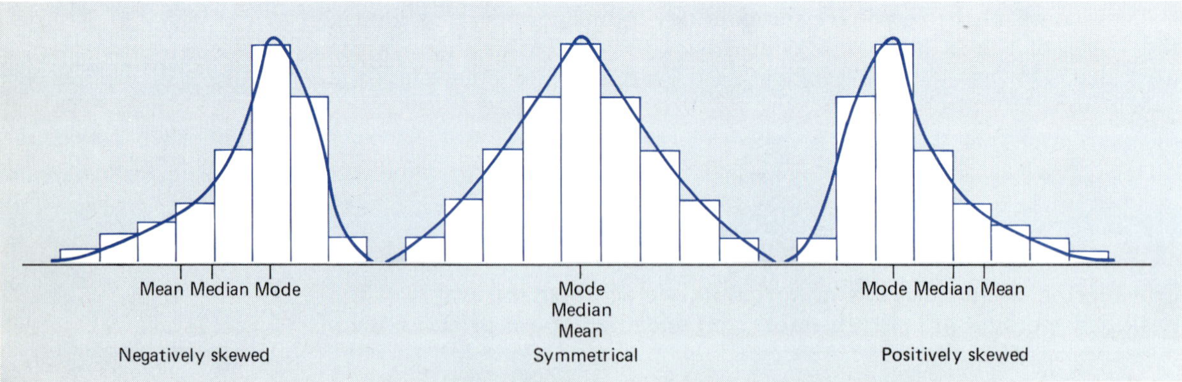

is greater than the median and Sk is a positive value. The opposite is true for negatively skewed distributions.

Example 5.9

Compute the coefficients of skewness for:

a. The distribution of parts produced by production employees of the Advance Machine Company.

$$\mu = 74.8, \qquad \text{Md} = 75.1, \qquad \sigma = 3.5$$

$$\text{(5.13)} \qquad \text{Sk} = \frac{3(\mu - \text{Md})}{\sigma} = \frac{3(74.8 - 75.1)}{3.5} = \frac{-0.9}{3.5} = -0.257$$

This distribution is negatively skewed. It is illustrated in Figure 5.3.

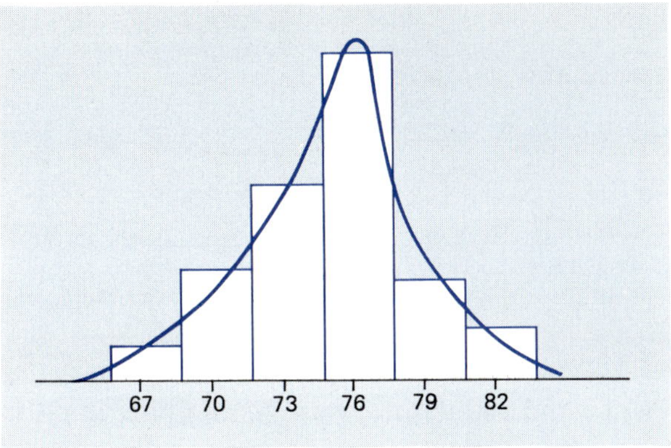

Figure 5.3
Distribution of parts produced (from Example 5.9).

b. The distribution of hours worked by production employees of the Advance Machine Company.

$$\mu = 40.6, \qquad \text{Md} = 40.5, \qquad \sigma = 2.30$$

$$\text{(5.13)} \qquad \text{Sk} = \frac{3(\mu - \text{Md})}{\sigma} = \frac{3(40.6 - 40.5)}{2.30} = \frac{0.3}{2.30} = 0.130$$

This distribution is positively skewed. It is illustrated in Figure 5.4.

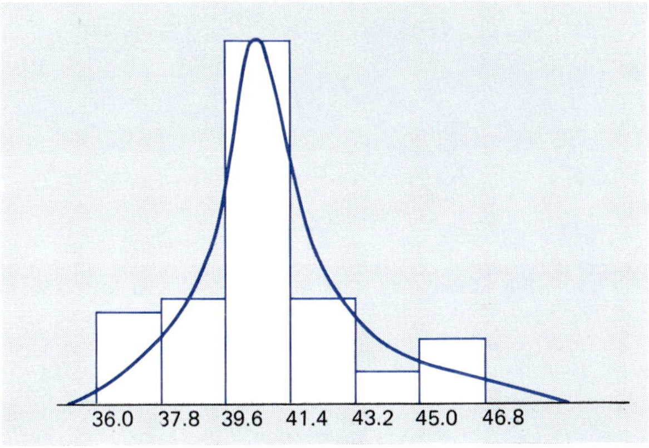

Figure 5.4
Distribution of number of hours worked (from Example 5.9).

Definitions

Coefficient of variation: A measure of relative variation. See Equation 5.12 below.

Range: A measure of variation determined by subtracting the lowest value in a set from the highest.

Skewness: Departure from symmetry.

Standard deviation: The standard deviation is best defined by Formula 5.3 below.

Variance: The variance is best defined by Formula 5.4 below.

Formulas

Range:

$$R = x_l - x_s \tag{5.1}$$

Variance and standard deviation:

$$\sigma^2 = \frac{\sum_{i=1}^{N}(x_i - \mu)^2}{N} \tag{5.2}$$

$$\sigma = \sqrt{\frac{\sum_{i=1}^{N}(x_i - \mu)^2}{N}} \tag{5.3}$$

Variance and standard deviation, computational form:

$$\sigma^2 = \frac{\sum_{i=1}^{N} x_i^2}{N} - \left(\frac{\sum_{i=1}^{N} x_i}{N}\right)^2 \tag{5.4}$$

$$\sigma = \sqrt{\frac{\sum_{i=1}^{N} x_i^2}{N} - \left(\frac{\sum_{i=1}^{N} x_i}{N}\right)^2} \tag{5.5}$$

Variance and standard deviation for grouped data:

$$\sigma^2 = \frac{\sum_{i=1}^{k}(x_i - \mu)^2 f_i}{N} \tag{5.6}$$

$$\sigma = \sqrt{\frac{\sum_{i=1}^{k}(x_i - \mu)^2 f_i}{N}} \tag{5.7}$$

Variance and standard deviation for grouped data, computational form:

$$\sigma^2 = \frac{\sum_{i=1}^{k} x_i^2 f_i}{N} - \left(\frac{\sum_{i=1}^{k} x_i f_i}{N}\right)^2 \tag{5.8}$$

(5.9) $$\sigma = \sqrt{\frac{\sum_{i=1}^{k} x_i^2 f_i}{N} - \left(\frac{\sum_{i=1}^{k} x_i f_i}{N}\right)^2}$$

Variance and standard deviation, step deviation method:

(5.10) $$\sigma^2 = c^2 \left[\frac{\sum_{i=1}^{k} u_i^2 f_i}{N} - \left(\frac{\sum_{i=1}^{k} u_i f_i}{N}\right)^2\right]$$

(5.11) $$\sigma = c\sqrt{\frac{\sum_{i=1}^{k} u_i^2 f_i}{N} - \left(\frac{\sum_{i=1}^{k} u_i f_i}{N}\right)^2}$$

Coefficient of variation:

(5.12) $$V = \frac{\sigma}{\mu}$$

Coefficient of skewness:

(5.13) $$\text{Sk} = \frac{3(\mu - \text{Md})}{\sigma}$$

Problems

5.1 For the following set of values of x,

10, 17, 25, 29, 33, 40, 49

a) Find the range.

b) Find the variance and standard deviation using Equations 5.2 and 5.3.

c) Find the variance and standard deviation using Equations 5.4 and 5.5.

5.2 Find the range, variance, and standard deviation for the following set of values of x,

59, 75, 27, 63, 27, 48, 56, 42.

5.3 Find the range, variance, and standard deviation for the following set of values of x,

150, 200, 125, 310, 330, 250, 225.

5.4 Find the range of the following frequency distribution:

Class	*Frequency*
0 and less than 3	4
3 and less than 6	3
6 and less than 9	5
9 and less than 12	6
12 and less than 15	7

5.5 Use the step deviation method to compute the variance and standard deviation of the frequency distribution in problem 5.4.

5.6 Use the step deviation method to compute the variance and standard deviation of the following frequency distribution:

Number of Accidents	*Number of Weeks*
0 to 4	6
5 to 9	25
10 to 14	11
15 to 19	7
20 to 24	1

5.7 Find the range of the frequency distribution in problem 5.6.

5.8 Determine the range, variance, and standard deviation of the following frequency distribution:

Days Past Due	*Number of Accounts*
30 to 44	40
45 to 59	45
60 to 74	40
75 to 89	25
90 to 104	25
105 to 119	20
120 to 134	5

5.9 Compute the variance and standard deviation of the following frequency distribution:

Life in Hours	*Number Failed*
1 to 50	2
51 to 100	8
101 to 150	15
151 to 200	20
201 to 250	25
251 to 300	20
301 to 350	10

5.10 Compute the standard deviation of the following frequency distribution:

Price (in thousands of dollars)	*Number of Houses*
25 and under 30	30
30 and under 35	40
35 and under 40	45
40 and under 45	35
45 and under 50	20
50 and under 55	30
55 and under 60	25
60 and under 65	15
65 and under 70	10

5.11 Compute the variance and standard deviation of the following frequency distribution:

Diameter (in centimeters)	*Number of Parts*
2.05 to 2.14	3
2.15 to 2.24	39
2.25 to 2.34	65
2.35 to 2.44	200
2.45 to 2.54	378
2.55 to 2.64	196
2.65 to 2.74	71
2.75 to 2.84	41
2.85 to 2.94	7

5.12 Compute the variance and standard deviation of the following frequency distribution:

Ages (in years)	*Frequency*
16 to 20	22
21 to 25	94
26 to 30	101
31 to 35	49
36 to 40	18
41 to 45	11
46 to 50	3
51 to 55	1
56 to 60	1

5.13 Compute the coefficient of variation of the distribution of lightbulb lifetimes in problem 5.9.

5.14 Compute the coefficient of variation of the distribution of housing prices in problem 5.10.

5.15 **a)** Which exhibits the greater amount of variation, the distribution of housing prices referred to in problem 5.14, or the distribution of lightbulb lifetimes referred to in problem 5.13?

b) Why could this same conclusion not be drawn by simply comparing the standard deviations of the two distributions?

5.16 Use coefficients of variation to compare the variation of the distribution of accidents in problem 5.6 with the variation of the distribution of diameters in problem 5.11.

5.17 Use coefficients of variation to compare the variation of the distribution of ages in problem 5.12 with the variation of the distribution of past-due accounts in problem 5.8.

5.18 Compute the coefficient of skewness for the distribution in problem 5.4.

5.19 Compute the coefficient of skewness for the distribution of housing prices in problem 5.10.

5.20 Compute the coefficient of skewness for the distribution of lightbulb lifetimes in problem 5.9.

5.21 Compute the coefficient of skewness for the distribution of accidents in problem 5.6 and for the distribution of ages in problem 5.12, and indicate which distribution is more asymmetrical.

5.22 Compute the coefficient of skewness for the distribution of past-due accounts in problem 5.8 and for the distribution of diameters in problem 5.11, and indicate which distribution is more asymmetrical.

A
$
B
$
C
$$
D
$
E
$
F
$

CHAPTER 6 INDEX NUMBERS

6.1
Construction of a simple relative

6.2
Some common business index numbers

6.3
Problems of index number construction

6.4
Price index numbers

6.5
Quantity index numbers

6.6
Other business applications of index numbers

In the previous chapters we studied methods for describing a large population of statistical data at a particular point in time. Studies such as these are often called cross-sectional studies. Although we made no attempt to describe any changes that occur with time, populations do change with time. There are descriptive measures that are used to summarize changes in certain population characteristics. The changing characteristics are usually economic ones such as prices, production quantities, and buying power. One type of measure used for describing these changes is an INDEX NUMBER.

An index number expresses the relationship between two other numbers, one of which is used as the basis for comparison and is called the *base number*. The index number is always expressed as a percentage of the base, even though the percent sign is not shown. Because the index number is a ratio of two other numbers, it is sometimes called a RELATIVE.

The concept of a relative is not a new one. We have discussed other kinds of relative values in previous chapters. For instance, the coefficient of variation is a relative because it expresses relative variation as the ratio of the standard deviation of a set of data to the mean of the same set. We also discussed relative frequencies, which are simply the ratios of the number of elements in a particular class of a frequency distribution to the total number of elements in the entire distribution.

Because index numbers are relatives, they are not limited to comparisons of conditions at different points in time. They are also useful for comparing different populations with respect to some characteristic at a single point in time. For example, a cost-of-living index at some particular time can be constructed for various United States cities. One city, such as Washington, D.C. or New York, is selected as the base and the cost relatives for the other cities are computed with respect to this base. In 1976, the United Nations Statistical Office constructed index numbers to compare the cost of living in various cities around the world. The cost of living based on 120 different goods and services was determined for New York City and used as the base. The cost of living index number for New York was 100. The cost of living index numbers for several other cities were London—91, Paris—124, Rome—94, and Vienna—108. These numbers can be interpreted as meaning that in 1976 the particular set of goods and services included in the index cost about 24% more in Paris than in New York. In London, the same set of items cost about 9% less than in New York.

6.1 Construction of a Simple Relative

Any series of numbers can be converted to a series of relatives in a few simple steps:

Step 1. Select a number to be used as the base. This is usually, but not necessarily, one of the numbers in the series.

Step 2. Divide every number in the series by the base number.

Step 3. Multiply the quotient by 100.

Example 6.1

Convert the following series of numbers to relatives, using 15 as the base: 3, 12, 16, 28, 30.

Step 1. The number 15 is selected as the base.

Step 2. Divide each number in the series by 15.

$$\frac{3}{15} = 0.200 \qquad \frac{12}{15} = 0.800 \qquad \frac{16}{15} = 1.067$$

$$\frac{28}{15} = 1.867 \qquad \frac{30}{15} = 2.000$$

Step 3. Multiply each decimal number by 100.

$$(0.200)(100) = 20.0 \qquad (0.800)(100) = 80.0$$
$$(1.067)(100) = 106.7 \qquad (1.867)(100) = 186.7$$
$$(2.000)(100) = 200.0$$

The series of relatives with base 15, corresponding to the original series of numbers, is

20.0 80.0 106.7 186.7 200.0

Example 6.2

If it cost $11,848 to purchase a particular package of goods and services in New York in 1976, and the same package of goods and services cost $10,640 in Los Angeles, convert the Los Angeles figure to a relative using New York as the base.

The cost figure for Los Angeles is divided by the base and multiplied by 100:

$$\frac{\$10{,}640}{\$11{,}848}(100) = 89.8$$

The cost index number for Los Angeles is 89.8. From this information, we can conclude that in 1976 the cost of living was 10.2% less in Los Angeles than in New York.

An index number constructed from a series of numbers as illustrated in Example 6.1 is called a SIMPLE INDEX NUMBER. The index numbers most commonly used to study economic activity are COMPOSITE INDEX NUMBERS because each value in the series is a summary of several other values. For example, the index numbers constructed by the United Nations Statistical Office to compare the cost of living in various cities are composite indexes. Each of the values for the individual cities that were used to construct the index is a summary of the prices of 120 different commodities and services in each of the cities. The dollar values used to construct the index number in Example 6.2 are also a summary of the prices of various goods and services. The construction of a simple index number is quite easy. The major problems of index number construction occur with composite indexes. Therefore, the discussion that follows is devoted to the construction of composite index numbers and the term "index number" will be used in this context unless stated otherwise.

6.2 Some Common Business Index Numbers

Even though a business manager may not be required to construct an index number in the course of routine activities, he or she may frequently be required to interpret and use index numbers constructed and reported by someone else. Many of the series of index numbers useful in business are developed and reported by agencies of the United States Government. In this section we briefly discuss three of the most important ones.

The Wholesale Price Index. The wholesale price index, abbreviated WPI, is compiled by the Bureau of Labor Statistics of the United States Department of Labor. This index shows the rate and direction of changes in the primary market price of approximately 2200 different commodities. It is based on prices paid by wholesalers to manufacturers and other producers. In addition to the overall index for the complete set of commodities, separate indexes are also compiled for various commodity groups and subgroups.

A manufacturer or wholesaler can use the WPI to compare changes in prices that he receives or pays, with the national average as represented by the index. By using the appropriate group and subgroup indexes, it is possible to compare price movements among various types of commodities.

The wholesale price index has been published continually since 1890 and can be found in the *Survey of Current Business* and the *Monthly Labor Review*. Appendix B.1 provides a sample of the WPI.

The Consumer Price Index. The consumer price index, or CPI, is also compiled and published by the Bureau of Labor Statistics. The purpose of the CPI is to measure changes in the prices of goods and services purchased by families of city-dwelling wage earners and clerical workers. The index is based on the prices of approximately 400 different items. Since it is based on the costs of purchases made by families of a specific income level, the CPI is not truly representative of the cost of living of all persons or families in the United States. Nevertheless, subject to this limitation, it is a reasonably good indicator of changes in the general cost of living.

In addition to the overall consumer price index, indexes are also constructed for various subgroups such as food, housing, apparel, and upkeep. Also, consumer price indexes are published for most of the larger cities in the United States. However, unlike the cost of living index published by the United Nations Statistical Office, the CPIs for two cities cannot be compared directly, because the base for each city's CPI is the cost of living figure for that city in the base year. The current CPI for a given city indicates the change in the cost of living in that city compared to the cost during the base year in that city.

The CPI has been published continuously since 1919. The base period is moved forward every ten years or so and the goods and services included in the index are revised periodically. The current base period for this index is 1967. The CPI is published in the *Survey of Current Business* and in the *Monthly Labor Review*. Appendix B.2 shows a sample CPI.

The Index of Industrial Production. The Index of Industrial Production, or IIP, is compiled and published by the Board of Governors of the Federal Reserve System. This index measures changes in the value or quantity of output of the manufacturing, mining, and utility industries. It does not include agricultural production, transportation, construction, or the trade and service industries.

The index of industrial production was first computed in the 1920s and includes about 100 components. Separate indexes are constructed for various groups and subgroups. Of all the current indexes, the IIP is the one most widely used as an indicator of general business conditions and the overall state of the economy. This index is reported monthly in the *Federal Reserve Bulletin*. Appendix B.3 is a sample.

6.3 Problems of Index Number Construction

The major problems encountered in the construction of index numbers can be grouped into four principal categories:

1. Selecting the items to be included in the composite index.
2. Choosing an appropriate base period.
3. Choosing an appropriate weighting system.
4. Determining the method for computing the index.

Selection of Items. The choice of the specific items to be included in any composite index must relate to the purpose of the index. An index of food prices, for example, must be based on a carefully selected set of food items representative of the intended index group. An index designed by a manufacturer to reflect changes in the costs of raw materials used in his operations has to include all of the materials purchased by the firm or a representative sample of those materials.

If an index is constructed from a sample rather than from a complete census of items, then the choice of the specific commodities included in the index becomes extremely important. This choice cannot be made by some chance selection process, but must relate to the purpose of the index and the nature of the items represented.

For example, the Bureau of Labor Statistics has determined that the typical, moderate-income, city-wage-earner family purchases about 2000 different commodities. However, only about 400 of these are included in the consumer price index. In determining which of the 2000 items should be included, consideration had to be given to the relative importance of each item and its effect upon the living costs of the type of family described above.

Also, if a sample of commodities is used to construct a price index for the raw material purchases of a particular manufacturing firm, the sample should include all the principal materials purchased but might exclude those materials that are purchased in such small quantities that their effect on the overall index is negligible.

Choice of Base Period. The choice of a base period for index numbers is influenced by two considerations. First, the period should be recent enough so that comparisons with the base are meaningful. Constructing an index of current prices using the year 1900 as a base is of little use in assessing recent price movements. Second, the period selected should be a typical period with respect to the activity of interest. Of course, the definition of typical is somewhat subjective. In general, a period should be selected that exhibits relative stability with respect to the variable being observed.

The use of several years as a base rather than a single year provides a moderating effect on year-to-year variations. Some indexes use a base period of three years' duration, such as 1947–1949 and 1957–1959.

Selection of a Weighting System. Since all commodities or items included in an index are usually not equally important, it is necessary to apply a system of weights, so the index will reflect the relative importance of its components. In constructing an index, the price of each commodity is generally weighted by the quantity sold or purchased during some period. The difference between the two most common weighting systems in use is that one, the Laspeyres system, uses base period values as weights, while the other, the Paasche system, uses the current period values as weights. We will use both systems in Section 6.4.

Selection of a Computational Method. Index numbers are usually based on one of two computational methods. They are either AGGREGATIVE INDEX NUMBERS or AVERAGE OF RELATIVES INDEX NUMBERS. Each of these may be either weighted or unweighted.

In the aggregative type of index, prices (or whatever variable is being observed) are summed for all items in both the base period and the current period, and the index is simply the ratio of these two aggregates. In the average of relatives index, a simple index is first computed for each commodity using current and base year values. These relatives are then summed and the average is determined. We will discuss both types of index numbers in Section 6.4.

Common Notation for Index Number Construction. Although variations do occur, the following notation is fairly typical of that used with reference to index numbers.

1. Capital letters are used to represent index numbers.

2. The subscript indicates the period for which the index applies. Therefore:

P_i = a price index for period i. The period may be any unit of time.

Q_i = a quantity index for period i.

3. Lowercase letters represent the price or quantity of individual commodities. Again, the subscript indicates the period to which that price or quantity applies.

p_i = the price of a particular commodity in period i.

q_i = the quantity of a particular commodity in period i.

4. The subscript 0 is used to represent the base period. However, when the periods are years, a subscript corresponding to the calendar year may be used. For example,

P_0, p_0, and q_0 represent the base-period price index number, the base year price, and the base year quantity, respectively.

P_1, p_1, and q_1 represent the price index number, the price, and the quantity for period one relative to the base period, respectively.

P_{79}, p_{79}, and q_{79} represent the index number, price, and quantity for the year 1979, compared to a specified base year.

6.4 Price Index Numbers

Price index numbers are probably the most common and useful business indexes. Therefore, we use them to provide general illustrations of index number construction. Other types of indexes are constructed using the same general procedures.

The four types of price index numbers we cover in this section are

1. The unweighted aggregative index

2. The unweighted average of relatives index

3. The weighted aggregative index
4. The weighted average of relatives index

Unweighted Aggregative Price Index. If you refer to the introduction to this chapter and to Example 6.1, you will recall that a relative is simply the ratio of two numbers expressed as a percent. A *price relative* is the ratio of two prices, a current period price, which we will designate p_i, and a base period price, p_0. If we construct a simple index number, that is, one involving the price of a single commodity, the index for period i is

$$P_i = (100)\frac{p_i}{p_0} \tag{6.1}$$

However, most index numbers are constructed for groups of commodities. The simplest composite index number involving groups of commodities is the unweighted aggregative price index. This index number is constructed by adding the prices of all the commodities in both the base period and the current period and obtaining the ratio of the two aggregates. The formula for this price index is

$$P_i = (100)\frac{\Sigma p_i}{\Sigma p_0} \tag{6.2}$$

Note that this differs from the simple index number of Equation 6.1 only by the summation signs, indicating that the ratio is one of groups of prices, not the ratio of a single price in each period.

Example 6.3 The president of the Eastboro Portland Cement Company has been receiving numerous complaints from customers because of a recent 10% increase in the price of cement. The company's justification for the increase is the continually rising costs of the raw materials used in the manufacture of Portland cement. In response to these complaints, the president decides to send a letter to all of the company's regular customers showing the price increases of all the principal raw materials over the past five years. However, the assistant to the president, a recent business school graduate, suggests that a series of index numbers that summarizes these price increases is much easier to read and a great deal more effective. The data pertaining to the prices of the basic components of Portland cement are provided in Table 6.1.

Using the data in Table 6.1 and the unweighted aggregative method, a series of price index numbers was constructed for the years 1975–1978 using 1975 as the base period. The procedure follows:

Step 1. Sum the columns of prices of the components for each year. These sums are

$$\Sigma p_0 = \$41.00,\ \Sigma p_1 = \$41.70,\ \Sigma p_2 = \$43.15,\ \Sigma p_3 = \$45.05$$

TABLE 6.1
PRICE PER TON OF PORTLAND CEMENT COMPONENTS, 1975–1978 (FROM EXAMPLE 6.3)

Component	*1975*	*1976*	*1977*	*1978*
Tricalcium silicate	$ 8.25	$ 8.55	$ 8.50	$ 8.75
Dicalcium silicate	7.25	7.50	7.50	7.95
Tricalcium aluminate	7.50	7.50	8.00	8.35
Ferric oxide	6.50	6.35	6.85	7.35
Magnesia	11.50	11.80	12.30	12.65
	$41.00	$41.70	$43.15	$45.05

Step 2. State the formula for the unweighted aggregative price index.

$$P_i = (100)\frac{\Sigma p_i}{\Sigma p_0} \tag{6.2}$$

Step 3. Evaluate the formula for $i = 0, 1, 2, 3.$

$$P_0 = (100)\frac{\$41.00}{\$41.00} = 100.0$$

$$P_1 = (100)\frac{\$41.70}{\$41.00} = 101.7$$

$$P_2 = (100)\frac{\$43.15}{\$41.00} = 105.2$$

$$P_3 = (100)\frac{\$45.05}{\$41.00} = 109.9$$

The series of index numbers indicates that the prices of the raw materials used by the Eastboro Portland Cement Company have increased by 9.9% since the base year, 1975, with the largest increase occurring from 1977 to 1978. Instead of using the subscripts 0, 1, 2, and 3 to indicate the years 1975, 1976, 1977, and 1978, we can use the last two digits of each year as subscripts, so that the series of price indexes appears as:

$$P_{75} = 100.0 \qquad P_{76} = 101.7 \qquad P_{77} = 105.2 \qquad P_{78} = 109.9$$

Example 6.4

The Tri-State Food Cooperative is a nonprofit organization that purchases food products directly from farmers and sells to its members at cost plus operating expenses. A recent annual membership report compared increases in the cooperative's prices to changes in the food component of the consumer price index over three years. Unweighted aggregative index numbers were constructed for 1976, 1977, and 1978 from the prices of the commodities which represented the bulk of the cooperative's sales. The data from which the index was constructed are provided in Table 6.2. The year 1976 was used as the base period.

The index numbers were constructed as follows:

Step 1. Sum the columns of commodity prices for each year:

$$\Sigma p_{76} = \$7.10 \qquad \Sigma p_{77} = \$7.50 \qquad \Sigma p_{78} = \$7.75$$

TABLE 6.2
PRICES OF FOODSTUFFS SOLD BY TRI-STATE FOOD COOPERATIVE—1976, 1977, AND 1978 (FROM EXAMPLE 6.4)

Foodstuff	*Measurement Unit*	*1976*	*1977*	*1978*
Beef	pound	$1.10	$1.05	$1.15
Bacon	pound	1.25	1.35	1.50
Eggs	dozen	0.95	0.95	0.90
Cheese	pound	1.15	1.25	1.20
Milk	gallon	1.55	1.70	1.75
Butter	pound	1.10	1.20	1.25
		$7.10	$7.50	$7.75

Step 2. State the formula for the unweighted aggregative price index.

$$P_i = (100)\frac{\Sigma p_i}{\Sigma p_0} \tag{6.2}$$

Step 3. Evaluate the formula for $i = 76, 77, 78$.

$$P_{76} = (100)\frac{\$7.10}{\$7.10} = 100$$

$$P_{77} = (100)\frac{\$7.50}{\$7.10} = 106$$

$$P_{78} = (100)\frac{\$7.75}{\$7.10} = 109$$

According to these indexes, this particular set of food items cost 9% more to purchase in 1978 than it did in 1976. However, this conclusion is misleading for two principal reasons. First, the prices of the various food items are reported in terms of different units: pounds, gallons, and dozens. The choice of the measurement unit affects the value of this index. For example, if milk is priced by the quart instead of by the gallon, the resulting index number has a different value. Second, in this index, no consideration is given to the fact that some items are purchased in greater quantities than others. Consequently, those food items have a much greater effect on the overall cost to the membership than items purchased in much smaller quantities. For example, if the membership consists primarily of young families with small children, milk would account for a much larger proportion of the food purchases than cheese. Increases in the cost of milk then would affect their food expenditures much more than increases in the cost of cheese. However, if the membership consists mostly of older couples, the reverse might be true.

This second defect also applies to Example 6.3. All five raw materials are not used in equal proportions in the manufacture of Portland cement and, therefore, should not have equal weight in the price index. The unweighted aggregative index should be used only in cases where all items are reported in the same units, and also where all items are of equal importance.

Unweighted Average of Relatives Price Index. The first defect attributed to the unweighted aggregative index is remedied by using the *average of relatives index*. This index is constructed by first converting each reported price for each commodity into a simple relative by dividing by the base period price of that commodity. These relatives are then summed for each year, the sum is multiplied by 100, and then divided by the number of commodities. By converting all prices to relatives and finding their averages, the effect of the measurement unit is eliminated. The formula for the unweighted average of relatives index is

$$P_i = \frac{(100)\Sigma(p_i/p_0)}{N} \tag{6.3}$$

where N represents the number of commodities entering into the index. Let's see how the average of relatives index affects the previous example.

Example 6.5

Because of the defects in the unweighted aggregative index, the Tri-State Food Cooperative decides to construct a new price index using the unweighted average of relatives method. For the data in Table 6.5, they proceed as follows.

Foodstuff	*1976*	*1977*	*1978*
Beef	1.10/1.10 = 1.000	1.05/1.10 = 0.955	1.15/1.10 = 1.045
Bacon	1.25/1.25 = 1.000	1.35/1.25 = 1.080	1.50/1.25 = 1.200
Eggs	0.95/0.95 = 1.000	0.95/0.95 = 1.000	0.90/0.95 = 0.947
Cheese	1.15/1.15 = 1.000	1.25/1.15 = 1.087	1.20/1.15 = 1.043
Milk	1.55/1.55 = 1.000	1.70/1.55 = 1.097	1.75/1.55 = 1.129
Butter	1.10/1.10 = 1.000	1.20/1.10 = 1.090	1.25/1.10 = 1.136
	6.000	6.309	6.500

Step 1. Each price in the table is expressed as a relative by dividing by the 1976 base period price. The simple indexes are then summed for each year.

Step 2. The summed relatives for each year are multiplied by 100.

$$(100)\Sigma(p_{76}/p_{76}) = (100)(6.000) = 600.0$$
$$(100)\Sigma(p_{77}/p_{76}) = (100)(6.309) = 630.9$$
$$(100)\Sigma(p_{78}/p_{76}) = (100)(6.500) = 650.0$$

Step 3. The results of step two are divided by $N = 6$, resulting in the following index numbers.

Unweighted Average of Relatives Price Index
$P_{76} = 100.0$
$P_{77} = 105.2$
$P_{78} = 108.3$

If you compare the average of relatives indexes with the aggregative indexes in the previous example, you will notice that they are different. The difference results because the unweighted average of relatives index eliminates the effect of the different measurement units for which the prices are quoted. However, it does not correct the other deficiency of an unweighted index. That is, all items in the index are weighted equally rather than according to their comparative importance.

Weighted Aggregative Price Index. The most common type of weight used in constructing a price index is quantity. This assumes that the quantity of a commodity sold or purchased in a given period is a valid measure of its comparative importance. In Example 6.3, the comparative importance of each component of Portland cement should be proportional to the amount of each component used in its manufacture. In Example 6.4, the purpose of the price index was to measure the effect of price changes on the amount spent for food by a typical family member of the cooperative. Since each typical-family member probably does not purchase equal quantities of beef and cheese, or attach the same importance to bacon as to milk, the effects of these price changes depend on the amounts purchased. However, simply deciding that prices should be weighted by quantities does not completely solve the problem. The question still remains, what quantities should be used as weights?

The two most common weighting systems are named for the individuals most influential in their development. When an index number is constructed using *base year* quantities as weights it is known as a LASPEYRES INDEX. The formula for a Laspeyres weighted aggregative price index is

$$P_i = (100)\frac{\Sigma p_i q_0}{\Sigma p_0 q_0} \tag{6.4}$$

This index measures the change in the total cost of a fixed quantity of goods that represent a consumption pattern typical of the base year. Exactly how long the consumption pattern of the base period should be used as a weighting factor is problematical. Consumption patterns change with time and when this pattern changes the base period should also be changed.

The other common weighting system uses *current year* quantities as weights and is known as a PAASCHE INDEX. The formula for a Paasche weighted aggregative index is

$$P_i = (100)\frac{\Sigma p_i q_i}{\Sigma p_0 q_i} \tag{6.5}$$

This index measures the change in the total cost of a fixed quantity of goods that represent a consumption pattern typical of the current year. The use of current year weights avoids the problem of changing consumption patterns; however, it requires that the weighting system be continually reestablished. Because of the difficulty involved in establishing a new set of weights and applying them each period, the Laspeyres method is by far the more commonly used.

Note that the only difference between these two indexes, Equations 6.4 and 6.5, is the subscript on q, the weighting factor.

Example 6.6

The board of directors of the Tri-State Food Cooperative realize that the unweighted aggregative price index numbers developed in Example 6.4 are not truly indicative of the actual price increases because of the defects in that index. They decide to construct a new index for 1978 relative to the base year, 1976. The Laspeyres method using base year weights is to be used. They proceed as follows using the data from Table 6.3.

Step 1. Compute p_0q_0 and p_iq_0 as shown in Table 6.3. These values are indicated in the table using subscripts which correspond to calendar years. In this case, $p_0 = p_{76}$, $p_i = p_{78}$, $q_0 = q_{76}$.

Step 2. Sum the columns of weighted prices, $p_{76}q_{76}$ and $p_{78}q_{76}$.

Step 3. State the formula for the Laspeyres weighted aggregative price index.

$$P_i = (100)\frac{\Sigma p_iq_0}{\Sigma p_0q_0} \tag{6.4}$$

Step 4. Substitute the values from Table 6.3 in the formula and evaluate.

$$P_{78} = (100)\frac{36510}{32715} = 111.6$$

This index number for 1978 of 111.6 indicates that the price of the fixed quantities of foodstuffs specified increased by 11.6% between 1976 and 1978. This compares with the unweighted aggregative index number, 109, and the unweighted average of relatives index number, 108.3.

TABLE 6.3
PRICES OF FOODSTUFFS SOLD BY TRI-STATE FOOD COOPERATIVE IN 1976 AND 1978 WITH QUANTITIES FOR 1976 (FROM EXAMPLE 6.6)

Foodstuff	*Quantity Sold Base Year* (q_{76})	*Price Base Year* (p_{76})	*Price Current Year* (p_{78})	$p_{76}q_{76}$	$p_{78}q_{76}$
Beef	3000 pounds	1.10	1.15	3300	3450
Bacon	2000 pounds	1.25	1.50	2500	3000
Eggs	1500 dozen	0.95	0.90	1425	1350
Cheese	800 pounds	1.15	1.20	920	960
Milk	15000 gallons	1.55	1.75	23250	26250
Butter	1200 pounds	1.10	1.25	1320	1500
				32715	36510

Not only does the weighted index give consideration to the comparative importance of the various quantities included in the index, it also eliminates the effects of the different measurement units that might be used. In Example 6.4, if the price of milk had been reported in terms of quarts instead of gallons, the value of the unweighted aggregative index would have been different. However, when every unit price is weighted by a quantity of the same unit, the choice of the unit becomes immaterial. This is easily illustrated using milk as an example. When we computed the weighted aggregative price index based on the data in Table 6.3, the price of milk per gallon was weighted by the number of gallons sold. This resulted in $p_{78}q_{76} = (1.75)(15{,}000) = 26{,}250$. If milk had been reported in terms of quarts, the price would have been \$0.4375, or $\frac{1}{4} \times \$1.75$, per quart and the quantity sold would have been 60,000 quarts, which is $4 \times 15{,}000$ gallons. Then the product of price and quantity would have been obtained from $p_{78}q_{76} = (0.4375)(60{,}000) = 26{,}250$ which is identical to the previous result.

Example 6.7 Some members of the board of directors of the Tri-State Food Cooperative think that it might be preferable to use the Paasche method of constructing the index, which weights the prices by current year quantities. The data and calculations required to obtain the price index for 1978 are based on Table 6.4 and Equation 6.5.

The Paasche weighted aggregative price index for 1978 is

$$P_i = (100)\,\frac{\Sigma p_i q_i}{\Sigma p_0 q_i} \tag{6.5}$$

$$P_{78} = (100)\,\frac{33{,}330}{30{,}005} = 111.1$$

TABLE 6.4
PRICES OF FOODSTUFFS SOLD BY TRI-STATE FOOD COOPERATIVE IN 1976 AND 1978 WITH QUANTITIES FOR 1978 (EXAMPLE 6.7)

Foodstuff	*Quantity Sold Current Year* q_{78}	*Price Base Year* p_{76}	*Price Current Year* p_{78}	$p_{76}q_{78}$	$p_{78}q_{78}$
Beef	3500 pounds	1.10	1.15	3850	4025
Bacon	1800 pounds	1.25	1.50	2250	2700
Eggs	1700 dozen	0.95	0.90	1615	1530
Cheese	1000 pounds	1.15	1.20	1150	1200
Milk	13000 gallons	1.55	1.75	20150	22750
Butter	900 pounds	1.10	1.25	990	1125
				30005	33330

The difference between the values of the index numbers computed by the two different weighting methods in Examples 6.6 and 6.7 is about 0.5%. The size of this difference depends on the relative amount of change in the quantities of the various commodities from the base period to the current period. If these quantity changes are slight, then the difference between the two indexes is also slight. However, if substantial changes occur between the two periods, the indexes also differ substantially. It is not correct to say that one of the weighting methods is better than the other. Each has its own interpretation. The Laspeyres index indicates that a particular set of commodities which represents a buying pattern typical to the base year costs 11.6% more to purchase in 1978 than it did in 1976. The Paasche index indicates that the price of a set of commodities which represents a buying pattern typical of 1978 has increased by 11.1% since 1976.

Weighted Average of Relatives Price Index. The unweighted average of relatives price index can be converted to a more satisfactory index with the application of a systematic weighting system. The weights used in constructing this index of relatives, however, are usually value weights rather than physical quantities. Note, however, that value is simply price times quantity, that is, pq. The formula for a weighted average of relatives index using base period values as weights is

$$P_i = \frac{\Sigma(100)(p_i/p_0)(p_0q_0)}{\Sigma p_0q_0}. \tag{6.6}$$

Note that Equation 6.6, after cancelling like terms, is identical to Equation 6.4, the weighted aggregative index. Since the weighted average of price relatives and the weighted aggregative price index are equivalent, it is reasonable to ask why Equation 6.6 should ever be used, since the arithmetic involved in using that equation seems to be more complicated. The answer lies primarily in the way the data are presented. When the data are available only in terms of price relatives or if information on value is easier to obtain than prices and quantities, then the use of Equation 6.6 involves less arithmetic. For example, if the data in Table 6.3 were expressed in terms of relatives rather than actual prices as shown in Table 6.5, it would be easier to use Equation 6.6 in computing the index number for 1978.

The weighted average of relatives price index for 1978 would be

$$P_{78} = \frac{\Sigma(100)(p_{78}/p_{76})(p_{76}q_{76})}{\Sigma p_{76}q_{76}} \tag{6.6}$$

$$P_{78} = \frac{3{,}650{,}630.5}{32{,}715} = 111.6$$

TABLE 6.5
PRICE INDEXES OF FOODSTUFFS SOLD BY TRI-STATE FOOD COOPERATIVE IN 1976 AND 1978 WITH QUANTITIES FOR 1976 (1976 = 100)

Foodstuff	*Quantity Sold Base Year* (q_{76})	*Price Index Base Year* $100(p_{76}/p_{76})$	*Price Index Current Year* $100(p_{78}/p_{76})$	*Weight* $p_{76}q_{76}$	$100(p_{78}/p_{76})p_{76}q_{76}$
Beef	3000 pounds	100.0	104.5	3300	344850.0
Bacon	2000 pounds	100.0	120.0	2500	300000.0
Eggs	1500 dozen	100.0	94.7	1425	134947.5
Cheese	800 pounds	100.0	104.3	920	95956.0
Milk	15000 gallons	100.0	112.9	23250	2624925.0
Butter	1200 pounds	100.0	113.6	1320	149952.0
				32715	3650630.5

which is exactly the same index number we obtained using the weighted aggregative method.

A comparable formula for a weighted average of relatives price index using current year quantities as weights is

$$P_i = \frac{\Sigma(100)(p_i/p_0)\,p_0q_i}{\Sigma p_0q_i} \tag{6.7}$$

Changing the Base Period. It is frequently necessary to change the base of an index number series from one period to another without recomputing the entire series from the original data. This occurs when a base period for a series is no longer considered normal or typical, or when a more recent period is more suitable as a basis for comparison. A change in base also becomes necessary when two or more series that were computed with reference to different bases are to be compared.

In all of the examples about the Tri-State Food Cooperative, the reason for constructing the index numbers was to enable the membership to compare the changes in food prices charged by the cooperative with changes in the food component of the CPI for the same period. Since Tri-State chose 1976 as its base period, and the present base period for the CPI is 1967, it is necessary to shift the base of one of the index number series in order to make them comparable.

The procedure for shifting the base of a series of index numbers is quite simple. It is accomplished by dividing each index number in the original series by the index number *expressed in decimal form* of the new base period. That is,

$$\text{New index number} = \frac{\text{Original index number}}{\text{Index number of new base period}}$$

Example 6.8

In order to compare changes in the price of food sold to members of the Tri-State Food Cooperative to general price changes as measured by the food component of the consumer price index, it was necessary to shift the base of the CPI series to the same base used in constructing Tri-State's index numbers. The base period used by Tri-State was 1976. The procedure is illustrated below.

Year	*Tri-State Food Index (1976 = 100)*	*Original CPI Food Index (1967 = 100)*	*CPI for New Base Period 1976 (in decimal form)*	*New CPI Food Index (1976 = 100)*
1967	—	100.0	1.701	58.8
1976	100.0	170.1	1.701	100.0
1978	111.6	190.7	1.701	112.1

The new series of index numbers can now be compared with Tri-State's index numbers as constructed from the data in Table 6.3 and Example 6.6. Note that shifting the base does not change the relationship between any two numbers in the series. For example, the index number for 1978 exceeds the number for 1976 by exactly the same percentage in both series.

It is now possible to conclude that although the prices charged by the Tri-State Food Cooperative increased by 11.6% from 1976 to 1978, the overall increase in food prices, as represented by the CPI food index, is 12.1%.

6.5 Quantity Index Numbers

Quantity indexes are constructed for measuring changes in physical quantity that occur over some period of time. The best known quantity index currently published in the United States is the Index of Industrial Production. The price index answers the question of how much we must spend in different periods if we buy the same assortment of goods in each period at different prices. The quantity index tells us how much we must spend in different periods if we buy varying assortments of specified goods at constant prices. In the first case, the difference in the amount spent each year is due to the price changes. In the second case, the difference is due to changes in the quantities purchased.

The construction of a quantity index involves the same problems with respect to weights and base periods and essentially the same methods as a price index. However, where we use quantities as weights in constructing a price index, we use prices as weights in constructing a quantity index. If base year prices are used as weights then we have a Laspeyres quantity index, whereas if current year prices are used as weights the resulting index is of the Paasche type.

The weighted aggregative quantity index using base year prices as weights is

$$Q_i = (100)\frac{\Sigma q_i p_0}{\Sigma q_0 p_0} \tag{6.8}$$

The weighted average of relatives quantity index using base year values as weights is

$$Q_i = \frac{\Sigma(100)(q_i/p_0)(q_0 p_0)}{\Sigma q_0 p_0} \tag{6.9}$$

The weighted aggregative quantity index using current year prices as weights is

$$Q_i = (100)\frac{\Sigma q_i p_i}{\Sigma q_0 p_i} \tag{6.10}$$

The weighted average of relatives quantity index using current year prices as weights is

$$Q_i = \frac{\Sigma(100)(q_i/q_0)(q_0 p_i)}{\Sigma q_0 p_i} \tag{6.11}$$

As with the comparable price indexes, the weighted average of relatives procedure results in the same index number as the weighted aggregative. The decision as to which procedure to use is dictated by the form of the data.

Example 6.9 An agency of the United States Department of Agriculture constructs a quantity index which indicates the change in the quantity of different grains produced from year to year. The quantities are weighted by 1975 prices. The index constructed for 1978 is based on the data provided in Table 6.6.

TABLE 6.6

GRAIN PRODUCTION IN 1975 AND 1978, WITH 1975 PRICES (EXAMPLE 6.9)

Grain	*1975 Price (per bushel)* (p_{75})	*1975 Production (millions)* (q_{75})	*1978 Production (millions)* (q_{78})	($q_{75}p_{75}$)	($q_{78}p_{75}$)
Rye	2.46	5767	6202	14186.8	15256.9
Oats	4.60	1521	2354	6996.6	10828.4
Barley	2.42	383	393	926.9	951.1
Wheat	3.52	2134	2615	7511.7	9204.8
				29622.0	36241.2

A weighted aggregative quantity index is constructed using these data and Equation 6.8 as follows.

$$(6.8) \qquad Q_{78} = (100)\frac{\Sigma q_{78}p_{75}}{\Sigma q_{75}p_{75}} = (100)\frac{(36{,}241.2)}{(29{,}622.0)} = 122.3$$

The index indicates that by 1978, the production of these four crops had increased by 22.3% over 1975.

6.6 Other Business Applications of Index Numbers

Deflation of Time Series. A time series is a series of numerical observations recorded with respect to time. The topic of time series analysis is covered in Chapter 17. A time series in which the observations are recorded in terms of dollars contains fluctuations due to changes in the overall level of prices in addition to actual changes of the value of the observed commodities. When the price of a commodity increases while the commodity itself is unchanged, we have inflation. The effect of inflation is to reduce the value of the dollar in terms of its purchasing power. Therefore, since 1978 dollars, for example, do not have the same purchasing power as 1974 dollars, a time series comparing dollars from these years is misleading. However, if all of the figures reported are converted to, say, 1967 dollars, then valid comparisons can be made. The process of converting a series of dollar figures recorded over time to a common constant value coinciding with the value of the dollar in the base year is called *deflation of a time series*. The value used as the basis of comparison is the common dollar.

Fluctuations in a time series due to price changes or inflation can and should be removed by dividing the figure reported for each time period by an appropriate price index for the same period. The resulting series shows actual changes in value in terms of constant dollars.

Example 6.10

Cinderella Clothes, Inc. operates a chain of children's clothing stores. The company management is interested in evaluating their sales performance for the past five years. The sales data are as follows:

Year	*Gross Sales (\$ thousands)*
1974	720
1975	765
1976	845
1977	900
1978	970

Since Cinderella Clothes is a retail chain, the appropriate index for them to use is the Consumer Price Index or CPI. Values of the CPI for these same years for the category of apparel and upkeep are as follows, where 1967 = 100:

1974—136.2 1975—142.3 1976—146.9
1977—151.0 1978—154.5

The procedure for deflating the Cinderella Clothes sales figures to constant 1967 dollars is illustrated in tabular form below. The actual dollars in thousands are divided by the value of the CPI for that period and the quotient is multiplied by 100. The results are the deflated sales figures in the last column.

Year	*Gross Sales (\$ thousands)*	*CPI (1967 = 100)*	*Deflated Sales (\$ thousands)*
1974	720	136.2	528.6
1975	765	142.3	537.6
1976	845	146.9	575.2
1977	900	151.0	596.0
1978	970	154.5	627.8

Using the original gross sales figures, the apparent increase in sales from 1974 to 1978 is (970 − 720)/720 = 0.347 or 34.7%. However, this increase includes the effect of inflation, that is, the generally increasing price level. When the inflationary effect is eliminated from the series, the sales increase in constant dollars is (627.8 − 528.6)/528.6 = 0.188 or 18.8%. Using this procedure and reporting gross sales in terms of constant dollars, we see that almost half of the apparent sales increase between 1974 and 1978 is actually due to inflation. Although an actual 18.8% increase in gross sales over that period might be considered quite satisfactory, it is certainly far less than the 34.7% increase that was indicated when the original gross sales figures were reported.

Long Term Lease or Purchase Contracts. It is not unusual for companies to enter into contracts that require payments to be made over a long period of time. A typical contract of this type might be for the leasing of property. If a property is leased for a period of 10, 15, or 20 years, and if the general level of prices increases during that period, the purchasing power of the lease payments declines substantially. Consequently, the owner of the property receives much less in terms of actual purchasing power than he may have intended at the time he entered into the lease agreement.

In order to avoid this decline in the purchasing power of the lease payments, contracts that involve future payments are often written with a clause that provides for payments to be adjusted periodically for changes in the price level. These changes are represented by an appropriate index number.

Example 6.11

In 1976, the management of the Advance Machine Company concluded that additional production facilities would be needed to accommodate anticipated future business. Since the company did not want to tie up large amounts of capital in real estate, they decided to take a ten-year lease on a suitable building near their present plant. The lease began on January 1, 1977. The owner of the property was willing to lease the building at a monthly rental of \$3000 but demanded a clause in the contract that adjusted the rent each year according to the value of the consumer price index as reported for the month of September of the previous year. In September of 1976, the CPI was reported to be 177.6. In September of 1977, the CPI was 188.2. Since the monthly rent for 1978 should have the same relationship to the CPI of September 1977 as the 1977 rent had to the CPI of September 1976, the 1978 rent was determined as follows.

The rent \$3000 is to 177.6 as the 1978 rent is to 188.2:

$$\frac{\$3000}{177.6} = \frac{1978\text{ rent}}{188.2}$$

$$1978\text{ rent} = \frac{(188.2)(\$3000)}{177.6}$$

$$1978\text{ rent} = \$3179\text{ per month}$$

Wage Contracts. Wage contracts frequently contain escalator clauses, which tie automatic wage increases to increases in the cost of living as measured by the CPI. These clauses usually provide that wages increase by some specified amount or percent for each point or percentage increase in the index.

Example 6.12

Metric Systems, Inc. recently negotiated a three-year contract with the Amalgamated Electrical Workers Union that included a cost-of-living escalator clause. The clause specified that wages were to be increased by 1% for every increase of one point in the CPI. The contract became effective on January 1, 1977, at which time the CPI was 180.1. In March the CPI went to 181.1. The personnel department of Metric Systems was required to make the 1% adjustment in the wages of all employees covered by the contract. An electrical worker earning \$5.40 per hour at the start of the contract received a wage increase of 5.4 cents/hour retroactive to March 1.

Definitions

Aggregative index number: A composite index number which is constructed by taking the ratio of two sets of aggregated numbers, either weighted or unweighted.

Average of relatives index number: A composite index number constructed by first converting pairs of values in the base and nonbase periods to relatives and then computing the arithmetic mean of the relatives.

Base period: The time period used as the base or basis of comparison when constructing an index number.

Composite index number: A relative which summarizes the relationship between two sets of prices, quantities, or other values.

Laspeyres method: A weighting system for the construction of index numbers which uses base period values as weights.

Paasche method: A weighting system for the construction of index numbers which uses current period values as weights.

Relative: The ratio of two numbers usually expressed as a percent.

Simple index number: A percentage which expresses the relationship between two numbers, one of which is called the base.

Formulas

Simple unweighted price index number:

$$P_i = (100)\left(\frac{p_i}{p_0}\right) \tag{6.1}$$

Unweighted aggregative price index number:

$$P_i = (100)\left(\frac{\Sigma p_i}{\Sigma P_0}\right) \tag{6.2}$$

Unweighted average of relatives price index number:

$$P_i = \frac{(100)\Sigma(p_i/p_0)}{N} \tag{6.3}$$

Weighted aggregative price index number using base period quantities as weights—Laspeyres method:

$$P_i = (100)\left(\frac{\Sigma p_i q_0}{\Sigma p_0 q_0}\right) \tag{6.4}$$

Weighted aggregative price index number using current period quantities as weights—Paasche method:

$$P_i = (100)\left(\frac{\Sigma p_i q_i}{\Sigma p_0 q_i}\right) \tag{6.5}$$

Weighted average of relatives price index number using base period values as weights—Laspeyres method:

$$P_i = \frac{\Sigma(100)(p_i/p_0)(p_0 q_0)}{\Sigma p_0 q_0} \tag{6.6}$$

Weighted average of relatives price index number using current period values as weights—Paasche method:

$$P_i = \frac{\Sigma(100)(p_i/p_0)(p_0 q_i)}{\Sigma p_0 q_i} \tag{6.7}$$

Weighted aggregative quantity index number using base period prices as weights—Laspeyres method:

$$Q_i = (100)\left(\frac{\Sigma q_i p_0}{\Sigma q_0 p_0}\right) \tag{6.8}$$

Weighted average of relatives quantity index number using base period values as weights—Laspeyres method:

$$Q_i = \frac{\Sigma(100)(q_i/p_0)(q_0p_0)}{\Sigma q_0p_0} \tag{6.9}$$

Weighted aggregative quantity index number using current period prices as weights—Paasche method:

$$Q_i = (100)\left(\frac{\Sigma q_i p_i}{\Sigma q_0 p_i}\right) \tag{6.10}$$

Weighted average of relatives quantity index number using current period values as weights—Paasche method:

$$Q_i = \frac{\Sigma(100)(q_i/q_0)(q_0p_i)}{\Sigma q_0p_i} \tag{6.11}$$

Problems

6.1 Convert the following series of numbers to a series of relatives using the number 250 as the base: 150, 200, 250, 300, 350, 400, 450.

6.2 Convert the series in problem 6.1 to a series of relatives using the number 350 as the base.

6.3 If a particular set of goods and services costs a typical family $13,747 in San Francisco, $12,962 in Philadelphia, and $14,440 in Boston, compute cost relatives for each city using San Francisco as the base.

6.4 For the data in problem 6.3, compute cost relatives using Philadelphia as the base.

6.5 For the data in problem 6.3, compute cost relatives using Boston as the base.

6.6 The following table provides the price per pound and the quantities purchased of peanuts, pecans, and cashew nuts during the years 1975, 1976, and 1977 in a particular locality. Construct an unweighted aggregative price index for these data using 1976 as the base.

	Price per Pound (in dollars)			*Quantities Purchased (in thousands of pounds)*		
Commodity	1975	1976	1977	1975	1976	1977
Peanuts	0.98	1.04	1.09	1200	1100	1150
Pecans	1.02	1.06	1.10	600	650	700
Cashews	1.04	1.05	1.12	650	650	675

6.7 The following table shows the relative prices per board foot of spruce, white pine, and cedar for the years 1976, 1977, and 1978. The base year for these relatives was 1977. Construct an unweighted average of relatives price index.

Commodity	*1976*	*1977*	*1978*
Spruce	98.4	100.0	102.5
White pine	96.2	100.0	105.7
Cedar	98.8	100.0	104.4

6.8 From the data concerning the Eastboro Portland Cement provided in Table 6.1, convert all prices to relatives using 1976 as the base year and construct an average of relatives price index.

6.9 From the data in problem 6.6, construct a Laspeyres weighted aggregative price index for each year, using 1975 as the base.

6.10 From the data in problem 6.6, construct a Paasche weighted aggregative price index for 1977, using 1975 as the base year.

6.11 From the data in problem 6.6, construct a Paasche weighted aggregative price index for 1977 using 1976 as the base year.

6.12 From the data in problem 6.6, construct a Laspeyres weighted average of relatives price index for 1976 and 1977 using 1975 as the base year.

6.13 For the data in Table 6.4, construct a Paasche weighted average of relatives price index for 1978, using 1976 as the base year.

6.14 In Example 6.3, index numbers that show the change in the price of Portland cement manufactured by the Eastboro Portland Cement Company were constructed for the years 1975–1978, with 1975 as the base. Shift the base of this index number series to 1977 and show the new index numbers for each year.

6.15 For the data in Example 6.3, shift the base of the index number series to 1978.

6.16 Use the production quantities in Table 6.6 and the following prices to construct a weighted aggregative quantity index for 1978 using 1978 prices as weights. The base year is 1975.

Grain	*1978 Price (in dollars per bushel)*
Rye	2.76
Oats	4.65
Barley	2.58
Wheat	3.65

6.17 The catch in millions of pounds and the price per pound of cod, flounder, and haddock is provided below for the years 1977 and 1978. Construct a Laspeyres weighted aggregative quantity index for the year 1978, using 1977 as the base year.

Fish	*Production 1977 (in millions of lbs.)*	*Production 1978 (in millions of lbs.)*	*Price 1977 (in dollars per lb.)*	*Price 1978 (in dollars per lb.)*
Cod	11	14	0.68	0.72
Flounder	22	36	0.75	0.77
Haddock	35	38	0.70	0.70

6.18 For the data in problem 6.17, construct a Paasche weighted aggregative quantity index for 1978, using 1977 as the base year.

6.19 The prices and quantities of the principle raw materials purchased by the Eastern Industrial Products Company for the years 1976 and 1977 are as follows.

	Quantity (in tons)		*Price per Ton (in dollars)*	
Material	*1976*	*1977*	*1976*	*1977*
A	140	150	27	30
B	55	65	75	75
C	90	85	15	18

Construct a Paasche weighted aggregative price index for 1977 using 1976 as the base year.

6.20 For the data in the previous problem, construct a Laspeyres weighted aggregative price index for 1977, using 1976 as the base year.

6.21 Gross sales for the Eastern Industrial Products Company and the appropriate CPI for the years 1975–1978 are listed below. Convert the gross sales to sales in constant 1967 dollars by deflating the series.

Year	*Gross Sales (in thousands of dollars)*	*CPI 1967 = 100*
1975	957.6	147.2
1976	1014.0	148.6
1977	1010.7	152.1
1978	1018.6	154.3

6.22 Profits after taxes of the Eastboro Portland Cement Company for the years 1975–1978 are provided below. Convert these profits to constant 1967 dollars using the Consumer Price Index numbers listed in problem 6.21.

Year	*After Tax Profits (in thousands of dollars)*
1975	103.4
1976	110.6
1977	112.1
1978	115.7

CHAPTER 7 PROBABILITY

7.1
The nature of probability

7.2
Some important definitions

7.3
Basic rules of probability

7.4
Conditional probability

7.5
Statistical independence

7.6
The rule of elimination

7.7
Bayes' rule

Before we can begin the study of statistical inference, we must become familiar with the basic concepts of probability. The previous chapters demonstrated that statistical data can be summarized and displayed in various ways. However, we did not attempt to make any inferences beyond the limits of the observed data. In order to make inferences, we must use probability concepts.

Probability involves the study of uncertain events, and as you will most certainly discover, uncertainty is a very common factor in business. The demand for a product in any future period is uncertain, and usually varies from period to period. Actions of competitors and reactions of potential customers to advertising appeals are never known in advance. Costs of materials change rapidly. Tomorrow's cost will not be known until tomorrow arrives. Unfortunately, business decisions must be made today in order to be effective tomorrow. Therefore, we must attempt to deal with uncertainty in some systematic way.

Since this is a textbook in business statistics, you may be surprised to see that many of the concepts and rules of probability are demonstrated with reference to gambling games that involve the tossing of coins, the rolling of dice, and the drawing of playing cards from a deck. The reason for using these types of examples is that they illustrate most clearly and most directly the operation of the rules of probability and the principles involved. In most cases, however, each gambling example is followed by a business-related example designed to illustrate the same principle. It is interesting to note that the mathematical theory of probability originated in the mid-seventeenth century as a result of attempts to solve problems involving games of chance.

7.1 The Nature of Probability

In its simplest sense, probability can be defined as a numerical measure of uncertainty. To be somewhat more specific, we will restate this general definition of probability and specify that *probability is a numerical measure of the likelihood of occurrence of an uncertain event.*

In searching for a useful measure of probability we require a numerical scale of values that encompasses the two extremes of "impossibility" and "absolute certainty." Any value between these two extremes represents the relative likelihood of the event occurring. We choose a value of zero corresponding to impossibility, and a value of one corresponding to certainty. Since there is no likelihood smaller than impossible and nothing is more certain than certainty, a probability measure can never have a value less than zero or greater than one. Furthermore, since any probability between zero and one expresses a relative likelihood, the probability measure must be a ratio or proportion. For example, if an event is just as likely to occur as it is not to occur, then its probability is $\frac{1}{2}$ or 0.5. If an event is three times as likely to occur as it is not to occur, its probability is 0.75. Probabilities are frequently expressed as percents, in which case the above probabilities are stated as 50% and 75%, respectively.

Simply stating the desirable characteristics of a probability measure does not indicate how such a measure should be determined. A more precise definition is required before we begin to compute probabilities of actual events. However, even mathematicians do not all agree on the true meaning of probability. Originally, probability was viewed as a purely *objective* measure of relative likelihood. That is, the probability of any event occurring was considered to be an exact value, which is determined by the state or condition of the universe. If these universal conditions are completely known, then the probability of an event can be precisely determined. However, in the absence of complete information about these conditions, we are sometimes required to estimate the probability on the basis of the information that is available.

A more recent development is the *subjective* view of probability. This view defines probability as an individual's state of belief regarding the likelihood of the occurrence of an event. However, regardless of whether a probability is determined objectively or subjectively, the value is always a number from zero to one. Furthermore, once the probability value is determined, regardless of the method, the same rules of probability always apply.

For most of the material in this book, we adopt the objective view of probability. However, in studying statistical decision theory, which is the subject of Chapter 18, we find it convenient to take the subjective approach.

7.2 Some Important Definitions

In the previous section, a general definition of probability provided that probability is a numerical measure of the likelihood of occurrence of an uncertain event. Although this definition tells us what probability is, it doesn't tell us how to compute it. To do that, we need a more precise mathematical definition. Since the theory of probability is closely related to set theory, we make extensive use of the concepts and notation of sets in defining and studying probability.

Before attempting to further define probability, we must define several terms that are necessary for the discussion. Note carefully the correspondence between some of these terms and those relating to sets.

Experiment: Any well-defined action that leads to a single well-defined result.

Outcome: Any single possible result of an experiment; an outcome corresponds to an element or elementary unit of a set.

Sample space: The sample space of an experiment is the set of all possible outcomes of the experiment; it is synonymous with the universal set. The sample space of an experiment is designated S.

Event: Any set of outcomes of an experiment; it is a subset of the sample space. Events are designated with capital letters such as A, B, and C, or occasionally with a descriptive word or phrase in boldface type, or set off by quotation marks.

Event space: That subset of outcomes which corresponds to some specific event.

Event set: Same as event space.

Example 7.1 Consider an experiment consisting of one roll of a single, six-sided die. The sample space of that experiment is

$$S = \{1,2,3,4,5,6\}$$

Each of the possible results of the experiment, that is, **a 1 appears, a 2 appears,** and so on, is an outcome. An event defined on this sample space is any subset consisting of any single outcome, or any other subset of outcomes including the entire sample space or universal set. The event, **odd number,** which we designate A, consists of the outcomes 1, 3, and 5. That is,

$$A = \{1,3,5\}$$

The Classical Definition of Probability. The classical definition of probability applies to finite sample spaces consisting of equally likely discrete outcomes. The probability of an event E is the ratio of the number of outcomes in the sample space corresponding to E to the total number of outcomes in the sample space. That is,

$$P(E) = \frac{n(E)}{n(S)} \tag{7.1}$$

where $P(E)$ stands for the probability of event E and $n(E)$ and $n(S)$ represent the number of outcomes in the respective sets.

Example 7.2 Suppose that, for the experiment described in Example 7.1, we are interested in determining the probability that the number one will appear on the uppermost face of the die.

$$P(\textbf{one}) = \frac{n(\textbf{one})}{n(S)} = \frac{1}{6} \tag{7.1}$$

The probability of an *odd* number appearing on the uppermost face is,

$$P(A) = \frac{n(A)}{n(S)} = \frac{3}{6} \quad \text{or} \quad 0.5 \tag{7.1}$$

The Relative Frequency Definition of Probability. In situations where the outcomes of an experiment are not all equally likely, or when it is not known whether they are equally likely, the application of the classical definition of probability may fail to yield the correct value. The probability of an event under these circumstances must be based on repetitive trials of the experiment.

The relative frequency definition of probability states:

> The probability of an event E is the limit, as n approaches infinity, of the ratio of the number of times that E occurs in n trials of an experiment.

That is,

$$P(E) = \lim_{n\to\infty}\left[\frac{n(E)}{n}\right] \tag{7.2}$$

where $n(E)$ now represents the number of occurrences of event E and n is the number of trials of the experiment. The symbol $\lim_{n\to\infty}$ is read: "The limit as n approaches infinity." This means that, according to the relative frequency definition, the true probability of E is the value of this ratio when the number of trials becomes infinitely large.

It is obviously impossible to conduct an infinite number of trials of an experiment. Therefore, the probability of an event is estimated from the relative frequency of occurrence of E in some finite number of trials, designated n. For practical purposes, then, the relative frequency definition of probability is

$$P(E) = \frac{n(E)}{n} \tag{7.3}$$

Example 7.3

In tossing an ordinary coin, it is usually assumed that the outcome "head" and the outcome "tail" are equally likely to occur on any single toss. A coin having this characteristic is called a fair coin. For a fair coin, the probability of a head occurring on a single toss is

$$P(H) = \frac{n(H)}{n(S)} = \frac{1}{2} \tag{7.1}$$

This value results from the classical definition of probability and Equation 7.1. Suppose, however, that we cannot assume the coin is fair. It may have been altered in some way to make one side or the other more likely to be up after a toss. Under these circumstances, the only way to determine the true probabilities of the possible outcomes is by experimentation, that is, by tossing the coin repeatedly and observing the number of times the outcome of interest occurs.

Suppose that in 1000 tosses of the coin a head occurs 570 times. The relative frequency estimate of the probability of a head on that coin is

570/1000, or 0.57. Since this is an estimate of the true probability of a head, another 1000 tosses of the same coin is unlikely to produce exactly 570 heads. However, in repeated experiments conducted under the same conditions, the observed relative frequencies tend to fluctuate around the true probability value. These fluctuations get smaller as the number of trials increases and tend to stabilize for a large number of trials. We defer our discussion of how large the number of trials should be until our discussion of estimation, in Chapter 10.

Example 7.4 Lance Clothiers, Inc. manufactures a wide variety of men's clothing, including dress shirts. After the shirts are manufactured, they are folded and packaged automatically by ten machines that are specially designed for this purpose. The output of each machine is packed in large cartons and shipped to retail stores. A routine inspection discloses that one of the packaging machines is not properly adjusted, and consequently is tearing each shirt it processes. Just prior to this inspection, a shipment of 100 cartons of shirts which includes 10 cartons from the defective machine is sent to customers. What is the probability that a customer receiving a carton from this shipment will find that it contains torn shirts?

Since each customer may get any one of the 100 cartons, the sample space of the experiment contains 100 elements. Therefore, since ten of the cartons belong to the event set $E = \{$**torn shirts**$\}$, by the classical definition of probability,

$$P(E) = \frac{n(E)}{n(S)} = \frac{10}{100} = 0.10 \tag{7.1}$$

Example 7.5 Metric Systems, Inc. produces integrated electronic circuits on a continuous production line. Occasionally, the process turns out a defective circuit. From time to time, the quality control department randomly selects 500 circuits from the production line and carefully inspects each one. From the latest sample of 500, 15 are found to be defective. What is the estimated probability that the process produces a defective circuit?

The random selection of a circuit from the production line corresponds to a single trial of an experiment. Five hundred selections represent 500 trials of the experiment, that is, $n = 500$. Let's designate the event, defective circuit, D. Since event D occurred 15 times, the estimated probability of a defective circuit by the relative frequency definition is

$$P(D) = \frac{n(D)}{n} = \frac{15}{500} = 0.03 \tag{7.3}$$

Probability for a Continuous Sample Space. When the elements or outcomes in the sample space of an experiment are countable, the sample space is said to be *discrete*. A discrete sample space can be finite or infinite. When the elements in the sample space are not countable, the sample space is *continuous*. The terms discrete and continuous were used similarly in Chapter 1 with reference to discrete and continuous variables.

A continuous sample space usually results from experiments

where observations are made in the form of measurements. The measurement of a length, an area, or any other measurable characteristic can have an infinite number of possible outcomes. Since the classical definition of probability requires that the sample space contains a finite number of discrete, equally likely outcomes, it is not directly applicable to determining probabilities from continuous sample spaces.

Continuous sample spaces are usually represented by areas, but can also be represented by intervals on a line. A continuous event space is then, an area or line segment encompassed by the area or line representing the sample space of the experiment. The probability of event E, that is, the probability that a randomly selected point belongs to E, is the ratio of the magnitude of the interval representing E to the magnitude of the interval representing S. If we let $m(E)$ and $m(S)$ represent the magnitudes of the intervals or areas corresponding to the event and sample spaces, then

$$P(E) = \frac{m(E)}{m(S)} \tag{7.4}$$

We discuss continuous sample spaces in more detail in Chapter 8.

Example 7.6

Consider the disk in Figure 7.1, which has a spinning pointer at the center. The disk is divided into four equal quadrants numbered one through four. If this disk is positioned horizontally on a perfectly level table and the pointer is spun, it can stop at any point on the circumference of the disk. Since there are an infinite number of points on the circumference, there are an infinite number of outcomes, and $n(S) = \infty$. However, if E is the event that the pointer stops on *any* point in the arc of quadrant one, and S is the circumference of the disk, then $P(E)$ is simple to determine. Let C be the magnitude of the circumference and $C/4$ be the magnitude of the arc of quadrant one. Then, by Equation 7.4,

$$P(E) = \frac{m(E)}{m(S)} = \frac{C/4}{C} = \frac{1}{4} = 0.25 \tag{7.4}$$

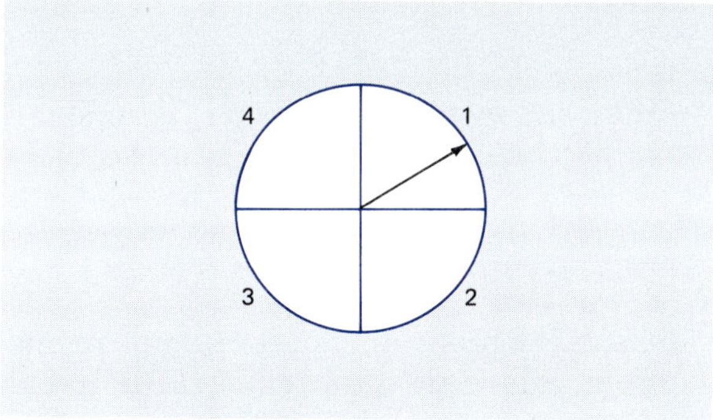

Figure 7.1
Disk with spinning pointer—a continuous sample space (Example 7.6).

7.3 Basic Rules of Probability

The first problem we encounter in beginning the study of probability is that of defining probability. However, having accomplished this in the previous section, we can now turn to the next problem, which involves determining a set of rules that govern the use of the probability measure.

The rules of probability are illustrated by the following experiment.

Consider an experiment consisting of one roll of a single six-sided die. The sample space of this experiment is

$$S = \{1,2,3,4,5,6\}$$

We define several events on this sample space. The event "an odd number occurs" is designated $A = \{1,3,5\}$. The event "a number less than four occurs" is designated $B = \{1,2,3\}$, while the event "a number greater than four occurs" is designated $C = \{5,6\}$. Of course, each of the individual outcomes of the experiment is also an event.

The probabilities of each of these events are

$$P(A) = \frac{n(A)}{n(S)} = \frac{3}{6}, \qquad P(B) = \frac{n(B)}{n(S)} = \frac{3}{6}, \qquad P(C) = \frac{n(C)}{n(S)} = \frac{2}{6}$$

by the classical definition of probability. Also, by the same definition,

$$P(\textbf{one}) = P(\textbf{two}) = P(\textbf{three}) = P(\textbf{four}) = P(\textbf{five}) = P(\textbf{six}) = \frac{1}{6}$$

Rule 1 The probability of an event is restricted to the range zero to one inclusive, where zero represents an impossible event and one represents a certain event. That is,

$$0 \leq P(E) \leq 1$$

Example 7.7 What is the probability of rolling a seven on a single roll of the die? $P(\text{SEVEN}) = 0$. A seven is an impossible event for this experiment, since $S = \{1,2,3,4,5,6\}$.

Rule 2 The probability of the entire sample space is 1. That is,

$$P(S) = 1$$

Rule 2 simply states that if an experiment is conducted, one of the outcomes in its sample space must occur.

Example 7.8 When the die is rolled, one of the numbers—1, 2, 3, 4, 5, or 6—must appear on the uppermost face. Therefore, $P(S) = 1$.

Rule 3 The probability of the complement of an event is equal to one minus the probability of the event. That is,

$$P(E') = 1 - P(E)$$

Rule 3 develops naturally from the concept of a set and its complement.

Since $n(E) + n(E') = n(S)$,

$$\frac{n(E)}{n(S)} + \frac{n(E')}{n(S)} = \frac{n(S)}{n(S)} = 1$$

and since $\dfrac{n(E)}{n(S)} = P(E)$, from Equation 7.1,

$$P(E) + P(E') = 1$$

Then, by simple algebra,

$$P(E') = 1 - P(E)$$

Example 7.9

Find the probability of the complement of C.

$$P(C') = 1 - P(C)$$

Since $P(C) = 2/6$,

$$P(C') = 1 - \frac{2}{6}$$

$$P(C') = \frac{4}{6}$$

Since event C is the occurrence of a number greater than four, C' must be the occurrence of a number that is four or less, that is, $C' = \{1,2,3,4\}$. Therefore,

$$P(C') = \frac{n(C')}{n(S)} = \frac{4}{6}$$

Rule 4 If A and B are *mutually exclusive* events defined on the sample space, then the probability of either A or B occurring is equal to the probability of A plus the probability of B. That is,

$$P(A \cup B) = P(A) + P(B) \tag{7.5}$$

Rule 4 is known as the *special additive rule of probability*. It results from Equation 2.5:

If

$$n(A \cup B) = n(A) + n(B),$$

then

$$\frac{n(A \cup B)}{n(S)} = \frac{n(A)}{n(S)} + \frac{n(B)}{n(S)}$$

and

$$P(A \cup B) = P(A) + P(B).$$

The special additive rule is not limited in application to two events, but can be extended to any number of mutually exclusive sets. In general, the probability of the union of any number of mutually exclusive events defined on a sample space is simply the sum of their individual probabilities.

Example 7.10 We are given:

$$B = \{1,2,3\}$$
$$C = \{5,6\}$$
$$(B \cup C) = \{1,2,3,5,6\}$$

Therefore,

$$P(B) = \frac{3}{6}$$

$$P(C) = \frac{2}{6}$$

$$P(B \cup C) = \frac{5}{6}$$

By Equation 7.5,

$$\begin{aligned} P(B \cup C) &= P(B) + P(C) \\ &= \frac{3}{6} + \frac{2}{6} \\ &= \frac{5}{6} \end{aligned}$$

Rule 5 If A and B are *any* two events defined on the sample space, then the probability of either A or B or both occurring is given by

(7.6) $$P(A \cup B) = P(A) + P(B) - P(A \cap B)$$

Rule 5 is known as the *general additive rule of probability* and is derived directly from Equation 2.6, which states that

$$n(A \cup B) = n(A) + n(B) - n(A \cap B)$$

therefore,

$$\frac{n(A \cup B)}{n(S)} = \frac{n(A)}{n(S)} + \frac{n(B)}{n(S)} - \frac{n(A \cap B)}{n(S)}$$

and

$$P(A \cup B) = P(A) + P(B) - P(A \cap B)$$

The general additive rule of probability can also be extended to apply to any number of events on a sample space.

Example 7.11 We are given:

$$A = \{1,3,5\}$$
$$B = \{1,2,3\}$$

Therefore,

$$P(A) = \frac{3}{6}$$

$$P(B) = \frac{3}{6}$$

$$(A \cap B) = \{1,3\}$$

$$P(A \cap B) = \frac{n(A \cap B)}{n(S)} = \frac{2}{6}$$

If $P(A \cup B) = P(A) + P(B) - P(A \cap B)$, then $P(A \cup B) = \frac{3}{6} + \frac{3}{6} - \frac{2}{6} = \frac{4}{6}$.

Example 7.12

Since Rule 5 involves a joint probability, we shall again illustrate its computation as follows.

If A and B are any two events defined on the sample space, then the probability of the intersection of A and B occurring is given by

$$P(A \cap B) = \frac{n(A \cap B)}{n(S)}$$

Since

$$A = \{1,3,5\}$$
$$B = \{1,2,3\}$$

and

$$n(S) = 6,$$

then

$$(A \cap B) = \{1,3\}$$
$$n(A \cap B) = 2$$

and

$$P(A \cap B) = \frac{n(A \cap B)}{n(S)} = \frac{2}{6}$$

The relationship among these three sets, A, B, and C, is illustrated in Figure 7.2.

7.4 Conditional Probability

There are many occasions when we would like to determine the probabilities of events under circumstances that place a restriction on the sample space of the experiment. The effect of this restriction or condition is to reduce the sample space of the experiment to one of its event spaces. The probability of an event subject to such a condition is called a CONDITIONAL PROBABILITY.

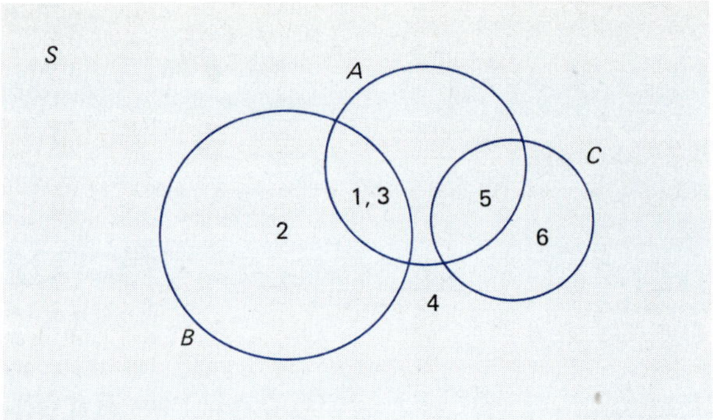

Figure 7.2
Venn diagram for illustrating rules of probability.

Suppose you agree to gamble on the outcome of an experiment in which a coin is tossed and then a die is rolled. Each outcome is identified by an ordered pair of symbols in which the first symbol represents the outcome of the coin toss and the second represents the result of the roll of the die. The terms of the gamble provide that you win every time a head appears on the coin and a number greater than four comes up on the die. We let H represent the event "head on the coin" and C represent the event "a number greater than four on the die." The sample space of this experiment consists of twelve equally likely outcomes as follows:

Coin Result	*Die Result*					
	1	2	3	4	5	6
H	$(H,1)$	$(H,2)$	$(H,3)$	$(H,4)$	$(H,5)$	$(H,6)$
T	$(T,1)$	$(T,2)$	$(T,3)$	$(T,4)$	$(T,5)$	$(T,6)$

The event "head on the coin" contains all of the outcomes in which the first symbol in the ordered pair is H. Since six out of twelve equally likely outcomes contain H, $P(H) = 6/12 = 1/2$. Note that this is the probability of getting a head on the coin regardless of what occurs on the die. Similarly, the event "number greater than four on the die" contains all of the outcomes in which the second symbol in the ordered pair is either a five or a six. Since four out of twelve equally likely outcomes correspond to this event, $P(C) = 4/12 = 1/3$.

The event "winning," however, requires the joint occurrence of H and C. The outcomes in this intersection are $(H \cap C) = \{(H,5), (H,6)\}$. The probability of winning in this game is the joint probability of H and C which is $P(H \cap C) = 2/12 = 1/6$.

Now suppose that before rolling the die you observe that a head has already come up on the coin. The probability of winning is no longer 1/6, since you have already satisfied one of the conditions that constitute the event "winning." What is the new probability?

Before tossing the coin, all 12 possible outcomes of the experiment have to be considered in computing the probability of winning. Now, however, given that the coin has come up heads, you need only be concerned with the subset of outcomes containing H, because the possibility of tails has already been eliminated. Therefore, the sample space of the experiment has been reduced from the original twelve outcomes to the six outcomes containing H. The restricted sample space contains the outcomes $\{(H,1), (H,2), (H,3), (H,4), (H,5), (H,6)\} = H$. The outcomes that correspond to winning are $\{(H,5), (H,6)\} = (H \cap C)$. The probability of winning is no longer simply the probability of the joint occurrence of H and C, but the probability of C *given* H. This is a conditional probability and is written $P(C\,|\,H)$. The sample space of the original experiment has been reduced to the event space of H, which contains six outcomes. Two of these outcomes correspond to event $(C\,|\,H)$, therefore,

$$P(C\,|\,H) = \frac{n(H \cap C)}{n(H)} = \frac{2}{6} = \frac{1}{3}$$

Conditional probability is usually defined in terms of other probabilities. Specifically, $P(C\,|\,H) = P(H \cap C)/P(H)$. Since $P(H \cap C)$ was previously determined to be 1/6 and $P(H)$ was 1/2, the conditional probability results from

$$P(C\,|\,H) = \frac{P(H \cap C)}{P(H)} = \frac{1/6}{1/2} = \frac{1}{3}$$

In general, for any two events A and B defined on a sample space, the conditional probability of one event given that the other occurs is obtained by dividing the joint probability of the two events by the probability of the conditional event. That is,

$$P(A\,|\,B) = \frac{P(A \cap B)}{P(B)} \tag{7.7}$$

$$P(B\,|\,A) = \frac{P(A \cap B)}{P(A)} \tag{7.8}$$

Probabilities such as $P(A)$, $P(B)$, $P(H)$, and $P(C)$ are generally called *marginal probabilities*. A *marginal probability* is the probability of a single event defined only with reference to the sample space of the experiment. A *joint probability* involves two or more events in the same sample space, and a *conditional probability* is defined with reference to a restricted sample space. These relationships are illustrated with a Venn diagram in Figure 7.3.

In the figure, the intersecting events A and B are defined on the

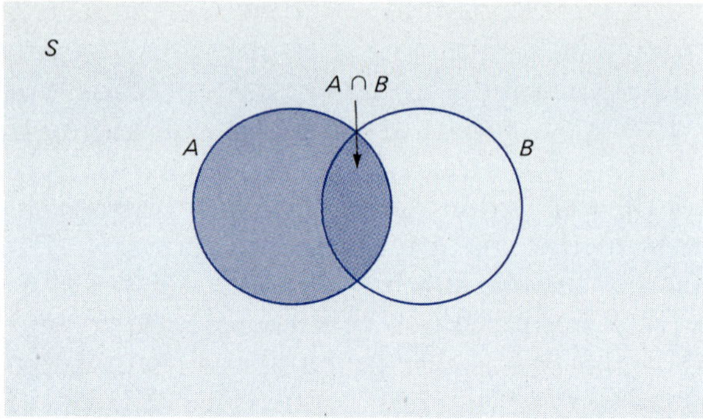

Figure 7.3
Venn diagram for illustrating marginal, joint, and conditional probabilities.

sample space S. The marginal and joint probabilities of A and B are defined with reference to S as:

$$(7.1)\quad P(A) = \frac{n(A)}{n(S)}, \qquad P(B) = \frac{n(B)}{n(S)}, \qquad P(A \cap B) = \frac{n(A \cap B)}{n(S)}$$

If we want to determine the probability of B given that A also occurs, the subset corresponding to the event of interest must be the intersection $(A \cap B)$, since both A and B must occur. The statement "that A also occurs" does not imply that A must occur before B, after B, or simultaneously with B, but simply that A must occur. However, since the occurrence of A is specified as a condition, the probability must be determined with respect to the event space of the conditional event. Therefore, the conditional probability of B given A is

$$\frac{n(A \cap B)}{n(A)} = \frac{P(A \cap B)}{P(A)} = P(B \mid A)$$

Note that this is simply the joint probability divided by the marginal probability of the conditional event.

Example 7.13 The marketing vice-president of Quality Appliances, Inc. demonstrated a newly developed kitchen appliance at the annual dealers' meeting. After the demonstration, he polled the audience of 200 dealers and obtained the following data.

140 of the dealers were men and 60 were women,

90 of the men indicated that they wanted to handle the new product, and

30 of the women indicated that they wanted to handle the new product.

If S represents the universal set consisting of all dealers, M represents the subset of dealers who are men, W represents the subset of dealers who are women, and E represents the subset of dealers who want to handle the new product, find $P(E \mid M)$ and $P(E \mid W)$.

By formulas 7.7 and 7.8,

$$P(E|M) = \frac{P(E \cap M)}{P(M)} \qquad \text{and} \qquad P(E|W) = \frac{P(E \cap W)}{P(W)}$$

From the data,

$$P(M) = \frac{n(M)}{n(S)} = \frac{140}{200} = 0.70 \qquad P(E \cap M) = \frac{n(E \cap M)}{n(S)} = \frac{90}{200} = 0.45$$

$$P(W) = \frac{n(W)}{n(S)} = \frac{60}{200} = 0.30 \qquad P(E \cap W) = \frac{n(E \cap W)}{n(S)} = \frac{30}{200} = 0.15$$

Therefore,

$$P(E|M) = \frac{0.45}{0.70} = 0.643 \qquad \text{and} \qquad P(E|W) = \frac{0.15}{0.30} = 0.50$$

We can now find an expression that gives the joint probability of two events. If we multiply both sides of Equations 7.7 and 7.8 by the marginal probability and rearrange the terms, we get

(7.9) $$P(A \cap B) = P(B)P(A|B)$$

and

(7.10) $$P(A \cap B) = P(A)P(B|A)$$

Equations 7.9 and 7.10 express the *general multiplicative rule of probability*. This rule simply states that the joint probability of two events is equal to the marginal probability of the first event multiplied by the conditional probability of the second event, given the first.

Example 7.14

The Advance Machine Company manufactures a particular part in large quantities for a customer. The tolerances specified by the customer are very tight, therefore the production of this part has been assigned to two experienced machinists whom we will call *A* and *B*.

At the end of each day, all parts produced are inspected and classified as either good or defective. Let *G* represent the event "good" and *D* represent "defective." The data on one day's production are summarized in Table 7.1. A table such as 7.1, in which data are classified according to two

TABLE 7.1
CLASSIFICATION OF PARTS BY CONDITION AND MACHINIST (EXAMPLE 7.14)

	Machinist		
Condition	*A*	*B*	Totals
G	80	88	168
D	20	12	32
Totals	100	100	200

categories, is called a contingency table. Here the categories are *machinist* and *condition of part.*

If all 200 parts are mixed together in a storage bin and a single part is randomly selected from the bin, find all the following probabilities.

a) Marginal probabilities: $P(A)$, $P(B)$, $P(G)$, and $P(D)$.

$$P(A) = \frac{n(A)}{n(S)} = \frac{100}{200} = 0.50 \qquad P(B) = \frac{n(B)}{n(S)} = \frac{100}{200} = 0.50$$

$$P(G) = \frac{n(G)}{n(S)} = \frac{168}{200} = 0.84 \qquad P(D) = \frac{n(D)}{n(S)} = \frac{32}{200} = 0.16$$

b) Joint probabilities: $P(A \cap G)$, $P(A \cap D)$, $P(B \cap D)$, $P(A \cap B)$.

$$P(A \cap G) = \frac{n(A \cap G)}{n(S)} = \frac{80}{200} = 0.40 \qquad P(A \cap D) = \frac{n(A \cap D)}{n(S)} = \frac{20}{200} = 0.10$$

$$P(B \cap G) = \frac{n(B \cap G)}{n(S)} = \frac{88}{200} = 0.44 \qquad P(B \cap D) = \frac{n(B \cap D)}{n(S)} = \frac{12}{200} = 0.06$$

$P(A \cap B) = 0$, since these subsets do not intersect.

c) Conditional probabilities: $P(G|A)$, $P(A|D)$, $P(D|A)$, $P(B|G)$, $P(G|B)$, $P(G|D)$.

$$P(G|A) = \frac{P(A \cap G)}{P(A)} = \frac{0.40}{0.50} = 0.80 \quad P(A|D) = \frac{P(A \cap D)}{P(D)} = \frac{0.10}{0.16} = 0.625$$

$$P(D|A) = \frac{P(A \cap D)}{P(A)} = \frac{0.10}{0.50} = 0.20 \quad P(B|G) = \frac{P(B \cap G)}{P(G)} = \frac{0.44}{0.84} = 0.524$$

$$P(G|B) = \frac{P(B \cap G)}{P(B)} = \frac{0.44}{0.50} = 0.88 \quad P(G|D) = \frac{P(D \cap G)}{P(D)} = \frac{0}{0.16} = 0$$

7.5 Statistical Independence

Whenever a question about probability involves two or more events, the concept of STATISTICAL INDEPENDENCE arises. Any two events are statistically independent if the occurrence or nonoccurrence of one of the events has absolutely no effect on the probability of the occurrence of the other.

Equations 7.9 and 7.10 provide a rule for determining the probability of the joint occurrence of two events. The equations contain a conditional probability, either $P(A|B)$ or $P(B|A)$. If the events A and B are statistically independent, then the condition specified should have no effect on the probability of the first event. In $P(A|B)$, for example, the condition that B must occur has no effect on the probability of A occurring if the two events are independent. Therefore, we can state that if A and B are statistically independent,

$$P(A|B) = P(A) \tag{7.11}$$

and

$$P(B|A) = P(B) \tag{7.12}$$

Conversely, if either Equation 7.11 or Equation 7.12 is true, then A and B must be statistically independent. Furthermore, if A and B are statistically independent, then

$$P(A \cap B) = P(A)P(B) \tag{7.13}$$

Equation 7.13 is called the *special multiplicative rule of probability* for independent events.

Example 7.15

In drawing cards from an ordinary deck of 52 playing cards, let A_1 represent the event "ace on the first draw" and A_2 represent the event "ace on the second draw."

a) If each card is replaced before the next card is drawn, what is the probability of getting two aces on the first two draws, that is, $P(A_1 \cap A_2)$? By the general multiplicative rule of probability, Equation 7.10, $P(A_1 \cap A_2) = P(A_1)P(A_2|A_1)$. Since there are four aces in the deck, $P(A_1) = 4/52$ by the classical definition of probability. After the first card is drawn and replaced, the probability of an ace on the second draw, given that an ace occurred on the first, is $P(A_2|A_1) = 4/52$. The composition of the deck for the second draw is the same as it was for the first; therefore, there are still four aces among 52 cards. By the same reasoning, the probability of getting an ace on the second draw if one was not drawn on the first is also 4/52. In other words,

$$P(A_2|A_1) = P(A_2|A_1') = P(A_2) = 4/52$$

Since the probability of getting an ace on the second draw is exactly the same regardless of whether an ace appears on the first, the events A_1 and A_2 are statistically independent.

b) If the first card is not replaced before the second card is drawn, what is the probability of getting two aces on the first two draws? The general multiplicative rule of probability states that $P(A_1 \cap A_2) = P(A_1)P(A_2|A_1)$. However, without replacement, $P(A_2|A_1) \neq P(A_2|A_1') \neq P(A_2)$. If the first card drawn is not replaced, then the composition of the deck for the second draw is changed. If the first card drawn is an ace, then three aces are left in the remaining 51 cards and $P(A_2|A_1) = 3/51$. If the first card drawn is not an ace, then $P(A_2|A_1') = 4/51$. Since the probability of an ace on the second draw depends on the result of the first draw, $P(A_2|A_1) \neq P(A_2)$ and the events A_1 and A_2 are not statistically independent.

Example 7.16

Consider once again Table 7.1. A production run of 200 parts was classified according to condition and machinist. It would be interesting to determine if the condition of a part is independent of the machinist who produces it. We can test for independence using various conditional and marginal probabilities obtained from the table. Most simply, however, if independence exists between these categories, then the events, "the part is good" and "the part was produced by A" are statistically independent. To test this we evaluate

$$P(G \cap A) \stackrel{?}{=} P(G)P(A)$$

From Table 7.1 we can determine that the joint probability $P(G \cap A) = 80/200 = 0.40$. Also, $P(G) = 168/200 = 0.84$ and $P(A) = 100/200 = 0.50$. Since $P(G)P(A) = (0.84)(0.50) = 0.42$ and $P(G \cap A) = 0.40$, A and G are not independent, and the condition of the part is not independent of the machinist who produced it.

We can also determine from the table that $P(A \mid G) = 80/100 \neq P(A) = 100/200$. The reader can also verify that $P(A \mid D) \neq P(A)$, $P(B \mid G) \neq P(B)$, and $P(D \mid B) \neq P(D)$.

7.6 The Rule of Elimination

We frequently use Venn diagrams to illustrate the relationships among subsets or event sets. Figure 7.4 illustrates a sample space that has been divided into four mutually exclusive subsets, A_1, A_2, A_3, and A_4. Another subset, B, has also been defined on the sample space so that it intersects with each of the A_i's.

Now consider the event B. B has also been partitioned into four mutually exclusive subsets that can be identified as $(A_1 \cap B)$, $(A_2 \cap B)$, $(A_3 \cap B)$, and $(A_4 \cap B)$. Since the subset B must be equal to the sum of all its parts,

$$B = (A_1 \cap B) + (A_2 \cap B) + (A_3 \cap B) + (A_4 \cap B)$$

and therefore,

$$P(B) = P(A_1 \cap B) + P(A_2 \cap B) + P(A_3 \cap B) + P(A_4 \cap B)$$

The right-hand side of the previous equation consists of the sum of several joint probabilities. From Equation 7.10, we know that

$$\begin{aligned} P(A_1 \cap B) &= P(A_1)P(B \mid A_1) \\ P(A_2 \cap B) &= P(A_2)P(B \mid A_2) \\ P(A_3 \cap B) &= P(A_3)P(B \mid A_3) \\ P(A_4 \cap B) &= P(A_4)P(B \mid A_4) \end{aligned}$$

Therefore, we can write

$$P(B) = P(A_1)P(B \mid A_1) + P(A_2)P(B \mid A_2) + P(A_3)P(B \mid A_3) + P(A_4)P(B \mid A_4)$$

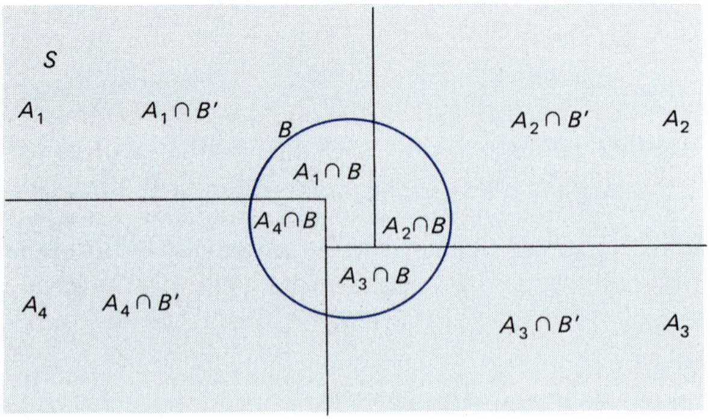

Figure 7.4
Partitioning of B.

In general, if an event B is partitioned into n mutually exclusive subsets, where each subset represents the intersection of B with A_i, then

$$P(B) = P(A_1)P(B|A_1) + P(A_2)P(B|A_2) + \cdots + P(A_i)P(B|A_i) + \cdots$$

or, more simply,

$$P(B) = \sum_{i=1}^{n} P(A_i)P(B|A_i) \qquad (7.14)$$

Equation 7.14 is frequently referred to as the *rule of elimination*. In simple terms, the rule of elimination states that if an event B is dependent upon the occurrence of any one of several other mutually exclusive events, the A_i's, none of which has zero probability, then the marginal probability of B is equal to the marginal probability of A_i times the conditional probability of B given A_i summed over all of the A_i's.

Example 7.17

The Essex Construction Company has ten bulldozers that are used on various construction projects. The bulldozers were purchased from three different manufacturers, namely: the A-1 Construction Equipment Corporation, which we designate A_1; Allstate Motors, which we call A_2; and Argonaut Tractor Company, designated A_3. Let B represent the event "breakdown of a bulldozer in the field."

It is estimated from performance records that each type of bulldozer has a different probability of breaking down. These conditional probabilities are

$$P(B|A_1) = 0.20 \qquad P(B|A_2) = 0.15 \qquad P(B|A_3) = 0.10$$

Suppose that a potential customer has asked Essex to determine the probability of a bulldozer breaking down in the field, resulting in a delay in the completion of a construction job. The probability of a breakdown is equal to the conditional probability of a breakdown, given that a specific type of bulldozer is in use, times the marginal probability that that type of bulldozer is in use. The products of these probabilities must then be added for all three types of bulldozers that Essex might use.

The marginal probability of each type of bulldozer being in use is proportional to the number of each type owned by Essex. Since Essex owns two bulldozers made by A-1, five made by Allstate, and three made by Argonaut, these probabilities are

$$P(A_1) = 0.20 \qquad P(A_2) = 0.50 \qquad P(A_3) = 0.30$$

Using the rule of elimination, the probability of a bulldozer breakdown occurring on any job is

$$\begin{aligned} (7.14) \quad P(B) &= P(A_1)P(B|A_1) + P(A_2)P(B|A_2) + P(A_3)P(B|A_3) \\ &= (0.20)(0.20) + (0.50)(0.15) + (0.30)(0.10) \\ &= 0.145 \end{aligned}$$

7.7 Bayes' Rule

Bayes' rule is frequently called the *rule of inverse probability*. This means that, given a conditional probability such as $P(B|A_1)$, it provides a method for determining $P(A_1|B)$. This is accomplished by using the following relationship established in Equations 7.9 and 7.10.

$$P(A_i \cap B) = P(B)P(A_i|B) = P(A_i)P(B|A_i)$$

If we equate only the last two parts of this expression and solve algebraically for $P(A_i|B)$, we get

$$P(B)P(A_i|B) = P(A_i)P(B|A_i)$$

$$P(A_i|B) = \frac{P(A_i)P(B|A_i)}{P(B)}$$

However, from the rule of elimination, Equation 7.14, we know that

$$P(B) = \Sigma P(A_i)P(B|A_i)$$

therefore,

$$P(A_i|B) = \frac{P(A_i)P(B|A_i)}{\Sigma P(A_i)P(B|A_i)} \tag{7.15}$$

Equation 7.15 is known as Bayes' rule.

Example 7.18 The dispatcher at the Essex Construction Company's headquarters receives a message that a bulldozer has broken down on an important construction job in Middletown. The message fails to specify which type of bulldozer is involved. In order to have the manufacturer provide service at the site, the make of the machine has to be known. However, all attempts to contact the supervisor at the construction site are unsuccessful. Since every minute of delay is costing the company money, the dispatcher decides to use Bayes' rule to determine the probability that, given a breakdown, the bulldozer involved is type A_1, A_2, or A_3. He then plans to contact the manufacturer for which the conditional probability is the highest.

The marginal and conditional probabilities from Example 7.17 are, respectively,

$$P(B|A_1) = 0.20 \qquad P(B|A_2) = 0.15 \qquad P(B|A_3) = 0.10$$
$$P(A_1) = 0.20 \qquad P(A_2) = 0.50 \qquad P(A_3) = 0.30$$

The inverse probabilities are obtained as follows.

$$\begin{aligned} P(A_1|B) &= \frac{P(A_1)P(B|A_1)}{P(A_1)P(B|A_1) + P(A_2)P(B|A_2) + P(A_3)P(B|A_3)} \\ &= \frac{(0.20)(0.20)}{(0.20)(0.20) + (0.50)(0.15) + (0.30)(0.10)} \\ &= 0.276 \end{aligned}$$

$$P(A_2|B) = \frac{P(A_2)P(B|A_2)}{P(A_1)P(B|A_1) + P(A_2)P(B|A_2) + P(A_3)P(B|A_3)}$$
$$= \frac{(0.50)(0.15)}{(0.20)(0.20) + (0.50)(0.15) + (0.30)(0.10)}$$
$$= 0.517$$

$P(A_3|B)$ can be calculated in exactly the same way. However, it is simpler to obtain this probability from the fact that the sum of the probabilities of all possible events must equal 1. Given that a breakdown occurred, it must have happened to one of the three types of bulldozers in use. Since $P(A_1|B) + P(A_2|B) + P(A_3|B) = 1$,

$$P(A_3|B) = 1 - [P(A_1|B) + P(A_2|B)]$$
$$P(A_3|B) = 1 - [0.276 + 0.517]$$
$$P(A_3|B) = 0.207$$

Given that a breakdown occurs, it is almost twice as likely to happen to an A_2 than it is to either of the other types. Therefore, the dispatcher calls Allstate Motors for service. The application of Bayes' rule to other types of problems involving decision making is covered in Chapter 18, Introduction to Statistical Decision Theory.

Definitions

Event: Any set of outcomes of an experiment; a subset of the sample space.

Event space—event set: The subset of outcomes corresponding to some designated event.

Experiment: Any well-defined action that leads to a single well-defined result.

Outcome: Any single, possible result of an experiment.

Probability:

General definition: A numerical measure of the likelihood of occurrence of an uncertain event.

Classical definition: The probability of an event E is the ratio of the number of outcomes in the event space of E to the total number of outcomes in the sample space of the experiment, provided that all outcomes are equally likely to occur.

Relative frequency definition: The probability of an event E is the limit, as n approaches infinity, of the ratio of the number of times that E is observed to occur in n trials of an experiment.

Sample space: The set of all possible outcomes of an experiment.

Statistical independence: Two events are statistically independent if the occurrence or nonoccurrence of one has no effect on the probability of the occurrence of the other.

Formulas

The classical definition of probability:

(7.1) $$P(E) = \frac{n(E)}{n(S)}$$

The relative frequency definition of probability:

(7.2) $$P(E) = \lim_{n\to\infty} \frac{n(E)}{n}$$

The relative frequency estimate of probability:

(7.3) $$P(E) = \frac{n(E)}{n}$$

Probability defined on a continuous sample space:

(7.4) $$P(E) = \frac{m(E)}{m(S)}$$

The special additive rule of probability:

(7.5) $$P(A \cup B) = P(A) + P(B)$$

The general additive rule of probability:

(7.6) $$P(A \cup B) = P(A) + P(B) - P(A \cap B)$$

Conditional probability:

(7.7) $$P(A|B) = \frac{P(A \cap B)}{P(B)}$$

(7.8) $$P(B|A) = \frac{P(A \cap B)}{P(A)}$$

The general multiplicative rule of probability:

(7.9) $$P(A \cap B) = P(B)P(A|B)$$

(7.10) $$P(A \cap B) = P(A)P(B|A)$$

Conditions satisfying statistical independence:

(7.11) $$P(A|B) = P(A)$$

(7.12) $$P(B|A) = P(B)$$

Special multiplicative rule of probability and condition satisfying statistical independence:

(7.13) $$P(A \cap B) = P(A)P(B)$$

Rule of elimination:

(7.14) $$P(B) = \Sigma P(A_i)P(B|A_i)$$

Bayes' rule:

(7.15) $$P(A_i|B) = \frac{P(A_i)P(B|A_i)}{\Sigma P(A_i)P(B|A_i)}$$

Problems

7.1 A single card is drawn at random from an ordinary deck of 52 playing cards. What is the probability that:

a) A club is drawn.

b) A deuce is drawn.

c) Either a heart or a club is drawn.
d) A red deuce is drawn.

7.2 For the experiment of problem 7.1, find the probability that:
a) Either a club or a deuce is drawn.
b) Either a black card or a picture card is drawn.
c) A black picture card is drawn.
d) Either a red ace or a black deuce is drawn.

7.3 A single die is rolled 1000 times. The frequency of each outcome is noted below. Estimate the probability of each possible outcome.

Outcome	*Frequency*
one	215
two	150
three	130
four	140
five	165
six	200

7.4 Describe the sample space of an experiment in which three coins are tossed simultaneously and each coin is observed as to whether a head or a tail is facing up. Hint: There are eight possible outcomes to this experiment. Identify the coins as coin 1, coin 2, and coin 3.

7.5 In the experiment of problem 7.4, find the probability of getting:
a) A tail on coins 1 and 3.
b) A head on coin 2.
c) A head on coin 1 and a tail on coin 3.
d) A tail on coins 1 and 3 and a head on coin 2.

7.6 In the experiment of problem 7.4, find the probability of getting:
a) An odd number of heads.
b) Exactly two tails.
c) At least two tails.
d) At most one head.

7.7 If a 5 or a 6 is counted as a success on a single roll of a die, and any number less than 5 is counted as a failure, then on three rolls find the probability of getting:
a) Two failures followed by a success.
b) A success on the third roll.
c) Exactly three successes.
d) At least two failures.

7.8 For the experiment in problem 7.7, find the probability of getting:
a) A failure on the first roll and a success on the third roll.
b) A failure on either the first or second roll.
c) A success on the third roll given that one failure occurs.
d) A success on the second roll given that a failure occurs on the first.

7.9 Describe the sample space of an experiment in which a blue die and a red die are rolled simultaneously. Hint: There are 36 outcomes to this experiment.

7.10 In the experiment of problem 7.9, what is the probability of getting:
a) A three on the red die?
b) A six on the blue die?

c) A three on the red die and a six on the blue die?
d) The sum of the numbers on both dice equal to nine?

7.11 In the experiment of problem 7.9, what is the probability that the sum of the numbers on both dice equals:
a) Seven?
b) One?
c) Two, three, or twelve?
d) Seven or eleven?

7.12 In drawing two cards from an ordinary deck of 52 playing cards without replacement, let $R = \{$**red card**$\}$, $M = \{$**picture card**$\}$, $C = \{$**club**$\}$, and let the subscript 1 or 2 designate the draw on which the event occurs. Find the following probabilities:
a) $P(R_1)$.
b) $P(C_1)$.
c) $P(R_1 \cap C_2)$.
d) $P(C_1 \cap R_2)$.

7.13 For the experiment of problem 7.12, find the following probabilities:
a) $P(M_2|M_1)$.
b) $P(R_2|C_1)$.
c) $P(C_2|R_1)$.
d) $P(M_2|R_1 \cap M_1)$.

7.14 For the experiment of problem 7.12, find the following probabilities:
a) $P(R_1 \cap C_1)$.
b) $P(R_1 \cap M_1)$.
c) $P(C_2)$.
d) $P(C_1 \cup C_2)$.

7.15 For the experiment of problem 7.12, find the following probabilities:
a) $P(C_2|C_1)$.
b) $P(R_2|R_1')$.
c) $P(C_1' \cap C_2)$.
d) $P(C_1' \cap C_2')$.

7.16 A lot of 5000 parts produced on three different assembly lines was inspected and each part graded into one of three classifications as shown below.

	Assembly Line			
Grade	1	2	3	*Totals*
Satisfactory	1295	1400	1505	4200
Rework	105	120	275	500
Scrap	30	50	220	300
Totals	1430	1570	2000	5000

If a part is selected at random from the lot, find the probability that:
a) The part was classified as scrap.
b) The part was produced on line 1 or line 2.
c) The part was produced on line 3 and was satisfactory.
d) The part was satisfactory.

7.17 For the data in problem 7.16, find the probability that:

a) The part was not satisfactory.

b) The part was satisfactory given that it was produced on line 2.

c) The part needed rework given that it was produced on line 3.

d) The part was produced on line 1 given that it needed rework.

7.18 For the data in problem 7.16, find the probability that:

a) The part was either satisfactory or needed rework given that it was produced either on line 1 or line 3.

b) Determine whether the grade of the part is independent of the assembly line on which it was produced.

7.19 In the experiment of problem 7.4, what is the probability of getting:

a) A head on coin 3 given exactly two heads?

b) A head on coin 3 given exactly two tails?

c) A tail on coin 2 given at least one head?

d) A tail on coin 2 given a head on coin 1?

7.20 In the experiment of problem 7.4, what is the probability of getting:

a) Exactly two heads given a head on coin 1?

b) Exactly two heads given a tail on coins 1 and 2?

c) At least two heads given a head on coin 1?

d) At least two heads given a tail on coin 1?

7.21 For the experiment of problem 7.9, what is the probability of getting:

a) A five on the blue die given a two on the red die?

b) A five or a six on the red die given a six on the blue die?

c) A one or a four on the red die given a two or a three on the blue die?

d) A one or a two on the blue die given a three, four, or five on the red die?

7.22 For the experiment of problem 7.9, find the probability that:

a) The sum of both dice is seven given a two on the blue die.

b) The sum of both dice is two, three, or twelve given a one on the red die.

c) The sum of both dice is ten given a six on the blue die.

d) The sum of both dice is two, three, or twelve given a three on the red die.

7.23 For the experiment of problem 7.7, find the probability of getting:

a) Exactly two successes or exactly two failures.

b) Exactly two successes given exactly one failure.

c) No successes given a failure on each of the first two rolls.

d) No failure given a success on the first roll.

7.24 The credit manager of a large department store performs an analysis on his charge account customers and finds that 5% of them have annual incomes of \$10,000 or less, 65% have incomes between \$10,000 and \$20,000, and the rest have incomes greater than or equal to \$20,000. He also finds that 20% of the accounts in the lowest income category are delinquent, 10% of those in the middle category are delinquent, but only 5% of those in the highest category are delinquent. If a charge account is selected at random, what is the probability that it is delinquent?

7.25 Suppose that the credit manager in the previous problem receives information that a particular charge account is delinquent, what is the probability that it belongs to a customer whose income is \$20,000 or more? What is the probability that it belongs to a customer who makes \$10,000 or less? What is the probability that it belongs to a customer in the middle income category?

7.26 The credit manager in the previous problem also found that 15% of all delinquent accounts are never collected. All nondelinquent accounts are eventually paid. Given that an account is paid in full, what is the probability that it was a delinquent account? What is the probability that any charge account will be paid in full?

7.27 A manufacturing company purchases a particular component part from three different suppliers, *A*, *B* , and *C*. Supplier *A* provides 25% of the manufacturer's requirements, supplier *B* provides 40%, and supplier *C*, 35%. From past experience it is known that the probability of getting a defective component from *A* is 10%, from *B* is 15%, and from *C* is 5%. What is the probability of the manufacturer receiving a defective component?

7.28 If the manufacturer in the previous problem receives a defective component, what is the probability that it came from *A*? From *B*? From *C*?

7.29 The Advance Machine Company produces an assembly consisting of three components designated component *A*, component *B*, and component *C*. Because of the differences in the complexities of the components, 2% of the *A*'s, 1% of the *B*'s, and 3% of the *C*'s that reach the assembly line are defective. What is the probability that any component is defective?

7.30 The three components used in the assembly by the Advance Machine Company are stored in different bins until they are needed in production. At that time, a component of each type is randomly selected and delivered to the assembly line. What is the probability that a completed assembly will contain three defective components?

7.31 Metric Systems, Inc. placed advertisements in several newspapers for individuals possessing certain technical qualifications. According to the Department of Health, Education, and Welfare, 22.5% of all individuals having these qualifications are women and 15% belong to minority groups. However, of the minority group individuals with the necessary qualifications, only 10% are women. Of the members of the nonminority group with these qualifications, 25% are women. Construct a Venn diagram illustrating these relationships.

7.32 From the data in problem 7.31, find the probability that:

a) A qualified applicant is not a member of a minority group.
b) A qualified applicant is a male member of a minority group.
c) A qualified applicant is a male who is not a member of a minority group.
d) A qualified applicant is female or a member of a minority group or both.

7.33 From the data in problem 7.31, find the probability that:

a) Given that a qualified applicant is a member of a minority group, the applicant is female.

b) Given that a qualified applicant is male, he is not a member of a minority group.

c) Given that a qualified applicant is male, he is a member of a minority group.

d) A qualified applicant is either female or a member of a minority group, but not both.

7.34 An automobile insurance company compiled statistics that indicated the probability of accidents for various classes of drivers living in the Middletown metropolitan area. The probabilities apply to the event that a driver will be involved in an automobile accident within a period of one year. These probabilities are

For male drivers over the age of 20, 1.75%.
For female drivers over the age of 20, 2.50%.
For male drivers 20 and under, 10.00%.
For female drivers 20 and under, 4.00%.

Records of the department of motor vehicles show that in Middletown, 45% of all licensed drivers are males over 20, 40% are females over 20, 8% are males 20 and under, and 7% are females 20 and under. From these data, determine the probability that a licensed driver will have an accident in Middletown during a one-year period.

7.35 For the data in problem 7.34, if an accident occurs in Middletown, what is the probability that:

a) A male 20 or under was driving the car?

b) A female 20 or under was driving the car?

CHAPTER 8 PROBABILITY DISTRIBUTIONS

8.1
Random variables

8.2
The probability function of a random variable

8.3
Expected value of a random variable

8.4
Variance of a probability distribution

8.5
Discrete probability distributions

8.6
Continuous probability distributions

Determining probabilities for various types of events is simplified if we can establish a general model for the sample space of the experiment. A model used to obtain probabilities is called a PROBABILITY DISTRIBUTION. A model is generally a simplified representation of some other thing. When a model uses mathematical symbols to represent real objects or processes, it is called a mathematical model. Probability distributions are mathematical models. They are also functions as defined in Chapter 2, and can be called probability functions. If we select a probability function that accurately represents the experiment of interest, then the function provides probabilities for the events in the sample space of the experiment. The accuracy of these probabilities depends on how well the model fits the situation. There are several well-known probability distributions that are suitable for a wide range of applications. We will study the most common of these in this chapter. However, before looking at specific distributions, we must first consider the meaning of the term *random variable*.

8.1 Random Variables

In Chapters 1 and 2 we discussed variables. We defined a variable as being some quantitative characteristic that varies over a set. We can continue to think of a variable in this way, but we should refine that concept slightly to include random variables. A RANDOM VARIABLE is a variable whose values are determined by the outcome of an experiment.

We will continue to use the notation adopted in previous chapters by designating random variables with lowercase letters, usually in the latter part of the alphabet. Specific values of a random variable are indicated by the use of subscripts such as x_1, x_2, and so on.

If a random variable is defined as a variable whose value is determined by the outcome of an experiment, then the values of a random variable are events. Suppose that an experiment consists of tossing a coin one time. The two possible outcomes of this experiment are "head" and "tail." If we let the random variable x represent the possible outcomes of this experiment, then x can assume two possible values, that is, $x = H$ or $x = H'$. We find that it is more convenient to specify values of a random variable in terms of numbers. Therefore, for this experiment, we will assign the value 1 to the random variable if a head occurs and the value 0 if a head fails to occur. Then, if we define the random variable x as representing the number of heads that occur on the toss of a coin, x can have the value 1 or 0. When all of the values of a random variable are numbers, we have a NUMERICAL RANDOM VARIABLE.

Suppose we define a new experiment consisting of four coin tosses. We can say that the new experiment consists of four trials of the original experiment. We still define the random variable x as representing the number of heads that occur during this experiment, that is, on four tosses. In this new experiment, x can have the value 0, 1, 2, 3, or 4. Each possible value of x corresponds to an event in the sample space of the experiment, but not necessarily to a single outcome. The possible outcomes in the sample space of this experiment and their correspondence to the possible values of x are shown in Table 8.1.

TABLE 8.1
OUTCOMES CORRESPONDING TO THE RANDOM VARIABLE *x* IN THE FOUR COIN-TOSS EXPERIMENT (*x* EQUALS THE NUMBER OF HEADS)

x	*Outcomes*
0	(T,T,T,T)
1	(H,T,T,T)(T,H,T,T)(T,T,H,T)(T,T,T,H)
2	(H,H,T,T)(H,T,H,T)(H,T,T,H)(T,H,H,T)(T,H,T,H)(T,T,H,H)
3	(H,H,H,T)(H,H,T,H)(H,T,H,H)(T,H,H,H)
4	(H,H,H,H)

In this experiment, the possible values of x are limited to certain numbers, specifically 0, 1, 2, 3, and 4. When the random variable is limited to certain values, it is called a DISCRETE RANDOM VARIABLE. Remember that in Chapter 1 we defined a discrete variable as one that is restricted to certain values within some range of values.

There are experiments in which the random variable is not limited to certain discrete values, but can assume any value within some range of values. Observations of outcomes in an experiment of this type are measurements. As an illustration, consider an experiment in which 16-ounce cans of fruit juice are weighed as they come off a production line. The weighing process assures that each can contains the correct amount of juice. The weight of each can is indicated by the position of a pointer on a very precise scale. Although it is necessary to read and record every weight in terms of discrete units, ounces, tenths of ounces, etc., the actual weight of the can is not limited to any of these discrete values. If the scale can be read to the hundredth of an ounce, a reading of 16.02 means that the true weight of the can may be anywhere between 16.015 and 16.025 ounces. The random variable corresponding to the outcome of this experiment is said to be a CONTINUOUS RANDOM VARIABLE. Remember that in Chapter 1 we defined a continuous variable as one that can assume any value within some range of values.

Example 8.1

To establish values for taxing purposes, the tax assessor of Buckingham County takes a survey of all single-family houses in the county. One criterion for the evaluation is the number of rooms in each house. If x is the random variable representing the number of rooms, then x must be a discrete random variable, since it is limited to values that are whole numbers. This is notwithstanding the fact that many real estate advertisements specify half-bathrooms.

Example 8.2 A study to determine the television viewing habits of the residents of Middletown requires information regarding the number of minutes each day that the television set is on. A device that records the "on" time is attached to sets in a representative sample of homes. If x represents the time that a television set is turned on in a particular home, is x a discrete or a continuous random variable?

Although the attachment records time in discrete units such as minutes and seconds, the actual time value is not so limited. If the attachment records a value of 127 minutes to the nearest minute, the set may have been on anywhere between 126.5 and 127.5 minutes. The random variable representing time is a continuous random variable.

8.2 The Probability Function of a Random Variable

In Chapter 2 we defined a function as a correspondence between two sets, one of which is called the range set and the other the domain set. We also state that the variable which corresponds to the range set is often called the dependent variable and the variable corresponding to the domain set is called the independent variable.

Consider a discrete random variable x whose values are determined by the outcome of some experiment. We are interested in a function of this random variable that assigns a probability to each possible value of x. Since a function as defined in Chapter 2 consists of a domain, a range, and a rule of association, the set of values of x is the domain of the function and the probabilities are the range. The function $f(x)$ is a probability function provided that $0 \leq f(x) \leq 1$ and $\sum_{x} f(x) = 1$.

Example 8.3 Recall the experiment in which a coin is tossed four times and the random variable x is defined as the number of heads that occur. The sample space of this experiment consists of the sixteen equally likely outcomes listed in Table 8.1. According to the classical definition, the probability of an event is the ratio of the number of outcomes that corresponds to that event, to the total number of outcomes of the experiment. Find the probability distribution of x.

The probability distribution of x consists of a set of probabilities corresponding to each value of x. Again, by the classical definition of probability and by counting outcomes:

$$P(x = 0) = 1/16 \qquad P(x = 1) = 4/16 \qquad P(x = 2) = 6/16$$
$$P(x = 3) = 4/16 \qquad P(x = 4) = 1/16$$

This probability distribution is provided in Table 8.2. Notice that the sum of all the probabilities is 1.

The probability function of a continuous random variable must be defined somewhat differently than that of a discrete random variable. Recall that in Chapter 7, continuous probabilities were

TABLE 8.2
PROBABILITY DISTRIBUTION OF THE NUMBER OF HEADS IN FOUR COIN TOSSES (EXAMPLE 8.3)

x (*Number of Heads*)	$f(x)$
0	$1/16$
1	$4/16$
2	$6/16$
3	$4/16$
4	$1/16$
	$16/16 = 1$

defined in terms of intervals instead of points in the sample space. Consequently, the probability function of a continuous random variable is a function that assigns probabilities to intervals defined on the sample space of an experiment. The probability function of a continuous random variable can be described by a smooth curve. The area under the curve between any two points corresponds to the probability that the random variable assumes a value in that interval. Discrete and continuous probability distributions are illustrated in Figures 8.1 and 8.2.

The probability distribution in Figure 8.1 is represented by a histogram, a graph we discussed in Chapter 3. The discrete random variable x can assume only those integer values that are labeled on the horizontal axis of the graph. Theoretically, the probability exists only at those values and is zero for all points between those values. The height of the bar centered at a specific value represents

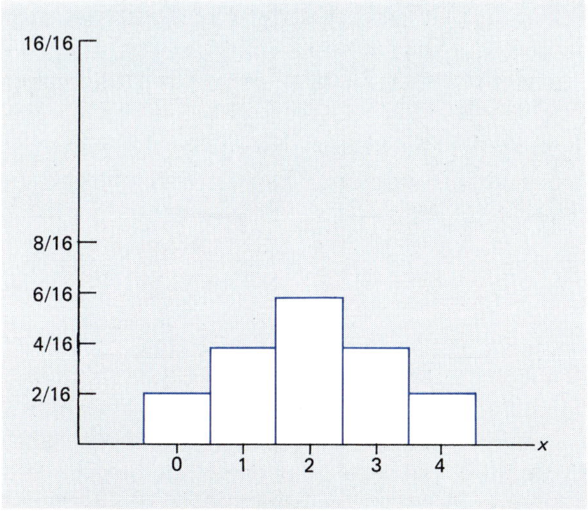

Figure 8.1
Probability distribution of a discrete random variable. x = number of heads in four coin tosses.

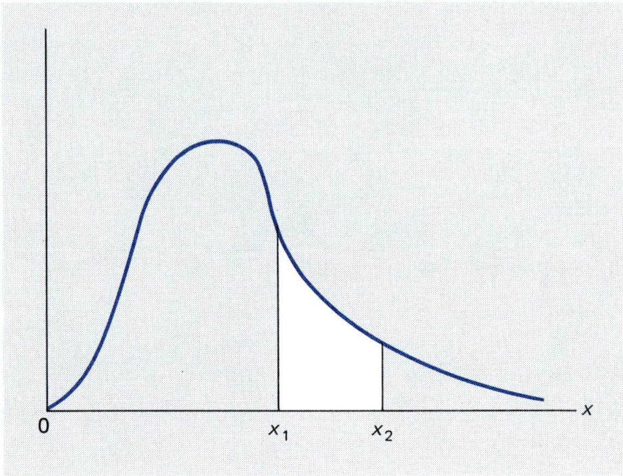

Figure 8.2
Probability distribution of a continuous random variable. x = television viewing time.

the probability that x assumes that value. However, by the nature of the histogram, the area in the bar also corresponds to this probability. Notice that the bar extends to a point halfway between two adjacent discrete values.

The probability distribution in Figure 8.2 is a smooth curve. Since the random variable in this experiment is continuous, x can assume any of an infinite number of values represented by the horizontal axis. Therefore, the probability that x assumes one specific value is zero, and the height of the curve at any point does not represent probability. For a continuous random variable, the probability that x assumes a value between any two points such as x_1 and x_2 is represented by the area under the curve between those two points.

Because a probability distribution is a function of a random variable, we use $f(x)$ to represent the probability that a random variable x is exactly equal to some value of x. That is, $f(x_i) = P(x = x_i)$. It is convenient to have notation that represents the probability that x is less than or equal to x_i. We use the cumulative probability distribution of x for this purpose. That is, $F(x_i) = P(x \leq x_i)$.

Much of this material about probability distributions is similar to the material on frequency distributions covered in Chapter 3. The relative frequency f_i/N is interpreted as representing the probability that some observed value x belongs to class *i*. In terms of probability for a discrete random variable, this is expressed as $f(x_i)$.

Probability distributions, like populations, are described by characteristics called PARAMETERS. If we consider the fact that probability distributions are used as models of populations, then it is logical that both populations and probability distributions have parameters. A parameter of a probability distribution is a constant value that characterizes the distribution.

8.3 Expected Value of a Random Variable

There is a special value of a random variable x called the EXPECTED VALUE and denoted by $E(x)$. If $f(x)$ is the probability function of a discrete random variable, then the expected value of x is defined as

$$E(x) = \sum_{x} xf(x) \tag{8.1}$$

Equation 8.1 simply means that the expected value of x is obtained by multiplying each value x by its probability of occurring and adding all of these products. If you compare this equation with Equation 4.5, which defines the mean of a frequency distribution, you will notice the similarity. If $f(x_i)$ is equated to the relative frequency f_i/N, then these formulas are identical and you can see that the expected value of a random variable is equal to the mean of its probability distribution. Consequently, we can state that

$$E(x) = \mu$$

Example 8.4

For the experiment of Example 8.3 in which four fair coins are tossed, find the expected value of x, the number of heads.

Using Equation 8.1 and Table 8.2,

$$\begin{aligned} E(x) = \sum_{x} xf(x) &= (0)(1/16) + (1)(4/16) + (2)(6/16) \\ &\quad + (3)(4/16) + (4)(1/16) \\ &= 0 + 4/16 + 12/16 + 12/16 + 4/16 \\ &= 32/16 = 2 \end{aligned}$$

The expected number of heads in four tosses of a fair coin is 2.

The symbol E is similar in a sense to the summation sign, Σ, since it indicates that a particular operation is to be performed. Also, like Σ, E has certain characteristics and rules that govern its use. These rules are stated below and should be compared with the rules for the use of the summation sign covered in Chapter 2.

1. The expected value of a constant is the constant itself.

$$E(c) = c$$

2. The expected value of a constant times a random variable is the constant times the expected value of the variable.

$$E(cx) = cE(x)$$

3. The expected value of a random variable plus a constant is equal to the expected value of the variable plus the constant.

$$E(x + c) = E(x) + c$$

4. The expected value of a constant times a random variable plus a different constant is equal to the first constant times the expected value of the variable plus the second constant.

$$E(c_1x + c_2) = c_1E(x) + c_2$$

5. The expected value of the sum of two random variables is equal to the sum of their expected values.

$$E(x + y) = E(x) + E(y)$$

Example 8.5 For the experiment of Example 8.3, if $c = 3$, use the above rules to find:

a) $E(cx)$, and verify the results.

From rule 2, $E(cx) = cE(x)$. Since $c = 3$ and, from Example 8.4, $E(x) = 2$, $E(3x) = 3E(x) = (3)(2) = 6$.

This is verified as follows.

x	c	cx	f(x)	cxf(x)
0	3	0	1/16	0
1	3	3	4/16	12/16
2	3	6	6/16	36/16
3	3	9	4/16	36/16
4	3	12	1/16	12/16
				96/16 = 6

b) $E(x + c)$ and verify the results.

From rule 3, $E(x + c) = E(x) + c$ and, therefore, $E(x + 3) = E(x) + 3 = 2 + 3 = 5$.

This is verified as follows.

x	c	x + c	f(x)	(x + c)f(x)
0	3	3	1/16	3/16
1	3	4	4/16	16/16
2	3	5	6/16	30/16
3	3	6	4/16	24/16
4	3	7	1/16	7/16
				80/16 = 5

Example 8.6 The probability distribution of the random variable y is as follows.

y	f(y)
1	1/6
2	2/6
3	3/6

a) Find $E(y)$

$E(y) = \sum_y yf(y)$ is computed as follows.

$yf(y)$
$(1)(1/6) = 1/6$
$(2)(2/6) = 4/6$
$(3)(3/6) = 9/6$
$\Sigma = 14/6 = 2\frac{1}{3}$

Therefore, $E(y) = 2\frac{1}{3}$.

b) Find $E(x + y)$, where x has the distribution provided in Example 8.5.

$E(x + y) = E(x) + E(y)$ by rule 5. Since $E(x) = 2$ from the previous example, $E(x + y) = 2 + 2\frac{1}{3} = 4\frac{1}{3}$.

Example 8.7

On any given day the demand for a particular brand of cigarettes carried by a retail tobacco store can be regarded as a function of a random variable. If x represents demand, the probability distribution for the demand is provided below. Find the *expected demand*, that is, $E(x)$.

Number of Cartons (x)	$f(x)$	$xf(x)$
0	0.05	0
1	0.10	0.10
2	0.15	0.30
3	0.25	0.75
4	0.20	0.80
5	0.10	0.50
6	0.10	0.60
7	0.05	0.35
	1.00	3.40

Since $E(x) = \sum_x xf(x) = 3.40$, the expected demand is 3.4 cartons per day.

8.4 Variance of a Probability Distribution

In Chapter 5 we discussed measures of variation for populations. The most useful of these measures are the *variance* and its positive square root, which is called the *standard deviation*. These same measures are also useful to help describe probability distributions of random variables. The variance of the distribution of a random variable is defined as

$$\text{Var}(x) = \sum (x - \mu)^2 f(x) = \sigma^2 \tag{8.2}$$

Notice the similarity between this formula and Equation 5.6. For comparison, Equations 4.5 and 8.1 for the population mean and the expected value of a random variable and Equations 5.6 and 8.2 for the respective variances are shown again below.

$$\mu = \frac{\Sigma xf}{N} \tag{4.5}$$

(5.6) $$\sigma^2 = \frac{\Sigma(x-\mu)^2 f}{N}$$

(8.1) $$E(x) = \mu = \sum xf(x)$$

(8.2) $$\mathrm{Var}(x) = \sigma^2 = \sum (x-\mu)^2 f(x)$$

If we equate the probability, $f(x)$, to the relative frequency, f/N, then Equations 4.5 and 8.1 are identical, as are 5.6 and 8.2.

A convenient shortcut formula for computing the variance of the distribution of a random variable is

(8.3) $$\mathrm{Var}(x) = \sigma^2 = \sum x^2 f(x) - \left[\sum xf(x)\right]^2$$

Remember that $\sum xf(x)$ is simply the mean or expected value of x and therefore Equation 8.3 can be written

(8.4) $$\mathrm{Var}(x) = \sigma^2 = \sum x^2 f(x) - \mu^2$$

Example 8.8 Find the variance of the probability distribution resulting from the experiment described in Examples 8.3 and 8.4 using Equation 8.2.

(8.2) $$\mathrm{Var}(x) = \sum (x-\mu)^2 f(x) = \sigma^2$$

x	$f(x)$	$xf(x)$	$x-\mu$	$(x-\mu)^2$	$(x-\mu)^2 f(x)$
0	1/16	0	−2	4	4/16
1	4/16	4/16	−1	1	4/16
2	6/16	12/16	0	0	0
3	4/16	12/16	1	1	4/16
4	1/16	4/16	2	4	4/16
		$\mu = 32/16 = 2$			$\sigma^2 = 16/16 = 1$

Example 8.9 Find the variance of the probability distribution of Example 8.7 using Equation 8.4.

x	$f(x)$	$xf(x)$	x^2	$x^2 f(x)$
0	0.05	0	0	0
1	0.10	0.10	1	0.10
2	0.15	0.30	4	0.60
3	0.25	0.75	9	2.25
4	0.20	0.80	16	3.20
5	0.10	0.50	25	2.50
6	0.10	0.60	36	3.60
7	0.05	0.35	49	2.45
		$\mu = 3.40$		14.70

$$(8.4) \qquad \begin{aligned} \sigma^2 &= \sum x^2 f(x) - \mu^2 \\ &= 14.70 - (3.40)^2 \\ &= 14.70 - 11.56 \\ &= 3.14. \end{aligned}$$

8.5 Discrete Probability Distributions

The Binomial Distribution. One of the most common discrete probability distributions is the BINOMIAL DISTRIBUTION. This distribution results from repeated INDEPENDENT TRIALS of an experiment in which there are only two possible mutually exclusive outcomes on each trial. For convenience, these outcomes are usually referred to as "success" and "failure." The random variable of interest is x, the total number of successes.

The concept of independence here is identical to that discussed in Chapter 7. Trials are independent if the probability of a specific outcome on any trial is unaffected by the outcome of any other trial.

Briefly then, in order for the binomial model to be applicable to some experiment or process, the process must have the following characteristics:

1) Only two mutually exclusive events or outcomes are possible on each trial of the experiment.

2) The repeated trials of the experiment are independent.

3) The probability of the event of interest occurring on any trial remains constant from trial to trial.

The binomial distribution results from a process of independent random sampling from a BERNOULLI process or population. The term Bernoulli population refers to an experiment which has only two possible outcomes, generally called SUCCESS and FAILURE. Such an experiment involves a single observation from a population divided into two mutually exclusive subsets according to the presence or absence of the attribute corresponding to the outcome, success. If the value 1 is assigned to an element having the attribute of interest and 0 to an element lacking that attribute, the population can be called a Bernoulli population.

The various coin tossing experiments we have used as examples have these characteristics. For example, on any single toss there are only two possible outcomes, "head" or "tail." The probability of getting a head on any toss of the coin is unaffected by the result of any other toss. And, of course, this probability remains constant no matter how many times the coin is tossed. Probabilities for this type of experiment can be determined by using the binomial distribution. The applicability of the binomial distribution is not limited to coin tossing. Many processes that occur in business and industry have similar characteristics. One of the most common types of these processes involves the selection of samples from infinite or very large populations.

The binomial distribution is

$$f(x) = \binom{n}{x} p^x (1-p)^{n-x} \tag{8.5}$$

where n represents the number of trials of the experiment, p represents the probability of a success on a single trial, and $\binom{n}{x} = n!/x!(n-x)!$ is the number of combinations of n things taken x at a time. The random variable x represents the number of successes and has the possible values 0, 1, 2, . . . , n.

The parameters of the binomial distribution are n and p. Once values are specified for the parameters, probabilities can be determined from the function. Although parameters of probability distributions are usually represented by Greek letters, the binomial distribution is an exception to this general rule.

Because the previous coin tossing experiments meet all of the requirements for the binomial distribution, it is possible to use the function specified as Equation 8.5 to determine the probabilities that we originally obtained by listing and counting outcomes in Tables 8.1 and 8.2. Let x be the random variable that represents the number of heads occurring in four coin tosses. If the coin is fair, the probability of a head occurring on any toss is $\frac{1}{2}$, therefore, $p = \frac{1}{2}$ and $1 - p = \frac{1}{2}$. The number of tosses is n, in this case, $n = 4$. Since we have specified the parameters of the binomial distribution for this experiment, we can now use the function to determine probabilities. Let us first determine the probability of getting exactly two heads in the four tosses.

$$\begin{aligned} f(x) &= \binom{n}{x} p^x (1-p)^{n-x} \qquad (8.5) \\ f(2) &= \binom{4}{2} (1/2)^2 (1/2)^2 \\ &= \frac{4!}{2!2!}(1/4)(1/4) \\ &= (6)(1/4)(1/4) \\ &= 6/16 \end{aligned}$$

The probabilities for all possible values of x from 0 to 4 are obtained in the same way:

$$P(x=0) = f(0) = \binom{4}{0}(1/2)^0(1/2)^4 = (1)(1)(1/16) = 1/16$$

$$P(x=1) = f(1) = \binom{4}{1}(1/2)^1(1/2)^3 = (4)(1/2)(1/8) = 4/16$$

$$P(x=2) = f(2) = \binom{4}{2}(1/2)^2(1/2)^2 = (6)(1/4)(1/4) = 6/16$$

$$P(x = 3) = f(3) = \binom{4}{3}(1/2)^3(1/2)^1 = (4)(1/8)(1/2) = 4/16$$
$$P(x = 4) = f(4) = \binom{4}{4}(1/2)^4(1/2)^0 = (1)(1/16)(1) = 1/16$$

Compare the probabilities here with those obtained by counting as shown in Table 8.2.

As with frequency distributions, it is often desirable to express the probability that x is less than or equal to some value, say c. This is provided by the CUMULATIVE BINOMIAL DISTRIBUTION, that is,

$$P(x \leq c) = F(c) = \sum_{x=0}^{c} \binom{n}{x} p^x (1-p)^{n-x} \tag{8.6}$$

where $0 \leq c \leq n$.

For this experiment, if $c = 2$, then the probability that $x \leq 2$ is the sum of the probabilities that $x = 0$, $x = 1$, and $x = 2$. Therefore, using the above results,

$$P(x \leq 2) = F(2) = 1/16 + 4/16 + 6/16 = 11/16$$

When n, the number of trials in the experiment, is small, determining probabilities by listing and counting outcomes is possible, though tedious. When n gets large this procedure is impractical if not impossible. In these situations the binomial distribution is useful.

Example 8.10

Consider an experiment consisting of rolling a fair die. A success is defined as the occurrence of a 1 or a 2. A failure corresponds to the occurrence of a 3, 4, 5, or 6. If the random variable x represents the number of successes in three rolls of the die, find the probability that $x = 0$, $x = 1$, $x = 2$, and $x = 3$. Also, find $F(1)$.

This experiment consists of three independent trials defined so that on each trial only one of two mutually exclusive events can occur, success or failure. If p represents the probability of a success on a single trial, then $p = \frac{1}{3}$ and $1 - p = \frac{2}{3}$. The probabilities, $P(x = x_i)$, are obtained from the binomial distribution with the parameters $n = 3$ and $p = \frac{1}{3}$ as follows.

$$P(x = 0) = f(0) = \binom{3}{0}(1/3)^0(2/3)^3 = (1)(1)(8/27) = 8/27$$
$$P(x = 1) = f(1) = \binom{3}{1}(1/3)^1(2/3)^2 = (3)(1/3)(4/9) = 12/27$$
$$P(x = 2) = f(2) = \binom{3}{2}(1/3)^2(2/3)^1 = (3)(1/9)(2/3) = 6/27$$
$$P(x = 3) = f(3) = \binom{3}{3}(1/3)^3(2/3)^0 = (1)(1/27)(1) = 1/27$$
$$F(1) = P(x \leq 1) = f(0) + f(1) = 8/27 + 12/27 = 20/27$$

Example 8.11

If one-fourth of the registered voters in the United States are independents, what is the probability that a random sample of five registered voters contains exactly three independents?

The random selection of elements from an infinite or very large population meets the independence requirement of the binomial distribution. Therefore, if one-fourth of all registered voters are independents, the probability of selecting an independent on a single trial is $\frac{1}{4}$. We use the binomial distribution with $p = \frac{1}{4}$ and $n = 5$ to determine this probability as follows.

$$P(x = 3) = f(3) = \binom{5}{3}(1/4)^3(3/4)^2 = (10)(1/64)(9/16)$$
$$= 90/1024 = 0.088$$

The mean and variance of any probability distribution can always be determined using Formulas 8.1, 8.2, 8.3, and 8.4. However, there is a simpler method for finding these values for the binomial distribution. The mean and variance of a binomial distribution are, respectively,

(8.7) $$E(x) = \mu = np$$

and

(8.8) $$\text{Var}(x) = \sigma^2 = np(1 - p)$$

In the experiment consisting of four coin tosses where x equals the number of heads, we found that $E(x) = 2$ and $\text{Var}(x) = 1$. The same results can be obtained more simply by using Equations 8.7 and 8.8 as follows.

(8.7) $$E(x) = np = (4)(1/2) = 2$$

(8.8) $$\text{Var}(x) = np(1 - p) = (4)(1/2)(1/2) = 1$$

Although binomial probabilities are easy to compute for small values of n, the computations become tedious when n is large. To avoid unnecessary computations, a table of the cumulative binomial probability distribution is included in the appendix for several values of n up to $n = 25$. Remember that a cumulative probability distribution provides the probability that the random variable is less than or equal to some specified value. Consequently, to find the probability that the random variable is exactly equal to some value c, it is necessary to subtract the tabulated value corresponding to $c - 1$ from the tabulated value corresponding to c. That is,

$$f(c) = F(c) - F(c - 1)$$

where

$$f(c) = P(x = c)$$

and

$$F(c) = P(x \leq c)$$

Example 8.12 For the binomial distribution with parameters $n = 10$ and $p = 0.20$, use Appendix A.1 to find:

a) $P(x \leq 4)$

b) $P(x = 4)$

c) $P(x = 2)$

d) $P(x > 4)$

e) $P(x > 2)$

a) 0.967 directly from the table

b) $0.088 = 0.967 - 0.879$

c) $0.302 = 0.678 - 0.376$

d) $0.033 = 1.000 - 0.967$

e) $0.322 = 1.000 - 0.678$

Example 8.13

If a production process produces 10% defective products, find the probability that a sample of fifteen items drawn randomly from the output of the process contains:

a) no defectives.

b) three or fewer defectives.

c) exactly three defectives.

The process of sampling from an infinite or very large population, in which each element selected is classified as either good or defective, can be modelled with a binomial distribution. The parameters in this case are $n = 15$ and $p = 0.10$. The probabilities are obtained from the binomial tables in the appendix.

a) 0.206 directly from the table

b) 0.944 directly from the table

c) $0.128 = 0.944 - 0.816$

Example 8.14

If 80% of the potential television viewing audience in a large metropolitan area watches television between the hours of 10:00 PM and 11:00 PM, find the probability that in a randomly selected sample of 25 potential viewers called during that hour:

a) all 25 are watching television.

b) 15 or fewer are watching television.

c) at least 15 are watching television.

d) exactly 15 are watching television.

a) $0.004 = 1.000 - 0.996$

b) 0.017 directly from the table

c) At least 15 means 15 or more. $P(x \geq 15) = 1 - P(x \leq 14) = 1.000 - 0.006 = 0.994$

d) $0.011 = 0.017 - 0.006$

The Hypergeometric Distribution. While the binomial distribution is applicable to repeated, independent trials of an experiment that has two mutually exclusive outcomes on each trial, a different model is required for situations in which the trials are not independent. For example, in drawing cards from a deck of playing cards, each draw or trial is independent of all others only if each card is replaced before the next card is drawn. When each card is not replaced, the complement of the deck changes after every draw. Consequently, the probability of the event defined for the experiment changes from one draw to the next, and that probability depends on which card or cards have already been removed from the deck.

Selecting a subset of elements from a population, whether the population consists of cards, manufactured parts, or people, is called *sampling*. When each element is not replaced before the next element is selected, the process is called sampling without replacement. The appropriate probability model for a process involving sampling without replacement from a finite population is the HYPERGEOMETRIC DISTRIBUTION.

Try to visualize a population consisting of N elements, C of which possess some characteristic of interest. It follows that $N - C$ do not possess that characteristic. Now suppose that we select a sample of n elements from this population without replacement. We want to know the probability that exactly x of the elements in the sample have the characteristic of interest. We let x represent the number of elements in the sample with the characteristic.

Let's list the various symbols and define them before we proceed.

N: the number of elements in the population.

C: the number of elements in the population with the characteristic of interest.

$N - C$: the number of elements in the population which do not have the characteristic of interest.

n: the number of elements in the sample.

x: a random variable which represents the number of elements in the sample having the characteristic of interest.

The probability of an event occurring has previously been defined as the ratio of the number of outcomes corresponding to that event to the total number of outcomes in the sample space of the experiment. We will construct a similar ratio to provide probabilities on the random variable x in a hypergeometric experiment. The total number of outcomes in the sample space is the number of different samples of size n that can be selected from a population of size N. This number is represented by the symbol $\binom{N}{n}$, which means the number of *combinations* of N things taken n at a time.

The event of interest, that the random variable x is equal to some specified value x_i, can be expressed as the joint occurrence in the sample of x elements with the characteristic and $n - x$ elements without it. The number of ways that this can occur is $\binom{C}{x}\binom{N-C}{n-x}$. The probability that $x = x_i$ is given by:

$$P(x = x_i) = f(x_i) = \frac{\binom{C}{x_i}\binom{N-C}{n-x_i}}{\binom{N}{n}} \tag{8.9}$$

Equation 8.9 is called the hypergeometric distribution.

Example 8.15

Suppose 40% of the people who respond to a magazine advertisement for health and accident insurance buy a policy. A salesman who receives ten responses selects three at random and arranges appointments. What is the probability that the salesman sells a policy to exactly one of these three respondents?

The population in this problem consists of the ten persons responding to the advertisement, therefore $N = 10$. The characteristic possessed by four of these is their tendency to purchase an insurance policy, thus $C = 4$. If the salesman randomly selects three responses, then $n = 3$. The probability that $x = 1$ is given by the hypergeometric distribution as:

$$P(x = 1) = f(1) = \frac{\binom{4}{1}\binom{6}{2}}{\binom{10}{3}}$$

Since

$$\binom{4}{1} = \frac{4!}{1!3!}, \quad \binom{6}{2} = \frac{6!}{2!4!}, \quad \text{and} \quad \binom{10}{3} = \frac{10!}{3!7!}$$

evaluating this expression results in $P(x = 1) = \dfrac{60}{120} = 0.50$.

Example 8.16

Suppose that a production process produces 30% defectives. The quality control department selects five parts at random from every lot of 20 parts produced and inspects each part in the sample. What is the probability that any sample contains: (a) exactly two defectives? (b) two or fewer defectives?

The population in this problem consists of the lot of 20 parts. Since 30% are known to be defective, each lot contains six defectives. The sample size is $n = 5$.

a) $P(x = 2) = f(2) = \dfrac{\binom{6}{2}\binom{14}{3}}{\binom{20}{5}} = 0.352$

b) $P(x \leq 2) = F(2) = f(2) + f(1) + f(0)$

$f(2) = 0.352$ from part (a)

$$f(1) = \frac{\binom{6}{1}\binom{14}{4}}{\binom{20}{5}} = 0.387$$

$$f(0) = \frac{\binom{6}{0}\binom{14}{5}}{\binom{20}{5}} = 0.129$$

$P(x \leq 2) = F(2) = 0.352 + 0.387 + 0.129 = 0.868$

The parameters of the hypergeometric distribution are N, n, and C. The mean is

$$\mu = n\left(\frac{C}{N}\right) \tag{8.10}$$

and the variance is

$$\sigma^2 = n\left(\frac{C}{N}\right)\left(1 - \frac{C}{N}\right)\left(\frac{N-n}{N-1}\right) \tag{8.11}$$

Since C/N represents the proportion of elements in the population that have the characteristic of interest, we can equate C/N to the binomial parameter p. If we compare the means and variances of the hypergeometric and binomial distributions, we find that the means are identical and the variances differ by the factor $(N - n)/(N - 1)$.

If N, the population size, gets very large, the hypergeometric distribution approaches the binomial distribution. Therefore, under certain circumstances, the binomial distribution may be used as an approximation to the hypergeometric distribution. A general rule is that when the sampling fraction, n/N, is less than or equal to 0.05, the binomial approximation may be used.

Example 8.17 At the end of 1977, the Emporium, a department store, had 200 delinquent charge accounts. Of these, 60 had balances over \$100. If an auditor takes a nonreplacement sample of 10 of the delinquent accounts, what is the probability that the sample contains exactly 5 accounts with balances greater than \$100? What is the probability that the sample contains at least 5 of these accounts?

Since this problem involves taking a nonreplacement sample from a finite population of 200 elements, the exact probability distribution is hypergeometric. To obtain the exact hypergeometric probability that $x = 5$, the following expression has to be evaluated.

$$f(5) = \frac{\binom{60}{5}\binom{140}{5}}{\binom{200}{10}}$$

However, since $n/N = 10/200 = 0.05$, this probability may be approximated using the binomial distribution. The binomial p is given by $C/N = 60/200 = 0.30$, and n is 10. Referring to the binomial tables in Appendix A.1,

$$P(x = 5) = 0.103$$
$$P(x \geq 5) = 0.150$$

The Poisson Distribution. There are situations where it is possible to observe the number of times an event occurs but it is impossible to observe the nonoccurrence of the event. We might, for example, examine a bolt of cloth to observe the occurrence of defects in the material. It is senseless, however, to attempt to observe the number of nondefects in the material. Or we might be interested in counting the number of customers that arrive at a teller's window in a bank or at a check-out counter in a supermarket. We can't possibly count the number of customers that don't arrive.

When discrete events occur in some continuous interval of space or time, an appropriate probability model may be the Poisson distribution. If x is the random variable that represents the number of occurrences in some interval of space or time, then the Poisson probability function is

$$P(x = x_i) = f(x_i) = \frac{e^{-\mu}\mu^{x_i}}{x_i!} \qquad (8.12)$$

where μ is the only parameter of the distribution and represents the mean or average number of occurrences for the specified interval. The letter e is an absolute constant with the approximate value 2.7183.

The variance of the Poisson distribution is also equal to μ. That is, $E(x) = \mu$ and $\text{Var}(x) = \mu$.

To use the Poisson distribution as a model, the following assumptions are necessary.

1. The occurrence of the event of interest within any interval is independent of its occurrence in any other nonoverlapping interval.

2. For a very small interval, the probability of one occurrence in the interval is proportional to the magnitude of the interval and independent of the location of the interval.

3. For a very small interval, the probability of more than one occurrence is negligible.

4. The average number of occurrences is constant for all intervals of the same size.

Tables of the cumulative Poisson distribution are provided in the Appendix. These tables provide the probability that the random variable x is less than or equal to some value *c*. That is,

$$P(x \leq c) = F(c) = \sum_{x=0}^{c} \frac{e^{-\mu}\mu^{x}}{x!} \tag{8.13}$$

Example 8.18 A citizens' group in Middletown petitioned the municipal traffic commission to install a traffic signal at a particularly dangerous intersection near an elementary school. Records kept by the group over a period of several years indicate that, on the average, three traffic accidents occur at the intersection every month.

What is the probability that in any given month:

a) no accidents occur at that intersection?

b) exactly two accidents occur?

c) two or fewer accidents occur?

d) three or more accidents occur?

This situation is one in which events are observed within some interval of time. In this case, the interval is one month. Assume that the required assumptions for the Poisson model are valid. Since the mean is given as 3, we refer to the Poisson tables to obtain the following probabilities.

a) $P(x = 0) = 0.049$, directly from the table.

b) $P(x = 2) = P(x \leq 2) - P(x \leq 1) = 0.423 - 0.199 = 0.224$.

c) $P(x \leq 2) = 0.423$, directly from the table.

d) $P(x \geq 3) = 1 - P(x \leq 2) = 1 - 0.423 = 0.577$.

Just as, under certain circumstances, the binomial distribution is used to approximate hypergeometric probabilities, there are also circumstances when the Poisson distribution can be used to approximate binomial probabilities. This approximation procedure produces satisfactory results when the binomial parameter n is large and p is small. As a general rule, the Poisson distribution is used as an approximation to the binomial distribution when $n \geq 20$ and $p \leq 0.05$. In using this approximation, the Poisson parameter μ is equated to the binomial np.

Example 8.19 The Reliance Insurance Company estimates that the probability that a particular type of automobile insurance policyholder will have an accident during a one year period is 0.02. If a random sample of 200 policyholders is taken, what is the probability that five or fewer had accidents during the past year?

If the random variable x represents the number of policyholders in the sample who were involved in an accident during the previous year, and if independence is assumed among policyholders, then x has a binomial distribution with $n = 200$ and $p = 0.02$. Note that if x of the policyholders included in the sample had an accident, then $n - x$ did not. This is not the same as the situation where accidents occurring at a particular intersection were counted.

The exact probability that x is less than or equal to 5 can be obtained from:

$$F(5) = \sum_{x=0}^{5} \binom{200}{x} (0.02)^x (0.98)^{200-x} \tag{8.6}$$

Since the binomial tables in the Appendix do not include values for n as large as 200 or values of p as small as 0.02, the evaluation of the above expression would be difficult. Fortunately, the Poisson approximation can be used here, since $n > 20$ and $p < 0.05$. The value of the Poisson parameter is $\mu = np = (200)(0.02) = 4$. Therefore, from the Poisson tables:

$$P(x \leq 5) = 0.785$$

The exact binomial probability is 0.788. In this case, the Poisson approximation is very good.

8.6 Continuous Probability Distributions

Probability distributions of continuous random variables are referred to as continuous probability distributions. As we discussed in Chapter 7, probability as defined for a continuous random variable refers to the probability that the variable assumes some value in a specified interval. The continuous probability distribution is pictured as a smooth curve. The area under the curve between two values represents the probability that the random variable assumes any value in that interval. Means and variances of continuous distributions are analagous to those measures as defined for discrete distributions.

The Normal Distribution. Probably the most useful probability distribution is the normal distribution. Because of its applicability as a model to a wide range of populations, we devote our entire discussion of continuous distributions in this chapter to this one distribution. Not only is the normal distribution useful as a model in its own right, it is also used as an approximation to many other probability distributions, including some of the discrete distributions covered in Section 8.5. A normal distribution is shown in Figure 8.3.

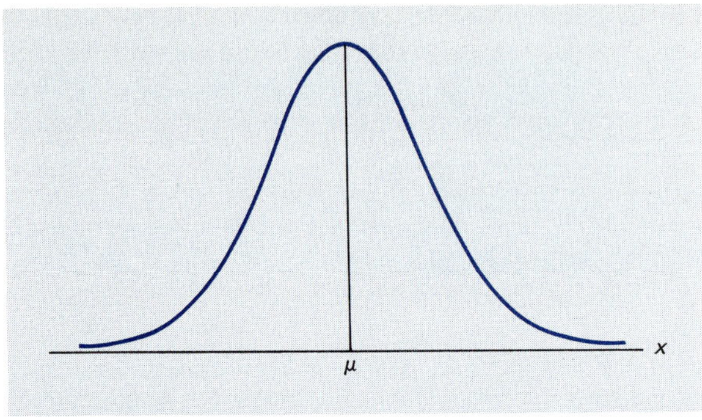

Figure 8.3
The normal distribution.

The normal distribution has certain characteristics worth noting. The most important of these are

1. It is a two parameter distribution. The parameters are μ, the mean, and σ, the standard deviation.

2. It is symmetrical about its mean, consequently the mean, median, and mode are identical.

3. It is infinite in extent, encompassing values from $-\infty$ to $+\infty$.

4. The total area under the curve is equal to 1.

The notation generally used to indicate that a random variable has a normal distribution with parameters μ and σ is $x : N(\mu, \sigma)$. Like the other distributions we have studied, the normal distribution is actually a family of distributions. Any specific member of this family of normal distributions is defined by specifying values for the two parameters. Therefore, $x : N(10, 50)$, $y : N(100, 15)$ means that the random variable x has a normal distribution with a mean of 10 and a standard deviation of 50, while the random variable y has a normal distribution with a mean of 100 and a standard deviation of 15.

To find areas under the normal distribution curve, we refer to Appendix A.3. Although there are many different normal distributions, only one is tabulated. That one is called the STANDARD NORMAL DISTRIBUTION. The standard normal distribution has a mean of zero and a standard deviation of one. The random variable that is distributed according to the standard normal distribution is usually called z, that is,

$$z : N(0, 1)$$

To find probabilities for some random variable x that has a normal distribution with parameters μ and σ, it is necessary to transform these values of x to corresponding values of z, a variable which has a standard normal distribution. Then the probabilities that are determined for z are identical to the probabilities for x.

Consider the random variable $x : N(100, 15)$. Suppose that we want to determine the probability that x has a value greater than 130, that is, $P(x > 130)$. Since, for continuous random variables, the probability at a specific point is always equal to 0, $P(x > 130) = P(x \geq 130)$. This is not true for discrete random variables.

The first step is to find a value of z that corresponds to $x = 130$. This is accomplished by the following transformation.

$$z = \frac{x - \mu}{\sigma} \tag{8.14}$$

Using Equation 8.14, the value of z that corresponds to $x = 130$ is

$$z = \frac{130 - 100}{15} = \frac{30}{15} = 2$$

Therefore, the probability that x is greater than 130 is equal to the probability that z is greater than 2. This last probability can be found in the appendix table. Although it may not be immediately apparent, z is the distance between the mean of the distribution and the specified value of x measured in units of the standard deviation. The value $z = 2$ simply means that the value of x, in this case 130, is two standard deviations above the mean of the distribution. If z equals -2, it represents a value of x two standard deviations below the mean. For $x : N(100, 15)$, Figure 8.4 shows the distribution with various values of x and the corresponding values of z. It also shows the percent of the area that falls within the

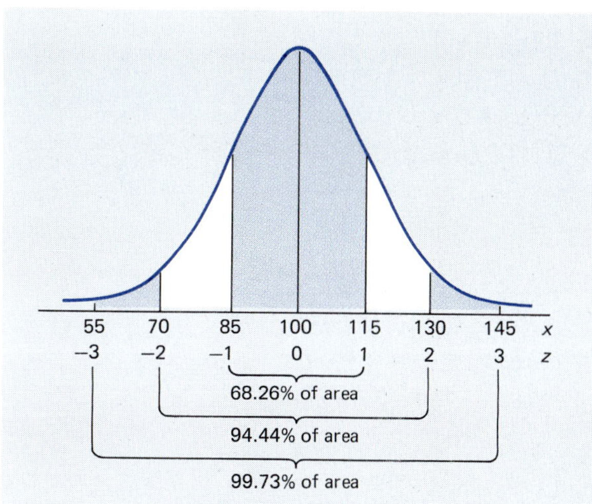

Figure 8.4
The distribution of $x:N(100, 15)$ and $z:N(0, 1)$.

interval defined by the mean plus and minus one, two, and three standard deviations for any normal distribution.

Since the normal distribution is symmetrical and the total area under the curve is equal to 1, then the area on each side of the mean must be equal to 0.5. Also, if the area between $z = -1$ and $z = +1$ is equal to 0.6826, then the area from the mean out to either $z = -1$ or $z = +1$ is equal to 0.3413.

To answer the question about the probability that $x > 130$, we simply find the probability that $z > 2$. From Figure 8.4 or the Appendix tables we can determine that the area between the mean, that is, $z = 0$ and $z = 2$, is 0.4722. Since the entire area to the right of the mean is equal to 0.5, that part of the area to the right of $z = 2$ must be $0.5 - 0.4722$, or 0.0278. Therefore, $P(x > 130) = 0.0278$.

The table of the standard normal distribution in the appendix provides areas or probabilities for values of z from $z = 0$ to $z = 4.0$. The probabilities in the body of the table correspond to areas between the mean and the specified value of z. Figure 8.5 illustrates the standard normal distribution. The shaded area corresponds to the probabilities provided in the table.

Probabilities in the table are provided only for positive values of z. Because of the symmetry of the distribution, all areas below the mean which correspond to negative values of z are identical to the corresponding areas above the mean. Therefore, to find the area between the mean and some negative value of z, look up the area between the mean and the corresponding positive value.

Example 8.20 If $z : N(0, 1)$, find the following probabilities.

a) $P(0 < z < 1.75)$

b) $P(z > 1.75)$

c) $P(z < -1.75)$

d) $P(-1.43 < z < 0.82)$

e) $P(1.75 < z < 2.28)$

f) $P(-2.28 < z < -1.75)$

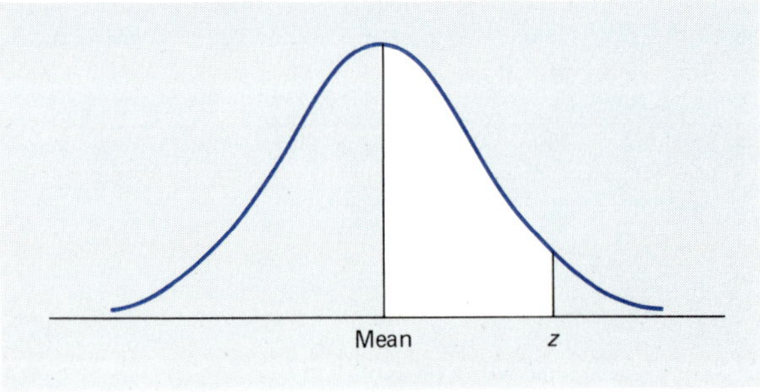

Figure 8.5
Area tabulated in the standard normal distribution.

Figure 8.6 shows the standard normal distribution, with the shaded areas corresponding to the probabilities required.

a) $P(0 < z < 1.75) = 0.4599$, directly from the table.

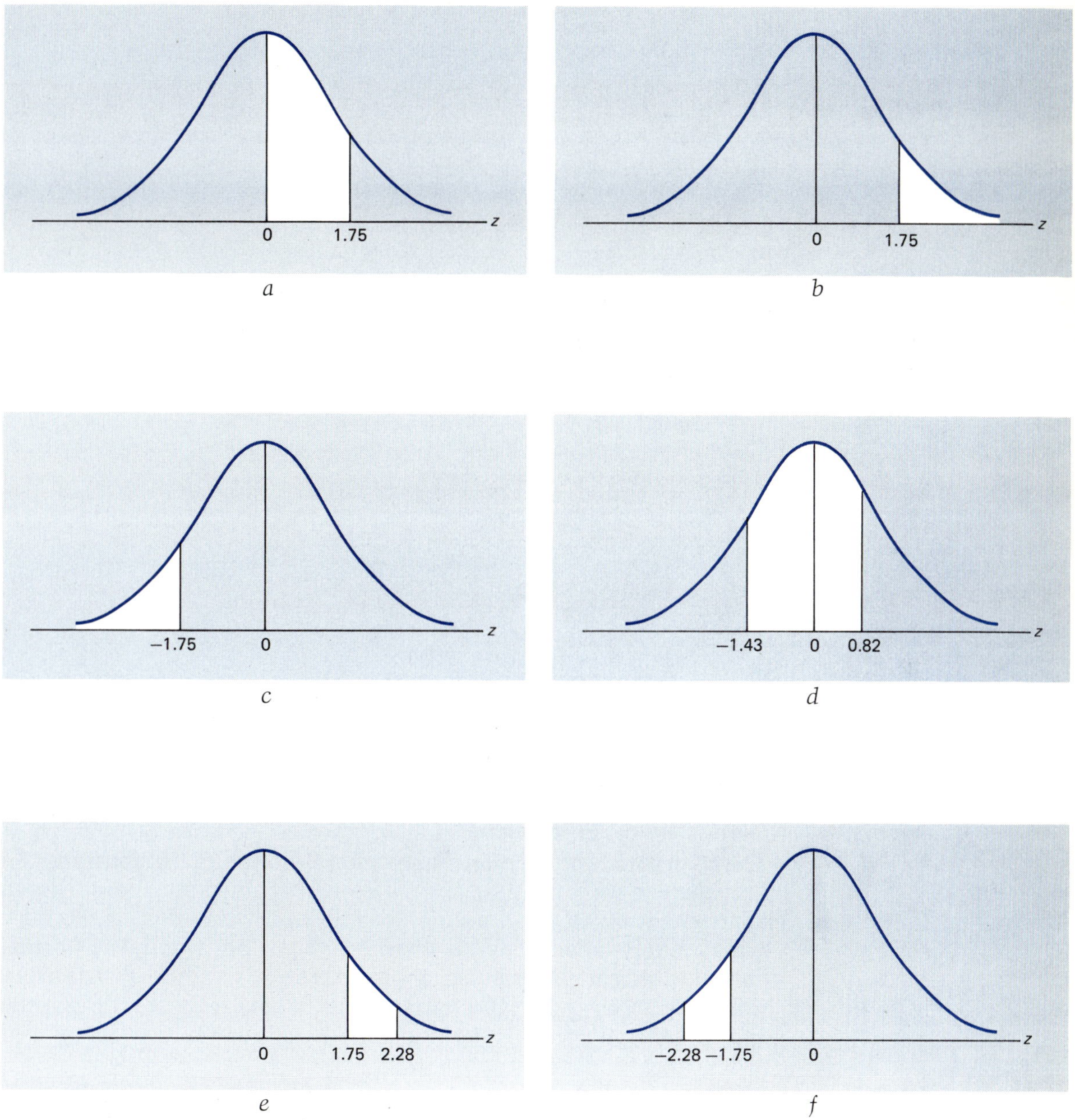

Figure 8.6
Areas under the normal distribution corresponding to Example 8.20.

b) $P(z > 1.75) = 0.5 - 0.4599 = 0.0401.$

c) $P(z < -1.75) = 0.0401$, by symmetry.

d) $P(-1.43 < z < 0.82) = P(-1.43 < z < 0) + P(0 < z < 0.82) = 0.4236 + 0.2939 = 0.7175.$

e) $P(1.75 < z < 2.28) = 0.4887 - 0.4599 = 0.0288.$

f) $P(-2.28 < z < -1.75) = 0.0288$, by symmetry.

Example 8.21 The lifetimes of Rollright automobile tires are known to be normally distributed with a mean life of 22,000 miles and a standard deviation of 2500 miles. Let the random variable x represent the lifetime in miles and find the probability that a randomly selected Rollright tire:

a) will not last for 20,000 miles.
b) can be driven for more than 26,000 miles.
c) will fail between 19,000 and 24,000 miles.

$x : N(22{,}000, 2{,}500)$

a) $$z = \frac{20{,}000 - 22{,}000}{2{,}500} = \frac{-2{,}000}{2{,}500} = -0.8$$

$$P(x < 20{,}000) = P(z < -0.8) = 0.2119$$

b) $$z = \frac{26{,}000 - 22{,}000}{2{,}500} = \frac{4{,}000}{2{,}500} = 1.60$$

$$P(x > 26{,}000) = P(z > 1.60) = 0.0548.$$

c) $$z_1 = \frac{19{,}000 - 22{,}000}{2{,}500} = \frac{-3{,}000}{2{,}500} = -1.20$$

$$z_2 = \frac{24{,}000 - 22{,}000}{2{,}500} = \frac{2{,}000}{2{,}500} = 0.80$$

$$P(19{,}000 < x < 24{,}000) = P(-1.20 < z < 0.80)$$
$$= 0.3849 + 0.2881 = 0.6730$$

The Normal Approximation to the Binomial Distribution. In Section 8.5 we saw that the Poisson distribution is used as an approximation to the binomial distribution when p is small and n is reasonably large. When n is large and p is *not* small, the normal distribution can be used to approximate a binomial distribution using the same mean and variance. As a general rule, the normal approximation to the binomial may be used when *both* np and $n(1 - p)$ are greater than 5.

Therefore, if x is a random variable with a binomial distribution, approximate probabilities can be obtained from the normal distribution with $\mu = E(x) = np$ and $\sigma = \sqrt{np(1 - p)}$. Figure 8.7 il-

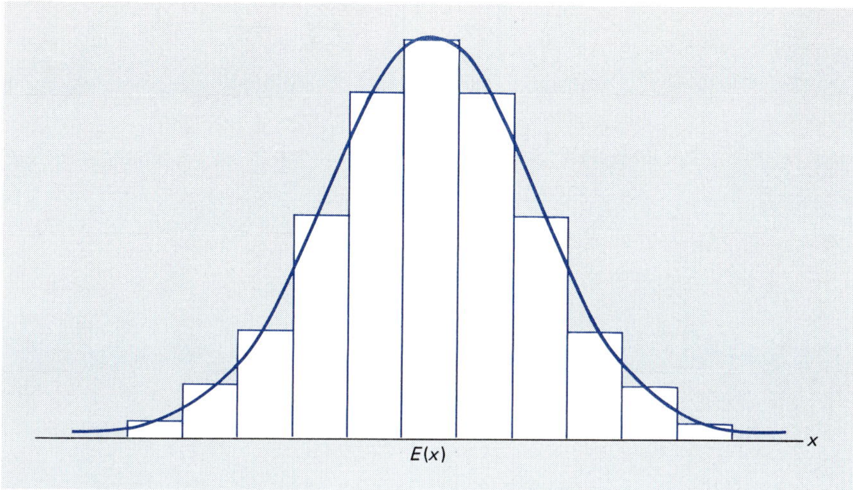

Figure 8.7
Binomial distribution approximated by a normal curve.

lustrates the approximation of a binomial distribution with a normal distribution.

Example 8.22

A market survey indicates that 70% of all adult shoppers who enter a department store actually make a purchase. To verify this finding, the management of the Emporium randomly selects 200 adult shoppers as they leave the store and asks each one if he or she has made a purchase in the store. If the survey findings are correct, what is the probability that at least 150 of the shoppers in the sample has made a purchase?

If x is the random variable representing the number of shoppers making purchases, then x has a binomial distribution with parameters $n = 200$ and $p = 0.70$. For this distribution, $E(x) = np = 140$ and σ, the standard deviation of the distribution of x, is $\sqrt{np(1-p)} = \sqrt{(200)(0.70)(0.30)} \doteq 6.5$.

To find the exact probability, it is necessary to evaluate:

$$P(x \geq 150) = \sum_{x=150}^{200} \binom{200}{x} (0.70)^x (0.30)^{200-x} \tag{8.6}$$

Without a computer, this may be difficult. However, since $np = 140$ and $n(1-p) = 60$, both of which are greater than 5, the normal approximation can be used. Since the binomial mean and standard deviation are 140 and 6.5, respectively, we will use the normal distribution having the same mean and standard deviation. Therefore, if x is approximately $N(140, 6.5)$, we find $P(x \geq 150)$ as follows.

1. $z = \dfrac{150 - 140}{6.5} = \dfrac{10.0}{6.5} = 1.54$
2. $P(z \geq 1.54) = 0.5 - 0.4382 \doteq 0.06$

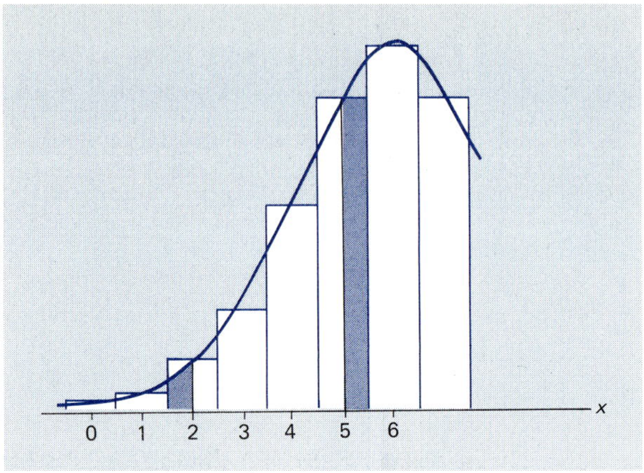

Figure 8.8
Normal approximation to the binomial distribution illustrating continuity correction.

The Continuity Correction Factor. Figure 8.8 illustrates a portion of a binomial distribution with an approximating normal distribution superimposed. The binomial distribution is represented by a histogram. The center of each bar in the histogram represents a discrete value of the random variable x, while the area in the bar represents the probability that x has that value.

Suppose we want to approximate the probability that x is less than or equal to 5. If we find that area under the approximating normal curve to the left of $x = 5$, we exclude the area in the right half of the bar, which corresponds to the value 5. This results in the estimated probability being too small. The area corresponding to the entire bar should be included. Since that bar actually extends from 4.5 to 5.5, the entire normal curve area to the left of 5.5 must be used to approximate the desired probability. Similarly, if we want to approximate the probability that x is greater than or equal to 2, the normal curve area to the right of $x = 1.5$ must be included. The addition or subtraction of half a unit is called a *continuity correction factor*. It is used when a discrete distribution is approximated by a continuous distribution.

In a problem where we want to approximate the probability that x is greater than or equal to 2 and less than or equal to 5, we have to include the normal curve area corresponding to the bars representing 2, 3, 4, and 5. This is the normal curve area that extends from $x = 1.5$ to $x = 5.5$. In this case, the half-unit correction factor must be applied at both ends of the interval. The significance of this correction factor decreases as n gets large. As a general rule, the continuity correction factor can be ignored whenever n is greater than 100.

Example 8.23

In Example 8.22, suppose that only 60 shoppers are surveyed as they leave the department store. What is the probability, assuming that the previous survey findings are correct, that at least 45 of the shoppers in the sample have made a purchase?

If x is the random variable representing the number of shoppers in the sample that make a purchase, then x has a binomial distribution with $n = 60$ and $p = 0.70$. $E(x) = np = 42$ and $\sigma = \sqrt{np(1-p)} = \sqrt{(60)(0.70)(0.30)} \doteq 3.5$.

The exact binomial probability could be obtained from

$$P(x \geq 45) = \sum_{x=45}^{60} \binom{60}{x} (0.70)^x (0.30)^{60-x} \tag{8.6}$$

However, since $np = 42$ and $n(1-p) = 18$, both of which are greater than 5, the normal approximation can be used. Also, since n is less than 100, the continuity correction factor must be applied. Therefore, if x is approximately $N:(42, 3.5)$, we proceed as follows.

1. Apply the continuity correction factor and find $P(x \geq 44.5)$.
2. $z = \dfrac{44.5 - 42.0}{3.5} = \dfrac{2.5}{3.5} = 0.71$
3. $P(z \geq 0.71) = 0.5 - 0.2612 \doteq 0.24$

Definitions

Bernoulli population: A population divided into two mutually exclusive subsets according to the presence or absence of an attribute of interest.

Continuous random variable: A numerical random variable that can assume any value in some range of values.

Discrete random variable: A random variable that is limited to certain values.

Expected value: The expected value of a random variable is equal to the mean of its probability distribution.

Independent trials: Trials of an experiment in which the probability of a specific outcome on any trial is unaffected by the outcome of any other trial.

Numerical random variable: A random variable whose values are numbers.

Parameter of a probability distribution: A constant value that characterizes the distribution.

Probability distribution: A function that assigns probabilities to values of a random variable.

Probability function: Same as a probability distribution.

Random variable: A variable whose values are determined by the outcome of an experiment.

Sampling fraction: The ratio of the sample size to the population size.

Formulas Expected value:

(8.1) $$E(x) = \sum xf(x)$$

Variance:

(8.2) $$\text{Var}(x) = \sum (x - \mu)^2 f(x)$$

(8.3) $$\text{Var}(x) = \sum x^2 f(x) - \left[\sum xf(x)\right]^2$$

(8.4) $$\text{Var}(x) = \sum x^2 f(x) - \mu^2$$

Binomial probability distribution:

(8.5) $$f(x) = \binom{n}{x} p^x (1 - p)^{n-x}$$

Cumulative binomial distribution:

(8.6) $$F(c) = \sum_{x=0}^{c} \binom{n}{x} p^x (1 - p)^{n-x}$$

Mean of a binomial distribution:

(8.7) $$E(x) = np$$

Variance of a binomial distribution:

(8.8) $$\text{Var}(x) = np(1 - p)$$

Hypergeometric probability distribution:

(8.9) $$f(x) = \frac{\binom{C}{x}\binom{N - C}{n - x}}{\binom{N}{n}}$$

Mean of a hypergeometric distribution:

(8.10) $$\mu = n\left(\frac{C}{N}\right)$$

Variance of a hypergeometric distribution:

(8.11) $$\sigma^2 = n\left(\frac{C}{N}\right)\left(1 - \frac{C}{N}\right)\left(\frac{N - n}{N - 1}\right)$$

Poisson probability distribution:

(8.12) $$f(x) = \frac{e^{-\mu}\mu^x}{x!}$$

Cumulative Poisson distribution:

(8.13) $$F(c) = \sum_{x=0}^{c} \frac{e^{-\mu}\mu^x}{x!}$$

The z transformation:

$$z = \frac{x - \mu}{\sigma} \quad (8.14)$$

Problems

8.1 Specify the random variable in each of the following experiments and indicate whether it is discrete or continuous.

a) The personnel manager of Metric Systems, Inc. reviews the records of all employees and notes the number of days each is absent from work.

b) The quality control manager of the Advance Machine Company records the number of defective parts produced on one particular production line each day for a period of 100 days.

c) Actuaries at the Reliance Insurance Company record the age at death of 10,000 customers.

8.2 Specify the random variable in each of the following experiments and indicate whether it is discrete or continuous.

a) One hundred light bulbs are life tested by the Northern Electric Company by turning on the bulbs and observing the time, to the nearest minute, when each burns out.

b) An agency of the Department of Health, Education, and Welfare surveys 1000 welfare mothers and records the number of children living with each.

c) A chain of children's clothing stores authorizes a survey of families in a particular marketing area. The age of the head of each household is observed and recorded to the nearest birthday.

8.3 Specify the random variable in each of the following experiments and indicate whether it is discrete or continuous.

a) To determine the number of switchboard operators needed during the noon hour, incoming calls during that hour are counted and recorded for twenty successive working days.

b) Inspectors measure and record a critical dimension of each part produced by a particular manufacturing process.

c) Each batch of punched cards processed at a computer center is verified and the number of errors on each card is recorded.

8.4 A fair coin is tossed three times. The number of heads observed = x. What is the probability distribution of x?

8.5 Consider an experiment in which a die is rolled twice. Let x = the number of **sixes** that occur. Show the probability distribution of x.

8.6 For the experiment defined in the previous problem, define a success as either a five or a six on the die. Let y = the number of successes. Provide the probability distribution of y.

8.7 Consider an experiment in which four playing cards—the ace of hearts, the ace of diamonds, the ace of clubs, and the ace of spades—are placed face down on a table. You select a card from among the four, note which one it is, and replace it. If you repeat this procedure three times, and if x represents the number of times you pick the ace of hearts, show the probability distribution of x.

8.8 Use Equation 8.1 to find $E(x)$ for problem 8.4.

8.9 Use Equation 8.1 to find $E(x)$ for problem 8.5.

8.10 Use Equation 8.1 to find $E(y)$ for problem 8.6.

8.11 Use Equation 8.1 to find $E(x)$ for problem 8.7.

8.12 Consider the probability distribution of problem 8.5. If $c = 3$, find the expected value of:
a) $x + c$
b) cx

8.13 Consider the probability distribution of problem 8.6. If $c = \frac{1}{2}$, find the expected value of:
a) $c + y$
b) cy

8.14 For the probability distribution of x which represents the demand for a particular brand of cigarettes in Example 8.7, let $y = x + 2$ and find $E(y)$.

8.15 For the probability distribution of x in problem 8.4, find Var(x) using Equation 8.2.

8.16 For the probability distribution of x in problem 8.5, find Var(x) using Equation 8.2.

8.17 For the probability distribution of y in problem 8.6, find Var(y) using Equation 8.3.

8.18 For the probability distribution of x in problem 8.7, find Var(x) using Equation 8.3.

8.19 In 20 independent trials of an experiment in which the constant probability of a success on a single trial is 0.6:
a) Determine the probability of getting exactly 14 successes.
b) Determine the probability of getting exactly 10 failures.
c) Determine the probability of getting at least 12 successes.
d) What is the name of the probability distribution that applies here?
e) Find the mean and variance of the random variable of this experiment.

8.20 In 15 independent trials of an experiment in which the constant probability of a success on a single trial is 0.8:
a) Determine the probability of getting 12 or fewer successes.
b) Determine the probability of getting more than 12 successes.
c) Determine the probability of getting 5 or more failures.
d) What is the name of the probability distribution that applies here?
e) Find the mean and variance of the random variable of this experiment.

8.21 A series of tests conducted by Metric Systems, Inc. indicates that a particular type of switch manufactured by the company fails to operate properly 5% of the time.
a) If a single switch is tested ten times, what is the probability that it will fail exactly once?

b) If ten switches are selected at random and each is tested once, what is the probability that exactly one will fail to operate?

8.22 If 40% of all adult males in the United States use electric shavers, find the probability that a random sample of 25 adult males taken by a razor-blade manufacturing company will contain eight or fewer electric-shaver users. What is the expected number of electric-shaver users in the sample?

8.23 In the previous problem, what is the probability that the sample will contain eight or fewer razor-blade users? What is the expected number of razor-blade users in the sample?

8.24 If a production process produces 10% defective products, what is the probability of finding one or more defectives among the first five parts produced on a shift?

8.25 If a box of ten fuses contains three that are defective, what is the probability that a fuse selected from the box will be defective? What is the probability that it will be good?

8.26 In the previous problem, if two fuses are selected from the box, what is the probability that:

a) both are good?
b) one is good?
c) both are bad?

8.27 The switches described in problem 8.21 are packed 20 to a box and shipped to customers. One customer, upon receiving his shipment, selects three switches at random from each box and tests them. If one or more of the switches tested fail to operate properly, the customer ships the entire box back to Metric Systems. What is the probability that any box of switches will be rejected by the customer and returned?

8.28 Out of a total of 15 executives in the main offices of Cinderella Clothes, Inc., four are women. What is the probability that a random sample of five of these executives contains:

a) no women?
b) one woman?
c) at least one woman?

8.29 In the previous problem, what is the expected number of women in the sample? What is the variance?

8.30 In problem 8.27, what is the expected value and variance of x, the number of defective switches in the sample of size 3?

8.31 If the average number of defects observed in a bolt of cloth is three per square yard, find the probability that any square yard of the material contains:

a) no defects.
b) exactly three defects.
c) exactly two defects.
d) two or fewer defects.
e) two or more defects.

8.32 From past experience, the Essex Construction Company estimates that a bulldozer will break down once in every 800 hours of operation. If a particular construction job requires 800 hours of bulldozer operation, what is the probability that:
a) there will be no breakdown?
b) there will be two or more breakdowns?
c) there will be less than three breakdowns?

8.33 If in the previous problem, the job required only 400 hours of bulldozer operation, find the probabilities asked for.

8.34 If, on the average, calls arrive at the switchboard of the Reliance Insurance Company at the rate of one every 20 seconds:
a) What is the probability that two calls will arrive in a 20-second period?
b) What is the probability that five or fewer calls will arrive in one minute?
c) What is the probability that at least five calls will arrive in one minute?

8.35 The annual dealers' meeting of Quality Appliances, Inc. was attended by 200 dealers of which 140 were men and 60 were women. If a random sample of 10 dealers is selected from among the attendees, what is the probability that:
a) three are women?
b) more than three are women?
Use an appropriate approximation.

8.36 If a production lot of 100 men's shirts produced by Lance Clothiers, Inc. contains 20 shirts with defects, find the probability that in four shirts selected from the lot at random:
a) exactly one contains a defect.
b) none contain a defect.
Use an appropriate approximation.

8.37 If the probability of a defective shirt being produced by Lance Clothiers is 5%, and if a sample of 100 shirts is randomly selected each day and inspected, what is the probability that:
a) there will be no defective shirts in the sample?
b) there will be more than five defective shirts in the sample?
Use an appropriate approximation.

8.38 If Cinderella Clothes, Inc. has determined that 0.5% of all incoming phone calls involve complaints, what is the probability that in 100 incoming calls there are
a) no complaints?
b) one complaint?
c) more than one complaint?
Use an appropriate approximation.

8.39 If $x : N(15,5)$, find:
a) $P(x > 17)$.
b) $P(x < 9)$.
c) $P(19 < x < 21)$.

d) $P(9 < x < 11)$.
e) $P(13 < x < 21)$.

8.40 If $x: N(125,50)$, find:
a) $P(x < 100)$.
b) $P(100 < x < 175)$.
c) $P(x > 175)$.
d) $P(50 < x < 100)$.
e) $P(125 < x < 200)$.

8.41 If the ages of the heads of households in a particular planned community have a normal distribution with a mean of 32 years and a standard deviation of four years, find the probability that the head of a randomly selected household is
a) over 40 years old.
b) less than 25 years old.
c) between 25 and 40 years old.

8.42 If 70% of Quality Appliance, Inc. dealers are men, what is the probability that a random sample of 50 dealers will contain:
a) more than 42 men?
b) less than 30 men?
c) between 30 and 42 men inclusive?
Use an appropriate approximation.

8.43 If a particular process produces 15% defective parts, what is the probability that in a day's production of 1000 parts, more than 175 are defective? Use an appropriate approximation.

CHAPTER 9 SAMPLING AND SAMPLING DISTRIBUTIONS

9.1
The rationale for sampling

9.2
Sample statistics

9.3
Random sampling

9.4
Sampling distributions

9.5
The sampling distribution of the mean

9.6
Sampling distributions and probabilities

9.7
Student's t distribution

9.8
The sampling distribution of the proportion and the sample number

The process of selecting a sample from a population is called SAMPLING. In this chapter we consider some common sampling methods as well as some distributions that result from sampling.

Information obtained from samples is the basis for all statistical inference. A sample is simply some part or subset of a population. When only a part of a population can be observed, all conclusions about the population characteristics or parameters must be inferred from the sample data.

9.1 The Rationale for Sampling

Samples are taken when it is either impractical or impossible to make observations on the entire population. Observing an entire population is impractical when the population is very large and such a procedure would be too costly in terms of both time and money. A company planning a survey to obtain information about the buying habits of adult Americans may find it extremely impractical to make observations on every member of the entire population. A census of the population is impossible if the population is infinite or in situations where the required observation results in the destruction of the element observed. For example, in industry it is sometimes necessary to test the strength of some manufactured item by applying a stress until the unit breaks. The amount of stress that results in breakage is the value of the observation that is noted. If this procedure is applied to an entire population, there would be nothing left. This type of testing is called destructive testing and requires that a sample be used.

If every member of a population is identical to every other member, at least with respect to the characteristic of interest, then all the information about the entire population with respect to that characteristic could be obtained by observing a single member. Such a population of identical elements is extremely unlikely, if not impossible. As discussed in Chapter 5, variation is inherent in nature. Elements differ from one another whether they are human beings, manufactured items, or pieces of paper that record information about business transactions.

If inferences about a population are based on data obtained from a sample, then the sample must resemble the population from which it is drawn, that is, it must be a *representative sample*. If conclusions are drawn about a population from a sample that is not representative, these conclusions are likely to be erroneous. On the other hand, if the sample is representative, inferences about the population based on this sample are likely to be correct.

Any inferences based on sample data will always include some error. This error is referred to as the sampling error. In statistics, the likelihood that the sampling error exceeds any specified magnitude must always be specified in terms of a probability value. It is for this reason that a study of probability and probability distributions precedes the study of statistical inference.

Because populations are variable, a representative sample must be large enough to exhibit the same, or a similar amount of, variation as the population. If the population exhibits no variation, a sample of size one is representative. The greater the variation in the population, the larger the sample must be to be representative.

Problems in sampling that relate to the representativeness of the sample generally fall into two categories: problems involving sample size and problems involving methods of sample selection. Problems regarding sample size are covered in Chapter 10. Some considerations regarding methods of sample selection are discussed in Section 9.3.

9.2 Sample Statistics

Values computed from samples are called STATISTICS. Statistics are typically used to make inferences about population parameters. In this section, we define and demonstrate computational methods for the most common statistics.

Inferences about a population are usually inferences about its parameters; that is, since populations are described by their parameters, these are the values that must be inferred from the sample data. Methods for computing parameters such as the population mean, variance, and standard deviation were covered in Chapters 4 and 5. While Greek letters such as μ and σ are usually used to represent parameters, Roman letters such as $\bar{x}$ and s are used to represent sample statistics. One exception to this convention that we have already encountered is the letter p, which represents a population proportion as well as a parameter of the binomial distribution.

The Sample Mean. The arithmetic mean of a sample of observations of a random variable is designated by placing a bar above the symbol for the random variable. For example, if the random variable is called x, then the mean of a sample of x values is designated $\bar{x}$.

The sample mean is computed like the population mean. The formulas for the sample mean for both ungrouped and grouped data are

$$\bar{x} = \frac{\sum_{i=1}^{n} x_i}{n} \tag{9.1}$$

and

$$\bar{x} = \frac{\sum_{i=1}^{k} x_i f_i}{n} \tag{9.2}$$

If these formulas are compared with Equations 4.3 and 4.6, the formulas for the population mean, notice that the only difference occurs in the denominators of the fractions. In the formula for the population mean, the denominator is N, which represents the population size, whereas in the formula for the sample mean, the denominator is n, which represents the sample size.

Sample Variance and Standard Deviation. The sample variance and the sample standard deviation are designated s^2 and s, respectively.

The formulas that define these two statistics for grouped data appear below. The comparable formulas for ungrouped data are omitted, since they are identical except for the elimination of the factor f_i.

$$s^2 = \frac{\sum_{i=1}^{k}(x_i - \bar{x})^2 f_i}{n - 1} \tag{9.3}$$

$$s = \sqrt{\frac{\sum_{i=1}^{k}(x_i - \bar{x})^2 f_i}{n - 1}} \tag{9.4}$$

In comparing Equations 9.3 and 9.4 with Equations 5.6 and 5.7, which define the population variance and standard deviation, notice that: In the sample statistic, $\bar{x}$ appears instead of μ in the numerator and the denominator is $n - 1$, one less than the sample size. This value, $n - 1$, in Equations 9.3 and 9.4 is called the "degrees of freedom."

More convenient computational formulas for the sample variance and sample standard deviation are, respectively,

$$s^2 = \frac{\sum_{i=1}^{k} x_i^2 f_i}{n - 1} - \frac{\left(\sum_{i=1}^{k} x_i f_i\right)^2}{n(n - 1)} \tag{9.5}$$

and

$$s = \sqrt{\frac{\sum_{i=1}^{k} x_i^2 f_i}{n - 1} - \frac{\left(\sum_{i=1}^{k} x_i f_i\right)^2}{n(n - 1)}} \tag{9.6}$$

Be sure to compare Equations 9.5 and 9.6 with Equations 5.8 and 5.9, which are the comparable formulas for parameters. Also remember that the step deviation shortcut method can also be used for computing sample statistics. The step deviation formulas for the sample variance and standard deviation are, respectively,

$$s^2 = c^2\left[\frac{\sum_{i=1}^{k} u_i^2 f_i}{n - 1} - \frac{\left(\sum_{i=1}^{k} u_i f_i\right)^2}{n(n - 1)}\right] \tag{9.7}$$

and

$$s = c\sqrt{\frac{\sum_{i=1}^{k} u_i^2 f_i}{n-1} - \frac{\left(\sum_{i=1}^{k} u_i f_i\right)^2}{n(n-1)}} \tag{9.8}$$

Example 9.1

A random sample of 25 sales slips is drawn from the records of one day's sales made by the cosmetics department at the Emporium Department Store. The amount of each sale is recorded and grouped into the following frequency distribution. Compute the mean, variance, and standard deviation of this sample. Use Equations 9.5 and 9.6 for the variance and standard deviation, respectively.

Class	*Class Mark* (x)	*Frequency* (*f*)	x*f*	x^2	x^2f
0 to 14	7	8	56	49	392
15 to 29	22	6	132	484	2904
30 to 44	37	4	148	1369	5476
45 to 59	52	4	208	2704	10816
60 to 74	67	3	201	4489	13467
		25	745		33055

$$\bar{x} = \frac{\Sigma xf}{n} = \frac{745}{25} = 29.8$$

$$\begin{aligned} s^2 &= \frac{\Sigma x^2 f}{n-1} - \frac{(\Sigma xf)^2}{n(n-1)} = \frac{33055}{24} - \frac{(745)^2}{(25)(24)} \\ &= \frac{33055}{24} - \frac{555025}{600} \\ &= 1377.29 - 925.04 \\ &= 452.25 \end{aligned}$$

$$s = \sqrt{452.25} \doteq 21.27$$

Example 9.2

For the data in the previous example, compute the variance and standard deviation using the step deviation method, Equations 9.7 and 9.8.

Letting a new variable, u, equal 0 at the midpoint of the middle class of the distribution, the new distribution and calculations are as follows.

Class	*u*	*f*	*uf*	u^2	u^2f
0 to 14	−2	8	−16	4	32
15 to 29	−1	6	−6	1	6
30 to 44	0	4	0	0	0
45 to 59	1	4	4	1	4
60 to 74	2	3	6	4	12
			−12		54

The class interval $c = 15$; therefore:

$$(9.7) \qquad s^2 = c^2\left[\frac{\Sigma u^2 f}{n-1} - \frac{(\Sigma uf)^2}{n(n-1)}\right] = (15)^2\left[\frac{54}{24} - \frac{(-12)^2}{(25)(24)}\right]$$

$$= 225\left[2.25 - \frac{(144)}{600}\right] = 225\,[2.25 - 0.24]$$

$$= (225)(2.01)$$

$$= 452.25$$

$$s = \sqrt{452.25} \doteq 21.27$$

The Sample Number. When dealing with a population of attributes in which all elements are classified into two mutually exclusive categories, the categories are usually referred to as success and failure. If a sample is drawn from a population, a success is counted if an element has the attribute of interest and a failure is counted if the element lacks that attribute. If X is a random variable that represents the total number of successes in the sample, X is also a statistic, and is called the SAMPLE NUMBER. If we assign each sample element the value 1 if it is a success, and the value 0 if it is a failure, then

$$(9.9) \qquad X = \sum_{i=1}^{n} \mathrm{x}_i \qquad \text{for x} = 0 \text{ or } 1$$

and each x_i is the observed value of a single sample element.

Previously, in discussing the binomial and hypergeometric distributions, the random variable was designated by x, which is consistent with our general notation for a random variable. However, now that we have defined a statistic which we call the sample number, we will use X to represent the number of successes in a sample.

The Sample Proportion. For a population of attributes as described in the previous paragraph, the parameter is the proportion of elements that have the attribute of interest. The population proportion is designated p. When a sample is drawn from a population, the sample proportion of successes is designated $\bar{p}$, which is defined as

$$(9.10) \qquad \bar{p} = X/n$$

where X is the sample number of successes and n is the sample size.

9.3 Random Sampling

All sample selection methods are divided into two general categories: JUDGMENT SAMPLING and PROBABILITY or RANDOM SAMPLING.

In judgment sampling, the choice of the specific members to be included in the sample is left to human judgment. If the individual making these decisions is knowledgeable about the population and has good judgment, then the resulting sample may be representative. Otherwise, the sample is not representative and any inferences based on that sample may be erroneous. Even if a judgment sample is reasonably representative, there is no objective method for determining the size or likelihood of sampling error.

Unlike judgment sampling, a measure of the sampling error and its probability *can* be obtained directly from the sample itself in random sampling.

The definition of a random sample is

> A RANDOM SAMPLE IS A SAMPLE DRAWN SO THAT EVERY ELEMENT IN THE POPULATION HAS A KNOWN AND NONZERO PROBABILITY OF BEING INCLUDED IN THE SAMPLE.

There are several common methods of obtaining random samples. Probably the most common sampling method is SIMPLE RANDOM SAMPLING. The definition of a simple random sample is

> A SIMPLE RANDOM SAMPLE IS A SAMPLE DRAWN SO THAT EVERY POSSIBLE COMBINATION OF n ELEMENTS IN THE POPULATION HAS THE SAME PROBABILITY OF BEING SELECTED.

Whenever we refer to a sample in this book, we mean a simple random sample unless it is stated otherwise.

Consider a population consisting of three elements, that is, $N = 3$. Suppose we want to select a simple random sample of two elements from this population. We can sample either with or without replacement. Sampling with replacement means that when each sample element is selected it is returned to the population before the next selection is made. In replacement sampling, it is possible for the same element to be selected for the sample more than once. In actual practice, replacement sampling is seldom if ever used.

Sampling without replacement simply means that when an element is drawn from the population it is not replaced. In nonreplacement sampling an element can be included in the sample only once. Since nonreplacement sampling is the only type that is important in business and industrial applications, we limit our discussion to sampling without replacement.

The total number of different nonreplacement samples of size n that can be selected from a population of size N is given by the combinatorial formula $\binom{N}{n} = N!/n!(N - n)!$. When $N = 3$ and $n = 2$, $\binom{3}{2} = 3$. That is, there are three distinct samples of size two that can be selected from a population of three elements. If the population elements are identified as **a, b,** and **c,** then the three possible samples are

sample 1: **a, b**

sample 2: **a, c**

sample 3: **b, c**

There are several procedures that could be used for selecting a simple random sample from this population. One method is to assign each of the possible samples a number as we did above, where the samples are numbered 1, 2, and 3. Each of the three numbers could then be written on separate slips of paper and a random selection made of one of the slips. Since each of the possible samples has the same probability of being selected—in this case the probability is $\frac{1}{3}$—we have a simple random sample. This procedure is extremely impractical in any real sampling situation. If the population of interest consists of as few as 30 elements and if samples of size 10 are to be selected, we must list over 30 million different possible samples, a virtual impossibility.

Another way to select a random sample is to assign numbers to each element in the population and then make a random selection of these numbers. For the population of three elements listed above, the elements **a, b,** and **c,** would be numbered 1, 2, and 3, respectively. On the first selection, each element has a probability of being selected of $\frac{1}{3}$. On the second selection, each of the remaining elements has a probability of being selected of $\frac{1}{2}$. Since on each selection all the remaining elements have the same probability of being selected, a simple random sample results.

In actual practice, the random selection of numbers corresponding to population elements is accomplished by using a table of random numbers. A portion of this table is included here as Table 9.1. All the digits in a table of random numbers have been generated by some random process. For convenience, the digits are grouped into sets of five.

To use this table it is necessary to have a numbered list of the elements in the population. This list is called a *sampling frame*. When numbers are assigned to the elements in the frame, a random selection of numbers from the table results in a random selection of the corresponding elements that constitute the sample.

Suppose a population consists of 700 elements and a random sample of 50 of these elements is to be selected. If the elements are listed in a frame, each element is assigned a three-digit number. Typically, these numbers run from 000 to 699. In selecting the sample, we would refer to a table of random numbers, and beginning on any page in the table and at any point on the page, proceed to select numbers in some systematic manner. Suppose that we happen to select the page reproduced as Table 9.1 and decide to

TABLE 9.1
RANDOM NUMBERS

23157	86952	02438	76520	97959	67422	69992	68479	10097	85017	84532	13618	91499	38631	79430	57421
05545	44109	05403	64894	35089	17691	89246	26940	37542	16719	82789	69041	80336	49172	16332	44670
14871	22115	86529	19645	62235	79655	81336	85157	08422	65842	27672	82186	44104	89232	57327	34679
38976	94324	43204	09376	58537	48274	81330	11100	99019	76875	20684	39187	12550	02844	15026	32439
97312	41548	93137	80157	08709	60523	2237	16505	12807	93640	39160	41453	63606	40387	65406	37920
11742	18226	29004	34072	08723	39323	05798	26457	66065	99478	70086	71265	61196	80240	44177	51171
43361	99436	42753	45571	56239	04595	10836	95270	31060	65119	26486	47353	15474	44910	99321	72173
93806	32584	21826	02051	44915	34823	51770	67897	85269	70322	21592	48233	94557	33683	86347	00926
49540	61777	67954	05325	41976	76559	24358	97344	63573	58133	41278	11697	42481	86430	19102	37420
36768	60452	38537	03529	15694	43438	36879	73208	73796	44555	81255	31133	23523	31379	68588	81675
07097	44673	61303	14905	14523	52494	24826	75246	98520	02295	13487	98662	04493	98086	32533	17767
43310	48897	48493	39808	94598	97654	16232	64051	11805	85035	54881	35587	00549	33185	04805	05431
61570	23350	65710	06288	81949	15307	00405	26898	83452	01197	86835	28021	35963	80951	68953	99634
31352	48625	44369	86507	73742	08391	49140	45427	88685	97907	19078	40646	59805	79752	02529	40200
57048	46359	74294	87517	49329	95236	32537	01390	99594	63265	96905	28797	46058	18633	99970	67348
09243	56092	84369	17468	90446	00597	45240	87379	65481	52841	59684	67411	32179	74029	74717	17674
07959	21225	13018	17727	45266	61406	41941	20117	80124	53722	71399	10916	69234	54178	10805	35635
93732	26958	02400	77402	28573	41430	96382	01758	74350	11434	51908	62171	19565	11664	77602	99817
72721	66995	07289	65252	16213	14938	71901	19476	69913	62375	99292	21177	45155	48324	32135	26803
61020	62841	31374	14225	78317	31994	98145	36168	09893	28337	20923	87929	94864	69074	45753	20505

start at the upper left-hand corner and read across the page. The first three-digit number is 231, and so the population element with the number 231 is the first member of the sample. The second three-digit number is 578 and the corresponding element becomes the second member of the sample. This process continues until 50 numbers are selected. If a number greater than 699, the highest number in the population, is read from the table, it is ignored and the next number is read. If a number is repeated, it is also ignored. By proceeding in this manner a random sample of 50 numbers is obtained.

Example 9.3 The owner of an automobile dealership in Middletown decides to sample customers of its service department to see if they are satisfied with the service they receive. A list of 500 of the past year's customers is prepared. This constitutes the sampling frame. A number from 000 to 499 is assigned to each element. Because of the large number of elements in the population, the owner decides to take a random sample of 15. Referring to the random number table (Table 9.1) he decides to begin with the three-digit number in the extreme lower left-hand corner of the table and read across from left to right. If the end of the row is reached before 15 valid numbers are obtained, he moves up to the next row and again begins reading across from left to right. Beginning at the lower left corner, the first number is 610. Since this number is not in the population, the owner goes to the next number, 206, which *is* in the population. The next 14 numbers are obtained in the same way, skipping any numbers larger than 499. The numbers the owner obtained for this sample are

206	284	131	374	142
257	453	283	372	092
387	445	205	057	272

Systematic Sampling. When the members of a population are already physically arranged in some order, a random sample is frequently obtained without preparing a separate list or frame. For example, folders containing records of customers' accounts might be filed in a cabinet according to the initial of the customer's last name or his account number. A day's production of a particular manufactured product might be stored in a bin in an order determined by the serial number of the product. Even houses along a street can constitute an ordered population. As long as the values of the characteristic being observed are unrelated to the order in which the elements are arranged, a random sample can be obtained by selecting every kth element in order, provided that the starting point is determined by some random method. The value k is called the *sampling interval.*

Suppose a population consists of 1000 accounts receivable and a random sample of 50 of these accounts is desired. The sampling interval is $k = N/n = 1000/50 = 20$. The sample is obtained by selecting every 20th account in the files. The starting point for the sample is determined by randomly selecting a number between 1 and 20. The numbers 1–20 could be written on separate slips of paper and drawn from a hat, or any other random method of selection could be used. Suppose, in this case, the number 8 is drawn. Then beginning with the 8th account in the file, every 20th account is selected until the sample of 50 is completed. The sample then consists of the 8th element, the 28th element, the 48th element, and so on.

Stratified Random Sampling. Another commonly used sampling procedure is called STRATIFIED RANDOM SAMPLING. In this procedure, the population is divided into subsets called *strata*. An attempt is made to define the strata so that the elements within a given stratum are relatively homogeneous compared to the population as a whole. A random sample, called a *subsample*, is then drawn from each stratum. The subsamples are combined to obtain the overall sample.

A stratified random sample of business firms might be stratified by number of employees, gross sales, type of business, or any other criteria considered to be of importance to the representativeness of the sample.

Cluster Sampling. Cluster sampling involves the random selection of entire groups, called *clusters*, from the population. Observations may then be made on every member of the cluster, in which case the procedure is called single-stage sampling, or a subsample may be taken from each cluster, which results in two-stage sampling.

Clusters differ from strata in that elements within a cluster are relatively heterogeneous compared to the overall population.

Most probability sampling methods are variations and/or combinations of the methods described above. The purpose of designing and using these apparently more complicated procedures is to reduce sampling error and sampling costs.

9.4 Sampling Distributions

We have defined the term "statistic" as a value computed from a sample. Since it is usually possible to draw many different samples of a given size from the same population, any particular statistic can have different values depending on which sample is selected. Therefore, if the same statistic is computed for each of the possible samples, the value of the statistic is likely to vary from sample to sample. Theoretically it would be possible to construct a frequency table showing the values assumed by the statistic and their frequency of occurrence. We would then have a distribution of values of a statistic. This distribution is called a *sampling distribution* because the values result from a process of sampling. Statistics computed from random samples are random variables.

Suppose we were to compute the means of all possible samples of size n that can be drawn from a population of size N. We would then have a set of $\binom{N}{n}$ sample means that can be arranged in the form of a distribution. This distribution would have its own mean, which we designate $\mu_{\bar{x}}$, and its own variance, $\sigma^2_{\bar{x}}$. The square root of the variance of a sampling distribution is called the *standard error* to distinguish it from the standard deviation of a population. The standard error of the mean is designated $\sigma_{\bar{x}}$.

We could follow this same procedure in computing the variance, the range, or any other statistic from all possible samples of a given size taken from a population. We would then have a sampling distribution of the variance, a sampling distribution of the range, or, in general, a sampling distribution of any statistic we choose. Each sampling distribution has its own mean, variance, and standard error. If we know the characteristics of the sampling distribution of a statistic, we can determine probabilities for the various possible values of the statistic.

At this point, we should relate the concept of a sampling distribution to the various probability distributions discussed in Chapter 8. The field of statistics concerns itself with distributions of random variables. When observations are made on all members of a population, we can obtain a frequency distribution of the population. Probability distributions are theoretical distributions that can be used as models of populations or of physical processes under certain specified conditions. Just as physical models, such as model airplanes or model cars, are not perfect representations of the real thing, probability distributions are not always exact representations of the populations they represent. They are, however, convenient methods for determining the probabilities of events when the populations or processes that generate these events meet certain criteria.

When events (that is, values of a random variable) are generated by a process of sampling, we have a sampling distribution. Some probability distributions are appropriate models for sampling distributions. For example, the binomial and hypergeometric distributions are theoretical distributions that provide the probabilities of events that result from a process of sampling. In that sense, they are actually sampling distributions. The Poisson and normal distributions are theoretical distributions that can be used as models for both population distributions and sampling distributions.

When we state that a population has a normal distribution, we imply that the particular mathematical model that describes the normal probability distribution provides a useful description of the distribution of population values. If we state that the means of samples are distributed normally, we imply that this same normal model is also useful for describing this particular distribution that results from a process of sampling. In the next few sections we describe the sampling distributions of several statistics and indicate the theoretical probability distribution that is most useful as a model.

9.5 The Sampling Distribution of the Mean

One of the most important sampling distributions in statistics is the distribution of the sample mean, $\bar{x}$. To determine the characteristics of the sampling distribution of the mean we must refer to two very important theorems. The first states:

> IF A POPULATION DISTRIBUTION IS NORMAL, THE SAMPLING DISTRIBUTION OF $\bar{x}$, THE MEAN OF RANDOM SAMPLES DRAWN FROM THE POPULATION, IS ALWAYS NORMAL WITH A MEAN OF μ AND A VARIANCE OF σ^2/n FOR SAMPLES OF ANY SIZE.

The second theorem is called the CENTRAL LIMIT THEOREM, which states:

> FOR ANY POPULATION HAVING A MEAN OF μ AND A VARIANCE σ^2, THE SAMPLING DISTRIBUTION OF THE SAMPLE MEAN APPROACHES A NORMAL DISTRIBUTION WITH A MEAN OF μ AND A VARIANCE OF σ^2/n AS n, THE SAMPLE SIZE, APPROACHES INFINITY.

In practice, the distribution of $\bar{x}$ approaches a normal distribution very rapidly, so that for samples as small as 30, the assumption of the normal distribution is quite valid. For sample sizes as great as 50, the shape of the population has little practical effect on the distribution of $\bar{x}$.

These two theorems apply when samples are drawn from infinite populations. When samples are drawn from finite populations, a correction factor must be applied to the variance and standard error of the mean. Therefore, when sampling from finite populations, the appropriate formulas for the variance and standard error of the mean are

$$\sigma_{\bar{x}}^2 = \frac{\sigma^2}{n}\left[\frac{N-n}{N-1}\right] \tag{9.11}$$

and

$$\sigma_{\bar{x}} = \frac{\sigma}{\sqrt{n}}\sqrt{\frac{N-n}{N-1}} \tag{9.12}$$

The factor $\sqrt{(N-n)/(N-1)}$ is called the *finite population correction* factor, frequently abbreviated fpc. Notice that as N, the population size, gets large relative to n, the sample size, the fpc approaches the value 1. Consequently, when the sampling fraction n/N is less than or equal to 0.05, the fpc can be ignored.

The reader is urged to refer to the discussion of the binomial approximation to the hypergeometric distribution in Chapter 8, where the finite population correction factor was also discussed.

Example 9.4 Consider a population consisting of five elements designated x_1, x_2, x_3, x_4, and x_5. The value of each x_i is

$$x_1 = 1, \qquad x_2 = 2, \qquad x_3 = 3, \qquad x_4 = 4, \qquad x_5 = 5$$

The mean and variance of this population can be computed directly from the x_i's, since all the population values are known. These parameters are

$$(4.3) \quad \mu = \Sigma x/N = 15/5 = 3$$

$$(5.2) \quad \sigma^2 = \frac{\Sigma(x-\mu)^2}{N}$$

$$= \frac{(1-3)^2 + (2-3)^2 + (3-3)^2 + (4-3)^2 + (5-3)^2}{5}$$

$$= \frac{(-2)^2 + (-1)^2 + (0)^2 + (1)^2 + (2)^2}{5}$$

$$= \frac{4+1+0+1+4}{5}$$

$$\sigma^2 = \frac{10}{5} = 2$$

$$\sigma = \sqrt{2} \doteq 1.41$$

Now suppose that samples of size $n = 3$ are drawn from this population of $N = 5$. The total number of different samples of size 3 is $\binom{N}{n} = \binom{5}{3} = 10$. Table 9.2 lists these ten samples and their means. In Table 9.3,

TABLE 9.2
LISTING OF SAMPLES OF SIZE $n = 3$ FROM POPULATION OF SIZE $N = 5$ (EXAMPLE 9.4)

Sample	Sample Elements	Sample Values	Sample Sum Σx	Sample Mean $\bar{x} = \Sigma x/n$
#1	$x_1x_2x_3$	1,2,3	6	6/3
#2	$x_1x_2x_4$	1,2,4	7	7/3
#3	$x_1x_2x_5$	1,2,5	8	8/3
#4	$x_1x_3x_4$	1,3,4	8	8/3
#5	$x_1x_3x_5$	1,3,5	9	9/3
#6	$x_1x_4x_5$	1,4,5	10	10/3
#7	$x_2x_3x_4$	2,3,4	9	9/3
#8	$x_2x_3x_5$	2,3,5	10	10/3
#9	$x_2x_4x_5$	2,4,5	11	11/3
#10	$x_3x_4x_5$	3,4,5	12	12/3

TABLE 9.3
FREQUENCY DISTRIBUTION OF MEANS OF SAMPLES OF SIZE $n = 3$ FROM POPULATION OF SIZE $N = 5$ (EXAMPLE 9.4)

Sample	$\bar{x}$	f	$\bar{x}f$
#1	6/3	1	6/3
#2	7/3	1	7/3
#3 and #4	8/3	2	16/3
#5 and #7	9/3	2	18/3
#6 and #8	10/3	2	20/3
#9	11/3	1	11/3
#10	12/3	1	12/3
		$\Sigma f = 10$	$\Sigma \bar{x}f = 90/3 = 30$

$$\mu_{\bar{x}} = \frac{\Sigma \bar{x}f}{\Sigma f} = \frac{30}{10} = 3$$

the data from Table 9.2 are arranged in the form of a frequency distribution. The mean of this distribution of sample means is computed in the same way as the mean of any frequency or probability distribution. Therefore, $\mu_{\bar{x}} = \Sigma \bar{x}f/\Sigma f = 30/10 = 3 = \mu$. Notice that the mean of this distribution of sample means is exactly equal to the population mean. That is, $\mu_{\bar{x}} = \mu = 3$.

The variance of this distribution of sample means is also computed from the data in the same way as any other variance. Table 9.4 repeats the frequency distribution of Table 9.3 and shows the calculations necessary for computing its variance.

Using the definitional formula for the variance, Equation 5.6, and substituting Σf in the denominator for the symbol N, its equivalent, we obtain $\sigma^2_{\bar{x}} = 1/3$. Of course, $\sigma_{\bar{x}}$, the standard error of the mean, is simply the square root of the variance, or $\sigma_{\bar{x}} = \sqrt{1/3} \doteq 0.58$.

TABLE 9.4
VARIANCE OF MEANS OF SAMPLES OF SIZE $n = 3$ FROM POPULATION OF SIZE $N = 5$

$\bar{x}$	f	$\bar{x} - \mu_{\bar{x}}$	$(\bar{x} - \mu_{\bar{x}})^2$	$(\bar{x} - \mu_{\bar{x}})^2 f$
6/3	1	(6/3 − 9/3) = −3/3	9/9	9/9
7/3	1	(7/3 − 9/3) = −2/3	4/9	4/9
8/3	2	(8/3 − 9/3) = −1/3	1/9	2/9
9/3	2	(9/3 − 9/3) = 0	0	0
10/3	2	(10/3 − 9/3) = 1/3	1/9	2/9
11/3	1	(11/3 − 9/3) = 2/3	4/9	4/9
12/3	1	(12/3 − 9/3) = 3/3	9/9	9/9
	10			30/9

$$\sigma^2_{\bar{x}} = \frac{\Sigma(\bar{x} - \mu_{\bar{x}})^2 f}{\Sigma f} = \frac{30/9}{10} = \left(\frac{30}{9}\right)\left(\frac{1}{10}\right) = \frac{30}{90} = \frac{1}{3}$$

It is also possible to obtain $\sigma_{\bar{x}}^2$ and $\sigma_{\bar{x}}$ by using Equations 9.11 and 9.12, since the variance of the population is known and is equal to 2. Because this example involves sampling from a very small, finite population, the use of the fpc is required. Since $\sigma^2 = 2$, $N = 5$, and $n = 3$,

$$(9.11)\quad \sigma_{\bar{x}}^2 = \frac{\sigma^2}{n}\left[\frac{N-n}{N-1}\right] = \frac{2}{3}\left[\frac{5-3}{5-1}\right] = \left(\frac{2}{3}\right)\left(\frac{2}{4}\right) = \frac{4}{12} = \frac{1}{3}$$

which agrees with the calculated result from Table 9.4.

The standard error can be obtained by taking the square root of the variance. It can also be calculated directly from Equation 9.12.

Example 9.5 Suppose the incomes of all wage earners in the United States are represented by the random variable x, which has a normal distribution with a mean of \$14,500 and a standard deviation of \$3000. If random samples of $n = 400$ are drawn from this population:

a) Find the mean, variance, and standard error of the sampling distribution of $\bar{x}$.

b) Is the distribution of $\bar{x}$ a normal distribution? Why?

a)

$$\mu_{\bar{x}} = \mu = \$14{,}500$$

$$(9.11)\quad \sigma_{\bar{x}}^2 = \frac{\sigma^2}{n} = \frac{(3000)^2}{400} = \frac{9{,}000{,}000}{400} = 22{,}500$$

$$(9.12)\quad \sigma_{\bar{x}} = \frac{\sigma}{\sqrt{n}} = \frac{3000}{20} = 150$$

b) Yes, the distribution of $\bar{x}$ is normal because the samples are drawn from a normal population.

Example 9.6 Suppose the random variable x represents the annual incomes of 500 production workers employed by Metric Systems, Inc. If the average annual income of these workers is \$13,200 with a standard deviation of \$2000, and if random samples of $n = 100$ are drawn from this population, find the mean, variance, and standard error of the sampling distribution of $\bar{x}$.

The finite population correction factor is required in this problem because the samples are drawn from a finite population and the sampling fraction $n/N = 100/500 = 0.20 > 0.05$.

$$\mu_{\bar{x}} = \mu = \$13{,}200$$

$$(9.11)\quad \sigma_{\bar{x}}^2 = \frac{\sigma^2}{n}\left[\frac{N-n}{N-1}\right] = \frac{(2000)^2}{100}\left[\frac{500-100}{500-1}\right]$$

$$= \left[\frac{4{,}000{,}000}{100}\right]\left[\frac{400}{499}\right] = 32{,}064$$

$$(9.12)\quad \sigma_{\bar{x}} = \frac{\sigma}{\sqrt{n}}\sqrt{\frac{N-n}{N-1}} = \frac{2000}{10}\sqrt{\frac{400}{499}} = 179.1$$

Example 9.7 Let the random variable y represent the diameters of 1000 special parts produced by the Advance Machine Company according to contract specifications. If the mean diameter of these parts is 2.5 centimeters and the standard deviation is 0.2 centimeters, and if samples of size $n = 36$ are drawn from this population of parts:

a) Find the mean, standard error, and variance of the sampling distribution of $\bar{y}$.

b) Is the fpc necessary? Why or why not?

c) Is the distribution of $\bar{y}$ a normal distribution? Why?

a)

$$\mu_{\bar{y}} = \mu = 2.5$$

(9.12)

$$\sigma_{\bar{y}} = \frac{\sigma}{\sqrt{n}} = \frac{0.2}{6} \doteq 0.033$$

$$\sigma_{\bar{y}}^2 = (0.033)^2 \doteq 0.001$$

b) Although this is a case of sampling from a finite population, the fpc is not necessary because the sample fraction $n/N = 36/1000 = 0.036 < 0.05$.

c) The sampling distribution of $\bar{y}$ is considered to be approximately normal by the Central Limit Theorem, since $n = 36$.

9.6 Sampling Distributions and Probabilities

Whenever a sample statistic is used to make an inference about a population parameter, some probability statement about the amount of error that might result should be included. One of the most common inferential procedures is estimation, a topic which is discussed thoroughly in Chapter 10.

In estimation, the value of the statistic is used as an estimate of the value of a population parameter. The difference between the value of the estimate and the value of the parameter is called the *sampling error*. For instance, in Example 9.3, the mean annual income of 500 production employees of Metric Systems was stated to be \$13,200. If an estimate of that mean income obtained from a sample is \$13,000, the sampling error is \$200.

Of course, if we know the value of the population parameter and then compute the value of the statistic from a sample, we can calculate the sampling error directly from the difference between the two values. In practice, however, the parameter value is not known. If it were, no estimate would be necessary. When the actual parameter value is not known, the sampling error cannot be calculated. However, knowledge about the sampling distribution of the statistic used as the estimator enables us to determine the probability that the sampling error is of any given magnitude.

To illustrate, refer to the sampling distribution of the means of samples of size $n = 3$ from the population of $N = 5$ elements in Table 9.3. We know that the mean of the $\bar{x}$'s is equal to the population mean, which is, in this case, equal to 3. If a random sample of size 3 is randomly selected from this population and $\bar{x}$ is used to estimate μ, the value of the sample statistic depends on which of the ten possible samples we happen to select. For example, if we select sample number 1, $\overline{x}$ has the value 2 and the sampling error is $|2 - 3| = 1$. We would get the same sampling error if we had selected sample number 10. The probability of selecting either sample number 1 or sample number 10 corresponds to the sum of their relative frequencies, that is, $1/10 + 1/10 = 2/10$. It is then proper to make the statement that in estimating the mean of this population with the mean of a randomly selected sample of size 3, the probability that the sampling error is exactly one unit is 0.2. The probability of sampling errors of other magnitudes can be determined directly from this distribution. In fact, if the characteristics of the sampling distribution of the statistic are known, the probability that the use of the statistic to make an inference about a parameter will lead to a sampling error of any magnitude can be determined.

Of course, in practical situations we are never able to list the complete sampling distribution of a statistic. However, we usually do know its characteristics. In Section 9.4, we found that the sampling distribution of means of samples drawn from a normal population has a normal distribution. We also know the mean and standard error of that sampling distribution. With this information, we can determine the probability that the sample mean will deviate from the population mean by any amount.

Example 9.8 In Example 9.5, the random variable x represents a large population of incomes which are normally distributed with a mean of \$14,500 and a standard deviation of \$3000. If a single random sample of $n = 400$ is drawn from this population, determine:

a) the probability that the sample mean is greater than \$14,750.

b) the probability that the sample mean is within \$250 of the population mean.

c) the probability that the sample mean deviates from the population mean by more than \$250.

In sampling from a normal population, the sampling distribution of the sample mean has a normal distribution. The mean or expected value of $\overline{x}$ is equal to μ, which is \$14,500. The standard error of $\overline{x}$, $\sigma_{\overline{x}} = \sigma/\sqrt{n} = 3000/20 = 150$. The required probabilities are obtained by transforming the specified values of $\overline{x}$ to values of z by the following relationship.

$$z = \frac{\overline{x} - \mu_{\overline{x}}}{\sigma_{\overline{x}}} \tag{9.13}$$

and finding the corresponding areas from the standard normal distribution.

a) The value of z corresponding to $\overline{x} = \$14,750$ is

$$z = \frac{14750 - 14500}{150} = \frac{250}{150} = 1.67$$

$$P(\bar{x} > 14750) = P(z > 1.67) = 0.0475$$

b) To determine the probability that $\bar{x}$ is within \$250 of the mean we must specify the endpoints of the interval within which $\bar{x}$ must lie. These endpoints are 14500 ± 250, resulting in the values 14250 and 14750. The values of z corresponding to these endpoints are -1.67 and $+1.67$ from part (a) above, and

$$P(14250 \leq \bar{x} \leq 14750) = P(-1.67 \leq z \leq +1.67) = 0.9050$$

c) The probability that the sample mean deviates from the population by more than \$250 is simply 1 minus the probability obtained in part (b). That is, $1 - 0.9050 = 0.0950$.

Example 9.9

In Example 9.6, the random variable x represents the annual incomes of 500 wage earners. If these incomes are normally distributed with a mean of \$13,200 and a standard deviation of \$2000, and if a single random sample of $n = 100$ is drawn from that population, determine:

a) the probability that the sample mean is less than \$13,000.

b) the probability that the sample mean is within \$150 of the population mean.

c) the probability that the sample mean deviates from the population mean by more than \$200.

In sampling from a normal population, the sampling distribution of $\bar{x}$ is always normally distributed with $E(\bar{x}) = \mu$. However, because this population consists of only 500 elements and the sampling fraction $n/N = 100/500 = 0.20 > 0.05$, the standard error of the mean is

$$\sigma_{\bar{x}} = \frac{\sigma}{\sqrt{n}}\sqrt{\frac{N-n}{N-1}}$$

$$= \frac{2000}{\sqrt{100}}\sqrt{\frac{500-100}{500-1}}$$

$$\doteq 179$$

a) The value of z corresponding to $\bar{x} = 13{,}000$ is

$$z = \frac{13{,}000 - 13{,}200}{179} = -1.12$$

$$P(\bar{x} < 13{,}000) = P(z < -1.12) = 0.1314$$

b) If a sample mean is within \$150 of the population mean, it must fall between the values 13,050 and 13,350. The corresponding values of z are

$$z_1 = \frac{13050 - 13200}{179} = -0.84$$

$$z_2 = \frac{13350 - 13200}{179} = 0.84$$

$$P(13{,}050 \leq \bar{x} \leq 13{,}350) = P(-0.84 \leq z \leq 0.84) = 0.5990$$

c) A value that deviates from the population mean by more than \$200 must either be greater than 13,400 or less than 13,000. The z value corresponding to 13,000 was determined to be -1.12 in part (a), and the area to the left of that value was 0.1314. The z value corresponding to 13,400 must, therefore, be $+1.12$, and the area to the right of that value must be 0.1314 due to the symmetry of the normal distribution. Therefore, the desired probability is $0.1314 + 0.1314 = 0.2628$.

9.7 Student's *t* Distribution

In Section 9.5 we saw that means of samples drawn from a normally distributed population always have a normal distribution. If $\bar{x}$ has a normal distribution, then the z statistic, $z = (\bar{x} - \mu_{\bar{x}})/\sigma_{\bar{x}}$, where $\sigma_{\bar{x}} = \sigma/\sqrt{n}$ also has a normal distribution, since z differs from $\bar{x}$ only by constant factors. However, the transformation from $\bar{x}$ to z requires a known value for σ, the population standard deviation. In most practical problems in statistical inference, σ is not known and has to be estimated by the sample standard deviation defined as $s = \sqrt{\Sigma(x - \bar{x})^2 f/(n - 1)}$ in Equation 9.4. Whenever it is necessary to use s as an estimate of σ, then $\sigma_{\bar{x}}$ must also be estimated by $s_{\bar{x}} = s/\sqrt{n}$. In this case, the ratio $(\bar{x} - \mu)/s_{\bar{x}}$, which is referred to as the t statistic, does not have a normal distribution. The distribution of this statistic is the Student's t distribution. Since values of the t statistic always result from a process of sampling, the t distribution is a sampling distribution. The t statistic is obtained from:

$$t = \frac{\bar{x} - \mu}{s_{\bar{x}}} \tag{9.14}$$

At first glance, the t distribution resembles the standard normal distribution. Both are symmetrical about a mean of zero and extend from $-\infty$ to $+\infty$. The t distribution is, however, more dispersed than the standard normal; that is, the area under the t distribution is usually less concentrated around the mean and more concentrated in the tails than in the normal distribution. Consequently, to find two points, equidistant above and below the mean, that encompass a specified area, we have to move further from the mean on the t distribution than on a normal distribution. A t distribution is compared with the standard normal distribution in Figure 9.1.

The t distribution has a single parameter called the *degrees of freedom*. Whenever the t statistic is computed from a single sample, the degrees of freedom are always one less than the sample size, that is, $n - 1$. As the sample size and, consequently, the degrees of freedom increase, the t distribution approaches the standard normal distribution. For most practical purposes, whenever the degrees of freedom are greater than 30 the two distributions are so similar that the normal distribution is used as an approximation to the t distribution with little significant error.

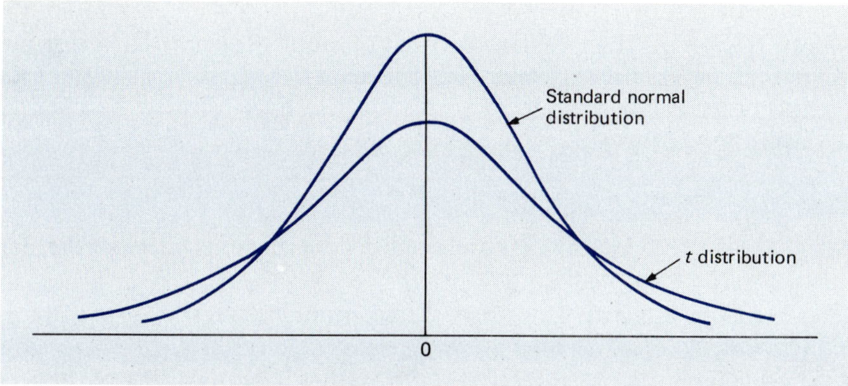

Figure 9.1
The t distribution and the standard normal distribution.

A table of the t distribution is provided in Appendix A.4. Notice that this table provides probabilities that correspond to the areas in the right tail of the distribution. This area is shaded in the diagram at the top of the table. The Greek letter α represents the area in the tail and values corresponding to t are tabulated in the body of the table for $\alpha = .100$, .050, .025, .010, and .005. The parameter *degrees of freedom* appears in the extreme left column of the table. Values are provided for degrees of freedom up to and including 30. The last row in the table shows the values of t when the degrees of freedom reach infinity. If you compare these values of t with the values of z, the standard normal variate, which cut off the same tail areas, you will find that they are identical. The t distribution is illustrated in Figure 9.2.

The t distribution is important in inferential procedures that involve the population mean when samples are drawn from a normal population but σ, the population standard deviation, is unknown. In those cases, σ must be estimated by s, the sample standard deviation, and the sampling distribution of $\bar{x}$ is treated as though it were a t distribution. The use of the t distribution instead

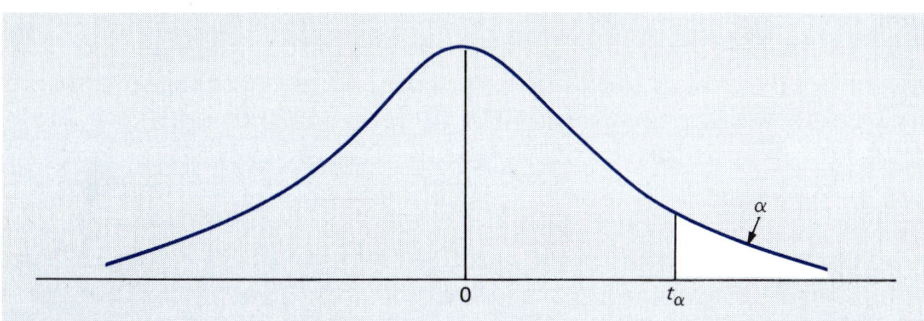

Figure 9.2
Area tabulated in the t distribution.

of the normal distribution is critical whenever the sample size is small. However, the t distribution approaches the standard normal distribution as the sample size increases. When the degrees of freedom are 30 or greater, the normal distribution may be used as an approximation.

Example 9.10

Find the following values of t:

a) The value of t with seven degrees of freedom that cuts off 1% of the area in the right tail of the distribution.

b) The values of t with 12 degrees of freedom that cut off 2.5% of the area in each tail of the distribution.

c) The value of t with four degrees of freedom that cuts off 5% of the area in the left tail of the distribution.

d) The values of t with 20 degrees of freedom that cut off 5% of the area in each tail of the distribution.

a) 2.998

b) $+2.179$ and -2.179

c) -2.132

d) $+1.725$ and -1.725

9.8 The Sampling Distribution of the Proportion and the Sample Number

In Chapter 8 we discussed distributions that result from a process of sampling from a population of attributes, that is, a population of elements which are classified into two mutually exclusive categories generally referred to as success and failure. Such a population is called a Bernoulli population or process. When the population is very large with respect to the sample size and when each observation is independent, the distribution of X is binomial. Since this random variable X is a value determined from a sample, the binomial distribution is actually a sampling distribution. The random variable X is called the SAMPLE NUMBER.

Since we know that the sampling distribution of X is binomial with parameters n and p, and $E(X) = np$ and $\text{Var}(X) = np(1 - p)$, we would like to determine the characteristics of the sampling distribution of $\bar{p}$, the sample proportion where $\bar{p} = X/n$.

Let's consider once more the sampling distribution of the number of heads in four tosses of a fair coin. This distribution is shown in Table 9.5. For every value of X there is a corresponding value of $\bar{p}$. In fact, $\bar{p}$ is simply X multiplied by the constant $1/n$. Therefore, the probability that $X = 1$ success in a sample of four trials is exactly equal to the probability that $\bar{p}$, the proportion of successes, is $\frac{1}{4}$. If the probabilities for X and the corresponding values of $\bar{p}$ are equal for all values, then the two variables must have the same probability distribution.

TABLE 9.5
SAMPLING DISTRIBUTION OF X AND $\bar{p}$ WHEN $n = 4$ AND $p = \frac{1}{2}$

X	$f(X) = f(\bar{p})$	$\bar{p}$
0	1/16	0
1	4/16	1/4
2	6/16	2/4
3	4/16	3/4
4	1/16	4/4

We can now state that the sampling distribution of $\bar{p}$, the sample proportion, is binomial with parameters n and p. However, since $\bar{p}$ and X are not identical random variables, their means and variances are not the same. These values for the two variables are

	X	$\bar{p}$
Mean (expected value)	np	p
Variance	$np(1-p)$	$\dfrac{p(1-p)}{n}$

When sampling from finite populations, such that the sampling fraction $n/N > 0.05$, the finite population correction factor must be used for the variances and standard errors. In this case:

$$\text{Var}(\bar{p}) = \left[\frac{p(1-p)}{n}\right]\left[\frac{N-n}{N-1}\right] \tag{9.15}$$

$$\sigma_{\bar{p}} = \sqrt{\frac{p(1-p)}{n}}\sqrt{\frac{N-n}{N-1}} \tag{9.16}$$

$$\text{Var}(X) = np(1-p)\left[\frac{N-n}{N-1}\right] \tag{9.17}$$

$$\sigma_X = \sqrt{np(1-p)}\sqrt{\frac{N-n}{N-1}} \tag{9.18}$$

Example 9.11

If 40% of all adult males prefer electric shavers to blade razors, and if a random sample of $n = 10$ is drawn from this population:

a) What is the expected value and variance of the proportion in the sample who prefer electric shavers?

b) What is the expected value and variance of the number in the sample who prefer electric shavers?

c) What is the probability that the number in the sample who prefer electric shavers is less than or equal to 2?

d) What is the probability that the proportion in the sample who prefer electric shavers is less than or equal to 0.2?

a) $E(\overline{p}) = p = 0.4$

$$\text{Var}(\overline{p}) = \frac{p(1-p)}{n} = \frac{(.4)(.6)}{10} = 0.024$$

b) $E(X) = np = (10)(.4) = 4$

$$\text{Var}(X) = np(1-p) = (10)(.4)(.6) = 2.4$$

c) $P(x \leq 2 | n = 10, p = .4) = 0.167$, from the binomial tables

d) $P(\overline{p} \leq 0.2 | n = 10, p = .4)$

$= P(x/10 \leq 2/10 | n = 10, p = .4)$

$= P(x \leq 2 | n = 10, p = .4)$

$= 0.167$, from the binomial tables

Definitions

Cluster sampling: A sampling method in which heterogeneous groups of elements are randomly selected from a population of such groups.

Judgment sampling: A sampling method in which the decision to include any population element in the sample is based on human judgment.

Probability sampling: A sampling method in which every element in the population has a known and nonzero probability of being included in the sample.

Random sampling: Same as probability sampling.

Sample number: The number of successes in a sample.

Sample proportion: The proportion of successes in a sample.

Sampling: The process of selecting samples from a population.

Simple random sampling: A sampling method which provides that every possible combination of n elements in the population has the same probability of being selected. A method of sampling so that on each selection every member of the remaining population has an equal probability of being selected.

Statistic: A value computed from a sample.

Stratified random sampling: A sampling method in which the population is divided into homogeneous subsets called strata, from which simple random samples are drawn and then combined into the overall sample.

Systematic sampling: A sampling method in which every kth element of the population is selected after the starting point is selected randomly.

Formulas

Sample mean:

(9.1)
$$\bar{x} = \frac{\sum_{i=1}^{n} x_i}{n}$$

(9.2)
$$\bar{x} = \frac{\sum_{i=1}^{k} x_i f_i}{n}$$

Sample variance (definition):

(9.3)
$$s^2 = \frac{\sum_{i=1}^{k} (x_i - \bar{x})^2 f_i}{n - 1}$$

Sample standard deviation (definition):

(9.4)
$$s = \sqrt{\frac{\sum_{i=1}^{k} (x_i - \bar{x})^2 f_i}{n - 1}}$$

Sample variance (computational):

(9.5)
$$s^2 = \frac{\sum_{i=1}^{k} x_i^2 f_i}{n - 1} - \frac{\left(\sum_{i=1}^{k} x_i f_i\right)^2}{n(n - 1)}$$

Sample standard deviation (computational):

(9.6)
$$s = \sqrt{\frac{\sum_{i=1}^{k} x_i^2 f_i}{n - 1} - \frac{\left(\sum_{i=1}^{k} x_i f_i\right)^2}{n(n - 1)}}$$

Sample variance (step deviation method):

(9.7)
$$s^2 = c^2 \left[\frac{\sum_{i=1}^{k} u_i^2 f_i}{n - 1} - \frac{\left(\sum_{i=1}^{k} u_i f_i\right)^2}{n(n - 1)} \right]$$

Sample standard deviation (step deviation method):

(9.8)
$$s = c\sqrt{\frac{\sum_{i=1}^{k} u_i^2 f_i}{n - 1} - \frac{\left(\sum_{i=1}^{k} u_i f_i\right)^2}{n(n - 1)}}$$

Sample number:

(9.9)
$$X = \sum_{i=1}^{n} x_i \qquad x = 0,1$$

Sample proportion:

$$\bar{p} = \frac{X}{n} \tag{9.10}$$

Variance of the sample mean:

$$\sigma_{\bar{x}}^2 = \frac{\sigma^2}{n}\left[\frac{N-n}{N-1}\right] \tag{9.11}$$

Standard error of the mean:

$$\sigma_{\bar{x}} = \frac{\sigma}{\sqrt{n}}\sqrt{\frac{N-n}{N-1}} \tag{9.12}$$

z transformation:

$$z = \frac{\bar{x} - \mu_{\bar{x}}}{\sigma_{\bar{x}}} \tag{9.13}$$

t transformation:

$$t = \frac{\bar{x} - \mu}{s_{\bar{x}}} \qquad \text{where } s_{\bar{x}} = \frac{s}{\sqrt{n}} \tag{9.14}$$

Variance of the sample proportion:

$$\text{Var}(\bar{p}) = \left[\frac{p(1-p)}{n}\right]\left[\frac{N-n}{N-1}\right] \tag{9.15}$$

Standard error of the proportion:

$$\sigma_{\bar{p}} = \sqrt{\frac{p(1-p)}{n}}\sqrt{\frac{N-n}{N-1}} \tag{9.16}$$

Variance of the sample number:

$$\text{Var}(X) = np(1-p)\left[\frac{N-n}{N-1}\right] \tag{9.17}$$

Standard error of the sample number:

$$\sigma_X = \sqrt{np(1-p)}\sqrt{\frac{N-n}{N-1}} \tag{9.18}$$

Problems

9.1 Specify and explain the factors that make sampling preferable to a complete census in a statistical investigation.

9.2 Which is more important in determining the size of a sample to assure that it is representative of the population, the population size or the population variation? Explain your answer.

9.3 Given the following frequency distribution:

a) Find the variance and standard deviation if the data are considered to be a population.

b) Find the variance and standard deviation if the data are a sample.

Class	*Frequency*
101 to 120	16
121 to 140	10
141 to 160	5
161 to 180	10
181 to 200	9

9.4 Using the following distribution of parts produced by employees of the Advance Machine Company, find the variance and standard deviation of the number of parts produced if the data are a sample.

Parts Produced	*Class Mark*	*Frequency (f)*
66 to 68	67	2
69 to 71	70	7
72 to 74	73	12
75 to 77	76	20
78 to 80	79	6
81 to 83	82	3
		50

9.5 Using the following distribution of hours worked by production employees of the Advance Machine Company, find the variance and standard deviation of the number of hours worked if the data are a sample.

Hours Worked	*f*
36.0 and under 37.8	6
37.8 and under 39.6	7
39.6 and under 41.4	24
41.4 and under 43.2	7
43.2 and under 45.0	2
45.0 and under 46.8	4
	50

9.6 A market research firm has been retained by several clients to conduct sample surveys. For the clients listed below, state whether the method described for selecting the sample will result in a random sample. Why or why not?

a) To obtain a random sample of customers of the Emporium Department Store, it was proposed that a numbered list be made of all charge account holders and the sample be selected from this list by using a table of random numbers.

b) To obtain a random sample of registered voters in Middletown for a citizens group, it was proposed that a list of all registered voters in the city be obtained and the sample selected by taking every

50th name on the list after the starting point was selected randomly.

9.7 Answer the question asked in problem 9.6 for the following situations.

a) To obtain a random sample of customers of the Emporium Department Store, interviewers are posted at all entrances to the store on three successive Wednesday afternoons. Each interviewer selects typical customers as they leave the store with packages.

b) To obtain a random sample of the adult residents of Middletown, the Middletown telephone directory is used as the sampling frame. Pages in the directory and names on the pages are selected using a table of random numbers.

9.8 If a random sample of $n = 400$ is selected from an infinite population having a standard deviation of 100, find $\sigma_{\bar{x}}$.

9.9 If the previous sample is taken from a population of $N = 2000$ with a standard deviation of 100, find $\sigma_{\bar{x}}$.

9.10 If a random sample of $n = 100$ is selected from a population of $N = 5000$ having a mean of 75 and a variance of 900, find the expected value of the sample mean and the variance of the sample mean.

9.11 If the random variable y has a normal distribution with a mean of 50 and a standard deviation of 15, find the probability that the mean of a sample of $n = 9$ drawn from the population has a value less than 42.

9.12 If the random variable x has a mean of 75 and a standard deviation of 21, find the probability that the mean of a random sample of $n = 36$ drawn from that population has a value greater than 76.

9.13 For the data in problem 9.11, find the probability that $\bar{y}$ is within six units of the population mean.

9.14 For the data in problem 9.12, find the probability that $\bar{x}$ deviates from the population mean by more than eight units.

9.15 If a random sample of $n = 25$ is drawn from the population described in problem 9.11, find a value of $\bar{y}$ such that:

a) the probability of getting a value that large or larger is 0.80.

b) the probability of getting a value that large or larger is 0.10.

9.16 If a random sample of $n = 15$ is drawn from an attributes population for which p, the proportion of successes in the population, is 0.40, find the exact probability that:

a) X, the number of successes in the sample, is 5 or less.

b) $\bar{p}$, the proportion of successes in the sample, is $\frac{1}{3}$ or less.

9.17 If a random sample of $n = 8$ is drawn from an attributes population for which p, the proportion of successes in the population, is 0.30, find the exact probability that:

a) the proportion of successes in the sample is exactly 0.25.

b) the proportion of successes in the sample is greater than 0.125.

c) the number of successes in the sample is greater than 3.

9.18 For the data in problem 9.16, find the expected value of $\bar{p}$, the standard error of $\bar{p}$, and the standard error of X, the number of successes in the sample.

9.19 For the data in problem 9.17, find $E(\bar{p})$, $\sigma_{\bar{p}}$, and σ_X.

9.20 If, in problem 9.16, the size of the population is $N = 200$, find $\sigma_{\bar{p}}$ and σ_X.

9.21 If, in problem 9.17, the size of the population is $N = 100$, find $\sigma_{\bar{p}}$ and σ_X.

9.22 If a random sample of $n = 200$ is drawn from an attributes population in which p, the proportion of successes in the population, is 0.20, use the normal approximation to find the probability that the proportion of successes in the sample is 0.15 or less.

9.23 A market research company has determined that 1% of all survey questionnaires sent out by mail are eventually returned. If, for a particular survey, a total of 5000 questionnaires are mailed, find the probability that:

a) 45 or less will be returned.

b) more than 1.25% will be returned.

c) between 1.10% and 1.25% will be returned.

9.24 In Example 9.7, the random variable y represents the diameters of a population of 1000 parts having a mean of 2.5 centimeters and a standard deviation of 0.2 centimeters. If a random sample of 36 parts is drawn from that population:

a) What is the probability that the sample mean is greater than 2.541 centimeters?

b) What is the probability that the sample mean is between 2.438 and 2.529 centimeters?

9.25 Last year, an audit of all charge accounts issued by the Southwest Oil Company disclosed that 4% were delinquent. This year, a random sample of 500 of these accounts will be taken. If the percentage of delinquent accounts has not changed since last year, what is the probability that there will be 14 or fewer delinquent accounts in the sample?

9.26 For the data of problem 9.25, what is the probability that the proportion of delinquent accounts in the sample is greater than 5%?

9.27 Suppose that in problem 9.25 the sample size is $n = 100$. What is the probability that the sample contains three or fewer delinquent accounts? Use the Poisson approximation.

9.28 The average outstanding balance of all of Southwest Oil Company's charge accounts is \$47.20, with a standard deviation of \$8.40. If these figures remain unchanged and a random sample of 225 of these accounts is selected, what is the probability that the sample mean is greater than \$48.50?

9.29 In problem 9.28, if the sample size is 100 accounts, what is the probability that the sample mean is greater than \$46.00?

9.30 If a single account is selected at random from the population described in problem 9.28, what is the probability that the outstanding balance on that account is \$57.00 or greater?

CHAPTER 10 ESTIMATION

10.1
The estimation process

10.2
Estimators and their properties

10.3
Point and interval estimates

10.4
One- and two-sided confidence intervals

10.5
Estimation of means

10.6
Estimation of proportions

10.7
Estimating variation

10.8
Estimating sample size

Statistical inference is a process that uses information obtained from samples to draw conclusions about the characteristics of an unknown population. One of the important inferential procedures is *estimation*. Most people associate the word "estimate" with the idea of an opinion or guess. This is not the meaning of the word in statistics. A statistical estimate is based on empirical evidence according to a set of well-established procedures. Even though an estimate based on incomplete information can never be free from error, there is still a great deal of difference between a statistical estimate and a guess or opinion. Since error is inherent in any estimate, it is usual to express all statistical estimates so that the likely error in the estimate is immediately evident.

Estimation is used frequently in all areas of business as a basis for decision making. Estimates are used to determine costs of production, the demand for a new product, the time necessary to complete a project, the effect of an advertising campaign on sales, and the size of the television viewing audience during some specified time period. In fact, most business decisions must be based on estimates simply because time and cost constraints usually make it impractical to obtain information about an entire population.

Of course, a business decision-maker likes to have complete information, if possible. The more information available for making a decision, the better that decision is likely to be. However, information is costly and a complete census is usually either impossible, impractical, or inefficient. These conditions make estimation a necessity.

10.1 The Estimation Process

Estimates are inferences about some characteristic or parameter of a population. Generally, the parameters that must be estimated most frequently are means, proportions, and standard deviations. Consider the following examples of situations where estimation is necessary for business decision making.

Example 10.1 Cinderella Clothes, Inc. is a company which operates a chain of children's clothing stores. The stores are generally located in suburban shopping centers. A decision must be made by the executive vice-president of the company regarding the opening of a store in a new shopping center now under construction. In the past, decisions of this type have been based on the average income of families living in the marketing area served by the shopping center. Information about the income level of families in the area is also useful in establishing pricing and credit policies and in determining the quality of the merchandise to be handled in the store. Cinderella Clothes contracts with a market research company to conduct a survey to estimate the mean income of the population.

Example 10.2

The Majestic Hotel, located in a large midwestern city, has 500 guest rooms. Because many travelers who make hotel reservations subsequently cancel them or fail to show up, it has been the practice of the hotel management to overbook. This means that more reservations are accepted and confirmed than there are rooms in the hotel. Overbooking has always been a matter of guesswork and on numerous occasions has led to serious altercations in the lobby with reservation-holding guests who cannot be accommodated. On other occasions, large blocks of rooms have been empty because the number of cancellations and no-shows have been underestimated. The hotel manager decides to use his knowledge of statistics to estimate the proportion of no-shows and cancellations and establish a standard reservations policy for the hotel.

Example 10.3

The Merit Manufacturing Company produces precision computer components according to specifications provided by their customers. A large order is received by the sales department for a particular component similar to other parts manufactured by the company. However, the specifications for this particular component are extremely precise. The tolerances, that is, the degree to which specified dimensions are allowed to vary, are very narrow. The production manager of the company is not sure if his machines and machinists are capable of producing the part to the required specifications. He decides to make a sample production run of these parts and use the data from the sample to estimate the variation of the production process. This is called a capability study. On the basis of this estimate of process variation, he will then make a decision as to whether or not to accept the order.

Example 10.4

The advertising manager of the Summit Record Company plans to advertise a new album on local television. A salesperson from UHF Channel 32 has recently called to try to convince the manager that Channel 32 is the place to advertise. The advertising manager is not convinced, however, and decides to conduct a survey for estimating the proportion of television sets in the local viewing area that are tuned to Channel 32 between the hours of 11 and 12 P.M.

Example 10.5

The president of The Emporium, a department store in the downtown area of a medium-sized eastern city, is disturbed at the losses incurred by the store due to shoplifting. The president believes that The Emporium can cope with the problem if it can be determined if losses are abnormally large or are consistent with those experienced by other department stores similar in size, type, and location. An administrative assistant is asked to look into the problem of estimating the average loss of this type for department stores similar to The Emporium.

All of these examples illustrate the need for estimation in business decision-making situations. A general procedure for making statistical estimates follows.

1. Define the estimation problem by specifying the parameter to be estimated and the population of interest.

2. Select the statistic that will be used to estimate the unknown parameter. This statistic is called an ESTIMATOR.

3. Draw a random sample from the population of interest.

4. Compute the value of the estimator from the sample data. The computed value of the estimator is called the estimate, or the POINT ESTIMATE.

5. Use the point estimate as a basis for constructing the INTERVAL ESTIMATE, which includes a statement about the size and likelihood of error.

Note that an estimator is a value computed from a sample for the purpose of providing an estimate of the unknown population parameter. An estimator is, therefore, a statistic as defined previously.

The specifics of the estimation procedure as applied to the previous examples are demonstrated in this chapter.

10.2 Estimators and Their Properties

An estimator is a statistic. For any specified parameter to be estimated, there may be more than one statistic that can be used as an estimator. The problem then arises as to which of the possible statistics makes the best estimator. For example, various measures of central tendency such as the sample mean, median, and midrange might be considered as possible estimators of the population mean. Some function of the sample range, as well as the sample standard deviation, might be used to estimate the population standard deviation. When faced with the problem of alternative estimators, it is necessary to have some criteria for selecting the best one. There are essentially four characteristics that a good estimator should have. They are unbiasedness, consistency, efficiency, and sufficiency. In comparing estimators, it is quite likely that one may be superior with respect to one or two of these characteristics, while another is superior with respect to the others. Consequently, it is not possible to provide a simple rule for selecting a best estimator. The selection of one estimator over another is frequently a matter of judgment. A brief, nonmathematical explanation of the desirable properties of estimators follows.

Unbiasedness. Suppose a statistic, v, is used for estimating some parameter θ. The statistic v has a sampling distribution. If the expected value or mean of the sampling distribution of v is equal to θ, v is said to be an unbiased estimator of θ. Stated symbolically, v is an unbiased estimator of θ if

$$E(v) = \theta$$

Sometimes, even though an estimator is biased, it is possible to identify the bias and remove it. For example, the sample variance, defined as

$$S^2 = \frac{\Sigma(x - \bar{x})^2}{n}$$

is a biased estimator of σ^2 because $E(S^2) \neq \sigma^2$. However, if this estimator is modified by replacing the n in the denominator with $n - 1$, we obtain

$$s^2 = \frac{\Sigma(x - \bar{x})^2}{n - 1}$$

which is the commonly used unbiased estimator of the population variance.

Consistency. If we define the error in an estimate as the absolute difference between the value of the estimator and that of the parameter being estimated, then

$$\text{Error} = |v - \theta|$$

It is desirable to make this error as small as possible. One procedure that should lead to a reduction in the size of the estimation error is to increase the size of the sample from which the estimate is made. A good estimator should produce estimates that get closer to the value of the parameter as the sample gets larger. An estimator that has this property is a *consistent estimator*. More precisely, v is a consistent estimator of θ if the probability that the estimation error becomes less than some arbitrarily chosen small number approaches 1, as n approaches infinity. This can be stated symbolically as

$$P[(|v - \theta|) < \epsilon] \rightarrow 1 \text{ as } n \rightarrow \infty$$

The arrow is read as "approaches" and the Greek letter epsilon, ϵ, represents a very small number.

Efficiency. We compare the efficiency of two estimators by comparing the variances of their sampling distributions. The more efficient estimator is the one that most frequently results in estimates which are close to the parameter being estimated. Since the variance of the sampling distribution of a statistic measures the amount of variation about the mean or expected value of the distribution, the estimator with the smaller variance is the more efficient.

Suppose we consider two different estimators of some parameter θ. If these two estimators, call them v_1 and v_2, are both unbiased,

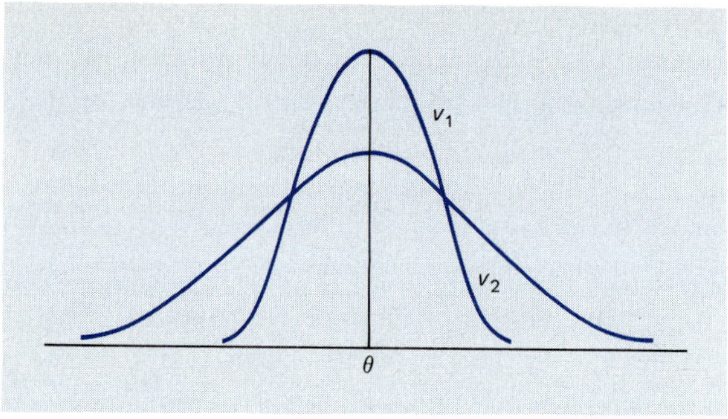

Figure 10.1
Sampling distributions of two estimators of θ.

their sampling distributions might look like the distributions of Figure 10.1. The sampling distribution of v_2 varies more widely than the distribution of v_1. We can obtain a numerical value for the relative efficiency of v_1 with respect to v_2 by finding the value of the ratio $\text{Var}(v_2)/\text{Var}(v_1)$. Relative efficiency (RE) is defined as

$$\text{RE}\left(\frac{v_1}{v_2}\right) = \frac{\text{Var}(v_2)}{\text{Var}(v_1)} \tag{10.1}$$

Example 10.6 Suppose the variances of two unbiased estimators of θ are

$$\text{Var}(v_1) = 27.30$$

and

$$\text{Var}(v_2) = 40.95$$

Find the relative efficiency of v_1 with respect to v_2.

$$\text{RE}\left(\frac{v_1}{v_2}\right) = \frac{\text{Var}(v_2)}{\text{Var}(v_1)} \tag{10.1}$$

$$\text{RE}\left(\frac{v_1}{v_2}\right) = \frac{40.95}{27.30}$$

$$\text{RE}\left(\frac{v_1}{v_2}\right) = 1.50$$

This result indicates that the estimator v_1 is $1\frac{1}{2}$ times as efficient as v_2 as an estimator of θ.

Example 10.7 One of the functions of the quality control department of a large manufacturer is to take periodic samples of the output of a particular production process to estimate the mean. Since the population of output is approximately normally distributed and the samples are large, the sampling distributions of both the sample mean and the sample median are also approximately normally distributed. Both statistics are unbiased estimators of the population mean. Since it requires less calculation to obtain the median from a set of data than it does to obtain the mean, an analyst suggests that the median be used as the estimator of the population mean. The quality control manager asks the analyst to determine the relative efficiency of the sample mean as an estimator with respect to the sample median. The variance of the sample mean is $\sigma_{\bar{x}}^2 = \sigma^2/n$, while the variance of the sample median is $\sigma_{md}^2 = 1.57\,\sigma^2/n$. The relative efficiency of the sample mean with respect to the sample median is, therefore,

$$\begin{aligned}\text{RE}\left(\frac{\text{mean}}{\text{median}}\right) &= \frac{\text{Var(median)}}{\text{Var(mean)}}\\ &= \frac{1.57\,\sigma^2/n}{\sigma^2/n}\\ &= 1.57\end{aligned}$$

Since the mean is 1.57 times as efficient as the median in estimating the population mean, a decision is made to continue using the sample mean as the estimator.

Sufficiency. The property of sufficiency is a difficult concept to define nonmathematically. Stated as simply as possible, a sufficient estimator is one that uses all the information about the population parameter contained in the sample. If the statistic v is a function of the sample values, then v is a sufficient estimator of θ if, having determined the value of v, no additional information about θ can be obtained from the sample.

For example, the sample mean, $\bar{x}$, is a sufficient estimator of the population mean since all the information in the sample is used in its computation. On the other hand, the sample midrange is not a sufficient estimator. Since the midrange is computed by averaging only the highest and lowest values in the sample, a great deal more information about the population mean can be obtained from the other $n - 2$ values in the sample.

10.3 Point and Interval Estimates

Whenever the value of an estimator is computed from the sample data, a single number is the result. This number is called the estimate. More precisely, however, it is a POINT ESTIMATE of the unknown parameter, because it represents a single point on some scale of values. If we assume that the estimator has the desirable properties discussed in the previous section, and that the random sample from which it was obtained is representative of the population, then the point estimate should be a value that is close to the true parameter value. However, it is very unlikely that any point estimate will be exactly equal to the unknown parameter value. If additional samples of the same size are drawn from the population in exactly the same way, the value of the estimator will vary from sample to sample. It is this variation among samples that results in sampling distributions and is the cause of error in estimation. Consequently, if a single sample is taken and the value of the point estimator is computed, we can be sure that some error is present in that estimate. However, nothing in the point estimate itself indicates how large that error is likely to be.

If the sampling error is defined as the difference between the point estimate and the value of the parameter, then the amount of error in an estimate can be described in terms of an interval of values about the point estimate. The likelihood that the error will be as large as or larger than the stated interval is expressed in terms of probability. The probability that a specified interval contains the unknown population parameter is called CONFIDENCE. The interval itself is called a CONFIDENCE INTERVAL or a CONFIDENCE INTERVAL ESTIMATE.

A confidence interval is generally based on the point estimate and is defined by an upper confidence limit, the UCL, and a lower confidence limit, the LCL. The confidence limits are usually obtained by adding or subtracting a factor, $k\text{SE}(v)$, to the point estimate. $\text{SE}(v)$ stands for the standard error of the estimator v, while k is called the *confidence multiplier*. The value of k depends on the sampling distribution of the estimator and the level of confidence desired.

The term "confidence" is usually interpreted by users of interval estimates as the probability that the unknown parameter lies in the interval. This is not precisely correct, since a parameter is a constant value and probability statements do not apply to constants. In estimation it is the interval that is variable and, therefore, the proper interpretation is that confidence represents the probability that the interval contains the unknown parameter. This is a theoretical point, and for practical purposes it makes little difference which way the user of the estimate interprets the probability statement.

Most interval estimates are constructed at the 95% or the 99% level of confidence, although any other confidence level can be used. The confidence level refers to the area or probability under the curve of the sampling distribution of the estimator, which lies between the upper and lower confidence limits. The value of k, the confidence multiplier, is generally designated in terms of the area or probability under the sampling distribution that lies *outside* of the confidence limits, that is, the tail area or areas.

For the estimation procedures covered in this chapter, only two sampling distributions are considered. These two are the normal distribution and the Student's t distribution. The confidence multipliers will therefore be either values of the standard normal variate, z, or values of the statistic t.

Example 10.8 The Colonial Canning Company uses an automatic machine to fill cans to a specified level. Since automatic machines have an unfortunate tendency to drift out of adjustment, Colonial's quality control department samples cans from its various product lines daily to estimate the average net weight of the cans. On one particular day, a random sample of 100 16-ounce cans of tomato juice is drawn from the day's production and weighed. The sample mean, which is used as the estimator, is computed to be $\bar{x} = 16.15$ ounces. The population is assumed to be normal with a known standard deviation equal to 0.55 ounces. Since any point estimate is subject to error, an interval estimate is to be constructed so that the company may have 99% confidence that the value of the true mean weight of that day's production is included in the interval obtained.

The important information necessary for the construction of this confidence interval is summarized as follows.

Population: normal

Population standard deviation: known, $\sigma = 0.55$ ounces

Estimator: $\bar{x} = 16.15$

Sample size: $n = 100$

Sampling distribution of $\bar{x}$: normal

Standard error of $\bar{x}$: $\sigma_{\bar{x}} = 0.55/\sqrt{100} = 0.055$

Confidence interval: two-sided, 99%

Confidence multiplier: $z_{(1-.99)/2} = 2.58$. Note that 2.58 is that value of the standard normal variate which, when both added to and subtracted from the mean, divides the central 99% of the area from the two tails.

The confidence interval is constructed as follows.

(10.2)
$$\begin{aligned} \text{CL} &= \bar{x} \pm z\,\sigma_{\bar{x}} \\ \text{CL} &= 16.15 \pm (2.58)\ (0.055) \\ \text{CL} &= 16.15 \pm 0.142 \\ \text{LCL} &= 16.01, \quad \text{UCL} = 16.29 \end{aligned}$$

This confidence interval estimate is summarized by the following statement:

The Colonial Canning Company can be 99% confident that the true mean net weight of all 16-ounce cans of tomato juice produced that day lies between 16.01 and 16.29 ounces.

Figure 10.2 illustrates this confidence interval defined on the sampling distribution of $\bar{x}$.

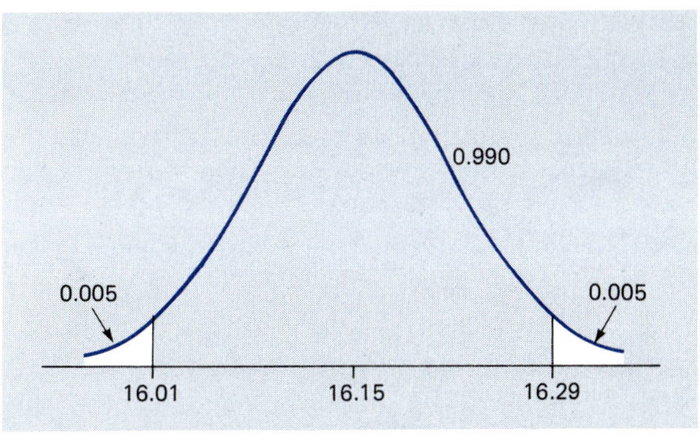

Figure 10.2
A two-sided 99% confidence interval on the mean weight (Example 10.8).

Example 10.9 In the previous example, the population of interest, one day's production, is assumed to be very large compared to the sample of 100 cans. Suppose, however, that the total production of 16-ounce cans of tomato juice for that particular day amounts to only 1000 cans. In this case, the sample consists of more than 5% of the population and, therefore, the use of the finite population correction factor is required. The data summarized in the previous example remains the same except for the following.

Population size: $N = 1000$

Standard error of $\bar{x}$:

(9.12)
$$\begin{aligned}\sigma_{\bar{x}} &= \frac{\sigma}{\sqrt{n}}\sqrt{\frac{N-n}{N-1}}\\ &= \frac{0.55}{\sqrt{100}}\sqrt{\frac{1000-100}{1000-1}}\\ &= \frac{0.55}{10}\sqrt{\frac{900}{999}}\\ &= 0.052\end{aligned}$$

The confidence interval is constructed as follows.

(10.2)
$$\begin{aligned}\text{CL} &= \bar{x} \pm z\sigma_{\bar{x}}\\ &= 16.15 \pm (2.58)(0.052)\\ &= 16.15 \pm 0.134\end{aligned}$$
$$\text{LCL} = 16.02, \qquad \text{UCL} = 16.28$$

10.4 One- and Two-Sided Confidence Intervals

In the previous section, the general procedure for constructing a confidence interval was explained and illustrated. The intervals constructed in those examples were two-sided intervals defined by both an upper and a lower confidence limit. There are many situations, however, in which interest is centered on only one of the confidence limits, either the upper or the lower. For example, the proprietor of a seafood restaurant on the West Coast might want an estimate of the average time it takes to ship fresh Maine lobster from the East Coast to her restaurant in Los Angeles via air freight. She has no interest in the lower confidence limit on this average time. She is primarily concerned with the maximum time, or upper limit. An estimate is needed that states, with 95% or 99% confidence, that the average shipping time is less than some value. In a situation like this, a one-sided confidence interval with a single upper limit is appropriate.

On the other hand, the production manager of the Colonial Canning Company in the previous examples might be primarily concerned that the automatic machine does not underfill the cans. In that case, he would prefer an interval estimate that assures him with some level of confidence that the average net weight is greater than some lower limit. He would then ask for a one-sided confidence interval with a single lower limit.

The procedure for the construction of a one-sided confidence interval is very similar to that of the two-sided interval. The primary difference is that for a two-sided interval a value equal to the confidence multiplier times the standard error is both added to and subtracted from the point estimate. For a one-sided interval, a comparable value is either added to or subtracted from the point estimate, but not both.

Since confidence intervals are defined on the sampling distribution of the estimator, the distance of the confidence limit or limits from the point estimate is determined by the form of the sampling distribution, the level of confidence desired, and the value of the confidence multiplier. If the sampling distribution of the estimator is normal, values of the confidence multiplier correspond to values of z, the standard normal variate. These values for various levels of confidence and for both one- and two-sided intervals are provided in Table 10.1.

TABLE 10.1

VALUES OF CONFIDENCE MULTIPLIERS FOR ONE- AND TWO-SIDED CONFIDENCE INTERVALS WHEN THE SAMPLING DISTRIBUTION IS NORMAL

Confidence Level	*One Sided* z_{1-C}*	*Two Sided* $z_{(1-C)/2}$*
90%	1.28	1.65
92%	1.41	1.75
94%	1.56	1.88
95%	1.65	1.96
96%	1.75	2.06
98%	2.06	2.33
99%	2.33	2.58
99.5%	2.58	2.81
99.9%	3.09	3.29

*The subscript on z refers to the area in one tail of the normal distribution. The tail is that portion of the distribution that lies further than z standard deviations from the mean. For example, when $C = 90$, a one-sided confidence limit divides the distribution so that 10% of the area is in one tail. A two-sided confidence interval divides the distribution so that 5% of the area is in each tail.

The reason for the different values of z for one- and two-sided intervals at the same level of confidence is attributable to the way the area under the curve of the sampling distribution is to be divided. In a two-sided interval, the probability or area associated with the C% level of confidence is generally centered about the point estimate and terminates at a finite distance in both directions as indicated by the two limits. The area or probability associated with $1 - C$ is divided between the two tails of the distribution, generally with $(1 - C)/2$ in each tail. In a one-sided confidence interval, the area associated with $C\%$ confidence extends infinitely far into one of the tails of the distribution. The single limit divides the area so that $C\%$ of the area lies on one side of the single limit and $(1 - C)\%$ lies on the other side. Therefore, the distance of the single limit either above or below the point estimate is not the same as that of the corresponding limit of a two-sided interval. The relationship between one-sided and two-sided confidence limits is illustrated in Figure 10.3. Note that the one-sided limits are closer to the center of the distribution than are either of the two-sided limits.

The equations for one-sided and two-sided confidence intervals on the mean, where σ is known, are given below.

Two-sided:

$$\bar{x} \pm z\,\sigma_{\bar{x}} \tag{10.2}$$

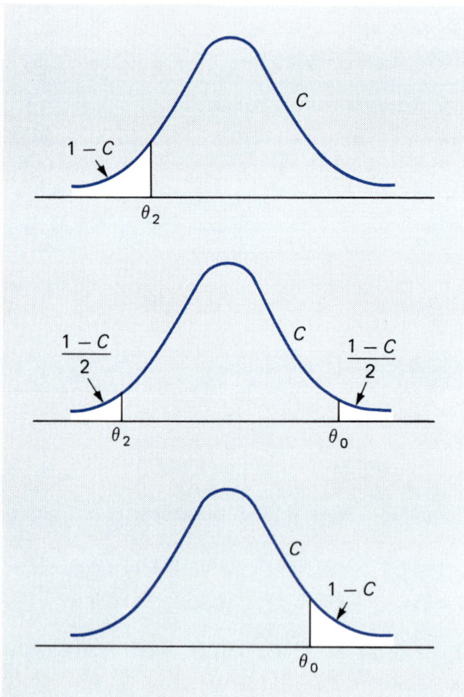

Figure 10.3
Relationship between limits in one-sided and two-sided confidence intervals.

where the value of the z multiplier corresponds to a tail area of $(1 - C)/2$.

One-sided, lower limit:

$$\bar{x} - z\,\sigma_{\bar{x}} \tag{10.3}$$

where the value of the z multiplier corresponds to a tail area of $1 - C$.

One-sided, upper limit:

$$\bar{x} + z\,\sigma_{\bar{x}} \tag{10.4}$$

where the value of the z multiplier corresponds to a tail area of $1 - C$.

Example 10.10

In Example 10.8, a 99% two-sided confidence interval was constructed on the mean net weight of 16-ounce cans of tomato juice. For the same data, it is decided to construct a 99% one-sided confidence interval with a single lower limit. The data for this problem are the same as that summarized in Example 10.8, except for the following.

Confidence multiplier: $z_{1-.99} = 2.33$

The confidence interval is constructed as follows.

$$\begin{aligned} \text{LCL} &= \bar{x} - z\,\sigma_{\bar{x}} \\ &= 16.15 - (2.33)\ (0.055) \\ &= 16.15 - 0.13 \\ &= 16.02 \end{aligned} \tag{10.3}$$

The quality control manager can now state that he is 99% confident that the true mean weight is 16.02 ounces or greater. This one-sided confidence interval is illustrated in Figure 10.4.

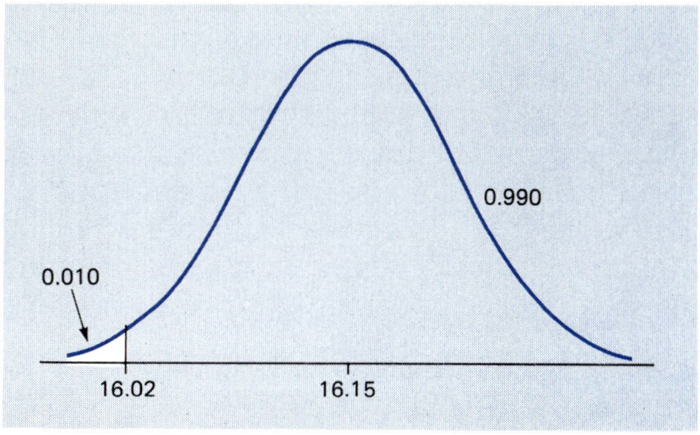

Figure 10.4
A one-sided 99% confidence interval on the mean weight (Example 10.10).

Example 10.11 If the quality control manager in the previous example is only concerned with the possibility of overfilling the cans, he might construct a one-sided confidence interval with a single upper limit. The data is the same as for Example 10.10. The confidence interval is constructed as follows.

$$\text{UCL} = \bar{x} + z\,\sigma_{\bar{x}} \tag{10.4}$$
$$= 16.15 + (2.33)(0.055)$$
$$= 16.15 + 0.13$$
$$= 16.28$$

The quality control manager can state on the basis of this interval that he is 99% confident that the mean weight is 16.28 ounces or less. This interval is illustrated in Figure 10.5.

Figure 10.5
A one-sided 99% confidence interval on the mean weight (Example 10.11).

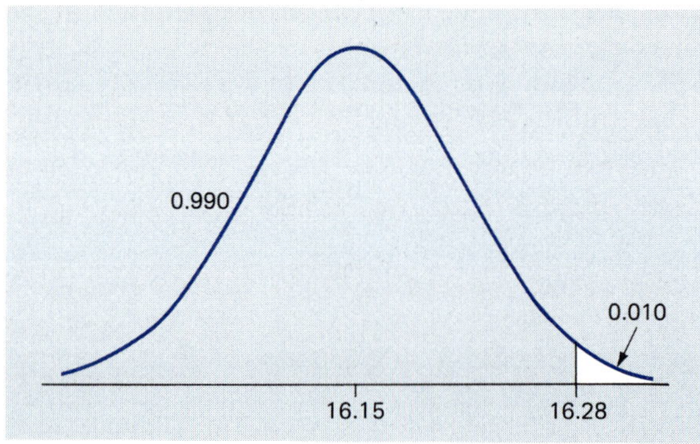

10.5 Estimation of Means

The most common type of problem in estimation involves population means. In fact, all the examples provided in the previous sections concern the estimation of means under conditions in which the population standard deviation is assumed to be known. This is the simplest situation that can be encountered. These cases of unknown means and known standard deviations are not very likely. If the population mean is unknown, it is most likely that the standard deviation is also unknown. In Example 10.8, the assumption of a known standard deviation is based on a great deal of experience and data accumulated on these particular automatic filling machines. These data lead to the conclusion that although the mean of the output is likely to shift over time, the variation, as measured by the standard deviation, remains constant at 0.55 ounces.

When the assumption of a known population standard deviation is valid, the procedure for the estimation of means is exactly that which was followed in the previous sections. However, if the population standard deviation is unknown, the procedure becomes somewhat more complicated. Since the standard error of $\bar{x}$ is dependent upon the standard deviation of the population, some estimate of σ must be used or it will be impossible to construct a confidence interval. It is common practice to use the sample standard deviation defined as

$$s = \sqrt{\frac{\Sigma(x - \bar{x})^2}{n - 1}}$$

as an estimate of σ. Then $\sigma_{\bar{x}}$ is estimated as

$$\hat{\sigma}_{\bar{x}} = s_{\bar{x}} = \frac{s}{\sqrt{n}} \qquad \text{or} \qquad \frac{s}{\sqrt{n}}\sqrt{\frac{N - n}{N - 1}}$$

The "hat" symbol, $\hat{}$, placed over the symbol for a parameter indicates an estimate.

When $s_{\bar{x}}$ is used as an estimate of $\sigma_{\bar{x}}$, the sampling distribution of $\bar{x}$ should no longer be treated as a normal distribution, nor should z, the standard normal variate, always be used as the confidence multiplier. We must now turn to the Student's t distribution and consider using the t variate as the confidence multiplier.

We will now discuss the estimation of means where the population is normally distributed but the population standard deviation is unknown. The estimation procedures fall into two categories, one for large samples and the other for small samples.

Large Samples For most practical purposes, large samples are considered to be samples that consist of more than thirty observations. For samples of this size, the distribution of the sample mean is approximately normal even if the population distribution is not precisely normal. Also, even though precise confidence intervals should be based on the t variate as the confidence multiplier when σ is unknown, for samples larger than 30 little loss in accuracy results if the interval is constructed using z as the multiplier. Under these circumstances, that is, $n > 30$, $\sigma_{\bar{x}}$ should be estimated with $s_{\bar{x}}$ and the confidence interval constructed in the same manner as it is when σ is known.

The equations for one- and two-sided confidence intervals on the mean, where σ is unknown, are given below.

	Large Samples: $n > 30$	Small Samples: $n \leq 30$
Two-sided:		
(10.5)	$\bar{x} \pm zs_{\bar{x}}$	$\bar{x} \pm ts_{\bar{x}}$

where the value of the multiplier corresponds to a tail area of $(1 - C)/2$.

One-sided lower limit:

(10.6)	$\bar{x} - zs_{\bar{x}}$	$\bar{x} - ts_{\bar{x}}$

where the value of the multiplier corresponds to a tail area of $1 - C$.

One-sided upper limit:

(10.7) $\bar{x} + zs_{\bar{x}}$ $\bar{x} + ts_{\bar{x}}$

where the value of the multiplier corresponds to a tail area of $1 - C$.

Example 10.12 In Example 10.1 Cinderella Clothes hired a marketing research company to estimate the average income of families living in the marketing area served by a shopping center under construction. No previous estimate of the population standard deviation was available, and therefore an estimation procedure for the mean had to accomodate the fact that the population standard deviation was unknown. The distribution of the population of incomes can, however, be assumed to be normal, although for large samples this is not a necessary condition. A 95%, two-sided confidence interval estimate is desired.

The marketing research company selected a random sample of 225 families residing in the area. The estimator of the population mean, $\bar{x}$, was computed from the sample to be \$14,250. The sample standard deviation, s, was \$985. The information about the problem is summarized as follows

Population: normal

Population standard deviation: unknown

Sample size: $n = 225$

Estimator: $\bar{x} = \$14,250$

Sample standard deviation: $s = \$985$

Sampling distribution of $\bar{x}$: normal

Standard error of $\bar{x}$: estimated by $s_{\bar{x}} = \dfrac{s}{\sqrt{n}} = \dfrac{\$985}{\sqrt{225}} = \$65.67$

Confidence interval: two-sided, 95%

Confidence multiplier: $z_{(1-.95)/2} = 1.96$

Although the sampling distribution of $\bar{x}$ is always normal when samples are taken from a normally distributed population, precise confidence intervals are constructed by treating the sampling distribution as a Student's t distribution and using the t variate as the confidence multiplier. However, for samples larger than 30, the z variate is used as the multiplier with little loss in accuracy. The confidence interval on mean income is then constructed as follows.

(10.5)
$$\begin{aligned} CL &= \bar{x} \pm zs_{\bar{x}} \\ &= \$14,250 \pm (1.96)\ (\$65.67) \\ &= \$14,250 \pm \$128.71 \\ LCL &= \$14,121, \qquad UCL = \$14,379 \end{aligned}$$

The executive vice-president of Cinderella Clothes can now be 95% confident that the average income of families in the marketing area is between \$14,121 and \$14,379.

Example 10.13

If the executive vice-president in the previous example is interested only in a lower 95% confidence limit on the mean income, a one-sided confidence interval may be constructed. The data is the same as that summarized in Example 10.12, except for the following.

Confidence interval: one-sided, 95%

Confidence multiplier: $z_{1-.95} = 1.65$

The confidence interval is constructed as follows.

(10.6)
$$\begin{aligned} \text{LCL} &= \bar{x} - zs_{\bar{x}} \\ &= \$14{,}250 - (1.65)\ (\$65.67) \\ &= \$14{,}250 - 108.36 \\ &= \$14{,}142 \end{aligned}$$

On the basis of this confidence interval, the executive vice-president can now be 95% confident that the average income of families in the area is greater than or equal to $14,142.

Example 10.14

In Example 10.5, the president of the Emporium wants an estimate of the average loss incurred due to shoplifting based on losses experienced by similar stores. The administrative assistant to the president decides that a 98% two-sided confidence interval estimate of the mean dollar value of such losses would be useful. Referring to various directories, he selects a random sample of 50 stores from a total population of 400 stores similar to The Emporium. Each of the 50 stores is contacted and asked to supply information regarding their losses during the previous year. From the data obtained, he computes the average loss, $\bar{x} = \$58{,}742$, and the sample standard deviation, $s = \$2170$. Since the ratio $n/N = 50/400 = 0.125$, which is greater than 0.05, he realizes that the finite population correction factor has to be used in computing $s_{\bar{x}}$. The information about this problem is summarized as follows.

Population: unknown

Population standard deviation: unknown

Population size: $N = 400$

Sample size: $n = 50$

Estimator: $\bar{x} = \$58{,}742$

Sample standard deviation: $s = \$2170$

Sampling distribution of $\bar{x}$: approximately normal by Central Limit Theorem

Standard error of $\bar{x}$: estimated by

$$\begin{aligned} s_{\bar{x}} &= \frac{s}{\sqrt{n}}\sqrt{\frac{N-n}{N-1}} \\ &= \frac{\$2170}{\sqrt{50}}\sqrt{\frac{400-50}{400-1}} = \$287.42 \end{aligned}$$

Confidence interval: two-sided, 98%

Confidence multipliers: $z_{(1-.98)/2} = 2.33$

The confidence interval is constructed as follows.

$$\text{(10.5)} \qquad \begin{aligned} CL &= \bar{x} \pm z s_{\bar{x}} \\ &= \$58{,}742 \pm (2.33)\ (\$287.42) \\ &= \$58{,}742 \pm 669.69 \\ LCL &= \$58{,}072 \qquad UCL = \$59{,}412 \end{aligned}$$

The president of the Emporium can now be 98% confident that the average loss due to shoplifting for stores like The Emporium lies between \$58,072 and \$59,412.

Small Samples The distinction between large and small samples for estimating means becomes important when the population standard deviation is unknown and must be estimated with the sample standard deviation. In order for these small sample techniques to be used, the population must be normally distributed or approximately so. In sampling from a normal population, the sampling distribution of $\bar{x}$ is also normal. However, when the sample size is less than or equal to 30 and the population standard deviation is unknown, the sampling distribution of $\bar{x}$ must be treated as a Student's t distribution for constructing confidence intervals. The t variate is used as the confidence multiplier in these cases.

In all cases of sampling from a normal population when σ is unknown, the statistic

$$\frac{\bar{x} - \mu}{s_{\bar{x}}}$$

has a Student's t distribution. Therefore, precise confidence limits on the mean require the use of the t multiplier regardless of the size of the sample used. The multiplier, z, is used in the large sample cases simply because it is more convenient and the error introduced is small enough to be inconsequential.

Example 10.15 If the sample size in Example 10.12 had been $n = 25$ instead of $n = 225$, the 95% two-sided confidence interval would be computed using the t multiplier. The data are the same except for the following.

Sample size: $n = 25$

Standard error of $\bar{x}$: estimated by

$$s_{\bar{x}} = \frac{s}{\sqrt{n}} = \frac{\$985}{\sqrt{25}} = \$197.00$$

Confidence multiplier: $t_{(1-.95)/2} = 2.064$

The confidence interval is constructed as follows.

(10.5)
$$\begin{aligned} CL &= \bar{x} \pm ts_{\bar{x}} \\ &= \$14{,}250 \pm (2.064)\ (\$197.00) \\ &= \$14{,}250 \pm \$406.61 \\ LCL &= \$13{,}843 \qquad UCL = \$14{,}657 \end{aligned}$$

Example 10.16

If a 95% one-sided confidence interval is required, the data are the same as for Example 10.15, except for the following.

Confidence interval: one-sided, 95%

Confidence multiplier: $t_{1-.95} = 1.711$

The confidence interval is constructed as follows.

(10.6)
$$\begin{aligned} LCL &= \bar{x} - ts_{\bar{x}} \\ &= \$14{,}250 - (1.711)\ (\$197.00) \\ &= \$14{,}250 - \$337.07 \\ &= \$13{,}913 \end{aligned}$$

Example 10.17

Referring to Example 10.14, suppose that although 50 stores are contacted, only 21 respond. Because this represents a small sample and the population standard deviation is unknown, it is necessary to make the assumption that the sample comes from a normally distributed population. The data for this example are the same as in Example 10.14, except for the following.

Population: normal

Sample size: $n = 21$

Standard error of $\bar{x}$: estimated by

$$\begin{aligned} s_{\bar{x}} &= \frac{s}{\sqrt{n}}\sqrt{\frac{N-n}{N-1}} \\ &= \frac{\$2170}{\sqrt{21}}\sqrt{\frac{400-21}{400-1}} = \$461.51 \end{aligned}$$

Confidence multiplier: $t_{(1-.98)/2} = 2.528$

The confidence interval is constructed as follows.

(10.5)
$$\begin{aligned} CL &= \bar{x} \pm ts_{\bar{x}} \\ &= \$58{,}742 \pm (2.528)\ (\$461.51) \\ &= \$58{,}742 \pm \$1166.70 \\ LCL &= \$57{,}575 \qquad UCL = \$59{,}909 \end{aligned}$$

10.6 Estimation of Proportions

When the population characteristic of interest is an attribute, the most usual parameter to be estimated is the population proportion, that is, the proportion of the population that possesses that attribute. If the population can be divided into two classes, those possessing the attribute and those that do not, recall that such a population is called a Bernoulli population. The parameter of a Bernoulli population, that is, the proportion of that population possessing a specified attribute, is generally designated p. Remember that p is also a parameter of the binomial distribution. The point estimator used for estimating p is called $\bar{p}$, and is equal to X/n, where X represents the number of elementary units in the sample that possess the attribute of interest and n is the total number of units in the sample.

The sampling distribution of $\bar{p}$ is binomial with parameters p and n. The mean or expected value of $\bar{p}$ is simply p, the population proportion. The standard error of $\bar{p}$ is

(9.16)
$$\sigma_{\bar{p}} = \sqrt{\frac{p(1-p)}{n}}$$

Since the standard error of $\bar{p}$ is dependent upon p, which is unknown, it is necessary to substitute the sample proportion in the above equation. This provides an estimate of the standard error:

(10.8)
$$\hat{\sigma}_{\bar{p}} = \sqrt{\frac{\bar{p}(1-\bar{p})}{n}}$$

Remember also that when sampling from a finite population in which the sample represents more than 5% of the population, the finite population factor must be used in computing $\hat{\sigma}_{\bar{p}}$. In those cases,

$$\hat{\sigma}_{\bar{p}} = \sqrt{\frac{\bar{p}(1-\bar{p})}{n}}\sqrt{\frac{N-n}{N-1}}$$

Confidence intervals on the population proportion can be obtained by using the binomial distribution. However, it may be impossible to find an interval corresponding to the exact level of confidence desired due to the discreteness of the binomial distribution. If samples are large, it is much more convenient to use the normal distribution as an approximation to the binomial.

Remember that as the sample size gets large, the binomial distribution approaches the normal distribution, which has the same mean and standard deviation. Of course, the size of the sample required for the approximation to be used varies depending on the value of p. The further p deviates from 0.5 in either direction, the larger the sample must be. We have previously provided a rule of thumb which specifies that both np and $n(1-p)$ must be greater than 5 to use this approximation.

The confidence interval for proportions is constructed like the confidence interval for means; that is, the interval is based on the point estimate plus and/or minus the confidence multiplier times the standard error of the estimator. Since only large samples will be used here for estimating proportions, the normal approximation procedure is used and the confidence multiplier is z. Consequently, confidence intervals on p are constructed from one of the following.

Two-sided interval:

$$\bar{p} \pm z_{(1-C)/2}\sigma_{\bar{p}} \tag{10.9}$$

One-sided, lower limit:

$$\bar{p} - z_{(1-C)}\,\sigma_{\bar{p}} \tag{10.10}$$

One-sided, upper limit:

$$\bar{p} + z_{(1-C)}\,\sigma_{\bar{p}} \tag{10.11}$$

Example 10.18

In Example 10.2, the manager of the Majestic Hotel wants to estimate the proportion of confirmed reservations that are never claimed. He obtains a random sample of 300 reservations requested and confirmed during the past year and determines that 75 of these were either cancelled or unclaimed. The point estimate of the population proportion is $\bar{p} = 75/300 = 0.25$. The estimated standard error of $\bar{p}$ is computed from

$$\hat{\sigma}_{\bar{p}} = \sqrt{\frac{\bar{p}(1-\bar{p})}{n}} = \sqrt{\frac{(0.25)\ (0.75)}{300}} = 0.025 \tag{10.8}$$

The information about this problem is summarized as follows.

Population: Bernoulli

Population size: very large

Sample size: $n = 300$

Estimator: $\bar{p} = 0.25$

Sampling distribution of $\bar{p}$: approximately normal for $n = 300$

Standard error of $\bar{p}$: estimated by

$$\hat{\sigma}_{\bar{p}} = \sqrt{\frac{(0.25)\ (0.75)}{300}} = 0.025$$

Confidence interval: two-sided, 99%

Confidence multiplier: $z_{(1-.99)/2} = 2.58$

A 99% two-sided confidence interval on the true proportion is

$$\begin{aligned} \text{CL} &= \bar{p} \pm z\,\sigma_{\bar{p}} \\ &= 0.25 \pm (2.58)\ (0.025) \\ &= 0.25 \pm 0.0645 \\ \text{LCL} &= 0.19 \qquad \text{UCL} = 0.31 \end{aligned} \tag{10.9}$$

On the basis of this sample, the manager can be 99% confident that the true proportion of cancellations and no-shows will be between 0.19 and 0.31. In terms of percentages, this is between 19% and 31%.

Having obtained the upper and lower confidence limits in terms of proportions and percentages, the manager must now relate these to the number of rooms in the hotel and use this information in establishing some policy regarding reservations. Since there are 500 rooms in the hotel, if exactly 500 reservations are confirmed, the manager can be 99% confident that between 95 and 155 of the rooms will not be claimed. What should his policy be on overbooking?

If the hotel accepts reservations for 595 rooms, based on the lower confidence limit, the manager can be almost certain that he will be able to accommodate all guests arriving with reservations. However, since the proportion of no-shows can be anywhere from 19% to 31%, it is still likely that he will have a considerable number of vacant rooms. On the other hand, if he accepts reservations for 655 rooms based on the upper confidence limit, he can be virtually certain that he will have no vacancies. However, he will still have to contend with some angry reservation holders for whom there are no rooms.

The estimation procedure has served to reduce the manager's uncertainty about the number of no-shows, but has not completely solved the problem for him by telling him exactly how many reservations to accept. The manager must decide which is more important, having no vacancies, in which case he would base his reservations policy on the upper confidence limit, or accomodating all reservation holders, in which case he would base his policy on the lower confidence limit. The manager concludes that the ill will that results from turning reservation holders away is more serious than having some vacancies in the hotel. He decides to establish a policy that provides assurance that he can accommodate all reservation holders. This means that a lower confidence limit is more important than an upper limit. Consequently, the manager should have constructed a one-sided interval with a single lower limit rather than the two-sided interval as in this example. A lower limit for a one-sided confidence interval is illustrated in the following example.

Example 10.19 The manager of the Majestic Hotel wants to establish a policy for overbooking that provides, with virtual certainty, that all holders of confirmed reservations will be accomodated. Therefore, he wants to base his reservations policy on the lower limit of a one-sided confidence interval. However, he feels that 99% confidence is not high enough. Since 100% confidence or absolute certainty is impossible to accomplish using samples, the manager decides to use 99.9% confidence. Therefore, his reservations policy is to be based on the lower limit of a 99.9% one-sided confidence interval on p.

The data are the same as in Example 10.18, except for the following.

Confidence interval: one-sided, 99.9%

Confidence multiplier: $z_{1-.999} = 3.09$

The lower confidence limit on p is computed as follows.

(10.10)
$$\begin{aligned}\text{LCL} &= \bar{p} - z\hat{\sigma}_{\bar{p}} \\ &= 0.25 - (3.09)\ (0.025) \\ &= 0.25 - 0.07725 \\ &= 0.17\end{aligned}$$

On the basis of this estimate, the manager can be almost certain that no less than 17% of the confirmed reservation holders will fail to claim their rooms. On the basis of 500 rooms, this means that 585 reservations can be made and confirmed with practically no risk of turning away holders of confirmed reservations.

Example 10.20

In Example 10.4, the Summit Record Company is considering placing spot advertisements for their new record album on local Channel 32 during a late-night talk show. Before making a decision, however, they decide to obtain a 98% two-sided confidence interval estimate of the proportion of television sets in the viewing area that tune to Channel 32 between the hours of 11 and 12 P.M. A random sample of 1500 families with television sets is taken and it is determined that 240 tune to Channel 32. Therefore, the point estimate of the population proportion is 240/1500 = 0.16. It is determined from other sources that there is a total of 25,000 families with television sets in the viewing area. The ratio of the families sampled to the total population is, therefore, 1500/25,000 = 0.06. The data for this problem are summarized as follows.

Population: Bernoulli

Population size: $N = 25{,}000$

Sample size: $n = 1500$

Sample to population ratio: $n/N = 0.06$

Estimator: $\bar{p} = 0.16$

Sampling distribution of $\bar{p}$: Approximately normal for $n = 1500$

Standard error of $\bar{p}$: estimated by

(10.8)
$$\begin{aligned}\hat{\sigma}_{\bar{p}} &= \sqrt{\frac{(0.16)(0.84)}{1500}}\sqrt{\frac{25{,}000 - 1500}{25{,}000 - 1}} \\ &= 0.00918\end{aligned}$$

Confidence interval: two-sided, 98%

Confidence multiplier: $z_{(1-.98)/2} = 2.33$

The confidence interval on p is computed as follows.

(10.9)
$$\begin{aligned}\text{CL} &= \bar{p} \pm z\hat{\sigma}_{\bar{p}} \\ &= 0.16 \pm (2.33)(0.00918) \\ &= 0.16 \pm 0.0214 \\ \text{LCL} &= 0.139, \qquad \text{UCL} = 0.181\end{aligned}$$

10.7 Estimating Variation

Variation is one of the important characteristics of a population. The most common measure of this characteristic is the standard deviation. Point and interval estimates of the standard deviation are frequently required, sometimes in conjunction with the estimation of means, and sometimes simply to provide information about the population variation itself.

The square root of the unbiased point estimator of the population variance is the sample standard deviation, defined as

$$s = \sqrt{\frac{\Sigma(x - \bar{x})^2}{n - 1}}$$

The procedure we use for obtaining a confidence interval requires large samples. Although we define large samples as samples that contain over 30 observations, for this procedure we require samples of at least 50 observations. We also require that the population is at least approximately normal.

When large samples are drawn from a normal population, the sampling distribution of s can be approximated by a normal distribution with a mean approximately equal to σ and a standard error of approximately

$$\sigma_s = \frac{\sigma}{\sqrt{2n}}$$

The procedure for obtaining a confidence interval on σ deviates somewhat from the familiar one used for means and proportions. In those cases, the interval resulted from the point estimate plus and/or minus the confidence multiplier times the standard error of the estimator. For standard deviations, an approximate $C\%$ two-sided confidence interval is expressed as

$$P\left[\frac{s}{1 + z/\sqrt{2n}} \leq \sigma \leq \frac{s}{1 - z/\sqrt{2n}}\right] \doteq C$$

In this inequality, the confidence limits are

$$\text{LCL} = \frac{s}{1 + z/\sqrt{2n}} \tag{10.12}$$

and

$$\text{UCL} = \frac{s}{1 - z/\sqrt{2n}} \tag{10.13}$$

The value of the z multiplier corresponds to a tail area of $(1 - C)/2$. Note that the upper and lower confidence limits are not equally distant from σ.

One-sided confidence intervals result from the following inequalities.

$$P\left[\sigma \geq \frac{s}{1 + z/\sqrt{2n}}\right] \doteq C$$

and

$$P\left[\sigma \leq \frac{s}{1 - z/\sqrt{2n}}\right] \doteq C$$

where

(10.14) $$\text{LCL} = \frac{s}{1 + z/\sqrt{2n}}$$

and

(10.15) $$\text{UCL} = \frac{s}{1 - z/\sqrt{2n}}$$

The value of the z multiplier corresponds to a tail area of $(1 - C)$.

Example 10.21

In Example 10.3, the production manager of the Merit Manufacturing Company finds it necessary to estimate the variation of a particular production process. Information about this variation is required to decide whether or not to accept a contract for the production of a certain part to very narrow tolerances. He orders a sample production run of 85 parts and computes the standard deviation from the sample:

$$s = \sqrt{\frac{\Sigma(x - \bar{x})^2}{n - 1}} = 0.050 \text{ centimeters}$$

The manager decides to construct a 99% two-sided confidence interval on the population standard deviation from these data. The population is assumed to be approximately normal. The information about this problem is summarized as follows.

Population: approximately normal

Population size: theoretically infinite

Sample size: $n = 85$

Estimator: $s = 0.050$

Sampling distribution of s: approximately normal for large samples

Confidence interval: two-sided, 99%

Confidence multiplier: $z_{(1-.99)/2} = 2.58$

The confidence interval is constructed as follows.

(10.12) $$\text{LCL} = \frac{s}{1 + z/\sqrt{2n}}$$

(10.13) $$\text{UCL} = \frac{s}{1 - z/\sqrt{2n}}$$

$$\text{LCL} = \frac{0.50}{1 + 2.58/\sqrt{170}}$$

$$\text{UCL} = \frac{0.50}{1 - 2.58/\sqrt{170}}$$

$$\text{LCL} = 0.042, \qquad \text{UCL} = 0.062$$

On the basis of this sample, the production manager can be 99% confident that the standard deviation of this particular production process is between 0.042 and 0.062 centimeters.

Example 10.22 In Example 10.21, the two-sided confidence interval provides the production manager with both an upper and a lower limit on the standard deviation. In considering the decision he has to make, the production manager realizes that he should be concerned with maximum rather than minimum variation. Therefore, a one-sided interval with an upper limit is more useful than the previously constructed two-sided interval. The manager decides to compute the upper confidence limit for a 99% one-sided confidence interval on the standard deviation of the process. The information for this problem is the same as for the previous example, except for the following.

Confidence interval: one-sided, 99%

Confidence multiplier: $z_{(1-.99)} = 2.33$

The confidence interval is constructed as follows.

$$\text{UCL} = \frac{s}{1 - z/\sqrt{2n}} \tag{10.15}$$

$$= \frac{0.05}{1 - 2.33/\sqrt{170}}$$

$$= 0.061$$

On the basis of this one-sided confidence interval, the production manager can be 99% confident that the true standard deviation of the process is no greater than 0.061 centimeters. With this information, the production manager is now able to make a decision about the contract. Since for a normal population approximately 99.7% of the values fall within three standard deviations of the mean, if the mean plus and minus three times the maximum standard deviation lies within the tolerance limits specified, the contract should be accepted. If this spread is greater than the specified tolerances, the production manager should reject the contract.

10.8 Estimating Sample Size

One of the most common questions asked about the estimation procedure is, "How large a sample is necessary to make an estimate?" Most of the examples in this chapter have indicated that a sample of a given size was taken without specifying how that particular sample size was determined.

The procedures for estimating the approximate sample size required in a given estimation procedure are fairly simple. The size of any confidence interval depends on several factors, including the size or amount of error that can be tolerated in the estimate. The error in an estimate is the absolute difference between the point estimate and the true value of the parameter being estimated. This error can always be expressed in terms of multiples of the standard error of the estimator. The size of the interval also depends on the level of confidence. The higher the confidence, the larger the interval. Confidence enters into this procedure through the value of the confidence multiplier. Finally, since the size of the interval varies inversely with the sample size, n is also an important factor. It is possible to express the size of a confidence interval in terms of error, the confidence multiplier, and the standard error of the statistic used as the estimator. That is,

$$E = k(\text{SE})$$

where E stands for error, k is the confidence multiplier, and SE represents the standard error of the estimator. Note that the value of the confidence multiplier used in sample size estimation must always correspond to a two-sided confidence interval.

So far, n, the sample size, does not appear in this equation. However, the standard error of a statistic always contains a factor related to the standard deviation of the population and a factor related to the size of the sample. The standard errors of the mean, proportion, and standard deviation for infinite or very large populations are, respectively,

$$\sigma_{\bar{x}} = \frac{\sigma}{\sqrt{n}}$$

$$\sigma_{\bar{p}} = \sqrt{\frac{p(1-p)}{n}}$$

$$\sigma_s = \frac{\sigma}{\sqrt{2n}}$$

By making the appropriate substitution for the standard error in the general formula stated above, the equation can always be solved for a value of n which meets the requirements established for the estimate. We will now consider the procedure for estimating the sample size for each of the three parameters of interest.

Sample Size for Means. In estimating the mean of a population, the error in the estimate is $E = |\bar{x} - \mu|$. The error can also be expressed as multiples of the standard error.

$$E = z_{(1-C)/2}\sigma_{\bar{x}}$$

Since $\sigma_{\bar{x}} = \sigma/\sqrt{n}$, by substitution we get

$$E = \frac{z\sigma}{\sqrt{n}}$$

This equation can be solved for n, resulting in

(10.16) $$n = \left[\frac{z\sigma}{E}\right]^2$$

Since E is the maximum error that can be tolerated in the estimate, and is specified in advance, and z is determined by the level of confidence desired, n and σ are the only unknowns in the sample size equation. It is, however, impossible to solve a single equation with two unknowns. Therefore, to solve for a value of n, σ must be either known or estimated.

Although it is not likely that the standard deviation of a population will be known if the mean is unknown, there are occasions when the standard deviation might be assumed to be known on the basis of other data, such as previous surveys. If there is no data available and σ is not known, then frequently a small pilot sample is taken for the purpose of obtaining a rough estimate of σ. This estimate of σ is used in the sample size equation for estimating n. When the sample size required for estimating the mean is determined, the data from the pilot sample can be incorporated as part of the larger sample.

Example 10.23 Suppose that in Example 10.12, the market research company retained by Cinderella Clothes needs an initial estimate of the size of the sample that has to be taken to estimate the mean income in the marketing area with 95% confidence. Cinderella Clothes has specified that the error in the estimate must not exceed \$135. From previous surveys of income levels of similar populations, the market researchers know that the standard deviation of incomes is generally about \$1000. This figure remains relatively stable even though the mean income can vary substantially. The data for this problem are summarized as follows.

Maximum error: \$135

Confidence level: $C = 95\%$

Confidence multiplier: $z = 1.96$

Population standard deviation: assumed known and equal to \$1000

The required sample size is estimated as follows.

(10.16) $$\begin{aligned} n &= \left[\frac{z\sigma}{E}\right]^2 \\ &= \left[\frac{(1.96)(\$1000)}{\$135}\right]^2 \\ &= 210.8 \end{aligned}$$

The minimum required sample size from this computation is 210.8. Since it is impossible to make an observation on 0.8 of an elementary unit, the next hgher whole number is used—in this case, a minimum sample of 211. In Example 10.12, the survey company actually took a sample of 225.

Sample Size for Estimating Proportions. When estimating a population proportion, the error in the estimate is $E = |\bar{p} - p|$. We can express the amount of error in terms of multiples of the standard error of the estimator.

$$E = z_{(1-C)/2}\sigma_{\bar{p}}$$

Since

$$\sigma_{\bar{p}} = \sqrt{\frac{p(1-p)}{n}} = \frac{\sqrt{p(1-p)}}{\sqrt{n}}$$

by substitution we get

$$E = \frac{z\sqrt{p(1-p)}}{\sqrt{n}}$$

Solving for n results in

$$n = \frac{z^2 p(1-p)}{E^2}$$

In this formula, the size of n depends on the value of p, which is unknown since it is that parameter which we are trying to estimate. In order to be conservative and not underestimate n, the simplest procedure is to substitute that value of p which will make n the largest. When $p = 0.5$, the product $p(1-p)$ is a maximum. Therefore, substituting 0.5 for p in the sample size equations results in the largest value of n. Making this substitution reduces the sample size equation to the following form.

$$\begin{aligned} n &= \frac{z^2 p(1-p)}{E^2} \\ &= \frac{z^2(.5)(.5)}{E^2} \\ &= \frac{.25z^2}{E^2} \\ &= \frac{z^2}{4E^2} \qquad (10.17) \end{aligned}$$

Example 10.24

In estimating the proportion of cancellations and no-shows in Example 10.19, the manager of the Majestic Hotel decides that the error in the estimate should not exceed eight percentage points, that is, 0.08, with a confidence level equal to 99%. The data for estimating sample size are

Maximum error: 0.08

Confidence level: $C = 99\%$

Confidence multiplier: $z = 2.58$

The sample size is estimated as follows.

$$n = \frac{z^2}{4E^2} \tag{10.17}$$

$$= \frac{(2.58)^2}{4(0.08)^2}$$

$$= 260.02$$

The minimum sample size should be 261. The manager actually took a sample of 300.

Sample Size for Estimating Standard Deviations. The same general procedure is followed to obtain an estimate of the sample size necessary for estimating a standard deviation. The error is equated to some multiple of standard errors of the estimator and the equation is solved for n. The sample size formula for standard deviations results from:

$$E = z_{(1-C)/2}\sigma_s$$

$$= \frac{z\sigma}{\sqrt{2n}}$$

$$n = \frac{1}{2}\left[\frac{z\sigma}{E}\right]^2 \tag{10.18}$$

The identical problem encountered in estimating the sample size for means also occurs here. The population standard deviation which appears in the formula must be known or estimated. Since σ cannot be assumed to be known, then the best approach is that of the small pilot sample. The data from the pilot sample can be used to estimate σ in the sample size formula and can then be combined with the larger sample in the final estimation procedure.

Example 10.25 In the case of the production manager of the Merit Manufacturing Company in Example 10.21, a 99% confidence interval on the standard deviation of a particular production process is required. The production manager, however, first has to estimate the size of the sample production run required to make this estimate. The manager wants a sample large enough to enable him to estimate the standard deviation of the process with an error not to exceed 0.05 centimeters, with 99% confidence. Referring to the sample size estimation formula, he sees that a value for σ is required. Since σ is the parameter to be estimated, it obviously is not known. He therefore decides to take a small pilot sample run of ten units to get an initial estimate of σ for use in the formula. After determining the sample size from the formula, he then uses the larger sample to make his final estimate of σ. The standard deviation of the small pilot sample is 0.25 centimeters. The data for this problem are summarized as follows.

Maximum error: 0.05 centimeters

Confidence level: $C = 99\%$

Confidence multiplier: $z = 2.58$

Population standard deviation: estimated from pilot sample to be 0.25 centimeters.

The required sample size is estimated as follows.

(10.18)
$$n = \frac{1}{2}\left[\frac{z\sigma}{E}\right]^2 = \frac{1}{2}\left[\frac{(2.58)(0.25)}{0.05}\right]^2 = 83.21$$

The minimum sample size that should be used is 84. In Example 10.21, the production manager actually used a sample of 85. However, since he already has data from the pilot sample of ten units, he needs only 74 or 75 additional units to meet the sample size requirements.

Definitions

Confidence: The probability that a specified interval contains the population parameter of interest.

Confidence interval estimate: A range of values which includes the population parameter with a stated probability.

Confidence limit: An endpoint of a confidence interval.

Estimator: A statistic used for estimating the value of a population parameter.

Point estimate: The numerical value of the estimator computed from the sample.

Unbiased estimator: An estimator whose expected value is equal to the value of the parameter to be estimated.

Formulas

Relative efficiency:

(10.1)
$$\text{RE}\left(\frac{v_1}{v_2}\right) = \frac{\text{Var}(v_2)}{\text{Var}(v_1)}$$

Confidence interval on the mean, σ known:

(10.2) Two-sided: $\bar{x} \pm z\sigma_{\bar{x}}$

where the value of the multiplier corresponds to a tail area of $(1 - C)/2$.

(10.3) One-sided, lower limit: $\bar{x} - z\sigma_{\bar{x}}$

where the value of the multiplier corresponds to a tail area of $(1 - C)$.

(10.4) One-sided, upper limit: $\bar{x} + z\sigma_{\bar{x}}$

where the value of the multiplier corresponds to a tail area of $(1 - C)$.

Confidence interval on the mean, σ unknown:

		$n \leq 30$	$n > 30$
(10.5)	Two-sided:	$\bar{x} \pm ts_{\bar{x}}$	$\bar{x} \pm zs_{\bar{x}}$

where the value of the multiplier corresponds to a tail area of $(1 - C)/2$.

		$n \leq 30$	$n > 30$
(10.6)	One-sided, lower limit:	$\bar{x} - ts_{\bar{x}}$	$\bar{x} - zs_{\bar{x}}$

where the value of the multiplier corresponds to a tail area of $(1 - C)$.

		$n \leq 30$	$n > 30$
(10.7)	One-sided, upper limit:	$\bar{x} + ts_{\bar{x}}$	$\bar{x} + zs_{\bar{x}}$

where the value of the multiplier corresponds to a tail area of $(1 - C)$.

Estimate of the standard error of the proportion:

(10.8) $$\hat{\sigma}_{\bar{p}} = \sqrt{\frac{\bar{p}(1 - \bar{p})}{n}}$$

Confidence interval on the proportion:

(10.9) Two-sided: $\bar{p} \pm z_{(1-C)/2}\sigma_{\bar{p}}$

(10.10) One-sided, lower limit: $\bar{p} - z_{(1-C)}\sigma_{\bar{p}}$

(10.11) One-sided, upper limit: $\bar{p} + z_{(1-C)}\sigma_{\bar{p}}$

Confidence interval on the standard deviation:

(10.12) Two-sided: $$\text{LCL} = \frac{s}{1 + z/\sqrt{2n}}$$

(10.13) $$\text{UCL} = \frac{s}{1 - z/\sqrt{2n}}$$

where the value of the multiplier corresponds to a tail area of $(1 - C)/2$.

(10.14) One-sided: $$\text{LCL} = \frac{s}{1 + z/\sqrt{2n}}$$

(10.15) $$\text{UCL} = \frac{s}{1 - z/\sqrt{2n}}$$

where the value of the multiplier corresponds to a tail area of $(1 - C)$.

Sample size for estimating means:

(10.16) $$n = \left[\frac{z\sigma}{E}\right]^2$$

Sample size for estimating proportions:

(10.17) $$n = \frac{z^2}{4E^2}$$

Sample size for estimating standard deviations:

(10.18) $$n = \frac{1}{2}\left[\frac{z\sigma}{E}\right]^2$$

Problems

10.1 If a population is normal with a standard deviation of 50, construct a 95% two-sided confidence interval on its mean if a random sample of $n = 100$ drawn from the population has a mean of 240.

10.2 For the data of problem 10.1, construct a 99% two-sided confidence interval on the population mean.

10.3 For the data of problem 10.1, construct a 99% one-sided confidence interval on the mean having a single lower limit.

10.4 An agency of the Department of Housing and Urban Development wants to determine the average purchase price of all new homes constructed in the United States during the year 1978. A random sample of 1000 records of house sales is drawn from the population. The mean of the sample is $39,100, with a standard deviation of $2000.

a) Construct a 98% two-sided confidence interval on the true mean price.

b) Construct a confidence interval on the mean so that we can state with 98% confidence that the true mean is equal to or greater than the stated limit.

10.5 A county housing agency wants to determine the average purchase price of new homes constructed in the county during 1978. A random sample of 25 records of house sales is drawn from the population of sales made in the county. The mean of this sample is $39,900, with a standard deviation of $2700. Assume that the population of sales prices is normally distributed.

a) Construct a 95% two-sided confidence interval on the true mean price.

b) Construct a confidence interval on the mean so that we can state with 95% confidence that the true mean is equal to or less than the stated limit.

10.6 Work problem 10.5 with the added assumption that a total of 200 houses were constructed and sold in the county during 1978.

10.7 Work problem 10.1 with the added assumption that the population size is 900.

10.8 A random sample of 625 families in a major metropolitan area indicates that each family averages 175 minutes of television viewing per day. The standard deviation of the sample is 40 minutes.

a) Construct a 96% two-sided confidence interval on the true average viewing time.

b) Find the lower 98% confidence limit on the average for a one-sided interval.

10.9 A large supermarket chain hires a market research company to estimate the average weekly expenditures per family on food in an area where company management is contemplating opening a new market. The researchers take a random sample of 225 families and find that the sample average is \$78.00, with a standard deviation of \$12.00.

a) Construct a 95% two-sided confidence interval on the population mean.

b) Construct a 99% two-sided confidence interval on the population mean.

c) Change the sample size to 625 and reconstruct the 95% two-sided interval.

d) Change the sample size to 25 and reconstruct the 99% two-sided interval. Assume that the population has a normal distribution.

10.10 Suppose that the supermarket management in the previous problem is interested in determining only a lower confidence limit on family food expenditures. Construct the appropriate one-sided intervals that correspond to parts (a) through (d) of the previous problem.

10.11 The credit manager of a large department store is interested in obtaining information about past-due accounts. A random sample of 100 past-due accounts indicates that the average number of days the accounts are delinquent is 35. The sample standard deviation is 10 days. Construct a 95% two-sided confidence interval on the average number of days past due for the population of delinquent accounts.

10.12 For the data in the previous problem, construct a 98% two-sided confidence interval.

10.13 Work problem 10.11 using a sample size of 16. Assume that the population is normally distributed.

10.14 From the sample drawn in problem 10.11, it was determined that 25 of the accounts are more than 60 days past due. Construct a 99% two-sided confidence interval on the true proportion of accounts that are more than 60 days past due.

10.15 A research team at the National Institute of Health wants to estimate the proportion of adults in the United States who smoke at least a package of cigarettes every day. A random sample of 1600 is drawn from the population of adult Americans. Of these 1600, 640 smoke at least one pack a day.

a) Construct a 99% two-sided confidence interval on the population proportion.

b) Construct a 97% two-sided confidence interval on the population proportion.

10.16 A county public health department wants to estimate the proportion of adults in the county who smoke at least a package of cigarettes every day. The adult population of the county is 2500. A random sample of 400 adults is selected and, of these, 100 smoke at least a pack a day.

a) Construct a 95% two-sided confidence interval on the population proportion.

b) Construct a 95% one-sided confidence interval on the proportion with a single lower limit.

10.17 A random sample of 250 parts taken from a continuous production line contains 50 defectives.

a) Construct a 99% one-sided confidence interval on the true proportion of defective parts produced, with a single lower limit.

b) Construct a 95% one-sided confidence interval on the true proportion, with a single upper limit.

c) Construct a 98% two-sided confidence interval on the true proportion.

10.18 A random sample of 250 parts is taken from a lot of 1000 parts that are being prepared for shipment to a customer. Forty parts from the sample are found to be defective.

a) Find the lower limit of a 99% one-sided confidence interval on the proportion of defective parts in the lot.

b) Find the upper limit of a 95% one-sided confidence interval on the proportion of defective parts in the lot.

c) Construct a 98% two-sided confidence interval on the proportion of defective parts in the lot.

10.19 As a result of the sample taken for the survey described in problem 10.4, it was determined that in 600 of the 1000 purchase transactions, the price of the house was greater than \$34,500. Construct a 90% two-sided confidence interval on the true proportion of houses sold in 1978 costing more than \$34,500.

10.20 Given that $n = 400$ and $\overline{p} = 0.25$:

a) Find the lower limit of a 99% one-sided confidence interval on the true proportion.

b) Construct a 99% two-sided confidence interval on the true proportion.

10.21 Given that $n = 625$ and $\overline{p} = 0.85$:

a) Find the upper limit of a 95% one-sided confidence interval on the true proportion.

b) Construct a 95% two-sided confidence interval on the true proportion.

10.22 Given that $n = 225$ and $\overline{p} = 0.35$:

a) Find the upper limit of a 90% one-sided confidence interval on the true proportion.

b) Construct a 90% two-sided confidence interval on the true proportion.

10.23 In problem 10.1 the population standard deviation was given. Suppose that that value was, in fact, an estimate from that sample of size 100. That is, $s = 50$. Construct a 95% two-sided confidence interval on the population standard deviation.

10.24 For the previous problem, find the upper limit of a 95% one-sided confidence interval on the population standard deviation.

10.25 Construct two 98% one-sided confidence intervals on the population standard deviation from the data provided in problem 10.4. One interval will have a single upper limit, the other will have a single lower limit. Assume that the population has a normal distribution.

10.26 Construct a 99% two-sided confidence interval on the population standard deviation using the data provided in problem 10.9. Assume that the population is approximately normal.

10.27 For the data in problem 10.11, assume that the population is approximately normal and construct the following confidence intervals on the true standard deviation.

a) 90% two-sided interval.

b) 90% one-sided interval with a single lower limit.

c) 90% one-sided interval with a single upper limit.

10.28 Assume that the Department of Housing and Urban Development is planning a new survey for 1980 similar to the one conducted in 1978 and described in problem 10.4. How large a sample should they take in order to have 98% confidence that they will be able to estimate the average cost of a new house with an error no greater than $100. They assume that the population standard deviation is $2000.

10.29 Find the required sample size for problem 10.28 if H.U.D. wants to be 99% confident that the error in their estimate will not exceed $200.

10.30 For the survey described in problem 10.8, how large a sample should be taken if the error in the estimate of average viewing time must not exceed 15 minutes with 99.5% confidence? Assume that the population standard deviation is 40 minutes.

10.31 For the supermarket survey described in problem 10.9, how large a sample must be taken to be 95% confident that the error in the estimate of average family expenditures for food will not exceed $2.00. Assume that the standard deviation of the population is $12.00.

10.32 How large a sample should the credit manager in problem 10.11 take in order to be 99% confident that the error in his estimate is no greater than two days. Assume that the population standard deviation is 10 days.

10.33 If the National Institute of Health wants to be 99% confident that their estimate of the proportion of heavy smokers in the population resulting from the survey described in problem 10.15 is in error by no more than two percentage points, how large a sample should have been taken?

10.34 In problem 10.17 how large a sample should be taken to assure that the error in the estimate of the truc proportion of defectives is

a) no greater than 0.03 with 95% confidence?

b) no greater than 0.05 with 95% confidence?

10.35 Determine how large a sample should be taken by H.U.D. in problem 10.19 if they want to be 98% confident that the error in their estimate of the proportion of houses costing more than $34,500 is no greater than:

a) 0.04.

b) 0.02.

10.36 How large a sample should be taken if the error in the estimate of the population standard deviation in problem 10.23 is not to exceed 4 with 99% confidence? Assume an initial estimate of $\widehat{\sigma} = 50$.

10.37 How large a sample should be taken if the error in the estimate of the population standard deviation described in problem 10.26 is not to exceed \$1.00 with 95% confidence. Assume an initial estimate of $\widehat{\sigma} = \$12.00$.

CHAPTER 11 HYPOTHESIS TESTING—ONE SAMPLE TESTS

11.1
The rationale for hypothesis testing

11.2
A general procedure for hypothesis testing

11.3
One-tailed and two-tailed tests

11.4
Errors in hypothesis testing

11.5
Tests about means—large samples

11.6
Tests about means—small samples

11.7
Tests about proportions

11.8
Calculating the probability of the type II error

11.9
Testing hypotheses with critical values of *z* and *t*

A hypothesis is an assumption to be tested. The statistical testing of hypotheses is probably the most important technique in statistical inference. Hypothesis tests are widely used in business and industry for making decisions. In fact, hypothesis testing is frequently referred to as statistical decision making.

Inferences about population parameters are generally made in one of two ways, either as an estimate of the parameter value or as a test of some assumption about the value of the parameter. However, although an estimate provides useful information to a decision maker, it does not clearly indicate exactly what action should be taken. On the other hand, a properly designed and executed test of a hypothesis can specify exactly what decision should be made and what action should be taken.

For example, in the previous chapter, Example 10.1 described a situation in which the executive vice-president of Cinderella Clothes had to make a decision about the opening of a store in a new shopping center. She intended to base this decision on an estimate of the average income of families living in the area served by the shopping center. Interval estimates of this average were obtained and although these estimates provided useful information on which the vice-president could base her decision, the estimates themselves did not indicate whether or not the store should be opened. However, a hypothesis test about some value of income that might be required before the store would be opened would lead to a specific decision.

Before turning to the specific procedures for hypothesis testing, however, consider a few additional examples of situations in business and industry where these tests might be used for making decisions.

Example 11.1 The Colonial Canning Company uses an automatic machine to fill cans to a specified weight and level. Since automatic machines have a tendency to drift out of adjustment, the quality control department of the company draws a random sample of 100 cans from each production line every day and weighs all the cans in the sample. The sample average, $\bar{x}$, is then compared with the desired mean weight of the population of cans filled on that particular production line. If the calculated value of $\bar{x}$ deviates significantly from the mean, μ, the assumption or hypothesis that the machine is properly adjusted is rejected. Consequently, production is halted and the machinery is readjusted.

Example 11.2 The advertising manager of the Summit Record Company plans to advertise a new album on local television. A salesperson for UHF Channel 32 claims that at least 20% of all television sets in the local viewing area are tuned to that channel between the hours of 11 and 12 PM daily. The advertising manager decides to test that claim by contacting a random sample of viewers between those hours. If the sample proportion verifies the claim, Channel 32 will be given the advertising contract. Otherwise, the contract will be placed elsewhere.

Example 11.3

The purchasing manager of the Global Office Machine Company purchases large quantities of a particular machine part from a supplier. This purchased part is used in one of Global's major products. One of the physical dimensions of the part is critical, for if it is too large or too small it cannot be assembled into the finished product. Because of the large quantities in which the part is purchased, it is impractical to inspect every part to determine if it meets the required specifications. Therefore, Global's quality control department established an acceptance procedure for all shipments. The procedure involves taking a sample of parts from every shipment and inspecting each part in the sample. The proportion of defective parts in the sample is compared with an acceptable hypothesized value of the population proportion of defectives. If the sample proportion is significantly greater than the acceptable value, the entire shipment is rejected and returned to the supplier.

In each of these examples, a statistical test of a hypothesis is established to make a specific decision. The decisions in these cases involve the following questions.

1. Should production be stopped and the machine readjusted?
2. Should the advertising contract be placed with Channel 32?
3. Should the shipment be returned to the supplier?

The considerations and procedures for establishing these tests are explained in the following sections.

11.1 The Rationale for Hypothesis Testing

If a random sample is drawn from a population, it is reasonable to assume that the sample will resemble the population. It is on the basis of this assumption that we make estimates of unknown parameters by using sample statistics. Now, if an assumption or hypothesis is made about the value of a population parameter, it is also reasonable to expect that an appropriate statistic computed from that sample will have a value that is reasonably close to the hypothesized parameter value. This is true only if the hypothesized value is correct and if the statistic has the properties that make it a good estimator of the pertinent parameter. The statistic used to test a hypothesis is called a *test statistic*.

Since statistics are random variables, they have sampling distributions that exhibit the property of variation. Consequently, even if the hypothesized value is correct, we do not expect the statistic to have exactly the same value as the parameter. Some difference is likely due to chance and to the particular sample actually drawn.

However, if the value of the statistic deviates significantly from the hypothesized parameter value, then there is reason to doubt that the hypothesis is correct. The greater the difference between the value of the statistic and the hypothesized value, the more doubt there is about the correctness of the hypothesis.

In statistics we cannot speak in vague generalities about large or small deviations. We must specify, in quantitative terms, exactly what constitutes a significant deviation. In statistical testing, significance is specified in terms of the probability that a particular deviation occurs by chance if the hypothesized value of the parameter is correct. If the properties of the sampling distribution of the test statistic are known, then the probability of a deviation of any magnitude is easily determined.

It is up to the decision maker to select the probability level that constitutes significance with regard to the particular hypothesis being tested. For example, if you were the decision maker, you might feel that a deviation that could occur by chance 10% of the time even if the hypothesis is correct, is not significant. On the other hand, a deviation so large that it could occur by chance less than 5% of the time might be significant. The probability level at which the decision maker concludes that an observed deviation cannot be attributed to chance is called the SIGNIFICANCE LEVEL of the test.

Example 11.4 As an illustration of what constitutes significance for you, suppose that you are asked to test the hypothesis that a particular coin is fair, that is, that there is a head on one side of the coin, a tail on the other, and the probability of either one occurring on a single toss is exactly one-half. You are not allowed to examine the coin, but you may observe any number of coin tosses and their results. At your request, the coin is tossed n times and you observe X consecutive heads. How many consecutive heads do you have to observe before you reject the hypothesis that the coin is fair?

A reasonable approach to the problem is to compare X, the number of heads observed in n tosses, with $E(X)$, the number expected if the coin is fair and determine when the difference is significant. Table 11.1 provides the probability of obtaining X consecutive heads in n tosses of a fair coin.

When this problem was posed to various groups of students, most would not reject the hypothesis that the coin is fair after observing only two or three consecutive heads. After all, three heads in three tosses can occur by chance with a fair coin $12\frac{1}{2}$% of the time. However, if you are prepared to reject the hypothesis of fairness after seeing four consecutive heads, your level of signficance for this test is 0.0625; that is, you consider an outcome that can occur by chance only $6\frac{1}{4}$% of the time sufficient evidence to reject the hypothesis of fairness. If you are not willing to reject the hypothesis after four consecutive heads, but reject it on observing five, your level of significance for this test is 0.03125. If you have to observe six consecutive heads before rejecting the hypothesis, your significance level for this test is 0.015625. In most statistical testing the significance levels used are 5% and

TABLE 11.1
PROBABILITIES OF CONSECUTIVE HEADS IN COIN TOSSING EXPERIMENT (EXAMPLE 11.4)

Number of Tosses (n)	*Number of Heads* (X)	*Expected Number of Heads if Coin is Fair* $E(X)$	$P(X = X_i)$
1	1	0.5	0.5
2	2	1.0	0.25
3	3	1.5	0.125
4	4	2.0	0.0625
5	5	2.5	0.03125
6	6	3.0	0.015625
7	7	3.5	0.0078125
8	8	4.0	0.0039062
9	9	4.5	0.0019531
10	10	5.0	0.0009765

1%. The symbol which represents the significance level is the Greek letter α, alpha.

11.2 A General Procedure for Hypothesis Testing

The first step in hypothesis testing is to establish the hypothesis to be tested. Since statistical hypotheses are usually assumptions about the value of some unknown parameter, the hypothesis specifies a numerical value or range of values for the parameter. These stated values, which are accepted or rejected on the basis of the test, constitute the *null hypothesis,* which is designated H_0. The null hypothesis corresponds to the hypothesis that we have been referring to in our discussion up to this point. If the condition specified under the null hypothesis is rejected by the test and therefore assumed to be false, then it follows that some other condition is assumed to be true. Therefore, rejection of the null hypothesis is equivalent to the acceptance of an *alternate hypothesis,* which is designated H_1. In any hypothesis test, there are always two hypotheses, the *null hypothesis* and the *alternate hypothesis.*

The null hypothesis can be expressed in either of two general forms. For instance, in testing a hypothesis about the value of an unknown population mean, the null hypothesis may state that the mean is equal to some single specific value. In this case, the null hypothesis is written:

$$H_0: \mu = k$$

where k represents some number. The alternate hypothesis states that the population mean is not equal to the value stated in the null hypothesis and is written:

$$H_1\colon \mu \neq k$$

For these statements of the null and alternate hypotheses, H_0 is rejected if the test statistic, in this case $\bar{x}$, deviates significantly from k in either direction, that is, if $\bar{x}$ is significantly larger or significantly smaller than k.

A second general form for testing hypotheses is to establish a null hypothesis that states that the unknown parameter is equal to or less than some value k or, alternatively, that it is equal to or greater than k. If the parameter is the mean, the null and alternate hypotheses are written, respectively,

$$H_0\colon \mu \leq k \qquad \text{and} \qquad H_1\colon \mu > k$$

in the first case, or

$$H_0\colon \mu \geq k \qquad \text{and} \qquad H_1\colon \mu < k$$

in the second.

It is common practice to establish the null hypothesis so that acceptance requires no positive action, while rejection leads to some positive act. This is not always possible, and sometimes there are circumstances where both acceptance and rejection of the null hypothesis lead to different positive actions.

Consider, once again, the case of the Colonial Canning Company described in Example 11.1. There a test was established which involves a hypothesis about the mean filled weight of 16-ounce cans of juice. The appropriate null and alternate hypotheses were, respectively,

$$H_0\colon \mu = 16$$

and

$$H_1\colon \mu \neq 16$$

In this case, acceptance of the null hypothesis would allow the process to continue. Rejection, however, would lead to a shutdown of the production line and readjustment of the machinery. In other words, acceptance of H_0 maintains the status quo, while rejection requires a positive action.

In Example 11.2, the test involved the proportion of television sets tuned to Channel 32 at a particular time of the day. The null and alternate hypotheses were, respectively,

$$H_0\colon p \leq 0.20$$

and

$$H_1\colon p > 0.20$$

In this case, in order to verify the claim made by Channel 32 and take the positive action of awarding them the advertising contract, the null hypothesis has to be rejected.

Once the null and alternate hypotheses are established, it is necessary to select the sample statistic that will be used as the test statistic. For the one-sample tests discussed in this chapter we will use $\overline{x}$ to test hypotheses about μ, and $\overline{p}$ to test hypotheses about p.

The third step in the hypothesis testing procedure is to establish a DECISION RULE. A decision rule is a statement that indicates the action to be taken, that is, accept H_0 or reject H_0, for all possible values of the test statistic. The decision rule divides the sample space of the experiment which corresponds to the area under the curve of the sampling distribution of the test statistic into two mutually exclusive regions. These regions are called the ACCEPTANCE REGION and the REJECTION or CRITICAL REGION.

Since hypotheses about parameters consist of assumptions about their numerical values, and because the test statistics assume numerical values, the decision rule indicates what set of values belongs to the acceptance region and what set belongs to the rejection region. The dividing line or lines between the acceptance and rejection regions correspond to some value or values of the test statistic which is called the CRITICAL VALUE. Decision rules for the hypotheses to be tested in Examples 11.1 and 11.2 might be stated as follows.

> In testing the hypothesis about the mean weight, take a random sample of $n = 100$ and compute $\overline{x}$ from the sample. If $\overline{x}$ has a value greater than or equal to 17 or less than or equal to 15, reject H_0; otherwise, accept H_0.

For this decision rule, the critical values of $\overline{x}$, designated $\overline{x}_c$, are 15 and 17. The rejection region consists of all values less than or equal to 15 and greater than or equal to 17. The acceptance and rejection regions are illustrated in Figure 11.1. The second decision rule might be

> In testing the hypothesis about the population proportion, take a random sample of $n = 400$ and compute $\overline{p}$ from the sample. If $\overline{p}$ has a value greater than or equal to 0.25, reject H_0; otherwise, accept H_0.

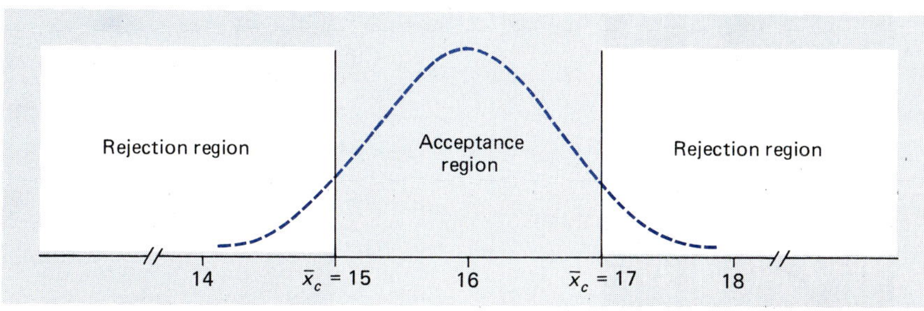

Figure 11.1
Diagrammatic representation of decision rule showing acceptance and rejection regions for a two-tailed test.

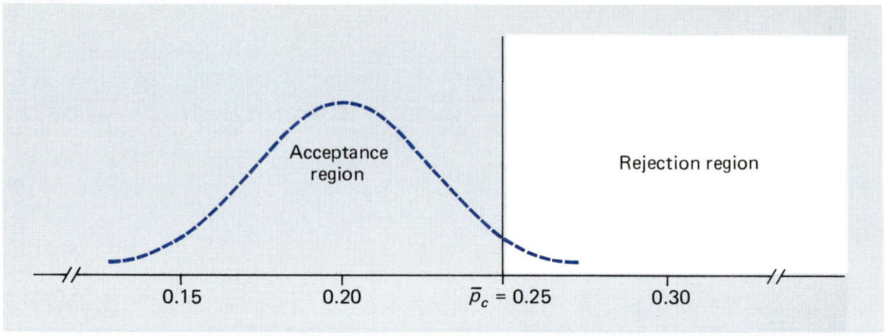

Figure 11.2
Diagrammatic representation of decision rules showing acceptance and rejection regions for a one-tailed test.

For this decision rule, the critical value of $\bar{p}$, designated $\bar{p}_c$, is 0.25. The rejection region consists of all values greater than or equal to 0.25. See Figure 11.2 for an illustration.

Notice that in Figure 11.1, for the first decision rule there are two rejection regions, while in Figure 11.2 there is only one rejection region. A decision rule which specifies two rejection regions and, therefore, two critical values of the test statistic is called a TWO-TAILED TEST. When only a single rejection region is specified we have a ONE-TAILED TEST. The form of the null and alternate hypotheses determines whether a one- or two-tailed test should be used. We will discuss one- and two-tailed tests in Section 11.3.

A general procedure for hypothesis testing is summarized below.

1. Establish the null and alternate hypotheses.
2. Specify the test statistic.
3. Select a significance level for the test and specify the size of the sample to be taken.
4. State the decision rule for the test.
5. Draw a random sample of size n from the population.
6. Compute the value of the test statistic from the sample.
7. Accept or reject the null hypothesis according to the decision rule.

11.3 One-Tailed and Two-Tailed Tests

In the previous section, we indicated that there are two general forms of the null and alternate hypotheses in statistical tests. The first form tests the null hypothesis that a parameter, such as μ, has a value that is exactly equal to some single specific value against the alternative that it is not equal to that value. Symbolically, the null and alternate hypotheses in this case are stated, respectively,

$$H_0\colon \mu = k$$

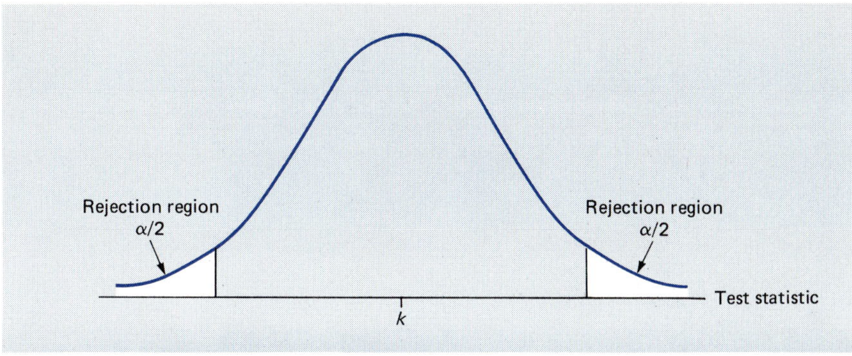

Figure 11.3
Rejection regions for two-tailed test.

and

$$H_1\colon \mu \neq k$$

In testing this hypothesis, a value of the test statistic equal to or close to k leads to acceptance of H_0. However, if the value of the test statistic deviates significantly from k in either direction, H_0 is rejected. Therefore, it is necessary to provide for a rejection region in both tails of the sampling distribution of the test statistic. This type of test is called a TWO-TAILED TEST. If the significance level for this test is α, the rejection regions are selected so that an area corresponding to $\alpha/2$ is placed in each tail of the sampling distribution, that is, half of the rejection region lies in each tail. This is illustrated in Figure 11.3.

The second form tests the hypothesis that the parameter value μ is equal to or less than k, or that it is equal to or greater than k against the appropriate alternative. That is,

$$H_0\colon \mu \leq k \qquad \text{versus} \qquad H_1\colon \mu > k$$

or

$$H_0\colon \mu \geq k \qquad \text{versus} \qquad H_1\colon \mu < k$$

These hypotheses result in a ONE-TAILED TEST, since the null hypothesis is rejected for a significant deviation from k in one

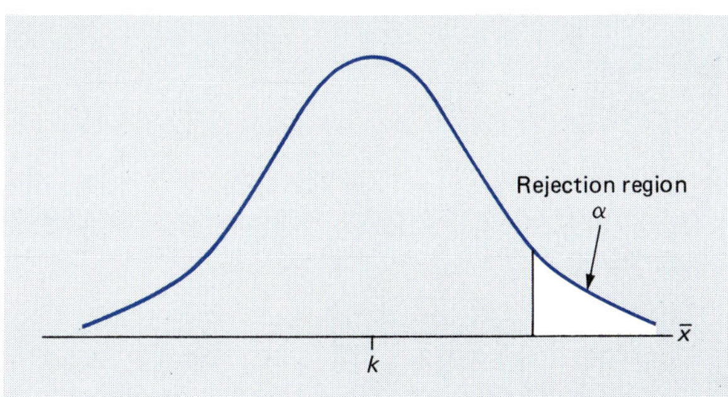

Figure 11.4
Rejection region for testing $H_0{:}\mu \leq k$ *vs* $H_1{:}\mu > k$.

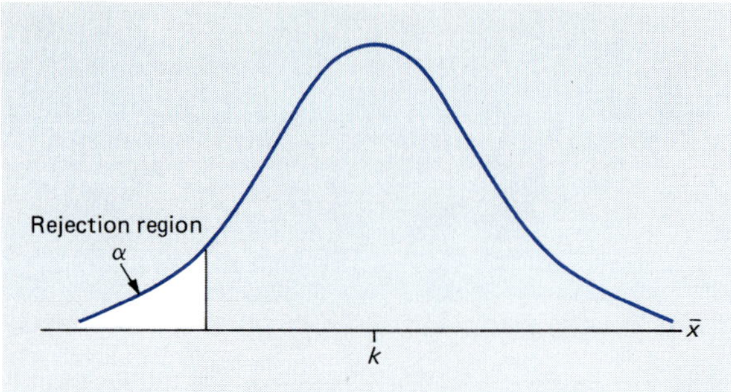

Figure 11.5
Rejection region for testing H_0: $\mu \geq k$ *vs* H_1: $\mu < k$

direction only. Therefore, the entire rejection region corresponding to α is located in only one tail of the sampling distribution of the statistic. The tail containing the rejection region is always in the direction of the inequality that appears in the alternate hypothesis. This is illustrated in Figures 11.4 and 11.5.

11.4 Errors in Hypothesis Testing

In testing any hypothesis, it would be desirable if the test always resulted in the acceptance of the null hypothesis when it is true and its rejection when it is not. However, as long as the decision to accept or reject a hypothesis is based on sample data, no testing procedure will always lead to the correct decision. An incorrect decision is an error, and we must now consider the two kinds of errors that are possible in any hypothesis test.

If the null hypothesis is true, a decision to reject it constitutes an error. An error of this kind is called a TYPE I ERROR. On the other hand, if the null hypothesis is false, a decision to accept it is also an error. This kind of error is called a TYPE II ERROR. Both kinds of error are illustrated in Table 11.2.

TABLE 11.2
TWO KINDS OF ERRORS IN HYPOTHESIS TESTING

	Condition	
Decision	H_0: *True*	H_0: *False*
Accept H_0	Correct Decision	Type II Error
Reject H_0	Type I Error	Correct Decision

Since a null hypothesis may be either true or false and the test can lead either to acceptance or rejection of the null hypothesis,

Table 11.2 shows that there are four possible results than can occur. These are

1. The null hypothesis is true and it is accepted. This is a correct decision.

2. The null hypothesis is true and it is rejected. This is an error of type I.

3. The null hypothesis is false and it is rejected. This is a correct decision.

4. The null hypothesis is false and it is accepted. This is an error of type II.

These errors are measured by their probability of occurring under some specified decision rule. The size of a type I error, that is, its probability, is referred to as α. Since α is also used to represent the significance level of the test, it follows that the significance level is the same as the probability of making a type I error. Therefore, whenever we select a significance level for a test of a hypothesis, we are specifying the risk we are willing to assume of rejecting a true null hypothesis.

The size or probability of making a type II error is called β, beta. The computation of values of β for a test are covered in Section 11.8.

An example of type I and type II errors in a situation that should be familiar to most of us occurs in our criminal justice system. Every time a defendant goes on trial, a null hypothesis is tested against an appropriate alternate hypothesis. The null hypothesis always presumes that the defendant is innocent, that is:

$$H_0\text{: The defendant is innocent}$$

and the alternate is

$$H_1\text{: The defendant is guilty}$$

Since the type I error is the error we incur when we reject a true null hypothesis, a guilty decision when the defendant is actually innocent constitutes an error of that type. On the other hand, finding the defendant innocent when he is guilty is a type II error. In this situation, the reader can determine which type of error has the more serious consequences.

Since the probability of making a type I error is the same as the significance level of the test which is always established in advance by the statistician or the decision maker, the type I error is controlled. The type II error is usually not controlled, but its value is computed, as we shall see later in this chapter.

11.5 Tests about Means—Large Samples

In hypothesis tests about population means, $\bar{x}$, the sample mean, is used as the test statistic. If the population from which the sample is drawn has a normal distribution, we know that the sampling distribution of $\bar{x}$ is also normal. Even if the population does not have a normal distribution, we can assume a normal distribution for $\bar{x}$ by the Central Limit Theorem, provided the sample is large. For our purposes, large samples are those for which $n > 30$.

When large samples are drawn from a population for testing a hypothesis about the mean, the critical values of $\bar{x}$ are determined from

$$\bar{x}_c = \mu_0 \pm z_{\alpha/2}\sigma_{\bar{x}} \tag{11.1}$$

for two-tailed tests, or

$$\bar{x}_c = \mu_0 + z_\alpha \sigma_{\bar{x}} \tag{11.2}$$

or

$$\bar{x}_c = \mu_0 - z_\alpha \sigma_{\bar{x}} \tag{11.3}$$

for one-tailed tests.

The zero subscript on μ refers to that value of the mean which is tested under H_0. The subscript on z refers to the area in the tail or tails of the sampling distribution corresponding to the rejection region, and z is a value of the standard normal variate. The standard error of the mean, $\sigma_{\bar{x}}$, is equal to $\sigma/\sqrt{n}$.

In situations where σ is not known, it must be estimated by $s_{\bar{x}} = s/\sqrt{n}$, where s is the positive square root of the unbiased estimator of the population variance. That is,

$$s = \sqrt{\frac{\Sigma(x - \bar{x})^2}{n - 1}}$$

Then the critical values of $\bar{x}$ are determined from

$$\bar{x}_c = \mu_0 \pm z_{\alpha/2}s_{\bar{x}} \tag{11.4}$$

for two-tailed tests or

$$\bar{x}_c = \mu_0 + z_\alpha s_{\bar{x}} \tag{11.5}$$

or

$$\bar{x}_c = \mu_0 - z_\alpha s_{\bar{x}} \tag{11.6}$$

for one-tailed tests.

Note that when samples are drawn from finite populations and the sampling fraction $n/N > 0.05$, the finite population correction, $\sqrt{[(N - n)/(N - 1)]}$ must be used in determining $\sigma_{\bar{x}}$.

Example 11.5

In Example 11.1, we described a situation in which a statistical test is to be used to decide whether or not an automatic machine should be readjusted. This decision is to be based on the average filled weight of cans coming off a production line. A null hypothesis is established that the mean weight is exactly 16 ounces. The alternate hypothesis states that the mean is not equal to 16 ounces. For these hypotheses, a two-tailed test is required.

It is decided that a random sample of 100 cans will be selected from a day's production and weighed, and $\bar{x}$, the sample mean, will be computed and used as the test statistic. A decision rule is formulated so that the probability of rejecting the null hypothesis incorrectly will be no greater than 0.01. From past experience, the population is known to be normal with a standard deviation of 0.55 ounces.

The important information pertaining to this test is summarized as follows.

Population: normal

Population standard deviation: known, $\sigma = 0.55$

Test statistic: $\bar{x}$

Sample size: $n = 100$

Sampling distribution of $\bar{x}$: normal

$$\sigma_{\bar{x}} = \frac{0.55}{\sqrt{100}} = 0.055$$

Significance level of test: $\alpha = 0.01$

Form of test: two-tailed

Hypotheses: H_0: $\mu = 16$ versus H_1: $\mu \neq 16$

Critical values of $\bar{x}$:

(11.1)
$$\begin{aligned} \bar{x}_c &= \mu_0 \pm z_{\alpha/2}\sigma_{\bar{x}} \\ &= 16 \pm (2.58)(0.055) \\ &= 16 \pm 0.142 \\ &= 15.858 \quad \text{and} \quad 16.142 \end{aligned}$$

Decision rule: A random sample of size $n = 100$ is drawn from the population and $\bar{x}$ is calculated. If $\bar{x}$ is less than or equal to 15.858 or greater than or equal to 16.142, H_0 is rejected; otherwise, H_0 is accepted.

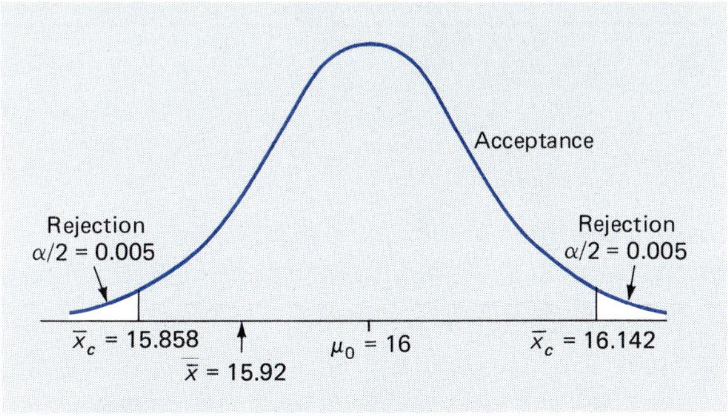

Figure 11.6
Hypothesis test of mean weight (Example 11.5).

When the sample is taken, $\bar{x}$ is calculated to be 15.92 ounces. Since 15.92 lies in the acceptance region, H_0 is accepted and the machine is not readjusted. This test is illustrated in Figure 11.6. The procedure and results would have been exactly the same if the form of the population had not been known, since the sampling distribution of $\bar{x}$ is assumed to be normal for a sample as large as $n = 100$.

Example 11.6 The executive vice-president of Cinderella Clothes, Inc. is required to make a decision regarding the opening of a store in a new shopping center. A criterion for this decision is based on the average annual income of families living in the area served by the shopping center. The criterion states that in order to open the store, the average family income in the area should be greater than \$14,500 per year. A null hypothesis that the mean income is less than or equal to \$14,500 is tested against the alternate hypothesis that the mean income is greater than \$14,500. The hypotheses are stated in this manner so that the sample evidence has to be strong enough to reject the null hypothesis before a decision can be made to open the store. This is a one-tailed test.

It is decided that the maximum acceptable risk of making a type I error is 0.05. A random sample of 225 families is drawn from the population. The sample mean is $\bar{x} = \$14,725$ and the sample standard deviation is $s = \$1000$. The distribution and standard deviation of the population are unknown.

The important information pertaining to this test is summarized as follows.

Population: unknown

Population standard deviation: unknown

Test statistic: $\bar{x}$

Sample size: $n = 225$

Sampling distribution of $\bar{x}$: normal, by Central Limit Theorem

Sample standard deviation: $s = 1000$

Standard error of $\bar{x}$: estimated by $s_{\bar{x}} = s/\sqrt{n} = 1000/15 = 66.67$

Significance level of test: $\alpha = 0.05$

Form of test: one-tailed

Hypotheses: H_0: $\mu \leq 14,500$ versus H_1: $\mu > 14,500$

Critical value of $\bar{x}$:

$$\begin{aligned} \bar{x}_c &= \mu_0 + z_\alpha s_{\bar{x}} \\ &= 14,500 + (1.65)(66.67) \\ &= 14,500 + 110.01 \\ &= 14,610.01 \end{aligned} \tag{11.5}$$

Decision rule: A random sample of size $n = 225$ is drawn from the population and $\bar{x}$ is calculated. If $\bar{x}$ is greater than or equal to 14,610.01, H_0 is rejected; otherwise, H_0 is accepted.

Since the sample mean is 14,725, which is greater than the critical value, H_0 is rejected and the decision is made to open the store. This test is illustrated in Figure 11.7.

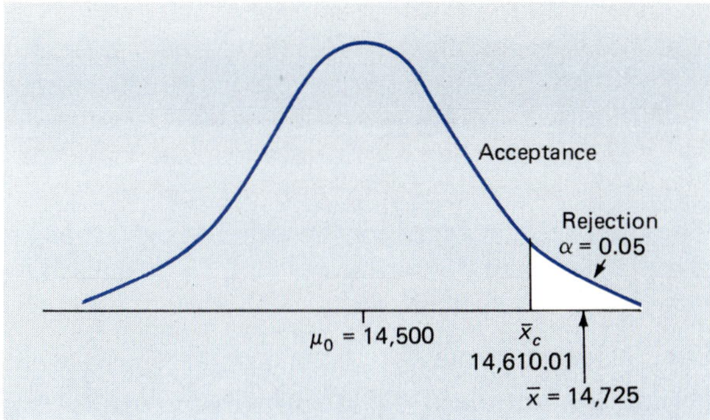

Figure 11.7
Test of hypothesis about mean income (Example 11.6).

11.6 Tests about Means—Small Samples

The distinction between large and small samples in statistics is not always clear-cut. In this book, we shall assume that samples greater than 30 are large samples and samples of 30 and less are small samples. Some of the reasons for selecting this criterion were discussed in Chapter 10.

For small samples, the Central Limit Theorem does not assure us that the sampling distribution of the mean is normal. Consequently, when testing hypotheses about means using small samples, we must assume that the samples come from a normally or approximately normally distributed population. Under these circumstances, the sampling distribution of $\bar{x}$ is normal but the method used for determining critical values of $\bar{x}$ depends on whether or not the population standard deviation is known. If σ is known, then the critical values of $\bar{x}$ are based on the known $\sigma_{\bar{x}}$ and the standard normal variate z. If σ is not known, the sample standard deviation must be used as an estimate and the standard error of the mean is estimated by $s_{\bar{x}} = s/\sqrt{n}$. In this case, the sampling distribution of the mean is treated as a Student's t distribution and the critical values of $\bar{x}$ are based on the estimate $s_{\bar{x}}$ and the t variate. These considerations are identical to those discussed in relation to estimating means with small samples in Section 10.5.

Example 11.7

If, in Example 11.5, samples of size 25 are used to test the hypothesis about the mean weight instead of samples of size 100, the data for the test is the same except for the following.

Sample size: $n = 25$

Standard error of $\bar{x}$: $\sigma_{\bar{x}} = 0.55/\sqrt{25} = 0.11$

Critical values of $\bar{x}$:

(11.1)
$$\begin{aligned}\bar{x}_c &= \mu_0 \pm z_{\alpha/2}\sigma_{\bar{x}} \\ &= 16 \pm (2.58)(0.11) \\ &= 16 \pm 0.284 \\ &= 15.716 \text{ and } 16.284\end{aligned}$$

Decision rule: A random sample of size $n = 25$ is drawn from the population and $\bar{x}$ is calculated. If $\bar{x}$ is less than or equal to 15.716 or greater than or equal to 16.284, H_0 is rejected; otherwise, H_0 is accepted.

When this sample was taken, $\bar{x}$ was computed to be 16.29 ounces. Since 16.29 is greater than the upper critical value of $\bar{x}$, H_0 is rejected.

Example 11.8 Consider the problem in the previous example if σ is unknown and the sample standard deviation is $s = 0.55$ ounces. The important information is summarized as follows.

Population: normal

Population standard deviation: unknown, estimated by $s = 0.55$

Test statistic: $\bar{x}$

Sample size: $n = 25$

Sampling distribution of $\bar{x}$: normal, but critical values are based on the Student's t distribution and the t variate

Standard error of $\bar{x}$: estimated by $s_{\bar{x}} = s/\sqrt{n} = 0.55/5 = 0.11$

Significance level of test: $\alpha = 0.01$

Form of test: two-tailed

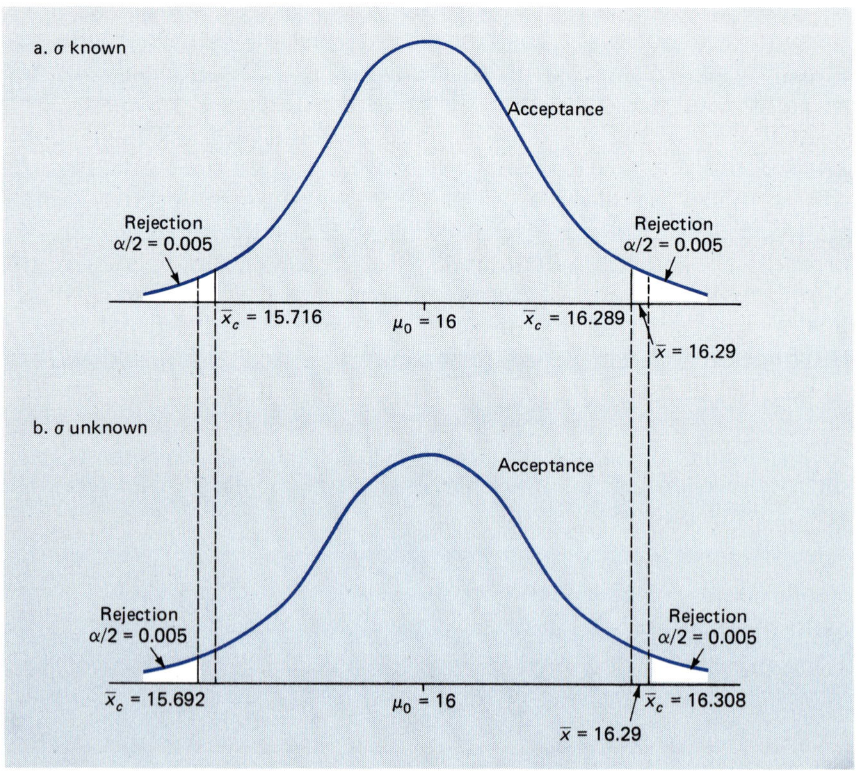

Figure 11.8 *Hypothesis tests for mean weights. a—σ known; b—σ unknown.*

Hypotheses: H_0: $\mu = 16$ versus H_1: $\mu \neq 16$

Critical values of $\bar{x}$:

(11.4)
$$\begin{aligned}\bar{x}_c &= \mu_0 \pm t_{\alpha/2}s_{\bar{x}}\\ &= 16 \pm (2.797)(0.11)\\ &= 16 \pm 0.308\\ &= 15.692 \text{ and } 16.308\end{aligned}$$

Decision rule: A random sample of size $n = 25$ is drawn from the population and $\bar{x}$ is calculated. If $\bar{x}$ is less than or equal to 15.692 or greater than or equal to 16.308, H_0 is rejected; otherwise, H_0 is accepted.

If the sample mean is computed to be $\bar{x} = 16.29$, under this decision rule the null hypothesis is accepted. Notice that the same calculated value of $\bar{x}$ results in the rejection of the null hypothesis in the previous example. These two examples differ only with respect to knowledge about the population standard deviation. Figure 11.8 illustrates these two cases.

Example 11.9

The Rollright Tire Company intends to base an advertising campaign on the claim that the average life of their new radial tire is more than 35,000 miles. Before proceeding, however, a life test is conducted using 20 of the new tires. The tires are placed on a special machine which simulates actual driving conditions and runs until all 20 of the tires fail. The time to failure of each tire is noted and the mean and standard deviation of the times are calculated to be $\bar{x} = 36{,}750$ and $s = 900$.

A null hypothesis is established that the mean life of these tires is less than or equal to 35,000. A significance level of 5% is chosen. Tire lifetimes are assumed to have a normal distribution. Should the null hypothesis be accepted or rejected, and what should the company do about the planned advertising campaign?

The important information for this test is summarized as follows.

Population: normal

Population standard deviation: unknown, estimated by $s = 900$

Test statistic: $\bar{x}$

Sample size: $n = 20$

Sampling distribution of $\bar{x}$: normal, but critical values are based on the Student's t distribution and the t variate

Standard error of $\bar{x}$: estimated by $s_{\bar{x}} = 900/4.47 = 201.3$

Significance level: $\alpha = 0.05$

Form of test: one-tailed

Hypotheses: H_0: $\mu \leq 35{,}000$ versus H_1: $\mu > 35{,}000$

Critical value of $\bar{x}$:

(11.5)
$$\begin{aligned}\bar{x}_c &= \mu_0 + t_{\alpha}s_{\bar{x}}\\ &= 35{,}000 + (1.729)(201.3)\\ &= 35{,}000 + 348.05\\ &= 35{,}348.05\end{aligned}$$

Decision rule: A random sample of size $n = 20$ is tested and $\bar{x}$ is calculated. If $\bar{x}$ is greater than or equal to 35,348.05, H_0 is rejected; otherwise, H_0 is accepted.

Since $\bar{x} = 36{,}750$, which is greater than $\bar{x}_c = 35{,}348.05$, the null hypothesis is rejected and the claim that the life of these tires is greater than 35,000 miles can be advertised.

11.7 Tests about Proportions

In testing hypotheses about population proportions, designated p, either the sample proportion, $\bar{p}$, or the sample number of successes, X, can be used as the test statistic. As discussed in Chapter 9, both $\bar{p}$ and X have binomial distributions with parameters n and p. The expected values and standard errors of these two statistics are

$$E(\bar{p}) = p \tag{11.7}$$

$$\sigma_{\bar{p}} = \sqrt{\frac{p(1-p)}{n}} \tag{11.8}$$

$$E(X) = np \tag{11.9}$$

$$\sigma_X = \sqrt{np(1-p)} \tag{11.10}$$

Don't forget that the finite population correction factor must be applied to the standard error formulas whenever the sampling fraction, n/N, is greater than 0.05.

Although the sampling distributions of $\bar{p}$ and X are binomial, it might not be possible to design a test of the population proportion having exactly the significance level desired because of the discreteness of these distributions. This is most frequently true when the sample size is small. We encountered a similar problem in constructing confidence intervals on the proportion in Chapter 10. However, when samples are large, the binomial distribution may be approximated by a normal distribution having the same mean and standard deviation. We shall consider only large sample tests of proportions under the assumption that the normal distribution provides an adequate approximation to the exact sampling distribution.

The procedure for testing hypotheses about proportions is identical to that followed for tests of means. Since the sampling distribution of the test statistic is assumed to be normal, critical values result from

$$\bar{p}_c = p_0 \pm z_{\alpha/2}\sigma_{\bar{p}} \tag{11.11}$$

for two-tailed tests or

$$\bar{p}_c = p_0 + z_{\alpha}\sigma_{\bar{p}} \tag{11.12}$$

or

$$\bar{p}_c = p_0 - z_{\alpha}\sigma_{\bar{p}} \tag{11.13}$$

for one-tailed tests. The symbol p_0 refers to the value of p specified in the null hypothesis.

Any hypothesis about p can also be tested using X, the sample number of successes, as the test statistic. Since X and $\bar{p}$ differ only by the constant factor $1/n$, their sampling distributions differ only in their means and standard errors, as shown in Equations 11.7 through 11.10. The critical values of X are obtained from

(11.14) $$X_c = np_0 \pm z_{\alpha/2}\sigma_X$$

for two-tailed tests or

(11.15) $$X_c = np_0 + z_\alpha\sigma_X$$

or

(11.16) $$X_c = np_0 - z_\alpha\sigma_X$$

for one-tailed tests.

Because the standard error of $\bar{p}$ or X is necessary for determining critical values of these test statistics, and these standard errors are dependent on a value of p, it is necessary to substitute some value for p in Equations 11.8 and 11.10. Since critical values for a test are based on the assumption that H_0 is true, p_0 is substituted for p in these standard error formulas.

Example 11.10

In Example 11.2, the advertising manager of the Summit Record Company plans to test a hypothesis about the proportion of television sets in the local viewing area that are tuned to Channel 32 between the hours of 11 and 12 PM daily. Channel 32 claims that more than 20% of the television sets in the area are tuned in to that channel at the specified time. A null hypothesis is established that the proportion tuned to Channel 32 is 20% or less and a significance level of 5% is selected. The sample size for this test is $n = 400$.

The important information for this test is summarized as follows.

Population: Bernoulli (attributes)

Test statistic: $\bar{p}$

Sample size: $n = 400$

Sampling distribution of $\bar{p}$: approximately normal

Standard error of $\bar{p}$:

$$\sigma_{\bar{p}} = \sqrt{\frac{p_0(1 - p_0)}{n}} = \sqrt{\frac{(.2)(.8)}{400}} = 0.02$$

Significance level: $\alpha = 0.05$

Form of test: one-tailed

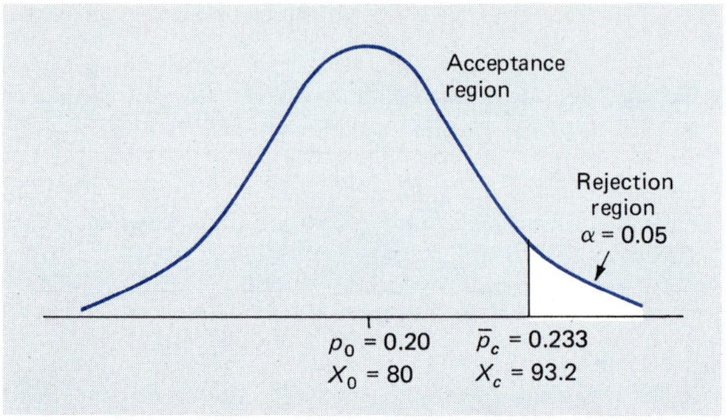

Figure 11.9
Hypothesis test of proportion of television viewers (Example 11.10).

Hypotheses: H_0: $p \leq 0.20$ versus H_1: $p > 0.20$

Critical values of $\overline{p}$:

(11.12)
$$\begin{aligned}\overline{p}_c &= p_0 + z_\alpha \sigma_{\overline{p}} \\ &= 0.20 + (1.65)(0.02) \\ &= 0.20 + 0.033 \\ &= 0.233\end{aligned}$$

Decision rule: A random sample of $n = 400$ is taken and $\overline{p}$ is calculated. If $\overline{p}$ is greater than or equal to 0.233, H_0 is rejected; otherwise, H_0 is accepted.

After the sample is taken, it is determined that the number of television sets in the sample tuned to Channel 32 is $X = 96$. The sample proportion is $X/n = 96/400 = 0.24$. Since 0.24 is greater than $\overline{p}_c = 0.233$, H_0 is rejected, Channel 32's claim is accepted, and Summit's advertising is given to Channel 32.

The decision rule for this test can also be stated in terms of X. Using the previous information, critical values of X are

(11.15)
$$\begin{aligned}X_c &= np_0 + z_\alpha \sigma_X \\ &= (400)(0.20) + (1.65)\sqrt{(400)(0.2)(0.8)} \\ &= 80 + (1.65)(8) \\ &= 80 + 13.2 \\ &= 93.2\end{aligned}$$

The decision rule in terms of X is: A random sample of $n = 400$ is taken and X, the number of successes, is determined. If X is greater than or equal to 93.2, H_0 is rejected; otherwise, H_0 is accepted.

Since there are 96 successes in the sample, the decision stands to reject H_0. This test is illustrated in Figure 11.9.

Example 11.11 In Example 11.3, the quality control department of the Global Office Machine Company wants to establish an acceptance procedure for shipments of a particular part that they purchase from a supplier. It is decided that shipments which contain 10% or fewer defectives should be accepted, but shipments containing more than 10% defectives should be returned to the supplier. A test is to be designed so that the probability of rejecting a good shipment, one with 10% or fewer defectives, is not to exceed 1%. A sample of size $n = 225$ is to be drawn from each shipment and the

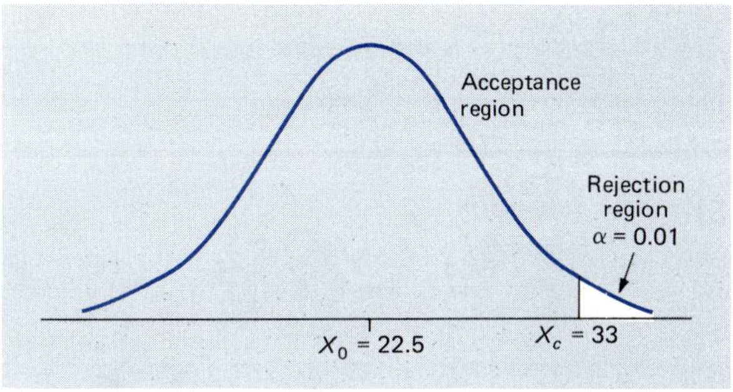

Figure 11.10
Hypothesis test of proportion of defectives (Example 11.11).

decision rule is formulated in terms of the number of defectives in the sample. The important information for this test is summarized as follows.

Population: Bernoulli (attributes)

Test statistic: X

Sample size: $n = 225$

Sampling distribution of X: approximately normal

Standard error of X:

(11.10)
$$\begin{aligned} \sigma_X &= \sqrt{np_0(1 - p_0)} \\ &= \sqrt{(225)(0.1)(0.9)} = 4.5 \end{aligned}$$

Significance level: $\alpha = 0.01$

Form of test: one-tailed

Hypotheses: H_0: $p \leq 0.10$ versus H_1: $p > 0.10$

Critical value of X:

(11.15)
$$\begin{aligned} X_c &= np_0 + z_\alpha \sigma_X \\ &= (225)(0.1) + (2.33)(4.5) \\ &= 22.5 + 10.485 \\ &= 32.985 \end{aligned}$$

Decision rule: A random sample of $n = 225$ is taken from each shipment. If X, the number of defectives in the sample, is greater than or equal to 32.985, the null hypothesis is rejected; otherwise, it is accepted.

It is obviously impossible to get 32.985 defectives in the sample. Therefore, the decision rule states that if 33 or more defectives are found, H_0 is rejected, but if there are 32 or fewer defectives in the sample, H_0 is accepted. This test is illustrated in Figure 11.10.

Example 11.12

If in the previous example it had been specified that each shipment consists of 2000 parts, the finite population correction factor would have to be applied to σ_X, since n/N is $225/2000 = 0.1125$. This results in the following change.

Standard error of X:

$$\sigma_X = \sqrt{np_0(1 - p_0)}\sqrt{\frac{N - n}{N - 1}}$$

$$= \sqrt{(225)(0.1)(0.9)}\sqrt{\frac{2000 - 225}{2000 - 1}}$$
$$= (4.5)(0.94)$$
$$= 4.23$$

Critical value of X:

(11.15) $$\begin{aligned} X_c &= np_0 + z_\alpha \sigma_X \\ &= (225)(0.1) + (2.33)(4.23) \\ &= 22.5 + 9.86 \\ &= 32.36 \end{aligned}$$

For practical purposes, the decision rule is the same as the one in the previous example.

11.8 Calculating the Probability of the Type II Error

In designing a statistical test of a hypothesis, the sample size and the significance level of the test are usually specified. This information, together with information about the population standard deviation and the form of the hypothesis to be tested, determines the decision rule for the test.

However, in Section 11.4 we pointed out that there are two types of errors associated with every hypothesis test. While the type I error results from a decision to reject the null hypothesis when it is true, a type II error results from a decision to accept the null hypothesis when it is false. Since α, the significance level, and also the probability of making a type I error, is usually specified, it becomes necessary to investigate β, the probability of making a type II error under the resulting decision rule.

A single value of α can always be associated with any test or decision rule. This value of α corresponds to the parameter value specified in the null hypothesis. For a two-tailed test in which the null hypothesis is of the form

$$H_0: \mu = k$$

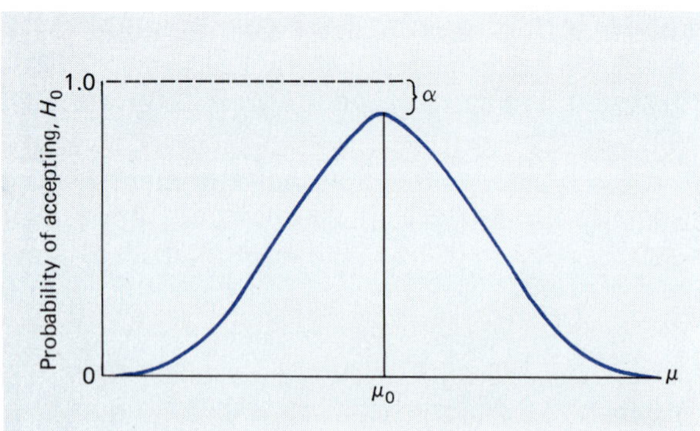

Figure 11.11
Operating characteristic curve for a two-tailed test.

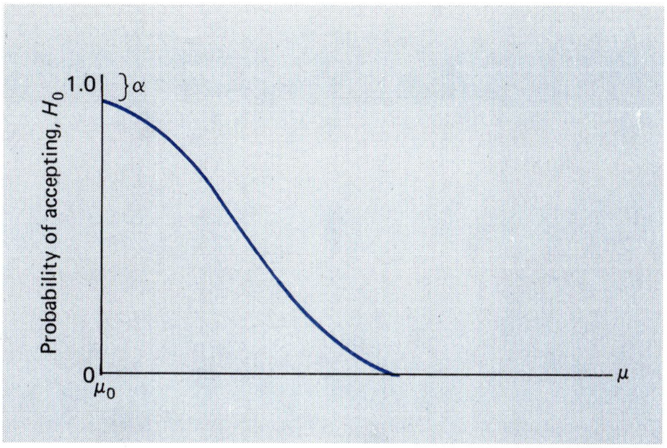

Figure 11.12
Operating characteristic curve for a one-tailed test.

α is the probability of rejecting H_0 when μ is exactly equal to k. For a one-tailed test of the form

$$H_0\colon \mu \leq k \qquad \text{or} \qquad H_0\colon \mu \geq k$$

the stated value of α is the maximum probability of rejecting the null hypothesis when it is true. This maximum probability applies when μ is exactly equal to k. For other values of μ for which H_0 is true, the probability of rejecting H_0 is less than this maximum.

On the other hand, there is no single value of β that can be associated with a test or decision rule. The probability of making a type II error varies depending on the true value of the parameter being tested, given that H_0 is false. Consequently, to determine a value for β, some alternate value of the unknown parameter must be specified. Usually, β is computed for a series of values of the parameter that make the null hypothesis false. These values of β are then plotted on a graph in which the horizontal scale represents alternate parameter values and the vertical scale represents the probability of accepting H_0. This graph is called an OPERATING CHARACTERISTIC CURVE, which is frequently abbreviated as "OC curve." Two operating characteristic curves are illustrated in Figures 11.11 and 11.12.

Example 11.13

Consider a one-tailed test of a hypothesis about a population mean. Assume that the null and alternate hypotheses are, respectively,

$$H_0\colon \mu \leq 75$$

and

$$H_1\colon \mu > 75$$

Population: normal

Standard deviation: $\sigma = 25$

Sample size: $n = 100$

Significance level: $\alpha = 0.05$

Test statistic: $\bar{x}$, $\sigma_{\bar{x}} = 25/\sqrt{100} = 2.5$

Sampling distribution of $\bar{x}$: normal

Critical value of $\bar{x}$:

$$\begin{aligned} \bar{x}_c &= \mu_0 + z_\alpha \sigma_{\bar{x}} \\ &= 75 + (1.65)(2.5) \\ &= 75 + 4.125 \\ &= 79.125 \end{aligned} \tag{11.2}$$

The decision rule for the test states that a random sample of $n = 100$ is drawn from the population, and if the sample mean is greater than or equal to 79.125, the null hypothesis is rejected; otherwise, it is accepted. The probability of making a type I error under this decision rule is 0.05.

The probability of making a type II error is determined under the assumption that H_0 is false, that is, that $\mu > 75$. However, to compute a specific value of β, an alternate value of μ must be indicated. Suppose, for example, we want to find a value of β if H_0 is false and μ is actually equal to 77. The relationship of the sampling distribution of $\bar{x}$ to the critical value of $\bar{x}$ under the decision rule stated above, and under two different assumptions about the population mean, is illustrated in Figure 11.13. The two assumptions about the population mean are $\mu = 75$ and $\mu = 77$. The dashed curve in the figure represents the sampling distribution of $\bar{x}$ under the first assumption, that is, $\mu = 75$. The portion of the right tail of the curve which falls in the rejection region is α. The unbroken-line curve represents the sampling distribution of $\bar{x}$ under the assumption that $\mu = 77$. The portion of the curve that lies in the acceptance region is β. The size of β is determined by finding the area under the curve which lies in the acceptance region, that is, to the left of the critical value of $\bar{x}$. This is easily accomplished by finding that value of z which corresponds to 79.125 with respect to the mean of 77, that is,

$$z = \frac{\bar{x}_c - \mu_1}{\sigma_{\bar{x}}} = \frac{79.125 - 77.0}{2.5} = 0.85$$

where μ_1 represents a value of the population mean corresponding to the alternate hypothesis. The area in the acceptance region under this as-

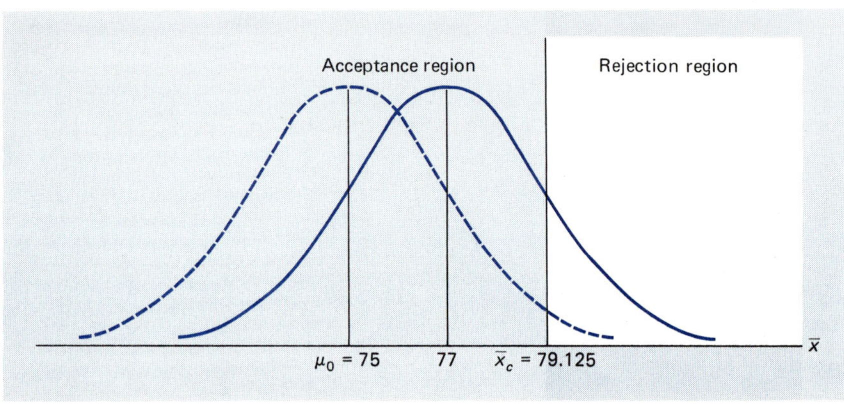

Figure 11.13
Probability of accepting or rejecting H_0 under different assumptions about the mean (Example 11.13).

TABLE 11.3
ACCEPTANCE PROBABILITIES FOR ALTERNATE VALUES OF THE MEAN (EXAMPLE 11.13)

Alternate Values of μ	*Probability of Accepting* H_0 *under the Decision Rule*	
75	0.95	$= 1 - \alpha$
76	0.8944	β
77	0.8023	β
78	0.6736	β
79	0.5199	β
80	0.3632	β
81	0.2266	β
82	0.1251	β

sumption corresponds to

$$P(z \leq 0.85) = 0.8023 = \beta$$

In order to construct an operating characteristic curve for this test, β is calculated for several other values of the parameter that make H_0 false. The procedure is exactly the same as that illustrated above, except for the values of μ_1. Probabilities of accepting H_0 under this test if μ has the values 76, 77, 78, 79, 80, 81, and 82 are provided in Table 11.3. The OC curve is shown in Figure 11.14.

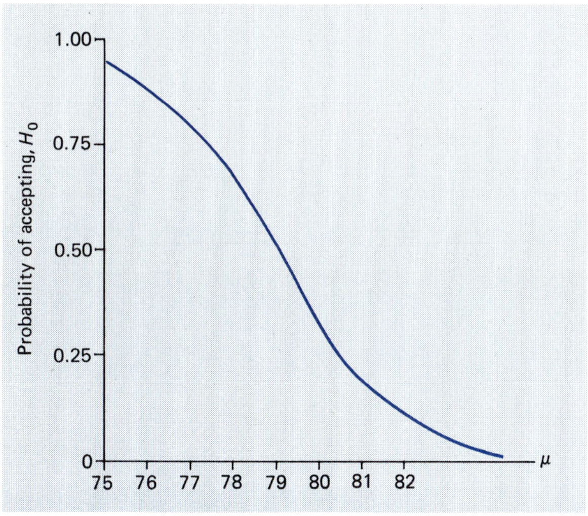

Figure 11.14
Operating characteristic curve for test of mean (Example 11.13).

Operating characteristic curves for proportions are constructed in the same way as for means. When the test is two-tailed, the OC curve resembles that in Figure 11.11. The construction of such a curve is left as an exercise for the reader.

TABLE 11.4
REJECTION PROBABILITIES FOR ALTERNATE VALUES OF THE MEAN

Alternate Values of μ	*Probability of Rejecting H_0 under the Decision Rule*	
75	0.05	$= \alpha$
76	0.1056	$1 - \beta$
77	0.1977	
78	0.3264	
79	0.4801	
80	0.6368	
81	0.7734	
82	0.8749	

The Power of a Test. The probability of making a type II error is defined as the probability of accepting the null hypothesis when it is false. Statisticians define the power of a test as the probability of *rejecting* the null hypothesis when it is false. In other words, the power of a test is the probability that it will lead to a correct decision when H_0 is false. For any specific value of the parameter which makes H_0 false, the power of the test is simply $1 - \beta$. Table 11.4 shows the power of the same test for which the values of β are given in Table 11.3. A graph of the power of the test for several values of the parameter that make H_0 false results in a POWER CURVE. This is illustrated in Figure 11.15.

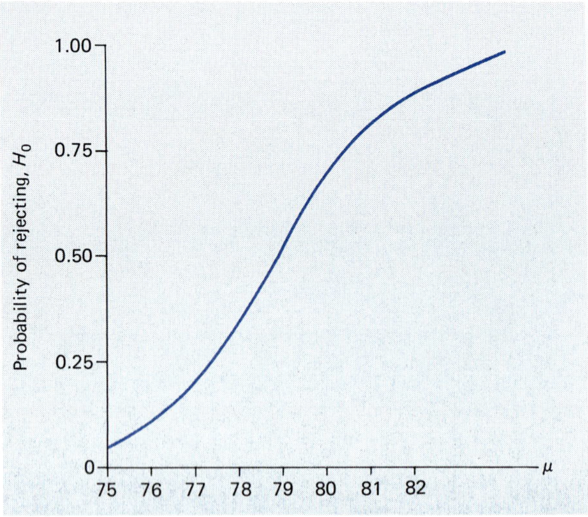

Figure 11.15
Power curve for test of mean.

Example 11.14

In Example 11.6, Cinderella Clothes designs a test of a hypothesis about the mean income of a particular population. The null and alternate hypotheses are, respectively,

$$H_0\colon\ \mu \leq 14{,}500$$

and

$$H_1\colon\ \mu > 14{,}500$$

The decision rule for this test states that a sample of $n = 225$ is drawn from the population and if $\bar{x}$, the mean of the sample, is greater than or equal to 14,610.01, H_0 is rejected; otherwise, H_0 is accepted. The standard error of the mean is estimated by $s_{\bar{x}} = 66.67$.

Find β and the power of the test if H_0 is false and μ is actually \$14,625.

If $\mu = 14{,}625$, then the mean of the sampling distribution of $\bar{x}$ is also 14,625. However, $\bar{x}_c$ is unchanged at 14,610.01. The probability of incorrectly accepting H_0 under these circumstances corresponds to the probability that $\bar{x}$ is less than 14,610.01. This probability is

$$\begin{aligned} P(\bar{x} < 14{,}610.01) &= P\left(z \leq \frac{14{,}610.01 - 14{,}625}{66.67}\right) \\ &= P(z \leq -0.22) = 0.413. \end{aligned}$$

Therefore, $\beta = 0.413$, and since the power of the test is simply $1 - \beta$, the power of this test if $\mu = 14{,}625$ is $1 - 0.413 = 0.587$.

Example 11.15

In Example 11.10, the Summit Record Company tested the hypotheses

$$H_0\colon\ p \leq 0.20$$

and

$$H_1\colon\ p > 0.20$$

with the following decision rule.

A random sample of $n = 400$ is taken and $\bar{p}$, the sample proportion, is calculated. If $\bar{p}$ is greater than or equal to 0.233, H_0 is rejected; otherwise, H_0 is accepted. For this test, $\sigma_{\bar{p}}$ is 0.02.

Find β and the power of the test if H_0 is false and p is actually 0.225.

If $p = 0.225$, the mean of the sampling distribution of $\bar{p}$ is 0.225. The critical value for the test, $\bar{p}_c$, is still 0.233. The probability of incorrectly accepting H_0 under this decision rule corresponds to

$$\begin{aligned} P(\bar{p} < 0.233) &= P\left(z \leq \frac{0.233 - 0.225}{0.02}\right) \\ &= P(z \leq 0.40) = 0.6554 \end{aligned}$$

Therefore, β is 0.6554. The power of the test is $1 - \beta = 0.3446$.

11.9 Testing Hypotheses with Critical Values of *z* and *t*

In all the hypothesis tests illustrated, the decision rule was defined with respect to the sampling distributions of either $\bar{x}$ or $\bar{p}$, depending on whether the test involves a mean or a proportion. However, it is always possible to test these hypotheses by defining the decision rule in relation to either the standard normal or the Student's t distribution; that is, instead of specifying critical values of $\bar{x}$ and $\bar{p}$, critical values of z and t can be used for the test. Then, after the sample is taken, the sample statistic is converted to a corresponding value of z or t which is compared to the critical value, z_c or t_c, to make the decision.

We have used the first procedure, that is, establishing the decision criteria in terms of $\bar{x}$ or $\bar{p}$, because it develops logically from the discussion of sampling distributions in Chapter 9 and closely resembles the procedure for constructing confidence intervals covered in Chapter 10. However, either procedure is valid, and the latter one is illustrated in the following examples.

Example 11.16 In Example 11.7, a test is established for the mean weight of cans. The sample size is $n = 25$ and the significance level of the test is 1%. The standard deviation of the population is equal to 0.55. The hypotheses are, respectively,

$$H_0: \mu = 16$$

and

$$H_1: \mu \neq 16$$

which implies a two-tailed test. The test statistic is $\bar{x}$.

Instead of defining the decision rule with respect to the sampling distribution of $\bar{x}$, we can define the decision rule with respect to the distribution of

$$z = \frac{\bar{x} - \mu_0}{\sigma_{\bar{x}}}$$

which has a standard normal distribution. The critical values of z are ± 2.58. The decision rule states:

A random sample of $n = 25$ is drawn from the population, $\bar{x}$, and the sample mean is converted to a value of z; if z is greater than or equal to 2.58 or less than or equal to -2.58, H_0 is rejected; otherwise H_0 is accepted. This test is illustrated in Figure 11.16.

The mean of the sample, as indicated in Example 11.7, was $\bar{x} = 16.29$. Therefore, the corresponding value of z was

$$z = \frac{16.29 - 16.00}{0.11} = 2.64$$

Since 2.64 is greater than the critical value of 2.58, H_0 is rejected.

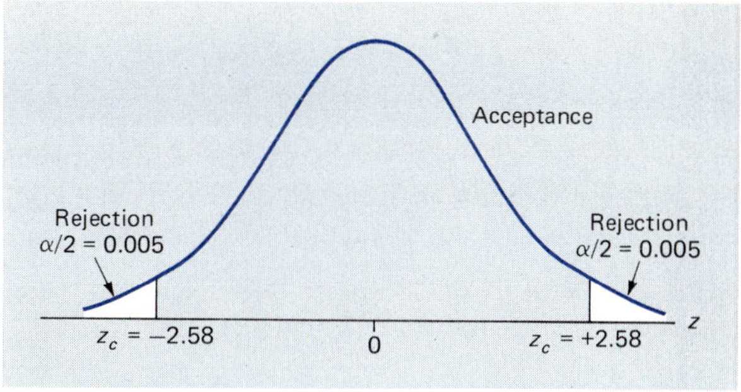

Figure 11.16
Critical value of z for test of mean (Example 11.16).

Example 11.17

The test in Example 11.8 is identical to the one in Example 11.7 except for the fact that σ is unknown and is estimated by the sample standard deviation, $s = 0.55$. Because of this and the small sample size, the critical values of $\bar{x}$ are based on the t distribution. The decision rule for this test can be based on critical values of the t statistic, as follows.

A random sample of $n = 25$ is drawn from the population and $\bar{x}$, the sample mean, is converted to a value of t from the transformation

$$t = \frac{\bar{x} - \mu_0}{s_{\bar{x}}}$$

If t is greater than or equal to 2.797 or less than or equal to -2.797, H_0 is rejected; otherwise, H_0 is accepted.

From Example 11.8, $\bar{x}$ is computed to be 16.29. Therefore,

$$t = \frac{16.29 - 16.00}{0.11} = 2.64$$

Since 2.64 is less than 2.797, H_0 is accepted. This test is illustrated in Figure 11.17.

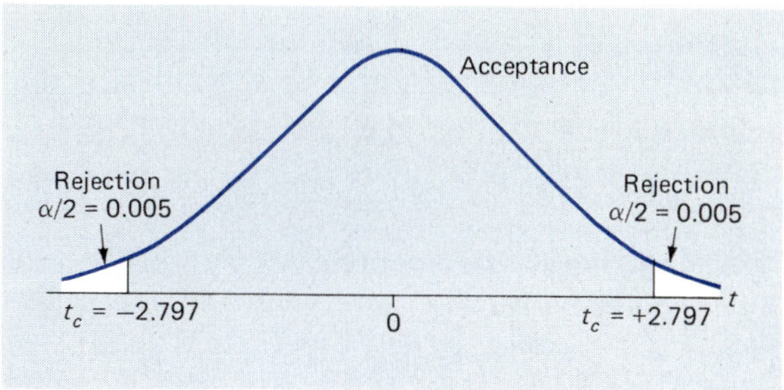

Figure 11.17
Critical value of t for test of mean (Example 11.17).

Example 11.18

In Example 11.11, a test of a hypothesis about the proportion of defectives in shipments is established. The significance level for the test is 1% and the hypotheses are, respectively,

$$H_0\colon p \leq 0.10$$

and

$$H_1\colon p > 0.10$$

A critical value for the test and the decision rule are specified with respect to the sampling distribution of X, the sample number of defectives, which is assumed to have a normal distribution. The critical value and the decision rule can be based upon z as follows.

A random sample of 225 parts is drawn from each shipment. The sample number, X, is converted to a value of z from

$$z = \frac{X - np_0}{\sigma_X}$$

If z is greater than or equal to 2.33, H_0 is rejected; otherwise, H_0 is accepted.

If a particular sample contains 34 defectives, the corresponding value of z is

$$z = \frac{34.0 - 22.5}{4.5} = 2.56$$

Since 2.56 is greater than 2.33, H_0 is rejected.

Definitions

Acceptance region: That portion of the area under the sampling distribution of a test statistic that corresponds to acceptance of the null hypothesis.

Alpha (α): The probability of making a type I error. The significance level of a test.

Beta (β): The probability of making a type II error.

Critical region: That portion of the area under the sampling distribution of a test statistic that corresponds to rejection of the null hypothesis.

Critical value: The value of a test statistic that divides the acceptance region from the critical region.

Decision rule: A rule that specifies the action to be taken with respect to the null hypothesis for all possible values of the test statistic.

One-tailed test: A hypothesis test in which the entire critical or rejection region is located in only one tail of the sampling distribution of the test statistic.

Operating characteristic curve: A graph of the probability of accepting the null hypothesis for alternate values of the parameter being tested.

Power: The probability that a test will reject the null hypothesis when it is false.

Power curve: A graph of the probability of rejecting the null hypothesis for alternate values of the parameter being tested.

Rejection region: Same as critical region.

Significance level: The probability level at which an observed deviation can no longer be attributed to chance.

Test statistic: The statistic used to test a hypothesis about a population parameter.

Two-tailed test: A hypothesis test in which the critical or rejection region for the test is divided between the two tails of the sampling distribution of the test statistic.

Type I error: Rejection of a true null hypothesis.

Type II error: Acceptance of a false null hypothesis.

Formulas

Critical values of the sample mean, two-tailed test, σ known:

(11.1) $$\bar{x}_c = \mu_0 \pm z_{\alpha/2}\sigma_{\bar{x}}$$

Critical value of the sample mean, one-tailed test, σ known:

(11.2) $$\bar{x}_c = \mu_0 + z_{\alpha}\sigma_{\bar{x}}$$

or

(11.3) $$\bar{x}_c = \mu_0 - z_{\alpha}\sigma_{\bar{x}}$$

Critical values of the sample mean, two-tailed test, σ unknown. Note that the t variate is used for small samples:

(11.4) $$\bar{x}_c = \mu_0 \pm z_{\alpha/2}s_{\bar{x}}$$

Critical value of the sample mean, one-tailed test, σ unknown. Note that the t variate is used for small samples:

(11.5) $$\bar{x}_c = \mu_0 + z_{\alpha}s_{\bar{x}}$$

or

(11.6) $$\bar{x}_c = \mu_0 - z_{\alpha}s_{\bar{x}}$$

Expected values and standard errors of binomial variates

(11.7) $$E(\bar{p}) = p$$

(11.8) $$\sigma_{\bar{p}} = \sqrt{\frac{p(1-p)}{n}}$$

(11.9) $$E(X) = np$$

(11.10) $$\sigma_X = \sqrt{np(1-p)}$$

Critical values of the sample proportion, two-tailed test

(11.11) $$\bar{p}_c = p_0 \pm z_{\alpha/2}\sigma_{\bar{p}}$$

Critical value of the sample proportion, one-tailed test

(11.12) $$\bar{p}_c = p_0 + z_{\alpha}\sigma_{\bar{p}}$$

or

(11.13) $$\bar{p}_c = p_0 - z_{\alpha}\sigma_{\bar{p}}$$

Critical values of the sample number, two-tailed test

(11.14) $$X_c = np_0 \pm z_{\alpha/2}\sigma_X$$

Critical value of the sample number, one-tailed test

(11.15) $$X_c = np_0 + z_\alpha \sigma_X$$

or

(11.16) $$X_c = np_0 - z_\alpha \sigma_X$$

Problems

11.1 Last year an audit of the accounts receivables of the Southwest Oil Company disclosed that the average account balance was \$62.73. This year, the auditors propose to take a sample in order to conclude whether or not that average has changed. Establish null and alternate hypotheses for this test and indicate whether it is a one- or two-tailed test.

11.2 The audit referred to in problem 11.1 also disclosed that 10% of all the accounts were more than 30 days past due. If a sample is taken this year to test the hypothesis that the percentage of delinquent accounts has not changed, state the null and alternate hypotheses. Is this a one- or two-tailed test?

11.3 A major car company makes the claim that its new small car, the Sprite, averages more than 42 miles per gallon of gas on the highway. A consumer group sets up a test of that claim. What are the null and alternate hypotheses for this test? Indicate whether it is a one- or two-tailed test.

11.4 The Green Brewery advertises that more people in Middletown drink their Green Label beer than all other beers together. In testing this claim, what are the null and alternate hypotheses? Is this a one- or two-tailed test?

11.5 If, in problem 11.1, it is concluded that the average account balance has not changed, when in fact the true mean is \$57.45, what type of error has been made?

11.6 Suppose that, in problem 11.2, it is concluded that the percentage of delinquent accounts is no longer exactly 10%, when in fact no change has occurred. What type of error has been made?

11.7 Explain the nature of a type I and type II error with reference to the test described in problem 11.3.

11.8 Explain the nature of the type I and type II error with reference to the test described in problem 11.4.

11.9 In testing the hypothesis that the mean of a normal population is equal to 25, the standard deviation of the population is assumed equal to 20. A sample of $n = 75$ is drawn from the population. The significance level of the test is 1%. Specify the null and alternate hypotheses and state the decision rule for the test.

11.10 Suppose that in the previous problem the hypothesis to be tested is that the mean is at least 25 and the distribution of the population is unknown. Specify the null and alternate hypotheses and state the decision rule. Does the lack of information about the population distribution make any difference in this case? Why or why not?

11.11 Work problem 11.9 with sample size $n = 25$.

11.12 Work problem 11.9 with sample size 25 and population standard deviation unknown. The sample standard deviation is $s = 30$.

11.13 Test the hypothesis established for problem 11.1 with a random sample of $n = 150$ at a significance level of 5%. State the decision rule and indicate the result of the test if $\bar{x} = \$60.75$ and $s = \$12.80$.

11.14 What is the probability that the sample mean in problem 11.13 is that small or smaller if the null hypothesis is true?

11.15 Consider the hypothesis tested in problem 11.9. Suppose that the decision rule for that test is as follows.

A random sample of $n = 75$ is drawn from the population. If $\bar{x}$ is less than or equal to 20 or greater than or equal to 30, H_0 is rejected; otherwise, H_0 is accepted.

What is the significance level of this test?

11.16 The Consumer Testing Bureau set up a test of a hypothesis about the gasoline mileage claimed for the Sprite automobile in problem 11.3. The test involves a random sample of five cars obtained from different dealers. Each car is started with a full gas tank and driven until all the gas is consumed. The sample mean is 46.3 miles per gallon and s is 9.0 miles. The significance level of the test is 5%.

a) State the decision rule for this test.

b) Should the car company's claim be accepted?

11.17 Suppose that the decision rule for the test in problem 11.16 is

A random sample of five cars is used in the test for gasoline mileage. If $\bar{x}$ is greater than or equal to 48.18, H_0 is rejected; otherwise, H_0 is accepted.

What is α for this test?

11.18 In testing the claim made for Green Label beer in problem 11.4, a random sample of 625 beer drinkers was taken. At a significance level of $2\frac{1}{2}$%, state the decision rule for the test.

11.19 In the previous problem, if 325 beer drinkers in the sample say that their brand is Green Label, what decision should be made about the null hypothesis? What is the probability of getting at least 325 Green Label drinkers in the sample if the true population proportion is actually 0.50?

11.20 The Lo-Voltage Electric Company claims that the average life of the television picture tubes that it manufactures is more than 2500 operating hours. A life test of a random sample of 100 of these tubes results in a sample mean of 2510 hours and a sample standard deviation of 50 hours.

a) Should their claim be accepted at $\alpha = 0.05$?

b) Should their claim be accepted at $\alpha = 0.01$?

11.21 Several years ago, losses incurred by Mountain Airlines due to no-shows prompted them to conduct a study to determine the average number of no-shows per flight. At that time, the average was 2.6. This year, a random sample of reservation records for 300 scheduled flights produced a mean of 2.1 with a standard deviation of 2.5. Can you conclude from this that the average number of no-shows per flight has changed at $\alpha = 0.01$?

11.22 The Reliance Insurance Company has data to show that the dollar value of automobile insurance claims has a normal distribution. The average value of all claims received last year was \$1050. A random sample of 20 claims received during the first quarter of this year has a mean of \$1150 and a standard deviation of \$200. At the 5% level of significance, can Reliance conclude that the average value of claims has increased since last year? Why or why not?

11.23 Work problem 11.22 if $\alpha = 0.01$.

11.24 Suppose in problem 11.20, the test had been conducted on a random sample of only 10 tubes. The sample mean and standard deviation are exactly the same. Lifetimes of television picture tubes have a normal distribution. Test the claim made by Lo-Voltage Electric at both the 5% and 1% levels of significance.

11.25 Metric Systems, Inc. purchases parts for one of its products from the Piedmont Metals Company. The production manager of Metric Systems has been complaining to the purchasing manager that the lead time required by Piedmont to make delivery is much too long. The purchasing manager disputes this, saying that the average time from submitting a purchase order until the shipment is delivered is no greater than 30 days. A random sample of 15 purchase orders is selected. The sample mean is 33 days with a standard deviation of seven days. Test the hypothesis that the average lead time is not greater than 30 days at $\alpha = 0.05$.

11.26 In testing the hypothesis that the value of a population proportion is at least 0.60 at a significance level of 0.10, a random sample of 250 elements is drawn from the population.

a) State the null and alternate hypotheses for this test.

b) State the decision rule in terms of $\overline{p}$.

c) State the decision rule in terms of X.

11.27 The president of Silver Suds, Inc., a manufacturer of household cleansers, is presented with a new dishwashing liquid that is supposed to be better than the competitive products on the market. After consulting with the marketing and financial vice-presidents, the president decides that the new product should be marketed if a test indicates that it is preferred by more than 25% of the people who try it. A market research firm was hired to randomly distribute 300 bottles of the new product. After a trial period, each person was asked whether the new dishwashing product is better than the previously used product.

Set up a test so that the probability of incorrectly concluding that the new product should be marketed is no greater than 2%. State the decision rule in terms of both the sample proportion, $\overline{p}$, and the sample number, X.

11.28 An agency of the federal government allocates funds to poverty programs in various municipalities according to the proportion of families in the area whose incomes are below the poverty level. When the program was started in Middletown, this proportion was determined to be 4%. This year, each municipality must determine whether or not that proportion has changed. A research firm is hired by the Middletown mayor's office to take a random sample of 1000 families and test the hypothesis that the proportion has not changed at a significance level of 1%. After the sample is taken, 57 of the families in the sample have incomes below the poverty level.

a) State the null and alternate hypotheses for this test.

b) State the decision rule.

c) Should the null hypothesis be accepted or rejected?

11.29 Suppose that the critical value of $\overline{p}$ for the test in problem 11.26 is 0.55. What is α for this test?

11.30 Suppose that the decision rule in problem 11.27 specifies that H_0 is rejected if X, the sample number, is greater than or equal to 88. What is α for this test?

11.31 Find α for the test in problem 11.28 if the decision rule had stated that H_0 would be rejected if $\overline{p}$ is less than or equal to 0.03 or greater than or equal to 0.05.

11.32 Suppose that the true population proportion in problem 11.27 is actually 0.33. Find β and the power of the test.

11.33 Suppose that the true population proportion in problem 11.26 is actually 0.54. Find β and the power of the test.

11.34 Suppose that the true population proportion in problem 11.28 is actually 0.03. Find β and the power of the test.

11.35 Consider the hypothesis test in problem 11.13. Suppose that the true population mean is actually \$62.00. Find β and the power of the test.

11.36 Restate the decision rule of problem 11.9 in terms of critical values of z.

11.37 State the decision rule for the hypothesis test of problem 11.13 in terms of critical values of z.

11.38 Restate the decision rule for the hypothesis test of problem 11.16 in terms of a critical value of t.

11.39 Restate the decision rule for the hypothesis test of problem 11.18 in terms of a critical value of z.

11.40 Restate the decision rule for the hypothesis test of problem 11.28 in terms of a critical value of z.

CHAPTER 12 HYPOTHESIS TESTING–TWO SAMPLE TESTS

12.1
The test statistic for tests of differences of means and proportions

12.2
Test of difference between means—large samples

12.3
Test of difference between means—small samples

12.4
Test of difference between proportions

12.5
Tests using paired observations

12.6
The F distribution

12.7
Test of the equality of variances

In Chapter 11, we considered tests about the value of a parameter of a *single* population. The decision to accept or reject the null hypothesis was based on data from a single sample from that population. There are, however, circumstances in which the focus of an investigation may be on two distinct populations. In many cases, interest is not in specific values of the population parameters, but in determining whether or not the values of the parameters of the two populations are equal. The null hypothesis to be tested is a hypothesis of *no difference* in parameter values. In terms of means and proportions, the two sample tests that we will study will test the following hypotheses.

$$H_0\colon \mu_1 = \mu_2 \qquad \text{versus} \qquad H_1\colon \mu_1 \neq \mu_2$$

and

$$H_0\colon p_1 = p_2 \qquad \text{versus} \qquad H_1\colon p_1 \neq p_2$$

where the subscripts 1 and 2 refer to populations 1 and 2.

In order to test hypotheses of this type, a random sample is drawn from each population. Consequently, we have a test based on two samples. For this type of test, the two samples must be *independent* of one another, that is, the value of any observation in one sample is unrelated to the value of any observation in the other sample. Consider two situations in which this type of test might be used.

Example 12.1 The Quality Appliance Company manufactures television receivers. The picture tubes installed in these sets are purchased from two other manufacturers, namely Lo-Voltage Electric Company and Midwestern Electronics. It has always been assumed that the average lifetimes of the tubes produced by both suppliers are the same. However, the new production manager of Quality Appliances decides that this assumption should be tested. Consequently, random samples of tubes produced by each manufacturer are placed on life test. The mean life of each sample is computed and the two means are compared. If the two sample means differ significantly, the null hypothesis that the population means are equal is rejected.

Example 12.2 The actuarial department of the Reliance Insurance Company planned a test to determine if there is any difference between the proportion of male and female teenage drivers involved in automobile accidents in which the damage to the driver's car exceeds $100. Independent random samples of male and female teenage drivers are drawn from the company's files of policyholders. The proportion of drivers in each sample who were involved in accidents during the previous year is determined. These two proportions are then compared. If the two sample proportions differ significantly, the null hypothesis that the population proportions are equal is rejected.

12.1 The Test Statistic for Tests of Differences of Means and Proportions

The null hypothesis in these tests is an assumption of *no difference* in parameter values and therefore states that $|\mu_1 - \mu_2| = 0$ and $|p_1 - p_2| = 0$. Recall that the two vertical lines designate absolute value.

Considering the test of means, if independent random samples are taken from each population and if the means of each sample are computed, then $\bar{x}_1$ is the mean of the first sample and $\bar{x}_2$ is the mean of the second. Since the two samples should be representative of their respective populations, we would expect $\bar{x}_1$ to be approximately equal to μ_1 and $\bar{x}_2$ to be approximately equal to μ_2. Also, $\bar{x}_1 - \bar{x}_2$ would be approximately equal to $\mu_1 - \mu_2$. If we define a statistic $d^* = \bar{x}_1 - \bar{x}_2$ and if $\mu_1 = \mu_2$, then the expected value of d^* is zero; that is, $E(d^*) = E(\bar{x}_1 - \bar{x}_2) = \mu_1 - \mu_2 = 0$. Read d^* as "d-star."

Furthermore, if $\bar{x}_1$ and $\bar{x}_2$ have normal distributions, either because the respective populations are normal or because of the Central Limit Theorem, then the distribution of d^* is also normal, with a mean equal to $\mu_1 - \mu_2$ and a variance of

$$\sigma^2_{d*} = \frac{\sigma_1^2}{n_1} + \frac{\sigma_2^2}{n_2} \tag{12.1}$$

where n_1 and n_2 are the sizes of the two samples. The standard error of d^* is the square root of the variance, that is,

$$\sigma_{d*} = \sqrt{\frac{\sigma_1^2}{n_1} + \frac{\sigma_2^2}{n_2}} \tag{12.2}$$

Sampling distributions of $\bar{x}_1$, $\bar{x}_2$, and d^* are illustrated in Figure 12.1.

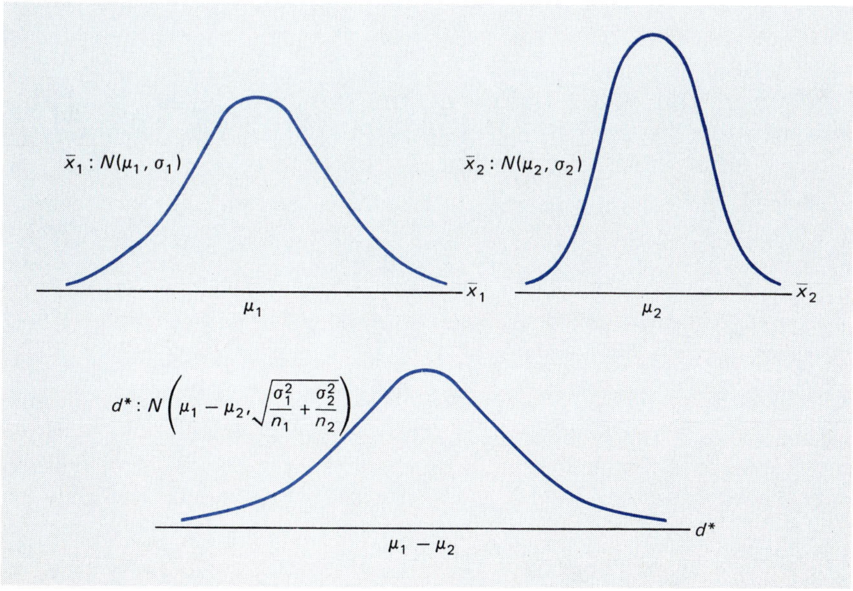

Figure 12.1
Sampling distributions of $\bar{x}_1$*,* $\bar{x}_2$*, and* d^**.*

In tests of differences between proportions, the test statistic is also denoted by d^*, where $d^* = \bar{p}_1 - \bar{p}_2$, the difference between the sample proportions. If the null hypothesis is true and there is no difference between the two population proportions, then $E(d^*) = 0$ and the standard error of d^* is

$$\sigma_{d*} = \sqrt{\frac{p(1-p)}{n_1} + \frac{p(1-p)}{n_2}} \tag{12.3}$$

Also, if $\bar{p}_1$ and $\bar{p}_2$ have distributions that are approximately normal, then the distribution of d^* is also approximately normal.

12.2 Test of Difference between Means—Large Samples

The null hypothesis for the test of the difference between two means is

$$H_0\colon \mu_1 = \mu_2$$

that is, the difference equals zero. The alternate hypothesis is then

$$H_1\colon \mu_1 \neq \mu_2$$

A random sample is drawn from each population, the mean of each sample is calculated, and the value of the statistic $d^* = \bar{x}_1 - \bar{x}_2$ is determined. If the null hypothesis is true, then $E(d^*) = 0$. Therefore, if d^* deviates significantly from 0 in either direction, the null hypothesis should be rejected. In order to establish a decision rule for this test, the significance level must be specified. Critical values of d^* are obtained from

$$d_c^* = 0 \pm z_{\alpha/2}\,\sigma_{d*} \tag{12.4}$$

where

$$\sigma_{d*} = \sqrt{\frac{\sigma_1^2}{n_1} + \frac{\sigma_2^2}{n_2}} \tag{12.2}$$

When σ_1 and σ_2 are unknown, the standard deviations of the samples must be used to estimate the population standard deviations. Under these circumstances, the standard error of d^* is estimated from

$$s_{d*} = \sqrt{\frac{s_1^2}{n_1} + \frac{s_2^2}{n_2}} \tag{12.5}$$

and critical values of d^* are obtained from

$$d_c^* = 0 \pm z_{\alpha/2}\,s_{d*} \tag{12.6}$$

Example 12.3 In Example 12.1, the production manager of the Quality Appliance Company tests the hypothesis that the mean lifetimes of television picture tubes manufactured by Lo-Voltage Electric and Midwestern Electronics are equal. If Lo-Voltage is designated company 1 and Midwestern company 2, the hypotheses are

$$H_0\colon \mu_1 = \mu_2$$

and

$$H_1\colon \mu_1 \neq \mu_2$$

Suppose that random samples of 100 tubes produced by each company are life tested, resulting in the following values.

$$\bar{x}_1 = 2560 \text{ hours}, \qquad s_1 = 210 \text{ hours}$$
$$\bar{x}_2 = 2640 \text{ hours}, \qquad s_2 = 300 \text{ hours}$$

The significance level for the test is $\alpha = 0.05$.

The important information concerning this test is summarized as follows.

Populations: unknown. Population standard deviations: unknown, estimated by the sample standard deviations.

Test statistic: $d^* = \bar{x}_1 - \bar{x}_2$. Sample sizes: $n_1 = 100$, $n_2 = 100$.

Sampling distribution of d^*: normal, by Central Limit Theorem. Standard error of d^*: estimated by

$$s_{d^*} = \sqrt{\frac{s_1^2}{n_1} + \frac{s_2^2}{n_2}} = \sqrt{\frac{(210)^2}{100} + \frac{(300)^2}{100}} = 36.62 \tag{12.5}$$

Significance level of test: $\alpha = 0.05$.

Form of test: two-tailed.

Hypotheses: $H_0\colon \mu_1 = \mu_2$ versus $H_1\colon \mu_1 \neq \mu_2$.

Critical values of d^*:

$$\begin{aligned} d_c^* &= 0 \pm z_{\alpha/2}\, s_{d^*} \\ d_c^* &= 0 \pm (1.96)(36.62) \\ d_c^* &= -71.77 \quad \text{and} \quad +71.77 \end{aligned} \tag{12.6}$$

Decision rule: Random samples of $n_1 = 100$ and $n_2 = 100$ are drawn from each population. The mean of each sample is calculated and the value of $d^* = \bar{x}_1 - \bar{x}_2$ is determined. If d^* is less than or equal to -71.77 or greater than or equal to $+71.77$, H_0 is rejected. Otherwise, H_0 is accepted.

Since the value of d^* obtained from the samples is

$$d^* = 2560 - 2640 = -80$$

which is less than -71.77, the null hypothesis is rejected. As a result of this test, the production manager of Quality Appliances concludes that the tubes produced by Midwestern have a longer mean lifetime than those produced by Lo-Voltage. This test is illustrated in Figure 12.2.

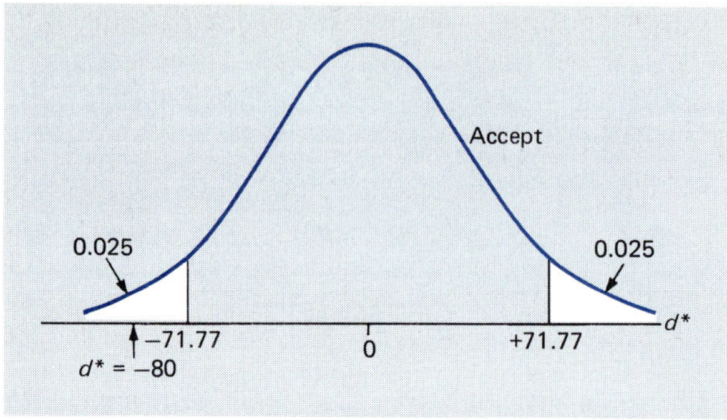

Figure 12.2
Test of difference between mean lifetimes of television picture tubes, large samples (Example 12.3).

12.3 Test of Difference between Means—Small Samples

We encounter similar problems in testing hypotheses about the difference between two population means with small samples as we do when the hypotheses involve only a single mean. Since the Central Limit Theorem does not apply to small samples, the populations from which the samples are drawn must be approximately normal. If the populations are normal, with known standard deviations, then the procedure is the same as for large samples. However, if the population standard deviations are not known, the procedure differs from the large sample case.

When small samples are used to test a hypothesis about the difference between the means of two normal populations with unknown standard deviations, it is necessary to assume that the two populations have *equal* standard deviations. If this assumption is valid, then an estimate of the common population standard deviation is made by pooling the data from both samples as follows.

$$s_p = \sqrt{\frac{(n_1 - 1)s_1^2 + (n_2 - 1)s_2^2}{n_1 + n_2 - 2}} \tag{12.7}$$

The *pooled estimate*, s_p, of the common population standard deviation, σ, is then used to find an estimate of the standard error of d^*, that is,

$$s_{d*} = \sqrt{\frac{s_p^2}{n_1} + \frac{s_p^2}{n_2}} \tag{12.8}$$

A simple computational form for estimating the standard error of d^* is

$$s_{d*} = s_p\sqrt{\frac{1}{n_1} + \frac{1}{n_2}} \tag{12.9}$$

Critical values of d^* are based on the t multiplier and s_{d*}. The sampling distribution of d^* must be treated as a t distribution because the population standard deviations are unknown. This is analagous to the situation where the sampling distribution of $\bar{x}$ is treated as a t distribution in single-sample tests when the population standard deviation is unknown. The degrees of freedom for a two-sample test are $n_1 + n_2 - 2$. The critical values of d^* are obtained from

$$d_c^* = 0 \pm t_{\alpha/2}\, s_{d*} \tag{12.10}$$

Whenever the degrees of freedom in a two-sample test of means are greater than 30, the z multiplier and the large sample methods may be used.

Example 12.4

Suppose in Example 12.3 that the lifetimes of television picture tubes have normal distributions and the two populations in that example have equal, though unknown, standard deviations. Also, suppose that only 10 Lo-Voltage tubes and 15 Midwestern Electronics tubes are available for testing. The sample means and standard deviations are

$$\begin{aligned} \bar{x}_1 &= 2560 \text{ hours}, \qquad & s_1 &= 210 \text{ hours} \\ \bar{x}_2 &= 2640 \text{ hours}, \qquad & s_2 &= 300 \text{ hours} \end{aligned}$$

The important information for this test is

Populations: normal. Population standard deviations: unknown but assumed to be equal; the common σ is estimated by

$$\begin{aligned} s_p &= \sqrt{\frac{(n_1 - 1)s_1^2 + (n_2 - 1)s_2^2}{n_1 + n_2 - 2}} \\ &= \sqrt{\frac{(9)(210)^2 + (14)(300)^2}{10 + 15 - 2}} \\ &= 268.4 \end{aligned} \tag{12.7}$$

Test statistic: $d^* = \bar{x}_1 - \bar{x}_2$. Sample sizes: $n_1 = 10$, $n_2 = 15$.

Sampling distribution of d^*: normal but critical values are based on the t distribution.

Standard error of d^*: estimated by

$$\begin{aligned} s_{d*} &= \sqrt{\frac{s_p^2}{n_1} + \frac{s_p^2}{n_2}} \\ &= \sqrt{\frac{(268.4)^2}{10} + \frac{(268.4)^2}{15}} \\ &= 109.57 \end{aligned} \tag{12.8}$$

Significance level of test: $\alpha = 0.05$.

Form of test: two-tailed.

Hypotheses: $H_0\colon \mu_1 = \mu_2$ versus $H_1\colon \mu_1 \neq \mu_2$.

Critical values of d^*:

$$\begin{aligned} d_c^* &= 0 \pm t_{\alpha/2}\, s_{d^*} \\ &= 0 \pm (2.069)(109.57) \\ &= -226.7 \quad \text{and} \quad +226.7 \end{aligned}$$

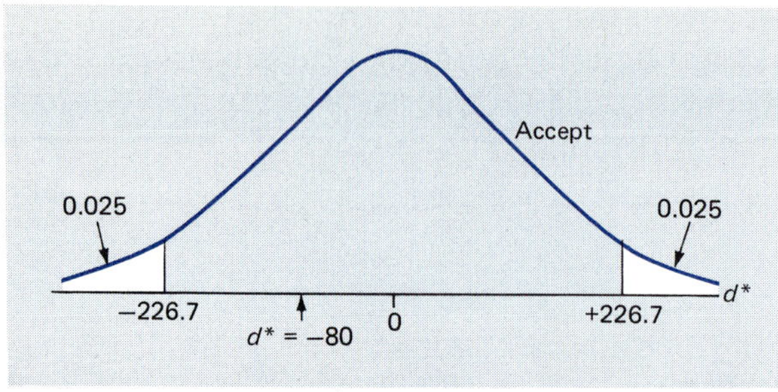

Figure 12.3
Test of difference between mean lifetimes of television picture tubes, small samples (Example 12.4).

Decision rule: Random samples of $n_1 = 10$ and $n_2 = 15$ are drawn respectively from each population. The mean of each sample is calculated and the value of $d^* = \bar{x}_1 - \bar{x}_2$ is determined. If d^* is less than or equal to -226.7 or greater than or equal to $+226.7$, H_0 is rejected. Otherwise, H_0 is accepted.

Since $d^* = 2560 - 2640 = -80$, the null hypothesis is accepted. It is not possible to conclude that there is a difference in the mean lifetimes on the basis of this test. Figure 12.3 illustrates this test.

12.4 Test of Difference between Proportions

The test of a difference between proportions is very similar to the test of a difference between means described in the previous sections. The null hypothesis in this test is H_0: $p_1 = p_2$, the hypothesis of no difference. The alternate hypothesis is H_1: $p_1 \neq p_2$. The test is two-tailed and the test statistic is $d^* = \bar{p}_1 - \bar{p}_2$. If the null hypothesis is true, then $E(d^*) = 0$.

Recall that $\bar{p}$, the sample proportion, has a binomial distribution. However, for large samples, the binomial distribution can be approximated by a normal distribution having the same mean and standard deviation. We shall only consider tests in which both samples are large enough for the sample proportions to be approximately normally distributed. Then, if $\bar{p}_1$ has a normal distribution and $\bar{p}_2$ has a normal distribution, $d^* = \bar{p}_1 - \bar{p}_2$ will also have a normal distribution with a mean equal to $p_1 - p_2$. The standard error of d^* is

$$\sigma_{d^*} = \sqrt{\frac{p_1(1 - p_1)}{n_1} + \frac{p_2(1 - p_2)}{n_2}} \tag{12.11}$$

If the null hypothesis is true and $p_1 = p_2$, then there is a single value of p which is common to both populations, and Equation 12.11 becomes

$$\sigma_{d^*} = \sqrt{\frac{p(1 - p)}{n_1} + \frac{p(1 - p)}{n_2}} \tag{12.3}$$

In order to evaluate σ_{d*} from the sample data, it is necessary to estimate this common value of p under the assumption that the null hypothesis is true. This is done by pooling the data from both samples. The *pooled estimate* of p is designated $\hat{p}$, p hat, and is

$$\hat{p} = \frac{n_1\bar{p}_1 + n_2\bar{p}_2}{n_1 + n_2} \tag{12.12}$$

The pooled estimate, $\hat{p}$, is substituted for p to obtain s_{d*}, the sample estimate of σ_{d*}. That is,

$$\hat{\sigma}_{d*} = s_{d*} = \sqrt{\frac{\hat{p}(1-\hat{p})}{n_1} + \frac{\hat{p}(1-\hat{p})}{n_2}} \tag{12.13}$$

A more convenient computational form for s_{d*} is

$$s_{d*} = \sqrt{\frac{\hat{p}(1-\hat{p})(n_1+n_2)}{n_1 n_2}} \tag{12.14}$$

Critical values of d^* are then obtained from

$$d_c^* = 0 \pm z_{\alpha/2}\, s_{d*} \tag{12.15}$$

Example 12.5

In Example 12.2, the actuarial department of the Reliance Insurance Company wants to test the hypothesis that there is no difference between the proportion of male and female teenage drivers involved in accidents in which the damage is over $100. Random samples of the records of 900 male and 750 female policyholders under the age of twenty are drawn from the company's files. During the previous year, 42 of the male drivers and 23 of the female drivers in the samples were involved in accidents. If the population of males is designated population 1 and the population of females population 2, then

$$\bar{p}_1 = \frac{42}{900} = 0.047$$

and

$$\bar{p}_2 = \frac{23}{750} = 0.031$$

The significance level for this test is $\alpha = 0.05$.

The important information for this test is summarized as follows.

Populations: Bernoulli (attributes).

Test statistic: $d^* = \bar{p}_1 - \bar{p}_2$. Sample sizes: $n_1 = 900$, $n_2 = 750$.

Sampling distribution of d^*: approximately normal.

Pooled estimate of p:

$$\begin{aligned} \hat{p} &= \frac{n_1\bar{p}_1 + n_2\bar{p}_2}{n_1 + n_2} \\ &= \frac{(900)(0.047) + (750)(0.031)}{900 + 750} \\ &= 0.039 \end{aligned} \tag{12.13}$$

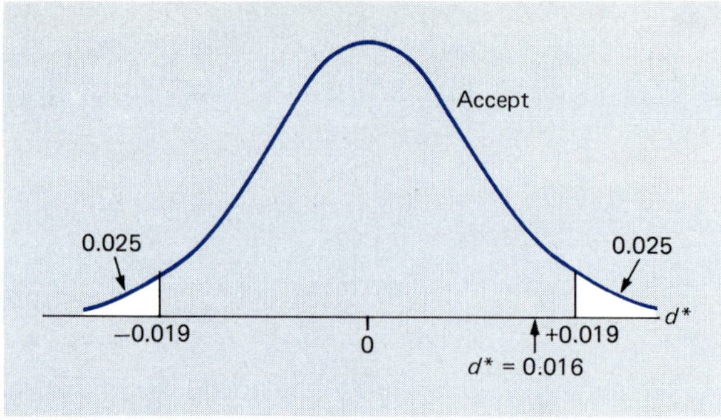

Figure 12.4
Test for difference in proportion of accidents between male and female teenage drivers (Example 12.5).

Standard error of d^*: estimated by

(12.14) $$s_{d*} = \sqrt{\frac{\hat{p}(1-\hat{p})(n_1+n_2)}{n_1 n_2}}$$

$$= \sqrt{\frac{(0.039)(0.961)(900+750)}{(900)(750)}}$$

$$= 0.0096$$

Significance level of test: $\alpha = 0.05$.

Form of test: two-tailed.

Hypotheses: H_0: $p_1 = p_2$ versus H_1: $p_1 \neq p_2$.

Critical values of d^*:

(12.15) $$d_c^* = 0 \pm z_{\alpha/2}\, s_{d*}$$
$$= 0 \pm (1.96)(0.0096)$$
$$= -0.019 \quad \text{and} \quad +0.019$$

Decision rule: Random samples of $n_1 = 900$ and $n_2 = 750$ are drawn respectively from each population. The sample proportion is determined for each sample and the value of $d^* = \bar{p}_1 - \bar{p}_2$ is calculated. If d^* is less than or equal to -0.019 or greater than or equal to $+0.019$, the null hypothesis is rejected; otherwise, H_0 is accepted.

Since $d^* = 0.047 - 0.031 = 0.016$, the null hypothesis is accepted. There appears to be no significant difference between the two proportions at the 5% level.

This test is illustrated in Figure 12.4.

12.5 Tests Using Paired Observations

In the previous two-sample tests, we specified that the two samples must be both *random* and *independent*. In other words, there must be no relationship between an observation obtained from one sample and a similar observation from the other. The examples in the previous sections appear to meet this requirement of independence. There are, however, circumstances in which two sets of observations may not be independent of one another. When some relationship can be shown between one member or observation of one sample and some corresponding member or observation from the other, then the related observations should be paired. A typical situation occurs when two observations are obtained from the same elementary unit, usually at different times. Under these circumstances, although we have two sets of sample values, the previous two sample tests are not applicable because the two samples are not independent. Consider the following example.

Example 12.6

After conducting a time and motion study on a particular manufacturing assembly operation, an industrial engineer on the staff of Metric Systems, Inc. suggests that a new procedure be adopted. He claims that the new procedure would reduce the average time required to perform this operation. The production manager, however, decides to conduct a test to see if the new procedure actually does reduce the average assembly time. He also does not want to ignore the possibility that the average time might actually increase under the new procedure.

A random sample of 10 assemblers is selected. Each assembler is timed using the old procedure. Then the same 10 assemblers are trained for two days in the new procedure and are timed again as they perform the operation. The two sets of assembly times are shown in Table 12.1, where

TABLE 12.1

DIFFERENCE BETWEEN ASSEMBLY TIMES USING TWO ASSEMBLY PROCEDURES (EXAMPLE 12.6)

Assembler	*Old Procedure Time in Minutes* (x)	*New Procedure Time in Minutes* (y)	*Difference* ($d = x - y$)	$d - \bar{d}$	$(d - \bar{d})^2$
1	14.6	14.0	0.6	−0.48	0.2304
2	10.7	11.4	−0.7	−1.78	3.1684
3	12.1	10.2	1.9	0.82	0.6724
4	13.9	10.4	3.5	2.42	5.8564
5	11.1	11.1	0	−1.08	1.1664
6	11.0	9.9	1.1	0.02	0.0004
7	10.4	9.5	0.9	−0.18	0.0324
8	10.9	10.9	0	−1.08	1.1664
9	13.0	11.6	1.4	0.32	0.1024
10	14.0	11.9	2.1	1.02	1.0404
			10.8		13.4360

$$\bar{d} = \frac{\Sigma d}{n} = \frac{10.8}{10} = 1.08$$

$$s_d^2 = \frac{\Sigma(d - \bar{d})^2}{n - 1} = \frac{13.4360}{9} = 1.493$$

$$s_d = \sqrt{1.493} = 1.22$$

$$s_{\bar{d}} = \frac{s_d}{\sqrt{n}} = \frac{1.22}{\sqrt{10}} = 0.386$$

the random variable x represents the times under the old procedure and y represents the times using the new method. The third column in the table provides values of a new random variable $d = x - y$, which is simply the difference between the two times for each assembler. A positive value of d represents a reduction in assembly time, while a negative value represents an increase. Be sure that you distinguish between the *random variable* d, the difference between two individual observations, and the *statistic* d^*, which is the difference between two sample means.

The hypothesis to be tested here is that there is no significant difference in the average time required to perform the operation between the two different procedures. The significance level selected for the test is $\alpha = 0.05$. If this hypothesis is true, then the average or mean difference should be zero. Therefore, the null and alternate hypotheses are, respectively,

$$H_0\colon \mu_d = 0$$

and

$$H_1\colon \mu_d \neq 0$$

where μ_d represents the mean of the distribution of a population of values of d. If the null hypothesis is true and if both the x's and y's have normal distributions, then $d = x - y$ also has a normal distribution with a mean of $\mu_d = 0$ and a standard deviation of σ_d. To test the hypothesis about the mean of the distribution, the test statistic should be $\bar{d}$, the mean of the sample differences. The standard deviation of the sample differences, s_d, is used as an estimate of σ_d. Both of these values are computed in Table 12.1.

Since d has a normal distribution with an unknown standard deviation which must be estimated by the sample standard deviation, the sampling distribution of $\bar{d}$ is treated as a t distribution with a mean equal to μ_d and a standard error of $s_{\bar{d}} = s_d/\sqrt{n}$. Critical values of $\bar{d}$ are then determined from

$$\begin{aligned} \bar{d}_c &= 0 \pm t_{\alpha/2} s_{\bar{d}} \\ &= 0 \pm (2.262)(0.386) \\ &= -0.873 \quad \text{and} \quad +0.873. \end{aligned} \tag{12.16}$$

Since $\bar{d} = 1.08$, which is greater than 0.873, the null hypothesis is rejected.

The reader should note that even though we have two sets of sample observations, this test is identical to a small sample test of a

Figure 12.5
Test of mean difference between assembly procedures (Example 12.6).

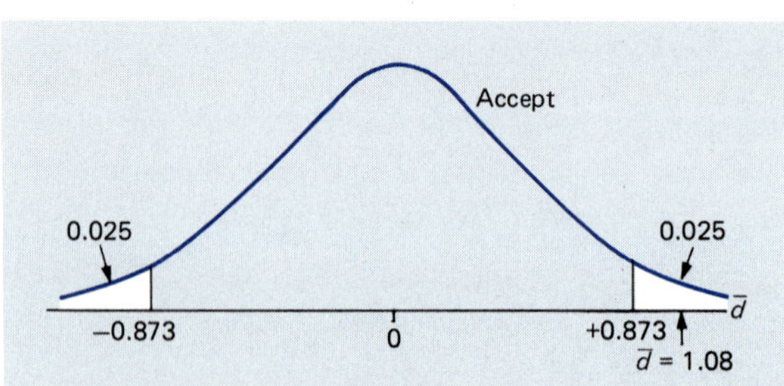

single mean. We use the t distribution to determine critical values of $\bar{d}$ because σ_d is unknown and is estimated by s_d. Of course, if the degrees of freedom are greater than 30, then z and the standard normal distribution may be used as an approximation.

This test is illustrated in Figure 12.5.

12.6 The *F* Distribution

As new statistical tests are introduced, it is sometimes necessary to introduce new sampling distributions for the test statistic. In Section 12.7, a test about the equality of two variances is described. The test statistic, designated F, is a ratio of two sample variances. The sampling distribution of this statistic is the F distribution. The statistic, F, is sometimes referred to as the *variance ratio*.

Like the t distribution, the F distribution is characterized by the number of degrees of freedom. However, unlike the t distribution, the F distribution is not symmetrical and has *two different* degrees of freedom, one for each of the sample variances that make up the ratio. These are designated df_1 and df_2, respectively, for the numerator and denominator of the ratio. The degrees of freedom, df_1 and df_2, are the two parameters of the F distribution.

If two independent random samples of size n_1 and n_2 are drawn from *normal* populations, then the statistic $(s_1^2/\sigma_1^2)/(s_2^2/\sigma_2^2)$ has an F distribution with $df_1 = n_1 - 1$ and $df_2 = n_2 - 1$. If the two population variances are equal, that is, $\sigma_1^2 = \sigma_2^2$, then this ratio reduces to s_1^2/s_2^2.

The F distribution is skewed to the right and extends from zero to plus infinity. Since it has two parameters, a three-way table is required to provide values of F which correspond to various values of α, the tail area. Abbreviated tables of this distribution are provided in Appendix A.6. There are separate pages for $\alpha = 0.01$ and $\alpha = 0.05$. Degrees of freedom for the numerator head the columns on each page and degrees of freedom for the denominator comprise the extreme left-hand column. The numbers in the body of the table are values of F which divide the distribution with α in the right tail. The distribution is illustrated in Figure 12.6. The unshaded area corresponds to α.

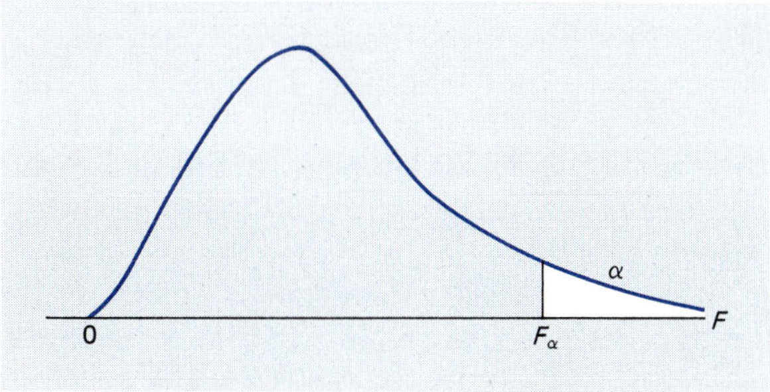

Figure 12.6
The F distribution.

Example 12.7 Two independent random samples are drawn from normal populations with variances σ_1^2 and σ_2^2. If $n_1 = 21$ and $n_2 = 30$, find the value of F that divides the distribution with 5% of the area in the right tail. Find the value of F that divides the distribution with 1% of the area in the right tail.

$$F_{.05,20,29} = 1.94.$$
$$F_{.01,20,29} = 2.57.$$

12.7 Test of the Equality of Variances

When testing the hypothesis that two normal populations have equal means using small samples, where the population standard deviations and variances are unknown, it is necessary to assume that these standard deviations are equal. However, before proceeding with the test, it is good practice to test that assumption. The hypotheses, in terms of the variances, are

$$H_0\text{: } \sigma_1^2 = \sigma_2^2$$

and

$$H_1\text{: } \sigma_1^2 \neq \sigma_2^2$$

If random samples are drawn from each of these populations and each sample variance is computed, the ratio

$$\frac{s_1^2/\sigma_1^2}{s_2^2/\sigma_2^2}$$

has an F distribution with $n_1 - 1$ and $n_2 - 1$ degrees of freedom. Now, if the null hypothesis is true and the two population variances are equal, then $\sigma_1^2 = \sigma_2^2 = \sigma^2$ and this ratio becomes

$$\frac{s_1^2}{\sigma_2} \div \frac{s_2^2}{\sigma_2} = \frac{s_1^2}{\sigma^2} \cdot \frac{\sigma^2}{s_2^2} = \frac{s_1^2}{s_2^2} = F \tag{12.17}$$

In other words, if H_0 is true, the ratio of the two sample variances has an F distribution with $n_1 - 1$ and $n_2 - 1$ degrees of freedom. Furthermore, if these samples are representative of their respective populations, then s_1^2 and s_2^2 should both be approximately equal to σ^2, the common variance, and the ratio s_1^2/s_2^2 should be approximately equal to 1. On the other hand, if the null hypothesis is false and $\sigma_1^2 \neq \sigma_2^2$, the sample variances tend to equal their respective population variances and s_1^2/s_2^2 is not expected to equal 1. Therefore, if we compute the variance ratio and its value is significantly different from 1, the null hypothesis should be rejected.

Because of the way in which the null and alternate hypotheses are stated, we would expect this to be a two-tailed test. This would require two critical values of F, one in each tail of the distribution. However, tables of the F distribution provide only right-tail areas.

Because of the skewed nature of the F distribution, the left-tail critical values are not simply the right-tail values prefixed by a minus sign. It is, therefore, much simpler to test these hypotheses with a one-tailed test using only the right tail of the distribution. This can be done if the larger of the two sample variances is *always* placed in the numerator of the variance ratio. Then, if the value of this ratio deviates significantly from 1, it will always be in the direction of the right tail, that is, greater than 1. Consequently, the entire rejection region for the test can be placed in the right tail of the distribution.

This testing procedure is valid as long as we can assume that if the null hypothesis is not true, the larger sample variance comes from the population with the larger variance. This is a reasonable assumption and in this case, the correct statement of the hypotheses is

$$H_0\colon \sigma_1^2 = \sigma_2^2$$
$$H_1\colon \sigma_1^2 > \sigma_2^2$$

Example 12.8

Suppose that in Example 12.4, where the hypothesis of equal mean lifetimes of two different populations of television picture tubes was tested using small samples, it was decided to first test the assumption that the two populations have equal variances. Recall that the small sample test of difference in means requires that the population standard deviations and variances be equal. The test of the equality of variances is to be constructed as a one-tailed test with a 1% level of significance. The important information for this test is summarized below.

Populations: normal.

Test statistic: $F = s_1^2/s_2^2$. Sample sizes: $n_1 = 10$, $n_2 = 15$.

Sampling distribution of F: F with $n_1 - 1$ and $n_2 - 1$ degrees of freedom.

Significance level: $\alpha = 0.01$.

Form of test: one-tailed.

Hypotheses: $H_0\colon \sigma_1^2 = \sigma_2^2$ versus $H_1\colon \sigma_1^2 > \sigma_2^2$.

Critical value of F: from the table, $F_{.01,9,14} = 4.03$.

Decision rule: Random samples of $n_1 = 10$ and $n_2 = 15$ are drawn from the two populations. The variance of each sample is computed and the ratio s_1^2/s_2^2 is formed with the larger sample variance in the numerator. If the value of this variance ratio is greater than or equal to 4.03, the null hypothesis is rejected; otherwise it is accepted.

From the data provided in Example 12.4, $s_1^2 = (300)^2 = 90{,}000$ and $s_2^2 = (210)^2 = 44{,}100$. Therefore,

$$s_1^2/s_2^2 = \frac{90{,}000}{44{,}100} = 2.04 \tag{12.17}$$

Since $2.04 < 4.03$, H_0 is accepted.

Definitions

Pooled estimate: A parameter estimate obtained from two or more samples by pooling or combining all of the sample data.

Variance ratio: The ratio of two sample variances.

Formulas

Variance and standard error of the difference between two sample means:

$$\sigma^2_{d*} = \frac{\sigma_1^2}{n_1} + \frac{\sigma_2^2}{n_2} \tag{12.1}$$

$$\sigma_{d*} = \sqrt{\frac{\sigma_1^2}{n_1} + \frac{\sigma_2^2}{n_2}} \tag{12.2}$$

Standard error of the difference between two sample proportions (equal population proportions):

$$\sigma_{d*} = \sqrt{\frac{p(1-p)}{n_1} + \frac{p(1-p)}{n_2}} \tag{12.3}$$

Critical values of the difference between sample means, large samples, standard deviations known:

$$d_c^* = 0 \pm z_{\alpha/2}\sigma_{d*} \tag{12.4}$$

Estimated standard error of the difference between sample means, large samples:

$$s_{d*} = \sqrt{\frac{s_1^2}{n_1} + \frac{s_2^2}{n_2}} \tag{12.5}$$

Critical values of the difference between sample means, large samples, standard deviations unknown:

$$d_c^* = 0 \pm z_{\alpha/2}s_{d*} \tag{12.6}$$

Pooled estimate of the common population standard deviation, two samples:

$$s_p = \sqrt{\frac{(n_1-1)s_1^2 + (n_2-1)s_2^2}{n_1+n_2-2}} \tag{12.7}$$

Estimated standard error of the difference between sample means, small samples:

$$s_{d*} = \sqrt{\frac{s_p^2}{n_1} + \frac{s_p^2}{n_2}} \tag{12.8}$$

Computational form for the estimated standard error of the difference between sample means:

$$s_{d*} = s_p\sqrt{\frac{1}{n_1} + \frac{1}{n_2}} \tag{12.9}$$

Critical values of the difference between sample means, small samples, standard deviations unknown:

$$d_c^* = 0 \pm t_{\alpha/2}s_{d*} \tag{12.10}$$

Standard error of the difference between two sample proportions:

$$\sigma_{d*} = \sqrt{\frac{p_1(1 - p_1)}{n_1} + \frac{p_2(1 - p_2)}{n_2}} \tag{12.11}$$

Pooled estimate of the common population proportion, two samples:

$$\hat{p} = \frac{n_1\bar{p}_1 + n_2\bar{p}_2}{n_1 + n_2} \tag{12.12}$$

Estimated standard error of the difference between sample proportions:

$$\hat{\sigma}_{d*} = s_{d*} = \sqrt{\frac{\hat{p}(1 - \hat{p})}{n_1} + \frac{\hat{p}(1 - \hat{p})}{n_2}} \tag{12.13}$$

Estimated standard error of the difference between sample proportions, computational form:

$$s_{d*} = \sqrt{\frac{\hat{p}(1 - \hat{p})(n_1 + n_2)}{n_1 n_2}} \tag{12.14}$$

Critical values of the difference between sample proportions:

$$d_c^* = 0 \pm z_{\alpha/2}s_{d*} \tag{12.15}$$

Critical values of the sample average difference:

$$\bar{d}_c = 0 \pm t_{\alpha/2}s_{\bar{d}} \tag{12.16}$$

Variance ratio:

$$\frac{s_1^2}{s_2^2} = F_{\alpha,df_1,df_2} \tag{12.17}$$

Problems

12.1 Test the following hypotheses at $\alpha = 0.01$

$$H_0: \mu_1 = \mu_2$$
$$H_1: \mu_1 \neq \mu_2$$

if two independent random samples of $n_1 = 125$ and $n_2 = 90$ have the following means and standard deviations.

$$\bar{x}_1 = 750 \qquad \bar{x}_2 = 725$$
$$s_1 = 140 \qquad s_2 = 180$$

12.2 Test the following hypotheses at $\alpha = 0.02$

$$H_0: \mu_1 = \mu_2$$
$$H_1: \mu_1 \neq \mu_2$$

if two independent random samples of $n_1 = 50$ and $n_2 = 60$ have the following means and standard deviations.

$$\bar{x}_1 = 980.7 \qquad \bar{x}_2 = 1010.2$$
$$s_1 = 75 \qquad s_2 = 80$$

12.3 Cinderella Clothes operates two stores in the Middletown area, one downtown and the other in a suburban mall. The executive vice-president of the company theorizes that lower gross sales in the downtown store might be due to a lower average dollar value per sales transaction. He decides to test the hypothesis that the average value of a sale in the downtown store is equal to that of the suburban store against the alternative that they are not equal. A random sample of 150 transactions is drawn from the sales records of the downtown store, store #1, and a second random sample of 200 transactions is drawn from the records of the suburban store, store #2. The means and standard deviations of these samples are

$$\bar{x}_1 = \$11.50 \qquad \bar{x}_2 = \$13.25$$
$$s_1 = \$3.25 \qquad s_2 = \$4.75$$

Test the hypothesis of no difference at $\alpha = 0.01$.

12.4 The controller of the Southwest Oil Company wants to test the hypothesis that the average outstanding balance on Southwest's credit card accounts is the same in the Pacific Coast region as it is in the Middle Atlantic region. Random samples of 400 accounts are selected from each region. The means and standard deviations of the samples are

$$\bar{x}_1 = \$44.20 \qquad \bar{x}_2 = \$41.50$$
$$s_1 = \$18.45 \qquad s_2 = \$15.25$$

Test the hypothesis of no difference at $\alpha = 0.05$.

12.5 Test the hypotheses in problem 12.1 if the two sample sizes are $n_1 = 19$ and $n_2 = 11$. Use the same sample means and standard deviations. What assumptions must be made to work this problem?

12.6 Test the hypotheses in problem 12.2 if the sample sizes are $n_1 = 15$ and $n_2 = 8$. What assumptions must be made to work this problem?

12.7 Test the hypotheses of problem 12.2 using the sample sizes of problem 12.6 but assume that $\sigma_1 = 75$ and $\sigma_2 = 80$. What assumption must be made to work this problem?

12.8 What would the result of the test in problem 12.3 be if the sample sizes are $n_1 = 16$ and $n_2 = 13$. What assumptions must be made to work the problem?

12.9 Repeat the test of problem 12.4 with the same data, except that the sample sizes are now $n_1 = 12$ and $n_2 = 12$. What assumptions must be made to work this problem?

12.10 Test the hypothesis that $p_1 = p_2$ against the alternate that $p_1 \neq p_2$ if two independent random samples of $n_1 = 400$ and $n_2 = 225$ produce 79 and 65 successes, respectively. Use $\alpha = 0.05$.

12.11 Test the hypothesis that $p_1 = p_2$ against the alternate hypothesis that $p_1 \neq p_2$, if two independent random samples of $n_1 = 120$ and $n_2 = 110$ produce 43 and 47 successes, respectively. Use $\alpha = 0.03$.

12.12 Test the hypotheses in Example 12.5 at $\alpha = 0.01$ if the sample sizes are $n_1 = 500$ and $n_2 = 400$. The number of accidents remains the same.

12.13 In a recent survey, 32% of a random sample of 250 drivers named Rollright tires as the best known brand of tires on the market. Last year, 38% of a sample of 150 drivers named Rollright. Test the hypothesis that both samples come from populations with the same percentage of drivers who consider Rollright the best known brand. Use $\alpha = 0.05$.

12.14 In a random sample of 500 California voters taken by an opinion polling company, 279 favor a new energy conservation bill being voted on in Congress. Only 235 of a random sample of 500 New York voters favor the bill. Can it be concluded at the 5% level of significance that the proportion favoring the legislation is the same in New York and California?

12.15 In order to test the effectiveness of a sales training program proposed by a firm of training specialists, the A&M Home Furnishings Company selects eight sales representatives at random to take the course. In the following table, y represents each representative's gross sales during the four weeks immediately preceding the course, while x represents the same representative's gross sales during the four-week period immediately following the course. Test the hypothesis that the training course has no effect on average sales at the 1% level of significance.

Sales Representative	*Gross Sales (in thousands of dollars) before Course* (y)	*Gross Sales (in thousands of dollars) after Course* (x)
1	90	97
2	83	80
3	105	110
4	102	110
5	110	118
6	78	86
7	97	93
8	75	81

12.16 A toothpaste manufacturer established a test of the effectiveness of a particular additive in reducing cavities in children's teeth. Twelve pairs of twins of the same sex were selected for the test.

One twin in each pair was given the toothpaste with the additive. The other twin was given the identical toothpaste without the additive. Both twins were instructed to brush in exactly the same way with the same frequency. The number of cavities discovered in periodic dental checkups during the year following the inception of the test were noted for each twin. Test the hypothesis that the additive had no effect in reducing cavities at the 5% level of significance. The data are provided in the following table. The variable y represents the number of cavities for the twin using the additive.

Pair	x	y
1	4	2
2	1	1
3	0	1
4	5	3
5	4	3
6	1	0
7	3	1
8	1	1
9	2	0
10	4	1
11	2	0
12	3	1

12.17 Repeat the test of problem 12.16 at the 1% level of significance.

12.18 To work problem 12.5, it was necessary to make the assumption that $\sigma_1^2 = \sigma_2^2$. Test this assumption at $\alpha = 0.01$ using the data given in that problem and in problem 12.1. State both the null and alternate hypotheses.

12.19 In problem 12.6, it was necessary to assume that $\sigma_1^2 = \sigma_2^2$. Test that assumption at $\alpha = 0.05$ using the data from that problem and problem 12.2. State both the null and alternate hypotheses.

12.20 Test the null hypothesis that $\sigma_1^2 = \sigma_2^2$ against the appropriate alternative at the 5% level of significance if random samples of $n_1 = 25$ and $n_2 = 30$ have variances of $s_1^2 = 1089$ and $s_2^2 = 361$, respectively. State the alternate hypothesis.

CHAPTER 13 ANALYSIS OF VARIANCE; MULTIPLE-SAMPLE TESTS OF MEANS

13.1
Notation for analysis of variance

13.2
The rationale for analysis of variance

13.3
Sums of squares and the ANOVA table

13.4
One-factor analysis of variance

13.5
Least significant difference

13.6
Two-factor analysis of variance

In Chapter 12, we discussed two-sample tests of population means; that is, we tested the hypothesis that the means of two populations were equal. The alternate hypothesis was that these population means were not equal. An extension of this situation leads us to tests that involve more than two populations and more than two samples. In other words, if there are three or more populations, we can test the null hypothesis that all the population means are equal against the alternate hypothesis that they are not all equal. This test is performed using a technique called *analysis of variance*.

At this point, it is important to examine the meaning of the alternate hypothesis in this test. The alternate hypothesis does not state that the population means are all different from one another. It simply states that they are not all equal. This means that the null hypothesis should be rejected if any one of the means is different from the others.

Although analysis of variance originated in the area of agricultural research, it is such a versatile technique that it has been easily adapted for tests in a wide variety of fields, including business. For example, in Chapter 1, a situation was described which concerned the effects of three different point-of-sale displays on the sale of a certain cosmetic product. In that situation, the populations in the problem were the potential purchasers of the product, who were exposed to each of the three different displays or *treatments*. These populations can also be referred to as *treatment populations*. The elementary units observed were the cosmetics departments, which used one of the displays. The random variable of the experiment was sales, in dollars or units, of the product.

The null hypothesis of equal means, in this case average sales, is a hypothesis of no difference or no treatment effect; that is, the null hypothesis states that the difference in displays has no effect on average sales. If this hypothesis is rejected, then it can be concluded that at least one of the treatments has an effect on average sales that is different from the others.

The word *treatment* in analysis of variance is used to refer to any factor in the experiment that is controlled at different levels or values. The treatments can be different point-of-sale displays, different drug dosages, assembly line techniques, sales training programs, or, in short, any controlled factor deliberately applied to the elementary units observed in the experiment.

The multiple-sample test of means by analysis of variance differs in one important respect from the two-sample tests discussed previously. In multiple-sample tests, variances are used to test the hypotheses about means. In all the previous tests of population means, sample means were used as the test statistics. In order to use this technique, two assumptions about the populations are required. First, each population must have a normal distribution, and second, the variances of these populations must all be equal.

13.1 Notation for Analysis of Variance

It will be helpful to describe the notation used in connection with analysis of variance before proceeding with the actual technique. Data are obtained from observations made on random samples of elementary units exposed to each treatment. These data can be displayed in a table similar to Table 13.1. Notice that the table has r rows and k columns. Each column corresponds to a different population or treatment. To be consistent with the usage common to the field, we will use the term "treatment."

Refer to Table 13.1 and consider an experiment in which random samples of elementary units are subjected to k different treatments. Each sample consists of $r_{.j}$ elementary units. The number $r_{.j}$ is the number of elementary units assigned to treatment j. An observation obtained from the ith elementary unit assigned to treatment j is designated x_{ij}. The subscript i runs from 1 to r, while the subscript j runs from 1 to k. That is, $i = 1,2,3,\ldots, r$ and $j = 1,2,3,\ldots, k$.

If all the x_{ij}'s in a single column are added, we get the sum of the values in that column, which is $\sum_i x_{ij}$. If a column sum is divided by the number of observations in that column, we get the column mean. For example, the mean of the jth column is $\bar{x}_{.j} = \sum_i x_{ij}/r_{.j}$. A dot used in place of a subscript, as in $\bar{x}_{.j}$, holds the place of a subscript which does not appear in that particular symbol. Since $\bar{x}_{.j}$ refers to the mean of column j, the row subscript is eliminated. The dot simply holds its place open and indicates that the j refers to the column. Conversely, a symbol like $\bar{x}_{i.}$ indicates the mean of row i and the dot indicates that for this particular symbol, the column subscript does not apply. The symbol $r_{.j}$ indicates the number of rows or observations contained in

TABLE 13.1
GENERAL ARRANGEMENT OF DATA FOR ANALYSIS OF VARIANCE

	Treatments					
Observations	1	2	3 . . .		. . . k	
1	x_{11}	x_{12}	x_{13} ·	x_{1j} · ·	x_{1k}	
2	x_{21}	x_{22}	x_{23} ·	x_{2j} · ·	x_{2k}	
3	x_{31}	x_{32}	x_{33} ·	x_{3j} · ·	x_{3k}	
·	·	·	· ·	· · ·	·	
·	x_{i1}	x_{i2}	x_{i3} ·	x_{ij} · ·	x_{ik}	
·	·	·	· ·	· · ·	·	
r	x_{r1}	x_{r2}	x_{r3} ·	x_{rj} · ·	x_{rk}	
Sums	$\sum_i x_{i1}$	$\sum_i x_{i2}$	$\sum_i x_{i3}$	$\sum_i x_{ij}$	$\sum_i x_{ik}$	$\sum_j \sum_i x_{ij}$
Means	$\bar{x}_{.1}$	$\bar{x}_{.2}$	$\bar{x}_{.3}$	$\bar{x}_{.j}$	$\bar{x}_{.k}$	$\bar{\bar{x}}$

the jth column. If the number of observations in all the columns are equal, that is, $r_{.1} = r_{.2} = \cdots = r_{.k}$, we can eliminate the subscripts entirely and simply use r to represent the number of observations in each column.

Since the data consist of r rows and k columns, the total number of observations is $rk = n$, providing of course, that each column contains the same number of observations. If the number of observations varies from column to column, then the total number is

$$r_{.1} + r_{.2} + \cdots + r_{.k} = \sum_j r_{.j} = n.$$

We illustrate the use of this notation with the following numerical example.

Example 13.1 An experiment consists of making observations on random samples of elementary units from three treatment populations. Each sample consists of five observations. The data are displayed in Table 13.2.

TABLE 13.2
EXAMPLE OF DATA IN ANALYSIS OF VARIANCE (EXAMPLE 13.1)

	Treatments			
Observations	1	2	3	
1	$x_{11} = 3$	$x_{12} = 3$	$x_{13} = 8$	
2	$x_{21} = 3$	$x_{22} = 8$	$x_{23} = 2$	
3	$x_{31} = 1$	$x_{32} = 3$	$x_{33} = 4$	
4	$x_{41} = 4$	$x_{42} = 7$	$x_{43} = 2$	
5	$x_{51} = 4$	$x_{52} = 4$	$x_{53} = 4$	
Sums	15 +	25 +	20	= 60
Means	$\bar{x}_{.1} = 3$	$\bar{x}_{.2} = 5$	$\bar{x}_{.3} = 4$	$\bar{\bar{x}} = 4$

Notice from the table that the same number of observations is obtained from each treatment population, $r_{.1} = r_{.2} = r_{.3} = r = 5$ and, therefore, $n = rk = (5)(3) = 15$.

The column sums are obtained by adding the values in each column, that is:

$$\sum_{i=1}^{5} x_{i1} = x_{11} + x_{21} + x_{31} + x_{41} + x_{51} = 3 + 3 + 1 + 4 + 4 = 15$$

$$\sum_{i=1}^{5} x_{i2} = x_{12} + x_{22} + x_{32} + x_{42} + x_{52} = 3 + 8 + 3 + 7 + 4 = 25$$

$$\sum_{i=1}^{5} x_{i3} = x_{13} + x_{23} + x_{33} + x_{43} + x_{53} = 8 + 2 + 4 + 2 + 4 = 20$$

The sum of the entire sample of n values in this table is

$$\sum_{j=1}^{3}\sum_{i=1}^{5} x_{ij} = \sum_{i=1}^{5} x_{i1} + \sum_{i=1}^{5} x_{i2} + \sum_{i=1}^{5} x_{i3} = 15 + 25 + 20 = 60$$

The column means are obtained by dividing each column sum by the number of observations in that column, that is:

$$\bar{x}_{.1} = \frac{\sum_i x_{i1}}{r_{.1}} = \frac{15}{5} = 3$$

$$\bar{x}_{.2} = \frac{\sum_i x_{i2}}{r_{.2}} = \frac{25}{5} = 5$$

$$\bar{x}_{.3} = \frac{\sum_i x_{i3}}{r_{.3}} = \frac{20}{5} = 4$$

The *grand mean*, designated $\bar{\bar{x}}$ and pronounced "x double-bar," is obtained by dividing the sum of all the values in the table by the total number of observations, that is,

$$\bar{\bar{x}} = \frac{\sum_j \sum_i x_{ij}}{n} = \frac{60}{15} = 4$$

13.2 The Rationale for Analysis of Variance

In testing a hypothesis of equal means, random samples are taken from normal populations having equal variances. These two assumptions, normality and equal variances, are essential to the analysis of variance procedure. Now, if the null hypothesis is true and all the treatment means are equal, then for all practical purposes we have essentially drawn a sample from a single population with mean μ and variance σ^2. In this case, the population from which the sample is taken would appear as in Figure 13.1a. However, if the null hypothesis is not true and the treatment populations have different means, these populations might appear as in Figure 13.1b. The horizontal scale in Figures 13.1a and 13.1b represents the values of the random variable x.

Now suppose we take a single sample of size n from the population in Figure 13.1a and compute the sample variance as an estimate of the population variance. This estimate is

$$s^2 = \frac{\Sigma(x - \bar{\bar{x}})^2}{n - 1}$$

On the other hand, suppose there are three different populations, as

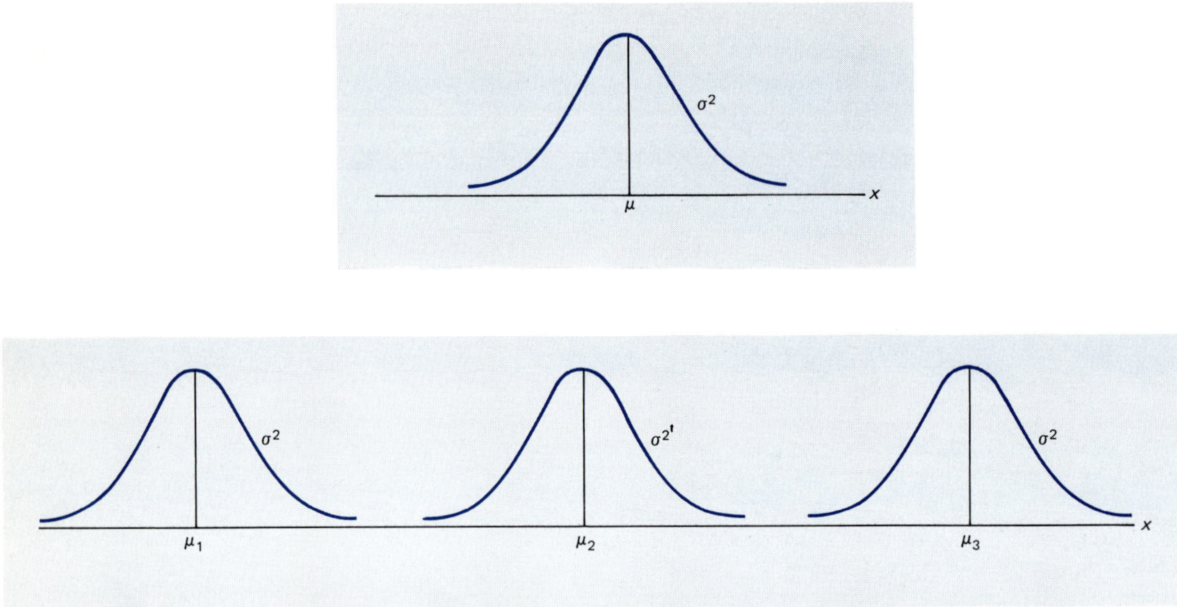

Figure 13.1
An illustration of two different situations in analysis of variance.

in Figure 13.1b, and three samples of size n_1, n_2, and n_3, respectively, are drawn, one from each population, and are used to estimate the common variance. One method of estimating the population variance is to pool the three individual sample variances. That is, an estimate of the common population variance is

$$\widehat{\sigma}^2 = \frac{(n_1 - 1)s_1^2 + (n_2 - 1)s_2^2 + (n_3 - 1)s_3^2}{n_1 + n_2 + n_3 - 3}$$

This formula is similar to Formula 12.7, which provides a pooled estimate of a population standard deviation obtained from two sample standard deviations.

Since each sample variance measures the variation *within* the sample, and this variation is due to sampling error, we will use the symbol s_E^2 to designate a variance estimate based on pooled sample variances. In general, for k samples, the pooled estimator for σ^2 is

$$(13.1) \qquad s_E^2 = \frac{(n_1 - 1)s_1^2 + (n_2 - 1)s_2^2 + \cdots + (n_k - 1)s_k^2}{n_1 + n_2 + \cdots + n_k - k}$$

Each individual sample variance is, by itself, a valid estimator of σ^2. However, the weighted average of all the sample variances, which is exactly what the pooled estimate is, is a better estimator of σ^2, since it is based on more data. Notice that by pooling the individual sample variances we now have an estimator based on

$n_1 + n_2 + \cdots + n_k - k$ degrees of freedom. Also notice that this pooled estimate of the population variance is based *only* on the variation measured *within* each sample.

There is another method for estimating the common population variance which is based on the variance among the sample means, $\sigma_{\bar{x}}^2$. We know that $\sigma_{\bar{x}}^2 = \sigma^2/n$; therefore, by algebra, σ^2 must be equal to $n\sigma_{\bar{x}}^2$. We can estimate $\sigma_{\bar{x}}^2$ from the several samples by subtracting each sample mean from the grand mean, squaring the difference, summing, and dividing by the degrees of freedom. That is, an estimate of $\sigma_{\bar{x}}^2$ is obtained from:

$$(13.2) \qquad s_{\bar{x}}^2 = \frac{(\bar{x}_{.1} - \bar{\bar{x}})^2 + (\bar{x}_{.2} - \bar{\bar{x}})^2 + \cdots + (\bar{x}_{.k} - \bar{\bar{x}})^2}{k - 1}$$

where k is the number of samples.

Since $\sigma^2 = n\sigma_{\bar{x}}^2$, we can now obtain an estimate of σ^2 by multiplying each term in the numerator of Equation 13.2 by $r_{.j}$. Therefore, the estimate of the common variance based on the variation among the sample means is

$$(13.3) \quad s_K^2 = \frac{r_{.1}(\bar{x}_{.1} - \bar{\bar{x}})^2 + r_{.2}(\bar{x}_{.2} - \bar{\bar{x}})^2 + \cdots + r_{.k}(\bar{x}_{.k} - \bar{\bar{x}})^2}{k - 1}$$

The $r_{.j}$ represents the size of the jth sample and k represents the number of samples. The subscript K in s_K^2 is used because this variance estimate is based on the variation among the k samples.

It is important to emphasize that this estimate is based on the variation *among* the k different sample means. Consequently, if a significant difference does exist among these means, the estimate s_K^2 will reflect this difference. On the other hand, if no significant difference exists, then s_K^2 will reflect the same variation as s_E^2. Therefore, if the null hypothesis of no difference in means is true, we would expect that $s_K^2 \doteq s_E^2$ and the ratio s_K^2/s_E^2 to be approximately equal to one. If a difference does exist, then s_K^2 will reflect this difference and will be larger than s_E^2 and the ratio s_K^2/s_E^2 should be greater than one. We learned in Chapter 12 that the ratio of two sample variances has an F distribution which can be used to test hypotheses about the equality of variances. The variance ratio and the F distribution can also be used to test the hypotheses about the equality of means because a significant difference in these variances implies a difference among the means.

Example 13.2

Using the data in Table 13.2 and the information that $s_1^2 = 1.5$, $s_2^2 = 5.5$, and $s_3^2 = 6.0$, compute values for s_E^2 and s_K^2 and use the variance ratio to test the hypothesis of no difference in treatment means at the 5% level of significance.

1. Compute the pooled estimate of the common population variance from Equation 13.1.

$$\begin{aligned} s_E^2 &= \frac{(4)(1.5) + (4)(5.5) + (4)(6.0)}{5 + 5 + 5 - 3} \\ &= \frac{6.0 + 22.0 + 24.0}{12} \\ &= \frac{52}{12} \\ &= 4.33 \end{aligned}$$

2. Compute an estimate of the common population variance based on the variation among the treatment means using Equation 13.3.

$$\begin{aligned} s_K^2 &= \frac{5(3-4)^2 + 5(5-4)^2 + 5(4-4)^2}{2} \\ &= \frac{(5)(1) + (5)(1) + (5)(0)}{2} \\ &= \frac{10}{2} \\ &= 5 \end{aligned}$$

3. Compute the variance ratio.

$$s_K^2/s_E^2 = \frac{5.00}{4.33} = 1.15$$

4. Determine the appropriate number of degrees of freedom for the numerator and the denominator of the variance ratio from the denominators of the expressions for s_K^2 and s_E^2.

$$df_K = 2$$
$$df_E = 12$$

5. Find the critical value of F at the 5% level of significance.

$$F_{2,12,0.05} = 3.89$$

Since the computed value of the variance ratio, 1.15, is less than the critical value of 3.89, there is insufficient evidence to reject the hypothesis of equal means at the 5% level of significance.

13.3 Sums of Squares and the ANOVA Table

The formula that defines the variance of any set of sample values is a fraction consisting of a numerator and a denominator. That is,

$$s^2 = \frac{\Sigma(x - \bar{x})^2}{n - 1}$$

This formula, modified for grouped data, appeared as Equation 9.3 in Chapter 9. The denominator of this fraction corresponds to degrees of freedom. The numerator, which is the sum of the squared deviations of the sample values about the sample mean, is called the sum of squares and is abbreviated by the letters SS. Consequently, the variance of any set of sample data is simply the sum of squares divided by the number of degrees of freedom in the sample. We can express this as

$$\text{sample variance} = \frac{SS}{df}$$

The sum of squares is a measure of variation. When the sum of squares is divided by the degrees of freedom, the result is frequently called the MEAN SQUARE, an alternate designation for the sample variance.

For the data in an analysis of variance problem, the sum of the squared deviations of all sample values from the grand mean is called the TOTAL SUM OF SQUARES and is abbreviated SS_T, where the subscript T indicates total. That is,

$$SS_T = \sum_{j=1}^{k} \sum_{i=1}^{r} (x_{ij} - \bar{\bar{x}})^2 \tag{13.4}$$

This total sum of squares is divided or partitioned into two parts. The portion attributed to the variation within the treatment samples is called the *error sum of squares*. The equation is

$$SS_E = \sum_{j=1}^{k} \sum_{i=1}^{r} (x_{ij} - \bar{x}_{.j})^2 \tag{13.5}$$

SS_E is the sum of the squared deviations of each observation within a treatment sample from the treatment mean. The sum of these values for each treatment is added over all the treatments. This error sum of squares is equivalent to the numerator of the expression for s_E^2 in Equation 13.1.

The remainder of the total sum of squares which is attributed to the variation among the treatments is called the *treatment sum of squares* and, since there are k treatments, is designated

$$SS_K = \sum_{j=1}^{k} r_{.j}(\bar{x}_{.j} - \bar{\bar{x}})^2 \tag{13.6}$$

Notice that SS_K is equivalent to the numerator of the expression for s_K^2 in Equation 13.3.

Since the total sum of squares has been partitioned into two parts, we can write the following relationship.

$$SS_T = SS_K + SS_E \tag{13.7}$$

However, the formulas for sums of squares, Equations 13.4, 13.5, and 13.6, are not the most convenient formulas for computing these quantities. As with the other variance formulas that you have encountered in this book, there are short-cut computational formulas. For the sums of squares, these computational forms are

$$\text{SS}_T = \sum_j \sum_i x_{ij}^2 - n\bar{\bar{x}}^2 \tag{13.8}$$

$$\text{SS}_E = \sum_j \sum_i x_{ij}^2 - \sum_j r_{.j}\bar{x}_{.j}^2 \tag{13.9}$$

$$\text{SS}_K = \sum_j r_{.j}\bar{x}_{.j}^2 - n\bar{\bar{x}}^2 \tag{13.10}$$

Notice that each of the computational formulas for a sum of squares consists of two terms. However, there are only three different terms in all three equations, that is,

$n\bar{\bar{x}}^2$ appears in SS_T and SS_K

$\sum_j \sum_i x_{ij}^2$ appears in SS_T and SS_E

and

$\sum_j r_{.j}\bar{x}_{.j}^2$ appears in SS_E and SS_K

Consequently, each of these terms needs to be computed only once to obtain all three sums of squares.

Example 13.3 Refer to the data in Table 13.3 to compute the sums of squares needed for an analysis of variance. The three different terms that appear in these sums can be obtained from the data in Table 13.3.

$$\sum_j \sum_i x_{ij}^2 = 302$$

$$n\bar{\bar{x}}^2 = (15)(4)^2 = (15)(16) = 240$$

$$\begin{aligned}\sum_j r_{.j}\bar{x}_{.j}^2 &= (5)(3)^2 + (5)(5)^2 + (5)(4)^2 \\ &= (5)(9) + (5)(25) + (5)(16) \\ &= 45 + 125 + 80 \\ &= 250\end{aligned}$$

Combining these terms in the proper order, we get:

$$\text{SS}_T = \sum_j \sum_i x_{ij}^2 - n\bar{\bar{x}}^2 = 302 - 240 = 62$$

TABLE 13.3
EXAMPLE OF COMPUTATION WITH DATA IN ANALYSIS OF VARIANCE (EXAMPLE 13.3)

	Treatments					
	1		2		3	
Observation	x_{i1}	x_{i1}^2	x_{i2}	x_{i2}^2	x_{i3}	x_{i3}^2
1	3	9	3	9	8	64
2	3	9	8	64	2	4
3	1	1	3	9	4	16
4	4	16	7	49	2	4
5	4	16	4	16	4	16
	$\Sigma x_{i1} = 15$	$\Sigma x_{i1}^2 = 51$	$\Sigma x_{i2} = 25$	$\Sigma x_{i2}^2 = 147$	$\Sigma x_{i3} = 20$	$\Sigma x_{i3}^2 = 104$
	$\bar{x}_{.1} = \frac{15}{5} = 3$		$\bar{x}_{.2} = \frac{25}{5} = 5$		$\bar{x}_{.3} = \frac{20}{5} = 4$	

$$\sum_j \sum_i x_{ij} = 15 + 25 + 20 = 60$$

$$n = rk = (5)(3) = 15$$

$$\bar{\bar{x}} = \frac{\sum_j \sum_i x_{ij}}{n} = \frac{60}{15} = 4$$

$$\sum_j \sum_i x_{ij}^2 = 51 + 147 + 104 = 302$$

$$SS_K = \sum_j r_{.j}\bar{x}_{.j}^2 - n\bar{\bar{x}}^2 = 250 - 240 = 10$$

$$SS_E = \sum_j \sum_i x_{ij}^2 - \sum_j r_{.j}\bar{x}_{.j}^2 = 302 - 250 = 52$$

After completing these computations, be sure to check that

$$SS_K + SS_E = SS_T$$

In this case,

$$10 + 52 = 62$$

Having computed the sums of squares, it is necessary to associate the appropriate number of degrees of freedom with each sum. The total number of degrees of freedom, df_T, is equal to the total number of observations minus one, that is,

(13.11) $$df_T = n - 1$$

For the data of Table 13.3, $df_T = 15 - 1 = 14$.

The number of degrees of freedom associated with the treatment sum of squares is simply one less than the number of treatments, that is,

(13.12) $$df_K = k - 1$$

For the data in Table 13.3, $df_K = 3 - 1 = 2$.

Finally, the number of degrees of freedom associated with the error sum of squares is equal to the total number of observations minus the number of treatments, that is,

$$df_E = n - k \tag{13.13}$$

For the data in Table 13.3, $df_E = 15 - 3 = 12$. As a check, be sure that $df_T = df_K + df_E$.

All the information about the analysis of variance is summarized in an analysis of variance table, generally abbreviated ANOVA. The ANOVA table is illustrated in Tables 13.4 and 13.5. Table 13.4 shows the general arrangement of an ANOVA table, while Table 13.5 provides the ANOVA table for the data in Table 13.3.

TABLE 13.4
GENERAL ARRANGEMENT OF THE ANOVA TABLE FOR A ONE-FACTOR ANALYSIS OF VARIANCE

Source of Variation	*Sum of Squares*	*Degrees of Freedom*	*Mean Square (SS/df)*	*Variance Ratio*
Among treatments (columns)	SS_K	$k - 1$	$MS_K = \frac{SS_K}{k-1}$	$\frac{MS_K}{MS_E} = F$
Error	SS_E	$n - k$	$MS_E = \frac{SS_E}{n-k}$	
Total	SS_T	$n - 1$		

TABLE 13.5
ANOVA TABLE (EXAMPLE 13.3)

Source of Variation	*Sum of Squares*	*Degrees of Freedom*	*Mean Square (SS/df)*	*Variance Ratio*
Among treatments	10	2	$10/2 = 5$	$\frac{5}{4.33} = 1.15$
Error	52	12	$52/12 = 4.33$	
Total	62	14		

13.4 One-Factor Analysis of Variance

The term *one-factor analysis of variance* refers to the fact that a single variable or factor of interest is controlled and its effect on the elementary units is observed. In the language of statistical experimentation, the treatments are referred to as different levels of the controlled factor. For example, if we are interested in the effect of different point-of-sale displays on the sales of a particular cosmetic product, the single factor of interest in this experiment is the set of point-of-sale displays. Each different display, or treatment, is a *level* of that factor. The elementary units are the cosmetics departments, and the response variable observed is sales of the product at each of the cosmetics departments observed.

A second example might relate to an experiment conducted by a manufacturer of pharmaceutical products who is interested in testing the effectiveness of a new drug in the treatment of obesity. In this case, the drug itself is the factor to be controlled; different dosages are the treatments or levels of the factor; the experimental units are subjects placed on the drug; and the observed response variable is the weight loss experienced by each subject.

The procedure for a one-factor analysis of variance is demonstrated by the following example.

Example 13.4

The marketing director of Ego Cosmetics, Inc., which manufactures and markets ladies' cologne and toilet water, wants to evalute the effectiveness of three different point-of-sale displays in promoting sales. These displays are placed at the cosmetics departments of the various department stores that sell these products. The marketing director decides to conduct a test. She selects 30 stores from the population of stores that handle Ego products and then randomly assigns ten of the selected stores to each of the three displays. The displays are placed in the test stores for a period of one week and the number of units sold during that week is noted for each store. These data are recorded in Table 13.6, where the x_{ij}'s represent sales in units of the product. The x_{ij}^2's appear in columns adjacent to the original observations.

The null hypothesis of this experiment is

$$H_0\colon \mu_1 = \mu_2 = \mu_3$$

that is, there is no difference in mean sales and, therefore, no treatment effect. The alternate hypothesis is simply that the means are not all equal and, therefore, at least one of the treatments has an effect that is significantly different from the others. The significance level chosen for this test is 5%.

The steps in the analysis of variance are listed below. An asterisk indicates a step in which one of the terms used in a sum of squares is computed.

1. Sum the x_{ij}'s in each column.

$$\sum_{i=1}^{10} x_{i1} = 170 \qquad \sum_{i=1}^{10} x_{i2} = 220 \qquad \sum_{i=1}^{10} x_{i3} = 150$$

2. Compute the column means.

$$\bar{x}_{.1} = \frac{170}{10} = 17 \qquad \bar{x}_{.2} = \frac{220}{10} = 22 \qquad \bar{x}_{.3} = \frac{150}{10} = 15$$

TABLE 13.6
DATA FOR THE TEST OF THREE POINT-OF-SALE DISPLAYS (EXAMPLE 13.4)

	Treatments						
	1		2		3		
Observation	x_{i1}	x_{i1}^2	x_{i2}	x_{i2}^2	x_{i3}	x_{i3}^2	
1	15	225	26	676	9	81	
2	21	441	22	484	15	225	
3	24	576	14	196	12	144	
4	10	100	29	841	18	324	
5	14	196	22	484	22	484	
6	16	256	14	196	17	289	
7	12	144	18	324	14	196	
8	19	361	25	625	14	196	
9	22	484	30	900	16	256	
10	17	289	20	400	13	169	
$r_{.j}$	10		10		10		$n = 30$
$\sum_i x_{ij}$	170		220		150		$\sum_j \sum_i x_{ij} = 540$
$\sum_i x_{ij}^2$		3072		5126		2364	$\sum_j \sum_i x_{ij}^2 = 10{,}562$
$\bar{x}_{.j}$	17		22		15		$\bar{\bar{x}} = 18$

$$\sum_j \sum_i x_{ij}^2 = 10{,}562$$

$$n\bar{\bar{x}}^2 = (30)(18)^2 = (30)(324) = 9720$$

$$\begin{aligned}\sum_j r_{.j}\bar{x}_{.j}^2 &= (10)(17)^2 + (10)(22)^2 + (10)(15)^2 \\ &= (10)(289) + (10)(484) + (10)(225) \\ &= 2890 + 4840 + 2250 \\ &= 9980\end{aligned}$$

3. Square each column mean.

$$\bar{x}_{.1}^2 = (17)^2 = 289 \qquad \bar{x}_{.2}^2 = (22)^2 = 484 \qquad \bar{x}_{.3}^2 = (15)^2 = 225$$

***4.** Compute

$$\begin{aligned}\sum_j r_{.j}\bar{x}_{.j}^2 &= r_{.1}\bar{x}_{.1}^2 + r_{.2}\bar{x}_{.2}^2 + r_{.3}\bar{x}_{.3}^2. \\ &= (10)(289) + (10)(484) + (10)(225) \\ &= 2890 + 4840 + 2250 \\ &= 9980\end{aligned}$$

5. Add all the column sums obtained in step 1 to obtain

$$\sum_{j=1}^{3}\sum_{i=1}^{10} x_{ij} = \sum_{i=1}^{10} x_{i1} + \sum_{i=1}^{10} x_{i2} + \sum_{i=1}^{10} x_{i3}$$
$$= 170 + 220 + 150$$
$$= 540$$

6. Compute the grand mean, $\bar{\bar{x}}$, by dividing the sum of all the observations by n.

$$\bar{\bar{x}} = \frac{\sum_j \sum_i x_{ij}}{r_{.1} + r_{.2} + r_{.3}} = \frac{\sum_j \sum_i x_{ij}}{n} = \frac{540}{30} = 18$$

***7.** Square $\bar{\bar{x}}$ and multiply by n.

$$n\bar{\bar{x}}^2 = (30)(18)^2 = (30)(324) = 9720$$

***8.** Sum the squares of all the original observations.

$$\sum_j \sum_i x_{ij}^2 = \sum_i x_{i1}^2 + \sum_i x_{i2}^2 + \sum_i x_{i3}^2$$
$$= 3072 + 5126 + 2364$$
$$= 10{,}562$$

9. Compute the sums of squares.

$$SS_T = \sum_j \sum_i x_{ij}^2 - n\bar{\bar{x}}^2 = 10{,}562 - 9{,}720 = 842$$

$$SS_K = \sum_j r_{.j}\bar{x}_{.j}^2 - n\bar{\bar{x}}^2 = 9{,}980 - 9{,}720 = 260$$

$$SS_E = \sum_j \sum_i x_{ij}^2 - \sum_j r_{.j}\bar{x}_{.j}^2 = 10{,}562 - 9{,}980 = 582$$

10. Determine the degrees of freedom associated with each sum of squares.

$$df_T = n - 1 = 30 - 1 = 29$$
$$df_K = k - 1 = 3 - 1 = 2$$
$$df_E = n - k = 30 - 3 = 27$$

11. Arrange the sums of squares and the degrees of freedom in an ANOVA table and compute the mean squares and variance ratio.

Sources of Variation	*Sum of Squares*	*Degrees of Freedom*	*Mean Square*	*Variance Ratio*
Treatments	260	2	130	6.02
Error	582	27	21.6	
Total	842	29		

12. Find the critical value of F with the appropriate degrees of freedom for the significance level of 5% and compare this value with the computed variance ratio.

$$F_{2,27,.05} = 3.35$$

Since the computed variance ratio of 6.02 is greater than the critical value of 3.35, the null hypothesis is rejected. It can be concluded that a treatment effect does exist, that is, that at least one of the displays has an effect on sales different from the others.

Although it is desirable to have samples of equal size assigned to each treatment, it is sometimes necessary to deal with unequal samples. This case is illustrated in the following example.

Example 13.5 A consumer testing organization plans to test four different brands of unleaded gasoline to determine if there is any difference among the brands in the mileage obtained. The level of significance for the test is 5%. All brands of gasoline are to be tested in randomly selected cars of the same make and model which are driven under similar conditions. However, the organization is able to obtain only 15 cars for the test. This means that one brand of gasoline can be used in three cars, while each of the others can be used in four cars. The data on the mileage obtained is shown in Table 13.7, where the treatments correspond to the brands of gasoline and the observations are in miles per gallon.

TABLE 13.7
DATA FOR TEST OF GASOLINE MILEAGE (EXAMPLE 13.5)

	Treatments							
	1		2		3		4	
Observation	x_{i1}	x_{i1}^2	x_{i2}	x_{i2}^2	x_{i3}	x_{i3}^2	x_{i4}	x_{i4}^2
1	22	484	21	441	15	225	27	729
2	19	361	26	676	19	361	19	361
3	23	529	22	484	20	400	23	529
4	24	576	21	441	14	196		
$r_{.j}$	4		4		4		3	
$\sum_i x_{ij}$	88		90		68		69	
$\sum_i x_{ij}^2$		1950		2042		1182		1619
$\bar{x}_{.j}$	22		22.5		17		23	

The terms required for the sums of squares are computed from the data of Table 13.7 as follows.

$$\sum_i r_{.j}\bar{x}_{.j}^2 = r_{.1}\bar{x}_{.1}^2 + r_{.2}\bar{x}_{.2}^2 + r_{.3}\bar{x}_{.3}^2 + r_{.4}\bar{x}_{.4}^2$$

$$= (4)(22)^2 + (4)(22.5)^2 + (4)(17)^2 + (3)(23)^2$$
$$= 1936 + 2025 + 1156 + 1587$$
$$= 6704$$

Note that $r_{.1} = r_{.2} = r_{.3} = 4$, but $r_{.4} = 3$.

$$n\bar{\bar{x}}^2 = (15)(21)^2 = 6615$$

$$\sum_j \sum_i x_{ij}^2 = 6793$$

The sums of squares are

$$\text{(13.8)} \qquad SS_T = \sum_j \sum_i x_{ij}^2 - n\bar{\bar{x}}^2 = 6793 - 6615 = 178$$

$$\text{(13.9)} \qquad SS_E = \sum_j \sum_i x_{ij}^2 - \sum_i r_{.j}\bar{x}_{.j}^2 = 6793 - 6704 = 89$$

$$\text{(13.10)} \qquad SS_K = \sum r_{.j}\bar{x}_{.j}^2 - n\bar{\bar{x}}^2 = 6704 - 6615 = 89$$

The degrees of freedom are

$$\text{(13.11)} \qquad df_T = n - 1 = 15 - 1 = 14$$

$$\text{(13.13)} \qquad df_E = n - k = 15 - 4 = 11$$

$$\text{(13.12)} \qquad df_K = k - 1 = 4 - 1 = 3$$

The ANOVA table is

Sources of Variation	*Sum of Squares*	*Degrees of Freedom*	*Mean Square*	*Variance Ratio*
Treatments	89	3	29.67	3.67
Error	89	11	8.09	
Total	178	14		

The critical value of F, for the appropriate degrees of freedom and a significance level of 5%, is

$$F_{3,11,.05} = 3.59$$

Since the computed variance ratio is greater than the critical value of F, the null hypothesis of no treatment effect is rejected at the 5% level, and we conclude that there is a difference in the mileage obtained from at least one brand of gasoline.

13.5 Least Significant Difference

When the null hypothesis in an analysis of variance problem is rejected, as it was in the previous two examples, it is proper to conclude that at least one of the sample means is significantly different from the others. However, we still do not know which means are different and which are not. When all the sample sizes are equal, a simple procedure called the least significant difference can provide the answer. If the sample sizes are not equal, more complicated procedures must be used.

The least significant difference, designated LSD, is the smallest difference between any two sample means that can be considered to be significant at the level selected for the test. Remember that a critical value of the difference between two sample means is computed from $d_c^* = ts_d^*$, where s_d^* is determined from a pooled estimate of the sample variances. A similar pooled estimate in analysis of variance is simply the mean square for error, or MS_E. Therefore, we obtain LSD from

$$\text{LSD} = t_{n-k,\alpha}\sqrt{\frac{2\text{MS}_E}{r}} \tag{13.14}$$

where $t_{n-k,\alpha}$ is the value of the statistic t which corresponds to the significance level of the test and the degrees of freedom for error, and r is simply the common sample size when the $r_{.j}$'s are all equal.

It can be shown that $t_{n-k,\alpha}^2 = F_{1,n-k,\alpha}$ and, therefore,

$$\text{LSD} = \sqrt{\frac{2(\text{MS}_E)F_{1,n-k,\alpha}}{r}} \tag{13.15}$$

Any two sample means are considered to be significantly different if their absolute difference is greater than or equal to the LSD.

Example 13.6 Since the null hypothesis in Example 13.4 was rejected, find which of the point-of-sale displays in that problem have effects that are significantly different from the others.

The MS_E from Example 13.4 is 21.6. The degrees of freedom for error are $df_E = 27$. The $r_{.j}$'s for each treatment are 10. Therefore,

$$\begin{aligned}\text{LSD} &= \sqrt{\frac{2(\text{MS}_E)F_{1,27.05}}{r}}\\ &= \sqrt{\frac{2(21.6)(4.21)}{10}}\\ &= 4.26\end{aligned}$$

Any two sample means with an absolute difference greater than or equal to 4.26 are considered to be significantly different. Since there are three sample means in the problem, the differences between each pair are

$|\bar{x}_{.1} - \bar{x}_{.2}| = |17 - 22| = 5 > 4.26$ significant difference

$|\bar{x}_{.1} - \bar{x}_{.3}| = |17 - 15| = 2 < 4.26$ no significant difference

$|\bar{x}_{.2} - \bar{x}_{.3}| = |22 - 15| = 7 > 4.26$ significant difference

Since the mean of treatment sample 2 is significantly greater than either of the other two means, the marketing director of Ego Cosmetics should use point-of-sale display 2 at all stores.

When the sample sizes are not all equal, as in Example 13.4, there is no simple procedure for determining which pair or pairs of treatments cause the significant differences in the analysis of variance. In this case, to avoid the use of complicated procedures, each pair of means can be tested individually by the test for the difference between two sample means, covered in Chapter 12.

13.6 Two-Factor Analysis of Variance

In a one-factor analysis of variance, the treatments constitute different levels of a single factor which is controlled in the experiment. There are, however, many situations in which the response variable of interest may be affected by more than one factor. For example, sales of Ego cosmetics, in addition to being affected by the point-of-sale display, might also be affected by the price charged, the size and/or location of the store, or the number of competitive products sold by the store. Gasoline mileage may be affected by the type of car driven, the way it is driven, road conditions, and other factors in addition to the brand of gasoline used.

When it is believed that two independent factors might have an effect on the response variable of interest, it is possible to design the test so that an analysis of variance can be used to test for the effects of the two factors simultaneously. Such a test is called a *two-factor analysis of variance*.

Suppose we have k levels of some factor K, the column factor, and r levels of factor R, the row factor. Each level of a factor corresponds to a treatment. A combination of one level of factor K with one level of factor R is frequently called a TREATMENT COMBINATION. There are then $rk = n$ treatment combinations. Each observation, an x_{ij}, constitutes a random sample of size one from one of the n populations, all of which are assumed to have normal distributions and a common variance, σ^2. The data for a two-factor analysis of variance is arranged in a table similar to Table 13.8. Notice the similarity of this table to Table 13.1. The only difference between the two tables is that in Table 13.8, each row constitutes a level of the row factor, and the row sums and means are computed in addition to the column sums and means.

With the two-factor analysis of variance, we can test two sets of hypotheses with the same data at the same time. First, we can test the null hypothesis that the column means are all equal against the alternate hypothesis that they are not all equal. This is expressed as

$$H_0\colon \mu_{.1} = \mu_{.2} = \mu_{.3} = \cdots = \mu_{.k}$$
$$H_1\colon \text{Not all the column means are equal}$$

Then we can test the null hypothesis that the row means are all

TABLE 13.8
GENERAL ARRANGEMENT OF DATA FOR A TWO-FACTOR ANALYSIS OF VARIANCE

Row Factor Levels	*Column Factor Levels* 1	2	3	·	·	·	·	*k*	*Row Sums*	*Row Means*
1	x_{11}	x_{12}	x_{13}	·	·	x_{1j}	·	x_{1k}	$\sum_j x_{1j}$	$\bar{x}_{1.}$
2	x_{21}	x_{22}	x_{23}	·	·	x_{2j}	·	x_{2k}	$\sum_j x_{2j}$	$\bar{x}_{2.}$
3	x_{31}	x_{32}	x_{33}	·	·	x_{3j}	·	x_{3k}	$\sum_j x_{3j}$	$\bar{x}_{3.}$
·	·	·	·	·	·	·	·	·		
·	·	·	·	·	·	·	·	·		
·	x_{i1}	x_{i2}	x_{i3}	·	·	x_{ij}	·	x_{ik}	$\sum_j x_{ij}$	$\bar{x}_{i.}$
·	·	·	·	·	·	·	·	·		
r	x_{r1}	x_{r2}	x_{r3}	·	·	x_{rj}	·	x_{rk}	$\sum_j x_{rj}$	$\bar{x}_{r.}$
Column sums	$\sum_i x_{i1}$	$\sum_i x_{i2}$	$\sum_i x_{i3}$			$\sum_i x_{ij}$		$\sum_i x_{ik}$	Grand sum $= \sum_j \sum_i x_{ij}$	
Column means	$\bar{x}_{.1}$	$\bar{x}_{.2}$	$\bar{x}_{.3}$			$\bar{x}_{.j}$		$\bar{x}_{.k}$	Grand mean $= \bar{\bar{x}}$	

equal against the alternate hypothesis that they are not all equal, that is,

$$H_0: \mu_{1.} = \mu_{2.} = \mu_{3.} = \cdots = \mu_{r.}$$
$$H_1: \text{Not all the row means are equal}$$

In order to test these hypotheses, the total sum of squares must now be partitioned into three parts:

$$SS_T = SS_K + SS_R + SS_E$$

where SS_R represents that portion of the total sum of squares which is attributed to the variation among the levels of the row factor. The computational forms for these sums of squares are

$$SS_T = \sum_j \sum_i x_{ij}^2 - n\bar{\bar{x}}^2 \tag{13.16}$$

$$SS_K = \sum_j r\bar{x}_{.j}^2 - n\bar{\bar{x}}^2 \tag{13.17}$$

$$SS_R = \sum_i k\bar{x}_{i.}^2 - n\bar{\bar{x}}^2 \tag{13.18}$$

The error sum of squares can most simply be obtained by subtraction. That is,

$$SS_E = SS_T - (SS_K + SS_R) \tag{13.19}$$

Notice that SS_T is obtained in the same way as in a one-factor analysis of variance. Also, since the number of levels of the row factor are the same across all the columns, the $r_{.j}$'s $= r$. This is reflected in Equation 13.17. Similarly, the number of levels of the column factor are the same across all of the rows and the $k_{i.}$'s $= k$.

In order to obtain all four sums of squares it is only necessary to compute four different terms. These are

$$\sum_j \sum_i x_{ij}^2 \quad \text{used in finding } SS_T$$

$$\sum_j r\bar{x}_{.j}^2 \quad \text{used in finding } SS_K$$

$$\sum_i k\bar{x}_{i.}^2 \quad \text{used in finding } SS_R$$

$$n\bar{\bar{x}}^2 \quad \text{used in finding } SS_T,\ SS_K, \text{ and } SS_R$$

Example 13.7

Consider the data presented in Table 13.9a, where the columns represent four levels of factor K and the rows represent three levels of factor R. Table 13.9b simply displays the squares of the observations provided in Table

TABLE 13.9

DATA FOR A TWO-FACTOR ANALYSIS OF VARIANCE (EXAMPLE 13.7)

A. TABLE OF x_{ij}.

	Column Factor Levels					
Row Factor Levels	*1*	*2*	*3*	*4*	*Row Sums*	*Row Means*
1	3	3	2	16	24	6
2	4	1	3	20	28	7
3	2	2	4	9	17	4.25
Column sums	9	6	9	45	Grand sum = 69	
Column means	3	2	3	15	Grand mean = 5.75	

B. TABLE OF x_{ij}^2.

	Column Factor Levels				
Row Factor Levels	*1*	*2*	*3*	*4*	*Row Sums*
1	9	9	4	256	278
2	16	1	9	400	426
3	4	4	16	81	105
Column sums	29	14	29	737	Grand sum = 809

13.9a. The terms needed for the sums of squares in a two-factor analysis of variance can be obtained from these data as follows, where $k = 4$, $r = 3$, and $n = rk = 12$.

$$\sum_j \sum_i x_{ij}^2 = 809$$

$$\begin{aligned}\sum_j r\bar{x}_{.j}^2 &= (3)(3)^2 + (3)(2)^2 + (3)(3)^2 + (3)(15)^2 \\ &= (3)(9) + (3)(4) + (3)(9) + (3)(225) \\ &= 27 + 12 + 27 + 675 \\ &= 741\end{aligned}$$

$$\begin{aligned}\sum_i k\bar{x}_{i.}^2 &= (4)(6)^2 + (4)(7)^2 + (4)(4.25)^2 \\ &= (4)(36) + (4)(49) + (4)(18.0625) \\ &= 144 + 196 + 72.25 \\ &= 412.25 \\ n\bar{\bar{x}}^2 &= (12)(5.75)^2 = (12)(33.0625) \\ &= 396.75\end{aligned}$$

Combining pairs of terms in the proper order, we get:

$$SS_T = \sum_j \sum_i x_{ij}^2 - n\bar{\bar{x}}^2 = 809 - 396.75 = 412.25$$

$$SS_K = \sum_j r\bar{x}_{.j}^2 - n\bar{\bar{x}}^2 = 741 - 396.75 = 344.25$$

$$SS_R = \sum_i k\bar{x}_{i.}^2 - n\bar{\bar{x}}^2 = 412.25 - 396.75 = 15.50$$

$$\begin{aligned}SS_E &= SS_T - (SS_K + SS_R) = 412.25 - (344.25 + 15.50) \\ &= 412.25 - 359.75 \\ &= 52.50\end{aligned}$$

The appropriate number of degrees of freedom for each of the sums of squares are

The total degrees of freedom are the total number of observations minus one, that is,

(13.20) $$df_T = n - 1$$

The degrees of freedom for the column sum of squares are the number of columns minus one, that is,

(13.21) $$df_K = k - 1$$

The degrees of freedom for the row sum of squares are the number of rows minus one, that is,

$$df_R = r - 1 \tag{13.22}$$

The degrees of freedom for the error sum of squares are simply

$$df_E = df_T - (df_K + df_R)$$
$$df_E = n - 1 - (k - 1 + r - 1)$$
$$df_E = n - k - r + 1$$
$$df_E = n - (k + r) + 1 \tag{13.23}$$

The ANOVA tables for a two-factor analysis of variance in general form and for the data in Table 13.9 are provided in Tables 13.10 and 13.11, respectively.

TABLE 13.10

GENERAL ARRANGEMENT OF THE ANOVA TABLE FOR A TWO-FACTOR ANALYSIS OF VARIANCE

Source of Variation	*Sum of Squares*	*Degrees of Freedom*	*Mean Square (SS/df)*	*Variance Ratio*
Among columns	SS_K	$k - 1$	$MS_K = SS_K/k - 1$	MS_K/MS_E
Among rows	SS_R	$r - 1$	$MS_R = SS_R/r - 1$	MS_R/MS_E
Error	SS_E	$n - (k + r) + 1$	$MS_E = SS_E/n - (k + r) + 1$	
Total	SS_T	$n - 1$		

TABLE 13.11

ANOVA TABLE (DATA OF TABLE 13.9)

Source of Variation	*Sum of Squares*	*Degrees of Freedom*	*Mean Square (SS/df)*	*Variance Ratio*
Among columns	344.25	3	114.75	114.75/8.75 = 13.11
Among rows	15.50	2	7.75	7.75/8.75 = 0.89
Error	52.50	6	8.75	
Total	412.25	11		

In testing the hypothesis of equal column means, that is, that different levels of the column factor have no effect, at $\alpha = 0.01$, the critical value of F is

$$F_{3,6,.01} = 9.78$$

Since the computed variance ratio for columns is 13.11, this null hypothesis is rejected and it can be assumed that at least one of the column means is different from the others.

In testing the hypothesis of equal row means, that is, that different levels of the row factor have no effect, at $\alpha = 0.01$, the critical value of F is

$$F_{2,6,.01} = 10.92$$

Since the computed variance ratio for rows is 0.89, this null hypothesis cannot be rejected and it is assumed that there is no row effect.

Because this analysis indicates that a significant column effect exists at the 1% level of significance, it is necessary to apply the least significant difference test to the column means. The LSD is not applied to the row means, since the hypothesis of equal means was not rejected.

The LSD for columns is

$$\text{LSD} = \sqrt{\frac{(2)(\text{MS}_E)F_{1,df_E,.01}}{r}} \tag{13.15}$$

$$= \sqrt{\frac{(2)(8.75)(F_{1,6,.01})}{3}}$$

$$= \sqrt{\frac{(2)(8.75)(13.75)}{3}}$$

$$= 8.96$$

Therefore, any two column means with an absolute difference greater than or equal to 8.96 are considered to be significantly different. Since there are four columns, the differences between each pair of means are

$|\bar{x}_{.1} - \bar{x}_{.2}| = 1 < 8.96$ not significant
$|\bar{x}_{.1} - \bar{x}_{.3}| = 0 < 8.96$ not significant
$|\bar{x}_{.1} - \bar{x}_{.4}| = 12 > 8.96$ significant
$|\bar{x}_{.2} - \bar{x}_{.3}| = 1 < 8.96$ not significant
$|\bar{x}_{.2} - \bar{x}_{.4}| = 13 > 8.96$ significant
$|\bar{x}_{.3} - \bar{x}_{.4}| = 12 > 8.96$ significant

We can conclude from this test that the mean of column four, that is, level four of the column factor, is significantly different from the other three.

Example 13.8 The Ego Cosmetics Company wants to test the effect of two different factors on sales of a newly developed ladies' cologne. Three different point-of-sale displays designed by the advertising department consititute three levels of the first factor. The other factor to be tested is selling price. Three prices are considered for the six-ounce bottle. The prices are \$4.75, \$5.50, and \$6.25. The level of significance for the test is 5%.

Nine department stores whose cosmetics departments carry Ego products are selected for the test and are randomly assigned to the treatment combinations. The stores are matched as closely as possible with respect to size, location, and other pertinent characteristics. The response variable observed is the number of bottles of the new cologne sold during the one-month test period. The data obtained at the end of that period are summarized in Table 13.12. The column factor is the point-of-sale advertising and the row factor is the price.

TABLE 13.12
ANALYSIS OF VARIANCE DATA POINT-OF-SALE DISPLAYS AND PRICE (NUMBER OF UNITS SOLD) (EXAMPLE 13.8)

Row Factor Levels Price	*Column Factor Levels Displays*		
	Display #1	*Display #2*	*Display #3*
Low price ($4.75)	80	60	40
Middle price ($5.50)	70	40	40
High price ($6.25)	30	20	10

The row and column means and the grand mean are obtained from the data as follows.

$$\sum_{i=1}^{3} x_{i1} = 80 + 70 + 30 = 180, \qquad \bar{x}_{.1} = \frac{180}{3} = 60$$

$$\sum_{i=1}^{3} x_{i2} = 60 + 40 + 20 = 120, \qquad \bar{x}_{.2} = \frac{120}{3} = 40$$

$$\sum_{i=1}^{3} x_{i3} = 40 + 40 + 10 = 90, \qquad \bar{x}_{.3} = \frac{90}{3} = 30$$

$$\sum_{j=1}^{3} x_{1j} = 80 + 60 + 40 = 180, \qquad \bar{x}_{1.} = \frac{180}{3} = 60$$

$$\sum_{j=1}^{3} x_{2j} = 70 + 40 + 40 = 150, \qquad \bar{x}_{2.} = \frac{150}{3} = 50$$

$$\sum_{j=1}^{3} x_{3j} = 30 + 20 + 10 = 60, \qquad \bar{x}_{3.} = \frac{60}{3} = 20$$

$$\sum_{j=1}^{3} \sum_{i=1}^{3} x_{ij} = 80 + 70 + 30 + 60 + 40 + 20 + 40 + 40 + 10 = 390$$

$$\bar{\bar{x}} = \frac{390}{9} = 43.3$$

The terms needed for the various sums of squares can be obtained from the data above, where $k = 3$, $r = 3$, and $n = rk = 9$.

$$\sum_{j=1}^{3}\sum_{i=1}^{3} x_{ij}^2 = (80)^2 + (70)^2 + (30)^2 + (60)^2 + (40)^2 + (20)^2 + (40)^2 + (40)^2 + (10)^2$$
$$= 6400 + 4900 + 900 + 3600 + 1600 + 400 + 1600 + 1600 + 100$$
$$= 21{,}100$$

$$\sum_{j=1}^{k} r\bar{x}_{.j}^2 = (3)(60)^2 + (3)(40)^2 + (3)(30)^2 = (3)(3600) + (3)(1600) + (3)(900)$$
$$= 10{,}800 + 4800 + 2700$$
$$= 18{,}300$$

$$\sum_{i=1}^{r} k\bar{x}_{i.}^2 = (3)(60)^2 + (3)(50)^2 + (3)(20)^2$$
$$= (3)(3600) + (3)(2500) + (3)(400)$$
$$= 10{,}800 + 7500 + 1200$$
$$= 19{,}500$$
$$n\bar{\bar{x}}^2 = (9)(43.3)^2 = (9)(1874.9) = 16{,}874.1$$

Combining pairs of terms in the proper order results in

$$SS_T = \Sigma\Sigma x_{ij}^2 - n\bar{\bar{x}}^2 = 21{,}100 - 16{,}874.1 = 4225.9$$
$$SS_K = \Sigma r\bar{x}_{.j}^2 - n\bar{\bar{x}}^2 = 18{,}300 - 16{,}874.1 = 1425.9$$
$$SS_R = \Sigma k\bar{x}_{i.}^2 - n\bar{\bar{x}}^2 = 19{,}500 - 16{,}874.1 = 2625.9$$
$$SS_E = SS_T - (SS_K + SS_R)$$
$$= 4225.9 - (1425.9 + 2625.9)$$
$$= 4225.9 - 4051.8 = 174.1$$

The analysis of variance table for these tests is shown in Table 13.13.

TABLE 13.13
ANOVA TABLE FOR TESTS OF POINT-OF-SALE DISPLAYS AND PRICE (EXAMPLE 13.8)

Source of Variation	*Sum of Squares*	*Degrees of Freedom*	*Mean Square*	*Variance Ratio*
Among columns (Displays)	1425.9	2	712.95	16.38
Among rows (Price)	2625.9	2	1312.95	30.16
Error	174.1	4	43.53	
Total	4225.9	8		

In testing the hypothesis that there is no display effect at $\alpha = 0.05$, the critical value of F is

$$F_{2,4,.05} = 6.94$$

Since the computed variance ratio for displays (columns) is 16.38, this null hypothesis is rejected and it can be assumed that at least one of the displays has a significant effect on sales at the 5% level.

In testing the hypothesis of no price effect at $\alpha = 0.05$, the critical value of F is also 6.94, since the degrees of freedom for rows happens to be the same as the degrees of freedom for columns in this example. Since the computed variance ratio for price (rows) is 30.16, this null hypothesis is also rejected and we can assume that price has a significant effect on sales at the 5% level.

Since both null hypotheses are rejected, it is necessary to determine which levels of the two factors are responsible for the differences. This is done by the method of the least significant difference. Since, in this problem $r = k$, the least significant difference is the same for both factors. That is,

$$\text{LSD}_{\text{Rows}} = \sqrt{\frac{2(\text{MS}_E)F_{1,df_E,\alpha}}{r}} = \text{LSD}_{\text{Columns}} = \sqrt{\frac{2(\text{MS}_E)F_{1,df_E,\alpha}}{k}}$$

$$\begin{aligned}\text{LSD} &= \sqrt{\frac{2(43.53)F_{1,4,.05}}{3}}\\ &= \sqrt{\frac{2(43.53)(7.71)}{3}}\\ &= 14.96\end{aligned}$$

Comparing all pairs of column and row means we get:

Column Means:

$|\bar{x}_{.1} - \bar{x}_{.2}| = 20 > 14.96$ significant
$|\bar{x}_{.1} - \bar{x}_{.3}| = 30 > 14.96$ significant
$|\bar{x}_{.2} - \bar{x}_{.3}| = 10 < 14.96$ not significant

Row Means:

$|\bar{x}_{1.} - \bar{x}_{2.}| = 10 < 14.96$ not significant
$|\bar{x}_{1.} - \bar{x}_{3.}| = 40 > 14.96$ significant
$|\bar{x}_{2.} - \bar{x}_{3.}| = 30 > 14.96$ significant

Therefore, with respect to the displays, we can conclude that display #1 has an effect on sales that is significantly different from displays #2 and #3. There is no significant difference between displays #2 and #3. Since sales are higher when display #1 is used, this display should be used to promote sales of the new cologne in all stores.

With respect to the price factor, we can conclude that the high price, \$6.25, affects the number of units sold. There is no significant difference in the effect on sales between the low and middle prices, \$4.75 and \$5.50, respectively. Since unit sales are less in stores where the high price was charged, the new cologne should not be priced at that level. However, since there is no significant difference in sales between the \$4.75 price and the \$5.50 price, it appears that the best price is \$5.50.

Definitions

ANOVA: Abbreviation for analysis of variance.

Factor: A variable of interest that is controlled in an analysis of variance.

Factor level: A value at which the factor is controlled; same as a treatment.

Least significant difference: The smallest difference between any two sample means in an analysis of variance that can be considered to be significant at the level selected for the test.

Mean square: The sum of squares divided by the number of degrees of freedom; a sample variance.

Sum of squares: The sum of the squared deviations of sample values about the sample mean.

Treatment: A value of a controlled factor; see factor level.

Treatment combination: A combination of one level of one factor with one level of another factor in a two-factor analysis of variance.

Treatment population: The theoretical population of elementary units that could potentially be exposed to a treatment.

Formulas

Pooled estimate of σ^2 from multiple samples:

$$s_E^2 = \frac{(n_1 - 1)s_1^2 + (n_2 - 1)s_2^2 + \cdots + (n_k - 1)s_k^2}{n_1 + n_2 + \cdots + n_k - k} \tag{13.1}$$

Estimate of $\sigma_{\bar{x}}^2$ from multiple samples:

$$s_{\bar{x}}^2 = \frac{(\bar{x}_{.1} - \bar{\bar{x}})^2 + (\bar{x}_{.2} - \bar{\bar{x}})^2 + \cdots + (\bar{x}_{.k} - \bar{\bar{x}})^2}{k - 1} \tag{13.2}$$

Alternate method of estimating σ^2 in analysis of variance:

$$s_k^2 = \frac{r_{.1}(\bar{x}_{.1} - \bar{\bar{x}})^2 + r_{.2}(\bar{x}_{.2} - \bar{\bar{x}})^2 + \cdots + r_{.k}(\bar{x}_{.k} - \bar{\bar{x}})^2}{k - 1} \tag{13.3}$$

Sums of squares and degrees of freedom for a one-factor analysis of variance:

$$SS_T = \sum_j \sum_i (x_{ij} - \bar{\bar{x}})^2 \tag{13.4}$$

$$SS_E = \sum_j \sum_i (x_{ij} - \bar{x}_{.j})^2 \tag{13.5}$$

$$SS_K = \sum_j r_{.j}(\bar{x}_{.j} - \bar{\bar{x}})^2 \tag{13.6}$$

$$SS_T = SS_K + SS_E \tag{13.7}$$

$$SS_T = \sum_j \sum_i x_{ij}^2 - n\bar{\bar{x}}^2 \qquad \text{Computational} \tag{13.8}$$

(13.9) $$SS_E = \sum_j \sum_i x_{ij}^2 - \sum_j r_{.j}\bar{x}_{.j}^2 \quad \text{Computational}$$

(13.10) $$SS_K = \sum_j r_{.j}\bar{x}_{.j}^2 - n\bar{\bar{x}}^2 \quad \text{Computational}$$

(13.11) $$df_T = n - 1$$

(13.12) $$df_K = k - 1$$

(13.13) $$df_E = n - k$$

Least significant difference:

(13.14) $$\text{LSD} = t_{n-k,\alpha}\sqrt{\frac{2(\text{MS}_E)}{r}}$$

(13.15) $$\text{LSD} = \sqrt{\frac{2(\text{MS}_E)F_{1,n-k,\alpha}}{r}}$$

Sums of squares and degrees of freedom for a two-factor analysis of variance:

(13.16) $$SS_T = \sum_j \sum_i x_{ij}^2 - n\bar{\bar{x}}^2 \quad \text{Computational}$$

(13.17) $$SS_K = \sum_j r\bar{x}_{.j}^2 - n\bar{\bar{x}}^2 \quad \text{Computational}$$

(13.18) $$SS_R = \sum_i k\bar{x}_{i.}^2 - n\bar{\bar{x}}^2 \quad \text{Computational}$$

(13.19) $$SS_E = SS_T - (SS_K + SS_R)$$

(13.20) $$df_T = n - 1$$

(13.21) $$df_K = k - 1$$

(13.22) $$df_R = r - 1$$

(13.23) $$df_E = n - (k + r) + 1$$

Problems

13.1 From the data in the following ANOVA table, test the hypothesis that there is no significant treatment effect at $\alpha = 0.01$.

ANOVA TABLE

Sources of Variation	*Sum of Squares*	*Degrees of Freedom*	*Mean Square*	*Variance Ratio*
Among treatments	151	9	16.8	3.23
Error	104	20	5.2	
Total	255	29		

13.2 From the data in the ANOVA table of problem 13.1, test for a significant treatment effect at $\alpha = 0.05$.

13.3 Given that $SS_T = 94$, $df_T = 34$, $SS_K = 30$, and $df_K = 5$, complete the ANOVA table and test for a significant treatment effect at $\alpha = 0.05$ and 0.01.

13.4 Given that $SS_T = 246$, $df_T = 27$, $SS_E = 120$, and $df_E = 20$, complete the ANOVA table and test for a significant treatment effect at $\alpha = 0.05$ and 0.01.

13.5 The following table shows values of a response variable x from three different samples each of size five. Test whether there is a significant difference in the sample means at $\alpha = 0.01$.

TREATMENTS

1	2	3
5	16	12
4	15	2
8	14	4
13	14	3
10	16	4

13.6 The following table shows values of a response variable x from four different samples each of size three. Test whether there is a significant difference in sample means at $\alpha = 0.01$.

TREATMENTS

1	2	3	4
9	12	3	4
10	10	4	2
8	8	2	6

13.7 The following table shows values of a response variable x from three samples of different sizes. Test whether there is a significant difference in the sample means at $\alpha = 0.01$.

TREATMENTS

1	2	3
5	10	2
4	6	2
3	11	1
4		3
		2

13.8 The following table shows values of a response variable x from four samples of unequal size. Test whether there is a significant difference in the sample means at $\alpha = 0.05$ and 0.01.

TREATMENTS

1	2	3	4
12	10	16	18
16	8	11	12
	16	9	9
	14		

13.9 The president of a large chain of supermarkets was prepared to order a large number of frozen food display cases for use in the markets. Before placing the order, he decided to test the products by storing half-gallon containers of milk in the cases made by different manufacturers and observing the spoilage time for each type of case. The display cases came from three different manufacturers designated, *A*, *B*, and *C*. Nine half-gallon containers of milk were randomly selected and assigned, three to each case. The response variable observed was the spoilage time in days. The data for this test is provided below. Test for a significant difference in the effect of the display cases at the 5% level of significance.

TREATMENTS

A	*B*	*C*
7	8	7
5	4	8
9	7	10

13.10 Four brands of tires were tested for durability and wear on specially designed machines which simulate road conditions. Four tires of each brand were subjected to the same test and the number of miles until wearout was noted for each tire. Test for a significant difference in tire mileage at the 5% level of significance. The data in thousands of miles is provided below.

TIRE BRAND

Rollright	*Extron*	*Goodheart*	*Durable*
24	19	22	15
14	10	15	9
20	12	18	14
21	14	20	12

13.11 Use the method of the least significant difference to determine which of the treatments in problem 13.5 have a significant effect at $\alpha = 0.01$.

13.12 Use the method of the least significant difference to determine which of the treatments in problem 13.6 have a significant effect at $\alpha = 0.01$.

13.13 Use the method of the least significant difference to determine which brand or brands of the frozen food display cases in problem 13.9 have a significant effect on spoilage time.

13.14 Use the method of the least significant difference to determine which of the tire brands in problem 13.10 had significantly different wear-out characteristics at $\alpha = 0.05$.

13.15 From the data in the following ANOVA table, test whether there is a significant difference in column means and row means at $\alpha = 0.01$.

ANOVA TABLE

Sources of Variation	*Sum of Squares*	*Degrees of Freedom*	*Mean Square*	*Variance Ratio*
Among columns	117.6	4	29.4	4.2
Among rows	142.8	3	47.6	6.8
Error	84.0	12	7.0	
Total	344.4	19		

13.16 From the data in the ANOVA table of problem 13.15, test whether there is a significant difference in column means and row means at $\alpha = 0.05$.

13.17 Given that $SS_K = 48.0$, $SS_E = 72.0$, $SS_T = 213.6$, $df_K = 4$, $df_E = 24$, and $df_T = 34$, complete the ANOVA table and test for a significant column and row factor effect at $\alpha = 0.05$.

13.18 For the data in problem 13.17, test for a significant column and row factor effect at $\alpha = 0.01$.

13.19 Given that $SS_R = 1852.5$, $df_R = 3$, $SS_E = 585.0$, $df_E = 9$, $SS_T = 3510.0$, and $df_T = 15$, complete the ANOVA table and test for a significant column factor effect at both $\alpha = 0.05$ and $\alpha = 0.01$.

13.20 For the data in problem 13.19, test for a significant row factor effect at both $\alpha = 0.05$ and $\alpha = 0.01$.

13.21 The advertising director of the Rollright Tire Company wanted to conduct a test for the effects of three different advertising slogans and three different advertising media on the sale of Rollright tires at the 5% level of significance. The slogans were simply referred to as slogan #1, slogan #2, and slogan #3. The three advertising media to be tested were radio spots, television spots, and newspaper ads. Nine test markets were selected, as similar as possible with respect to population, socioeconomic structure, and other pertinent factors, and randomly assigned to the treatment combinations. The response variable is the number of tires sold in the test market during the month of the test. The data appear in the following table. The x_{ij}'s are in hundreds. Test both row factors and column factors for significance at the 5% level.

Row Factor (Medium)	*Column Factor (Slogans)* #1	#2	#3
1. Radio	20	14	20
2. Television	10	10	15
3. Newspaper	21	30	25

13.22 When the new marketing manager took over at the Herbert Vacuum Cleaner Company, he was primarily concerned with methods of increasing sales. Two methods for accomplishing this goal were considered. One was a training course for new sales representatives who, in the past, received no formal training. The new manager considered the possibility of continuing as before, but he also considered having each new sales representative take a three- or five-day training course. These three possibilities with regard to training constitute three levels of this factor. The second factor was a possible change in the method of compensating sales representatives. In the past, all sales representatives were paid on a straight commission basis. The new manager wanted to consider the possible alternatives of salary plus commission or straight salary and no commission. We will refer to the training procedures as the levels of the column factor and the three compensation plans as the levels of the row factor.

Nine new representatives who were hired by the new marketing manager were randomly assigned to each treatment combination. The data resulting from this test are provided in the following table, where the response variable is the number of vacuum cleaners sold by each sales representative during his first month in the field. Test for the existence of a significant row factor effect and column factor effect at the 5% level of significance.

	Column Factor		
Row Factor	*1 No Formal Training*	*2 3-Day Training*	*3 5-Day Training*
1. Straight Commission	15	27	18
2. Salary and Commission	18	36	24
3. Straight Salary	12	21	15

13.23 For problem 13.22, if either the row factor or the column factor or both has a significant effect on sales at the 5% level of significance, use the method of the least significant difference to determine which levels of the significant factor were different from the others.

13.24 For the analysis of problem 13.21, if either or both factors had a significant effect at the 5% level of significance, use the LSD to determine which levels of the significant factor were different from the others.

X

CHAPTER 14 CHI-SQUARE TESTS

14.1
The chi-square distribution

14.2
The chi-square statistic

14.3
Tests of goodness-of-fit

14.4
Tests of independence—contingency tables

In the previous chapters on statistical inference it was often necessary to make certain assumptions about the populations from which the samples were drawn. In many of the statistical tests, for example, we had to assume that the samples came from normal populations. In other tests involving two or more samples, the added assumption of equal variances was required. When these assumptions cannot be justified it is necessary to use procedures that do not require that these conditions be met. These procedures are generally referred to as nonparametric methods. In this chapter, we discuss two types of nonparametric tests which belong to a category called *chi-square tests*. Several other nonparametric tests are covered in Chapter 15.

14.1 The Chi-Square Distribution

The statistic used in all chi-square tests is called chi-square, χ^2. The symbol χ is the Greek letter chi. The sampling distribution of χ^2 is called the chi-square distribution, with the single parameter degrees of freedom. The chi-square distribution is a continuous probability distribution which has the value zero at its lower limit and extends to infinity in the positive direction. Negative values of χ^2 are impossible. The distribution is skewed to the right, as illustrated in Figure 14.1. Values of χ^2 that divide the curve with a proportion of the area equivalent to α in the right tail are provided in Appendix Table A.5.

In the table of the chi-square distribution, degrees of freedom appear in the extreme left-hand column. The proportions of the area in the right tail, the unshaded area in Figure 14.1, are specified in the other column headings. The numbers in the body of the table correspond to the values of χ^2 that place the specified proportions in the tail.

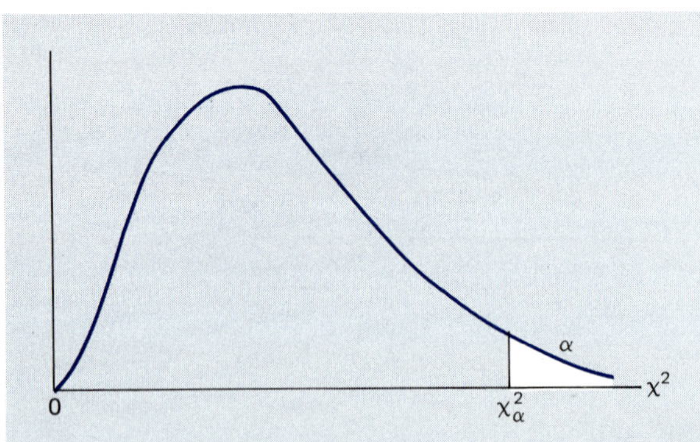

Figure 14.1
A chi-square distribution.

Example 14.1

Find the following.

a) the value of χ^2 with 10 degrees of freedom which cuts off 5% of the area in the right tail of the distribution.

b) the value of χ^2 with 40 degrees of freedom which cuts off 1% of the area in the right tail of the distribution.

c) the value of χ^2 with 27 degrees of freedom which cuts off 2.5% of the area in the right tail of the distribution.

d) the value of χ^2 with 19 degrees of freedom which cuts off 10% of the area in the right tail of the distribution.

a) 18.31 **b)** 63.69 **c)** 43.19 **d)** 27.20

14.2 The Chi-Square Statistic

Chi-square tests deal with frequencies or counts, that is, the number of observations that fall into certain categories. In each test, there is a set of frequencies obtained from a random sample. These frequencies are designated f. This is consistent with the notation used in Chapter 3 on frequency distributions. We then calculate a second set of expected frequencies based on the assumption that the null hypothesis to be tested is true. These expected frequencies or expectations are designated e. The chi-square statistic compares the differences between the two sets of frequencies and expresses them with a single numerical value. If these differences, as represented by the value χ^2, are significant at the level selected for the test, the null hypothesis is rejected.

The χ^2 statistic is computed from the data as follows.

$$\chi^2 = \sum_i \frac{(f_i - e_i)^2}{e_i} \tag{14.1}$$

where χ^2 has an approximate chi-square distribution. Critical values of χ^2 are obtained from the tables of the chi-square distribution. The degrees of freedom for a chi-square test depend on the type of test and certain other factors, which will be discussed as each particular test is illustrated.

Example 14.2

The Emporium department store planned to remain open until 9:30 P.M. one night each week. In considering which night, Monday through Friday, that the store should stay open, one executive suggested that the number of customers shopping was probably the same regardless of the night selected. The executive board decided to conduct a test to determine whether or not this assumption was true at the 5% level of significance. The null and alternate hypotheses for the test were

H_0: The number of shoppers is the same on all weeknights.

H_1: The number of shoppers is not the same on all weeknights.

Five different weeks were selected during which no holidays occurred and no special promotions were scheduled. This was done so that if differences in the number of shoppers did occur, they could be attributed to the night of the week on which the store was open and not to either of the other factors. During each of these five weeks, the store remained open on a different night and the number of shoppers entering the store was recorded. The data obtained were

Night	*Number of Shoppers* (f)
Monday	175
Tuesday	205
Wednesday	195
Thursday	210
Friday	215
	1000

These data represent the observed frequencies. We also need a set of expected frequencies for comparison. The expected frequencies are calculated under the assumption that the null hypothesis is true. Therefore, if the number of shoppers is the same on all five nights, we expect 1/5 of the total to shop on each night. Since the sample consisted of a total of 1000 shoppers, the expected frequency is 200 per night.

The observed and expected frequencies and the calculations of χ^2 are shown in Table 14.1. In this problem, the variable used for classifying the data is simply the night of the week on which the store was open. Each of these nights corresponds to a cell in the table.

The procedure for calculating χ^2 and testing the hypothesis is outlined as follows.

Step 1: The observed and expected frequencies are listed in columns 2 and 3 of Table 14.1.

Step 2: Each expected frequency is subtracted from the corresponding observed frequency and the difference is entered in column 4.

TABLE 14.1
DATA FOR CHI-SQUARE TEST OF SHOPPING NIGHTS (EXAMPLE 14.2)

1	2	3	4	5	6
Cells	*Observed Frequencies* (f)	*Expected Frequencies* (e)	$f - e$	$(f - e)^2$	$(f - e)^2/e$
Monday	175	200	−25	625	3.125
Tuesday	205	200	5	25	0.125
Wednesday	195	200	−5	25	0.125
Thursday	210	200	10	100	0.500
Friday	215	200	15	225	1.125
					5.000

Step 3: The differences in column 4 are squared and the values are entered in column 5.

Step 4: Each value in column 5 is divided by the expected frequency in the corresponding cell and the result is entered in column 6.

Step 5: The values in column 6 are summed, resulting in $\chi^2 = 5.000$.

Step 6: The number of degrees of freedom is determined. For this test, the degrees of freedom are one less than the number of cells, or four. In general, for tests involving a single variable of classification, the degrees of freedom are equal to the number of cells minus one. If we let k equal the number of cells, then $df = k - 1$.

Step 7: The critical value of χ^2 with four degrees of freedom for a significance level of 5% is obtained from the table of the chi-square distribution in the Appendix. This value is $\chi^2_{4,0.05} = 9.49$.

Step 8: The calculated value of χ^2 is compared with the critical value and, since $5.000 < 9.49$, the null hypothesis is accepted. We conclude that the number of shoppers is the same regardless of which night the store is open.

14.3 Tests of Goodness-of-Fit

The *goodness-of-fit test* is a test of a hypothesis about a population distribution. Information about the form or shape of a population can be as important as information about the value of its parameters. For example, in order to perform an analysis of variance, it is necessary to assume that the populations sampled have normal distributions. A goodness-of-fit test can be made on these observations to determine if this assumption is valid before beginning the analysis of variance. In many business situations, accurate information about the probability of occurrence of different types of events is of great value to an executive or manager. If the form of the population on which the events are defined is known, these probabilities can be determined. Remember, however, that even if the population distribution is known, parameter values must also be known or estimated in order to determine these probabilities.

In order to test a hypothesis about the form or distribution of a population, a random sample is taken. The observed frequencies are grouped into cells, similar to the classes in a frequency distribution, and are compared with a set of expected frequencies in the same cells. The expected frequencies are computed under the assumption that the population is of the type specified in the null hypothesis. The comparison is accomplished by computing a value of χ^2 using Equation 14.1. If the computed χ^2 is less than the critical value of χ^2 at the significance level selected with the appropriate degrees of freedom, the null hypothesis is accepted; otherwise, it is rejected.

The test illustrated in Example 14.2 is a goodness-of-fit test. The hypothesized distribution is a uniform distribution, although we do not state this fact in the example. A uniform distribution is a distribution in which all possible occurrences have exactly the same probability. In the example, the occurrences or events correspond to the arrival of a shopper on a specified night. If this distribution is uniform, then the probability of a shopper arriving on a Monday night is the same as the probability of a shopper arriving on a Tuesday night, and so on for the five nights. Since there are only five different nights and, therefore, five possible occurrences, the probability is 1/5 for each occurrence if the distribution is uniform. It was on the basis of the assumption that shopper arrivals are distributed uniformly over the five nights that we determined the expected frequencies of 200 per night. Stating the hypotheses in the form of a goodness-of-fit test, they are

H_0: The distribution of shoppers is uniform.

H_1: The distribution of shoppers is not uniform.

The procedure to be followed for a goodness-of-fit test is summarized as follows.

1. Establish the null hypothesis that the sample data come from a specified distribution. The alternate hypothesis is simply that the sample data do not come from the specified distribution.

2. Group the observed data into cells and note the frequency in each cell.

3. Determine the expected frequency in each cell under the assumption that the sample data came from the hypothesized distribution.

4. The minimum expected frequency in each cell must be five. If any expected frequency is less than five, then that cell must be combined with an adjacent cell or cells until the combined cells have an expected frequency of five or more.

5. Compute χ^2 from the observed and expected frequencies according to Equation 14.1 and compare this value with a critical value obtained from the chi-square table.

6. If the computed value of χ^2 is less than the critical value, accept the null hypothesis; otherwise, reject it.

Degrees of Freedom for the Goodness-of-Fit Test. The degrees of freedom for a goodness-of-fit test, where data are grouped into k cells, cannot be greater than $k - 1$. However, whenever the observed data are used to estimate the population parameters necessary for computing the expected frequencies, an additional degree of freedom must be deducted for each estimate. For example, if we are testing for goodness-of-fit to a Poisson distribution, μ, the parameter of the Poisson distribution, must be known in order to compute the expected frequencies. If the value of μ is not specified, then it must be estimated from the sample data. If there are k cells and μ is estimated from the data, the degrees of freedom for the test are $k - 1 - 1$, or $k - 2$.

If we test for goodness-of-fit to a normal distribution, two parameter values, μ and σ, are needed to compute the expected frequencies. If it is necessary to estimate both μ and σ from the data, then the degrees of freedom for the test are $k - 1 - 2$, or $k - 3$.

In general, for goodness-of-fit tests, if there are k cells and m parameters to be estimated from the data, the degrees of freedom for the test are $k - 1 - m$.

Example 14.3

The field service department of Metric Systems, Inc. keeps accurate records regarding failure of their products from data reported to their various field offices. The breakage of one particular component which is used in all their products appears to be the most common cause of product failure. It is theorized that the occurrence of breakage in the field can be modeled by a Poisson distribution. If this can be established, then the probabilities of failures can be determined and used to set levels of spare-parts inventories for the field offices and also to estimate warranty costs resulting from product failure.

A company statistician is assigned the problem of conducting a goodness-of-fit test to a Poisson distribution. His first step is to obtain data from a sample of field-office failure reports for a period of 200 working days. From these data he determines the frequency, that is, the number of days on which various numbers of breakages occur. He assumes that the average number of breakages per day is $\mu = 2.5$. The observed frequencies and the calculation of expected frequencies are provided in Table 14.2. Table 14.3 illustrates how cells are combined and the χ^2 statistic is calculated. If we

TABLE 14.2

PART BREAKAGE DATA—OBSERVED AND EXPECTED FREQUENCIES (EXAMPLE 14.3)

Cells (x = *number of breakages per day*)	*Observed Frequency* (f)	*Poisson Probability* [$P(x = x_i \mid \mu = 2.5)$]	*Expected Frequency* [$e_i = 200P(x = x_i)$]
0	6	0.082	16.4
1	32	0.205	41.0
2	48	0.256	51.2
3	52	0.214	42.8
4	25	0.134	26.8
5	14	0.066	13.2
6	9	0.028	5.6
7	4	0.010	2.0
8	3	0.003	0.6
9	4	0.001	0.2
10	3	0.001	0.2
	200		200

TABLE 14.3
A. PART BREAKAGE DATA—COMBINING CELLS (EXAMPLE 14.3)

x	f	e
0	6	16.4
1	32	41.0
2	48	51.2
3	52	42.8
4	25	26.8
5	14	13.2
6	9	5.6
7	4	2.0
8	3 } 23	0.6 } 8.6
9	4	0.2
10	3	0.2
	200	200

B. PART BREAKAGE DATA—REVISED CELLS AND FREQUENCIES AND CALCULATION OF χ^2 (EXAMPLE 14.3)

x	f	e	$f - e$	$(f - e)^2$	$(f - e)^2/e$
0	6	16.4	−10.4	108.16	6.60
1	32	41.0	−9.0	81.00	1.98
2	48	51.2	−3.2	10.24	0.20
3	52	42.8	9.2	84.64	1.98
4	25	26.8	−1.8	3.24	0.12
5	14	13.2	0.8	0.64	0.05
6 or more	23	8.6	14.4	207.36	24.11
	200				$\chi^2 = 35.04$

let x represent the number of breakages per day, then each cell in the table corresponds to a different value of x. The observed frequencies, f, are the number of days on which breakages equal x. The significance level for this test is 1%.

In Table 14.2, the expected frequencies are obtained by multiplying the total number of observations, 200, by the probability of x failures, assuming that the distribution is Poisson with $\mu = 2.5$. Remember that the average or mean number of failures is assumed by the statistician to be known and equal to 2.5. Under these circumstances, the appropriate hypotheses for this test are

H_0: The part breakages have a Poisson distribution with $\mu = 2.5$.

H_1: The part breakages do not have a Poisson distribution with $\mu = 2.5$.

It is important to be aware that if this null hypothesis is rejected, we cannot be certain whether it was due to the fact that the distribution is not Poisson, or that the mean is not 2.5, or both.

Since the expected frequencies in several of the cells in Table 14.2 are less than five, it is necessary to combine adjacent cells until all expected frequencies are five or more. This is done in Table 14.3a. The calculation of χ^2 is shown in Table 14.3b. The computed χ^2 value is 35.04. Since the revised number of cells is seven and the value of the parameter μ is given and does not have to be computed from the data, the number of degrees of freedom for this test is $7 - 1 = 6$. The critical value of χ^2 with six degrees of freedom at $\alpha = 0.01$ is 16.81. Since the computed χ^2 of 35.04 is greater than the critical value, the null hypothesis is rejected.

Example 14.4

In the previous example, rejection of the null hypothesis led to the conclusion that the distribution was not Poisson with $\mu = 2.5$. However, it is still possible that the distribution of breakages is Poisson with some other mean. The statistician decides to test the data again for goodness-of-fit to a Poisson distribution, but without assuming that the mean is known. Instead, he decides to estimate the mean from the data and use this estimate to determine the expected frequencies. The null and alternate hypotheses for this new test are

H_0: The part breakages have a Poisson distribution.

H_1: The part breakages do not have a Poisson distribution.

The mean is estimated from the data using Equation 9.2 as follows.

Cells x	*f*	*xf*
0	6	0
1	32	32
2	48	96
3	52	156
4	25	100
5	14	70
6	9	54
7	4	28
8	3	24
9	4	36
10	3	30
	200	626

The estimated mean is $\mu = \bar{x} = \Sigma xf/n = 626/200 = 3.13$, or approximately 3.1.

Table 14.4 provides the probabilities of the various values of x from the Poisson distribution with $\mu = 3.1$ and shows the expected frequency in each cell. It also indicates which of the cells must be combined to have the expected frequencies in all cells equal to or greater than five. Table 14.5 regroups the data into the revised cells and shows the calculation of χ^2. The degrees of freedom for this test are $8 - 1 - 1 = 6$. Notice that the additional degree of freedom was subtracted because the sample data were used to estimate the mean. The critical value of χ^2 with six degrees of freedom at $\alpha = 0.01$ is 16.81. Since the computed value of 14.34 is less than the critical value of 16.81, the null hypothesis is accepted. We can conclude that breakages of this particular part do follow a Poisson distribution.

TABLE 14.4
PART BREAKAGE DATA—OBSERVED AND EXPECTED FREQUENCIES (EXAMPLE 14.4)

Cells (x = *number of breakages per day*)	*Observed Frequency* (*f*)		*Poisson Probability* [$P(x = x_i \mid \mu = 3.1)$]	*Expected Frequency* (*e*)	
0	6		0.045	9.0	
1	32		0.140	28.0	
2	48		0.216	43.2	
3	52		0.224	44.8	
4	25		0.173	34.6	
5	14		0.108	21.6	
6	9		0.055	11.0	
7	4	14	0.025	5.0	7.6
8	3		0.009	1.8	
9	4		0.004	0.8	
10	3		0.000	0.0	
	200			200	

TABLE 14.5
PART BREAKAGE DATA—REVISED CELLS AND FREQUENCIES AND CALCULATION OF χ^2 (EXAMPLE 14.4)

x	f	e	$f - e$	$(f - e)^2$	$(f - e)^2/e$
0	6	9.0	−3.0	9.00	1.00
1	32	28.0	4.0	16.00	0.57
2	48	43.2	4.8	23.04	0.53
3	52	44.8	7.2	51.84	1.16
4	25	34.6	−9.6	92.16	2.66
5	14	21.6	−7.6	57.76	2.67
6	9	11.0	−2.0	4.00	0.36
7 or more	14	7.6	6.4	40.96	5.39
	200	200			$\chi^2 = 14.34$

The rejection of the null hypothesis in the previous example was the result of an incorrect assumption about the mean by the statistician.

Example 14.5 The director of an agency of the United States government believed that a general model of urban family income would be useful in helping the agency allocate funds for urban renewal. She hypothesized that the most appropriate model would be the normal distribution and she directed a staff member to test that hypothesis.

The staff member selected a random sample of 1000 urban families and recorded the annual income for each. The mean and standard deviation of the sample data were computed. The values obtained were: $\bar{x} = \$16{,}600$, or 16.6 thousands of dollars; and $s = \$6360$, or 6.36 thousands of dollars. The student can verify these values using the methods of Chapter 9 and the data provided in Table 14.6.

TABLE 14.6

URBAN FAMILY INCOME DATA—OBSERVED AND EXPECTED FREQUENCIES (EXAMPLE 14.5)

1 *Income* *(in thousands of dollars)*	2 *Midpoint* (*x*)	3 f	4 *Upper* *Boundary*	5 *z value* *at Boundary*	6 *Cumulative Probability* *at Boundary*	7 *Probability* *in Cell*	8 e
0– 3.9	1.95	20	3.95	−1.99	0.0233	0.0233	23.3
4– 7.9	5.95	50	7.95	−1.36	0.0869	0.0636	63.6
8–11.9	9.95	130	11.95	−0.73	0.2327	0.1458	145.8
12–15.9	13.95	330	15.95	−0.10	0.4602	0.2275	227.5
16–19.9	17.95	190	19.95	0.53	0.7019	0.2417	241.7
20–23.9	21.95	140	23.95	1.16	0.8770	0.1751	175.1
24–27.9	25.95	90	27.95	1.78	0.9625	0.0855	85.5
28–31.9	29.95	40	31.95	2.41	0.9920	0.0295	29.5
32–35.9	33.95	10	35.95	3.04	0.9988	0.0068	6.8
36–39.9	37.95	0	39.95	3.67	1.0000	0.0012	1.2
		1000				1.0000	1000

The hypotheses to be tested are

H_0: The data on family income come from a normally distributed population.

H_1: The data on family income do not come from a normally distributed population.

The significance level selected for the test is $\alpha = 0.01$.

The first three columns in Table 14.6 show the cells or classes, the cell midpoint, and the observed frequency for each cell. Column 4 provides the upper boundary of each cell. This is the point that is halfway between adjacent cell limits as defined in Chapter 3. The entries in the remaining columns are explained as follows.

Column 5. This is the z value corresponding to the upper boundary of each cell. This value is simply the deviation of any value from the mean in units of the standard deviation. Since the sample mean and standard deviation are $\bar{x} = 16.6$ and $s = 6.36$, the z value for the first cell is $(3.95 - 16.6)/6.36 = -1.99$. The other z values were calculated the same way.

Column 6. These are the cumulative probabilities at z obtained from the tables of the standard normal distribution in Appendix Table A.3. They represent the total area to the left of each cell boundary.

Column 7. The probability corresponding to each cell is obtained from the cumulative probabilities by subtraction. For example, the probability in the fourth cell is obtained by subtracting the cumulative probability at the upper boundary of the third cell from the cumulative probability at the upper boundary of the fourth cell. That is,

$$0.4602 - 0.2327 = 0.2275.$$

Column 8. The expected frequency in each cell is obtained by multiplying the probability in that cell by the total number of observations, in this case 1000.

TABLE 14.7
URBAN FAMILY INCOME DATA—CALCULATION OF χ^2 (EXAMPLE 14.5)

Cell	f	e	$f - e$	$(f - e)^2$	$(f - e)^2/e$
1	20	23.3	−3.3	10.89	0.47
2	50	63.6	−13.6	184.96	2.91
3	130	145.8	−15.8	249.64	1.71
4	330	227.5	102.5	10506.25	46.18
5	190	241.7	−51.7	2672.89	11.06
6	140	175.1	−35.1	1232.01	7.04
7	90	85.5	4.5	20.25	0.24
8	40	29.5	10.5	110.25	3.74
9–10	10	8.0	2.0	4.00	0.50
					$\chi^2 = 73.85$

The calculations for χ^2 are provided in Table 14.7. Notice that the last two cells were combined so that the expected frequency would be five or greater. The degrees of freedom for this test are $9 - 1 - 2 = 6$, since two degrees of freedom were lost because the two population parameters had to be estimated from the sample data. The critical value of χ^2 with six degrees of freedom at $\alpha = 0.05$ is 16.81. Since the calculated χ^2 of 73.85 far exceeds the critical value, the null hypothesis is rejected. It cannot be concluded that urban family incomes are normally distributed.

14.4 Tests of Independence—Contingency Tables

In a goodness-of-fit test, a set of observed frequencies is assigned to k cells according to a single classification variable. In Example 14.2, that single classification variable is a night of the week. In Examples 14.3 and 14.4, the variable is the number of part breakages per day. In Example 14.5 it is income. However, it is possible to classify observations according to two or more variables, as illustrated in Chapter 13 in the discussion of the two-way analysis of variance. For example, voters can be classified according to sex and political affiliation. Wage-earners can be classified according to education and income; families according to geographical location and home ownership. When observations are classified according to two variables and arranged in a table, the display is called a *contingency table*.

The chi-square statistic can be used to test whether the two variables of classification are independent of one another. Independence, in this context, has a similar meaning to statistical independence, defined in Chapter 7. If two classification variables are independent, then membership in a particular class of one variable has no effect on membership in a particular class of the other variable. With respect to the examples given above, if sex and political affiliation are independent, then a female is just as likely to be a Republican as a Democrat. If education and income are independent, then a person with only a high school education is just as likely to be earning a high income as a college graduate.

In tests of independence, the null hypothesis is always that the two classification variables are independent. The alternate hypothesis is that they are not independent. The chi-square statistic is computed in the same way as the goodness-of-fit test, except that the summation is over both rows and columns, that is,

$$\chi^2 = \sum_j \sum_i \frac{(f_{ij} - e_{ij})^2}{e_{ij}} \tag{14.2}$$

Of course, the expected frequencies are determined under the assumption of independence. The procedure is best explained through examples.

Example 14.6 In preparation for the coming election, the Sligh for Senator campaign committee hired Opinion Surveys, Inc. to determine if geographical location within the state has any effect on political affiliation. A random sample of 420 registered voters is obtained by OSI and each voter is classified according to his or her registration as a Republican, Democrat, or Independent, and according to his or her location in the state—east, central, or west. The data obtained are provided in Table 14.8.

TABLE 14.8
CROSS-CLASSIFICATION OF VOTERS BY POLITICAL AFFILIATION AND GEOGRAPHICAL LOCATION (EXAMPLE 14.6)

	Location			
Affiliation	*East*	*Central*	*West*	*Totals*
Republican	45	20	60	125
Democrat	105	35	60	200
Independent	50	25	20	95
Totals	200	80	140	420

The hypotheses to be tested at the 1% level of significance are

H_0: Location and political affiliation are independent.

H_1: Location and political affiliation are not independent.

In order to determine the expected frequencies, we proceed under the assumption that the two classification variables are independent. If this is true, then we expect that the proportion of Republicans, Democrats, and Independents in the population is the same for all locations. Conversely, the different locations are represented in the same proportions in all the political categories.

We might begin by obtaining the best estimate of the proportion of voters in each political category from the data. If we let p_R, p_D, and p_I represent the proportion of Republicans, Democrats, and Independents in the population, then these proportions can be estimated from the data as follows.

$$\hat{p}_R = \frac{125}{420} = 0.298, \qquad \hat{p}_D = \frac{200}{420} = 0.476, \qquad \hat{p}_I = \frac{95}{420} = 0.226$$

The expected frequency under the assumption of independence in each area is simply the number in the sample from the area multiplied by these estimated proportions; that is, the expected number of Republicans in the east, central, and west are, respectively:

$$\begin{aligned} (0.298)(200) &= 59.5 \\ (0.298)(80) &= 23.8 \\ (0.298)(140) &= 41.7 \end{aligned}$$

The expected number of Democrats in the east, central, and west are, respectively:

$$(0.476)(200) = 95.2$$
$$(0.476)(80) \quad = 38.1$$
$$(0.476)(140) = 66.7$$

The expected number of Independents in the east, central, and west are, respectively:

$$(0.226)(200) = 45.3$$
$$(0.226)(80) \quad = 18.1$$
$$(0.226)(140) = 31.6$$

A simple alternative method for obtaining the expected frequency in any cell of a contingency table is provided by the following rule.

> The expected frequency in any cell, e_{ij}, is simply the product of the total of the ith row and the jth column, divided by n, the sample size. That is,
>
> (14.3) $$e_{ij} = \frac{(t_{i.})(t_{.j})}{n}$$
>
> where $t_{i.}$ is the total of the ith row and $t_{.j}$ is the total of the jth column.

The subscript notation used here is consistent with that introduced in Chapter 13, Analysis of Variance. The expected frequencies using Equation 14.3 are provided in Table 14.9, while the calculation of χ^2 is shown in Table 14.10.

As with the goodness-of-fit tests, the calculated value of χ^2 must be compared with a critical value for the significance level selected with the appropriate number of degrees of freedom. The degrees of freedom for a chi-square test of independence will always be $(r - 1)(k - 1)$, where r represents the number of rows and k the number of columns in the contingency table. For this test of independence between political affiliation and location, there are three rows and three columns. Therefore, the number of degrees of freedom is $(3 - 1)(3 - 1) = (2)(2) = 4$.

TABLE 14.9

CROSS-CLASSIFICATION OF VOTERS BY POLITICAL AFFILIATION AND GEOGRAPHICAL LOCATION—CALCULATION OF EXPECTED FREQUENCIES (EXAMPLE 14.6)

	Location			
Affiliation	*East*	*Central*	*West*	*Totals*
Republican	$\frac{(125)(200)}{420} = 59.5$	$\frac{(125)(80)}{420} = 23.8$	$\frac{(125)(140)}{420} = 41.7$	125
Democrat	$\frac{(200)(200)}{420} = 95.2$	$\frac{(200)(80)}{420} = 38.1$	$\frac{(200)(140)}{420} = 66.7$	200
Independent	$\frac{(95)(200)}{420} = 45.3$	$\frac{(95)(80)}{420} = 18.1$	$\frac{(95)(140)}{420} = 31.6$	95
Totals	200	80	140	420

TABLE 14.10
TEST OF INDEPENDENCE: POLITICAL AFFILIATION VERSUS GEOGRAPHICAL LOCATION—CALCULATION OF χ^2 (EXAMPLE 14.6)

Cell	f	e	$f - e$	$(f - e)^2$	$(f - e)^2/e$
Republican—east	45	59.5	−14.5	210.25	3.53
Republican—central	20	23.8	−3.8	14.44	0.61
Republican—west	60	41.7	18.3	334.89	8.03
Democrat—east	105	95.2	9.8	96.04	22.09
Democrat—central	35	38.1	−3.1	9.61	47.61
Democrat—west	60	66.7	−6.7	44.89	134.56
Independent—east	50	45.3	4.7	22.09	0.49
Independent—central	25	18.1	6.9	47.61	2.63
Independent—west	20	31.6	−11.6	134.56	4.26
	420	420			$\chi^2 = 21.48$

The critical value of chi-square with four degrees of freedom at $\alpha = 0.01$ is 13.28. Since the computed value of 21.48 is greater than the critical value, H_0 is rejected and it can be assumed that political affiliation and geographical location are not independent.

Example 14.7 The Fireside Restaurant serves four fixed-price dinners at $8, $10, $12, and $14. For several years, the restaurant has been issuing its own Fireside credit cards, as well as accepting those issued by Eaters Club. A management consultant retained by the restaurant suggested that it might be advantageous to change the credit card policy. Before recommending a new policy, however, he decided to conduct a test to determine if the method of payment and the price of the dinner ordered were independent. He established the following hypotheses for the test.

H_0: Dinner price and payment method are independent.

H_1: Dinner price and payment method are not independent.

The significance level for the test is $\alpha = 0.05$.

For a period of one week, records were obtained showing the number of dinners of each price ordered and the method of payment. Tables 14.11a and 14.11b show the contingency tables with observed and expected frequencies. The calculation of the chi-square statistic is provided in Table 14.12. The degrees of freedom for this test are $(r - 1)(k - 1) = (4 - 1)(3 - 1) = (3)(2) = 6$.

The critical value of χ^2 with six degrees of freedom at $\alpha = 0.05$ is 12.50. Since the calculated χ^2 is 18.94, the null hypothesis is rejected, and it can be concluded at the 5% level of significance that the price of the dinner ordered and the method of payment are not independent.

Yates Correction Factor. When there are only two rows and two columns in a contingency table, chi-square will have only one degree of freedom. Under these circumstances, an adjustment

TABLE 14.11

A. DINNER PRICE VERSUS PAYMENT METHOD—OBSERVED FREQUENCIES (EXAMPLE 14.7)

	Payment Method			
Dinner Price	*Cash*	*Fireside Card*	*Eaters Club Card*	*Totals*
$8	200	70	130	400
$10	220	100	180	500
$12	190	80	130	400
$14	120	20	60	200
Totals	730	270	500	1500

B. DINNER PRICE VERSUS PAYMENT METHOD—EXPECTED FREQUENCIES (EXAMPLE 14.7)

	Payment Method			
Dinner Price	*Cash*	*Fireside Card*	*Eaters Club Card*	*Totals*
$8	195	72	133	400
$10	243	90	167	500
$12	195	72	133	400
$14	97	36	67	200
Totals	730	270	500	1500

TABLE 14.12

TEST OF INDEPENDENCE: DINNER PRICE VERSUS PAYMENT METHOD—CALCULATION OF χ^2 (EXAMPLE 14.7)

Cell[a]	f	e	$f - e$	$(f - e)^2$	$(f - e)^2/e$
11	200	195	5	25	0.13
12	70	72	−2	4	0.06
13	130	133	−3	9	0.07
21	220	243	−23	529	2.18
22	100	90	10	100	1.11
23	180	167	13	169	1.01
31	190	195	−5	25	0.13
32	80	72	8	64	0.89
33	130	133	−3	9	0.07
41	120	97	23	529	5.45
42	20	36	−16	256	7.11
43	60	67	−7	49	0.73
	1500	1500			$\chi^2 = 18.94$

[a] The first digit is the row number and the second digit is the column number. For example: Cell number 21 is the figure from the second row and the first column.

called the *Yates correction factor* should be used. This is a correction for continuity to compensate for the fact that while the observed frequencies are discrete values, the theoretical chi-square distribution is continuous. The correction is accomplished by subtracting 0.5 from the absolute value of the difference between f and e before squaring; that is, the corrected value of χ^2 is obtained from

$$\chi^2 = \Sigma \frac{(|f - e| - 0.5)^2}{e} \tag{14.4}$$

Example 14.8 In testing the hypothesis at $\alpha = 0.05$ that manual dexterity and sex are independent, the personnel department of Metric Systems, Inc. administered a test to a random sample of 100 production workers whose job involved the assembly of intricate electronic devices. The test scores were simply categorized as high and low and each worker was classified according to test score and sex. The observed and expected frequencies are provided in Tables 14.13a and 14.13b, respectively.

TABLE 14.13
A. MANUAL DEXTERITY VERSUS SEX—OBSERVED FREQUENCIES (EXAMPLE 14.8)

	Test Score		
Sex	*High*	*Low*	*Totals*
Male	37	34	71
Female	18	11	29
Totals	55	45	100

B. MANUAL DEXTERITY VERSUS SEX—EXPECTED FREQUENCIES

	Test Score		
Sex	*High*	*Low*	*Totals*
Male	39	32	71
Female	16	13	29
Totals	55	45	100

Since there are only two categories of each classification variable, we obtain a 2×2 contingency table. Therefore, it is necessary to use the Yates correction factor in calculating χ^2. These calculations are provided in Table 14.14.

The critical value of χ^2 with one degree of freedom at $\alpha = 0.05$ is 3.84. Since the computed value of χ^2 is 0.44, the null hypothesis is accepted and we conclude that manual dexterity and sex are independent.

TABLE 14.14
TEST OF INDEPENDENCE: MANUAL DEXTERITY VERSUS SEX—CALCULATION OF CHI-SQUARE (EXAMPLE 14.8)

Cell	f	e	$\lvert f-e \rvert$	$\lvert f-e \rvert -0.5$	$(\lvert f-e \rvert -0.5)^2$	$(\lvert f-e \rvert -0.5)^2/e$
11	37	39	2	1.5	2.25	0.06
12	34	32	2	1.5	2.25	0.07
21	18	16	2	1.5	2.25	0.14
22	11	13	2	1.5	2.25	0.17
	100	100				$\chi^2 = 0.44$

Definitions

Classification variable: Any variable characteristic of data which is used to divide individual observations into categories or cells.

Contingency table: A table in which data are classified according to two classification variables.

Goodness-of-fit test: A test of the hypothesis that data observed from a sample belongs to a stated population distribution.

Test of independence: A test of the hypothesis that membership in a particular category of one classification variable has no effect on membership in a particular category of another classification variable.

Formulas

Chi-square statistic for goodness-of-fit test:

(14.1) $$\chi^2 = \sum_i \frac{(f_i - e_i)^2}{e_i}$$

Chi-square statistic for test of independence:

(14.2) $$\chi^2 = \sum_j \sum_i \frac{(f_{ij} - e_{ij})^2}{e_{ij}}$$

Expected cell frequency for test of independence:

(14.3) $$e_{ij} = \frac{(t_{i.})(t_{.j})}{n}$$

Yates correction factor for 2×2 contingency tables:

(14.4) $$\chi^2 = \sum \frac{(|f - e| - 0.5)^2}{e}$$

Problems

14.1 The following data represent sets of observed and expected frequencies. Compute the value of χ^2 from these data.

Cell:	1	2	3	4	5	6	7	8
f:	22	27	20	12	6	6	4	3
e:	20	30	25	20	15	12	10	5

14.2 The following data represent sets of observed and expected frequencies. Compute the value of χ^2 from these data.

Cell:	1	2	3	4	5	6	7	8	9	10
f:	6	10	14	18	16	12	10	8	8	6
e:	10	10	10	10	10	10	10	10	10	10

14.3 The following data represent sets of observed and expected frequencies. Compute the value of χ^2 from these data.

Cell:	1	2	3	4	5	6	7
f:	9	11	12	7	5	3	1
e:	5	9	14	6	4	2	0

14.4 For the data in problem 14.1, test the hypothesis at $\alpha = 0.05$ that the observed data come from a population with the distribution represented by the expected frequencies. Assume that one population parameter was estimated from these data.

14.5 For the data in problem 14.2, test the hypothesis at $\alpha = 0.05$ that the observed data come from a population with the distribution represented by the expected frequencies. Assume that no parameters were estimated from these data.

14.6 For the data in problem 14.3, test the hypothesis at $\alpha = 0.01$ that the observed data come from a population with the distribution represented by the expected frequencies. Assume that one parameter was estimated from these data.

14.7 The marketing manager of the Apex Specialty Company wants to test the hypothesis that the number of sales calls made by its sales representatives during one week has a uniform distribution. During a randomly selected week, the number of calls made by each representative were as follows.

Sales representatives:	1	2	3	4	5	6	7	8
Number of calls:	14	26	31	30	14	16	29	28

Test this hypothesis at $\alpha = 0.01$

14.8 In order to keep check-out lines as short as possible, the manager of a Saveright Supermarket collected data on the pattern of arrivals for check-out for a period of 100 hours. The data are presented below, where x represents the number of arrivals per hour and f represents the observed frequency. Test the hypothesis that these arrivals have a Poisson distribution at $\alpha = 0.01$. (Use the closest value of the parameter estimate that appears in the Poisson tables.)

x:	0	1	2	3	4	5	6	7	8	9	10 or more
f:	0	4	8	15	20	25	15	10	2	1	0

14.9 In 50 random samples of $n = 5$ drawn from a continuous production line, the number of samples containing X defectives is noted below. Test whether these data have a binomial distribution with $p = 0.30$ at $\alpha = 0.05$.

X:	0	1	2	3	4	5
f:	4	13	17	12	3	1

14.10 Estimate p from the data given in problem 14.9. Then test the hypothesis that these data have a binomial distribution at $\alpha = 0.05$.

(Use the closest value of the parameter estimate that appears in the binomial table.)

14.11 A large midwestern state university publishes a daily newspaper which is read by most of its 30,000 students enrolled in undergraduate, graduate, and evening programs. An advertising agency in the area, which is considering this paper as an advertising medium for several of its clients, compiled the following data on the age distribution of its readership from a random sample of 100 student subscribers. Test the hypothesis, at $\alpha = 0.01$, that these data come from a normally distributed population with a mean of 25 years and a standard deviation of 5 years.

Age	*f*
16–18	3
19–21	20
22–24	30
25–27	15
28–30	13
31–33	11
34–36	8
	100

14.12 For the data in problem 14.11, test the hypothesis that these data come from a normally distributed population at $\alpha = 0.01$.

14.13 To determine whether the number of mistakes per page of typewritten material has a Poisson distribution, a random sample of 100 typewritten pages is selected from the output of a typing pool. The number of mistakes per page, x, and the frequency of occurrence in the sample, *f*, are

x	*f*
0	7
1	14
2	36
3	24
4	12
5	7
	100

Estimate the necessary parameter from the sample data and test the hypothesis that the number of mistakes per page has a Poisson distribution at $\alpha = 0.05$.

14.14 Use the data of Example 14.7 to test the hypothesis that dinner price and payment method are independent at $\alpha = 0.01$. Compare your conclusion here with the result of that example and explain the difference.

14.15 From the following data, obtained from a random sample of 200 families and arranged in a contingency table, test the hypothesis that the number of children in the family and the number of automobiles owned are independent at the 1% level of significance.

Number of Children	*Number of Automobiles* 0	1	2	3	*Totals*
0	3	10	12	0	25
1	6	22	27	0	55
2	5	29	28	3	65
3	0	12	17	1	30
4 or more	1	15	9	0	25
Totals	15	88	93	4	200

14.16 Test for independence between the level of schooling completed and income from the following data. Use the 1% level of significance.

Income Level	*Schooling Completed* High School	College	Graduate School	*Totals*
Low	165	110	25	300
Middle	225	250	125	600
High	10	40	50	100
Totals	400	400	200	1000

14.17 The Sligh for Senator campaign committee wanted information to assist them in formulating Mr. Sligh's position on several major issues, one of which is compulsory medical insurance. To determine whether attitudes on this issue were independent of the educational level of the voter, a random sample of 1000 voters was cross-classified as follows.

Education	*Attitude* For Compulsory Medical Insurance	Against Compulsory Medical Insurance	*Totals*
High School	150	325	475
College	150	300	450
Graduate School	50	25	75
Totals	350	650	1000

Test for independence at $\alpha = 0.01$.

14.18 The personnel department of Metric Systems, Inc. was reevaluating their policies on employment and retirement. As part of this reevaluation, they conducted a test to determine whether or not a relationship existed between age and productive output. The following data were obtained from a random sample of 100 production employees. Test the hypothesis that age and productivity are independent at $\alpha = 0.05$.

	Output			
Age	*Low*	*Medium*	*High*	*Totals*
21–35	13	15	12	40
36–50	11	13	11	35
51–65	7	10	8	25
Totals	31	38	31	100

14.19 The Reliance Insurance Company collected data from a random sample of 250 drivers. The drivers were cross-classified according to sex and whether or not an accident claim was filed by the driver during the previous year. Test the hypothesis that sex and accident involvement are independent at the 5% level of significance.

	Sex		
Accidents	*Male*	*Female*	*Totals*
None	13	6	19
One or more	137	94	231
Totals	150	100	250

14.20 The Reliance Insurance Company wanted to test for a relationship between accidents and age. A sample of 250 drivers was cross-classified according to age and accident occurrence. Test for independence between these two classification variables at the 5% level of significance.

	Accidents		
Age	*None*	*One or More*	*Totals*
Under 21	52	8	60
21 and over	179	11	190
Totals	231	19	250

CHAPTER 15 NONPARAMETRIC STATISTICAL METHODS

15.1
The runs test for randomness

15.2
The sign test

15.3
The rank sum test for two samples

15.4
The rank sum test for three or more samples: the Kruskal–Wallace test

15.5
Rank correlation

Most of the methods of statistical inference that we have discussed so far require certain assumptions about either the distribution of the population sampled or its parameters or both. These assumptions and the restrictions they place on the use of parametric tests were discussed in the introduction to Chapter 14 on chi-square tests. It was noted there that chi-square tests are nonparametric since they are not restricted by any assumptions about the distribution of the population sampled. For example, in the test of the hypothesis that two classification variables are independent, it was not necessary to stipulate that the variables involved come from a normally distributed population or, for that matter, from any particular type of population at all.

You will frequently see the terms *nonparametric* and *distribution-free* used with reference to these kinds of methods. These two terms are not actually synonymous, although they are often used as though they were. The term nonparametric describes procedures that make no hypothesis about population parameters, while "distribution-free" refers to tests that require no assumptions about the form of the population distribution. Despite this difference in meaning, we will use the term "nonparametric" in referring to both types of procedures.

Nonparametric tests are important in statistics for two reasons. First, since there are few restrictions on their use, they can be applied in many situations where comparable parametric tests cannot be used. Second, they are usually quite easy to apply. However, it is necessary to exercise some caution in using these tests in situations where all the conditions for using a comparable parametric test can be satisfied. In these situations, the nonparametric test will be less accurate and less efficient than the comparable parametric test. For this reason, nonparametric tests should be used only when parametric tests are not applicable to the problem, when the conditions required for the parametric test are not satisfied, or when the savings in time and effort gained by using a nonparametric test are of greater importance than the precision that might be sacrificed.

In this chapter we discuss several of the most widely used nonparametric tests.

15.1 The Runs Test for Randomness

Throughout this book we have referred to random samples in connection with all the inferential procedures covered. In the chapter on sampling, we defined the term "random sample" and discussed several methods for obtaining one. We also explained the importance of randomness in sampling and emphasized the point that in order to make probability statements about a population parameter, randomness in sampling is necessary. We will now consider a test for determining if some sequence of observations has been obtained in a random order.

To use the *runs test for randomness,* a sample of observations must be obtained from a population of attributes. That is, each observation must belong to one of two mutually exclusive categories. These categories may exist naturally in the population sampled or they may be defined artificially. An example of the first case is a population divided into categories like male and female, black and white, or good and defective. If the population is such that the observations assume various numerical values, then some arbitrary division can be established. A commonly used classification scheme is for each value to be categorized with respect to its position in relationship to some middle value such as the median, that is, above or below the median. A further requirement for this test is that the order of the observations must be maintained.

A *run* is simply a sequence of identical observations preceded and followed by a different observation. For example, in the following sequence of A's and B's, there are eight runs. Each run is underlined and separated from the others by a space: AAA B A BBBB AA BB AAA BB. Notice that a run can consist of a *single* observation if it is preceded and followed by a different observation.

The question that can be asked about such a sequence is, do the A's and B's appear in a random order? Notice that we are not questioning if the proportion of A's or B's is consistent with some hypothesis about the true population proportion.

The answer to the question about the randomness of the above sequence is difficult to answer intuitively. In any random sequence we would expect to see runs of varying lengths. However, too few runs or too many runs implies a lack of randomness. For example, the following two sequences appear not to be random for the reasons of too few and too many runs, respectively.

AAAAAAAAA BBBBBBBBB

AB AB AB AB AB AB AB AB AB

The first sequence above has only two runs in 18 observations, while the second sequence has nine runs in 18 observations. The runs test provides a criterion for determining, for any sequence of observations, what constitutes too few runs and what constitutes too many.

Having observed a sequence of observations of which n_1 are of one kind and n_2 are of the other, we can test the following hypotheses by counting the number of runs, R.

H_0: The sequence is random.

H_1: The sequence is not random.

If the sampling is random and if either n_1 or n_2 is greater than 20, then the statistic R, the number of runs, is approximately normally distributed with

$$\mu_R = \frac{2n_1n_2}{n_1 + n_2} + 1 \tag{15.1}$$

and

$$\sigma_R = \sqrt{\frac{2n_1n_2(2n_1n_2 - n_1 - n_2)}{(n_1 + n_2)^2(n_1 + n_2 - 1)}} \tag{15.2}$$

where

μ_R = the mean of the sampling distribution of the number of runs;
σ_R = the standard error of the sampling distribution of the number of runs;
n_1 = the number of observations of one kind;
n_2 = the number of observations of the other kind.

Given the statistic and the mean and standard error of its distribution, we can easily determine the value of z, the standard normal variate, from:

$$z = \frac{R - \mu_R}{\sigma_R}$$

We can compare this computed value of z with the critical values of z obtained from the tables of the standard normal distribution for the significance level chosen for the test. Since too few or too many runs lead to rejection of the hypothesis of randomness, this is a two-tailed test.

Example 15.1 An interviewer for a consumer testing organization was stationed at a particular store-front location in the downtown shopping area of Middletown. He was there to conduct interviews with shoppers leaving the store. In reviewing the completed forms submitted by the interviewer, his supervisor noted the following sequence of male and female shoppers interviewed. The supervisor decided to test the sequence for randomness at the 5% level of significance.

$\underline{\text{FF}}\ \underline{\text{M}}\ \underline{\text{F}}\ \underline{\text{MMMM}}\ \underline{\text{FFFFFF}}\ \underline{\text{M}}\ \underline{\text{FFF}}\ \underline{\text{MMMM}}\ \underline{\text{F}}\ \underline{\text{M}}\ \underline{\text{FFFFFFF}}\ \underline{\text{M}}\ \underline{\text{F}}\ \underline{\text{MMMMMM}}$

For this sequence we note

number of females, $n_1 = 21$

number of males, $n_2 = 18$

number of runs, $R = 14$

Computing the mean and standard error, we obtain:

$$\mu_R = \frac{2n_1n_2}{n_1 + n_2} + 1 \tag{15.1}$$
$$= \frac{(2)(21)(18)}{21 + 18} + 1$$
$$= \frac{756}{39} + 1$$
$$= 20.38$$

$$\sigma_R = \sqrt{\frac{2n_1n_2(2n_1n_2 - n_1 - n_2)}{(n_1 + n_2)^2(n_1 + n_2 - 1)}} \tag{15.2}$$
$$= \sqrt{\frac{(2)(21)(18)[(2)(21)(18) - 21 - 18]}{(21 + 18)^2(21 + 18 - 1)}}$$
$$= \sqrt{9.378}$$
$$= 3.06$$

The z statistic is computed from $z = (R - \mu_R)/\sigma_R = (14 - 20.38)/3.06 = -2.08$. The critical value of z for a two-tailed test at the 5% level of significance is $z_c = \pm 1.96$. Since $-2.08 < -1.96$, the null hypothesis is rejected. It appears that there are too few runs for a random sequence. The test is illustrated in Figure 15.1.

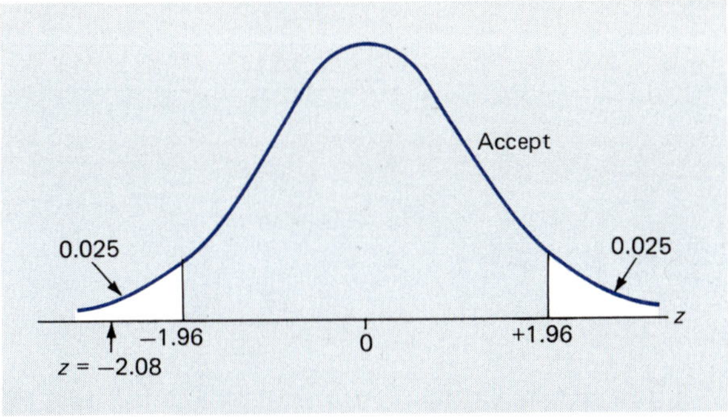

Figure 15.1
Runs test of randomness of interview data (Example 15.1).

Example 15.2

Sheet metal is rolled to a specified thickness on an automatic machine. The thickness of the sheet is a critical dimension that must be controlled within strict limits. Since variation from sheet to sheet is unavoidable, it is necessary to watch for machine drift, that is, a tendency for the machine setting to drift off-center in one direction. The following coded measurements were made on a sequence of 45 sheets as they left the machine. Test for randomness of the sequence at the 1% level of significance. The sequence reads down each column and then across the columns.

19	22	22	20	21	19	22	21	20
18	21	21	19	23	19	19	19	19
19	19	23	20	24	23	17	16	23
17	21	21	21	22	16	18	18	21
22	22	24	19	18	21	19	18	18

In order to apply the runs test, the data must be classified into two mutually exclusive categories. These categories are ABOVE the median and BELOW the median. To find the value of the median, the data must be rearranged in order of magnitude. The following array results from this rearrangement.

16	18	19	19	19	21	21	22	23
16	18	19	19	20	21	21	22	23
17	18	19	19	20	21	21	22	23
17	18	19	19	20	21	22	22	24
18	18	19	19	21	21	22	23	24

The median, or middle item of this set, is 20, with 21 values below the median and 21 values above it. Eliminating the three values that are equal to the median, that is, neither below nor above it, we obtain the following sequence of 42 observations from the original sequence of 45. The letter A represents values above the median and the letter B represents values below.

$\underline{\text{BBBB}}\ \underline{\text{AAA}}\ \underline{\text{B}}\ \underline{\text{AAAAAAA}}\ \underline{\text{B}}\ \underline{\text{A}}\ \underline{\text{B}}\ \underline{\text{AAAA}}\ \underline{\text{BBB}}\ \underline{\text{A}}\ \underline{\text{B}}\ \underline{\text{AA}}\ \underline{\text{BBBB}}\ \underline{\text{A}}\ \underline{\text{BBBBB}}\ \underline{\text{AA}}\ \underline{\text{B}}$

There are 17 runs in this sequence. Both n_1 and n_2 are equal to 21. The mean and standard error of R are computed from Equations 15.1 and 15.2, respectively.

$$\mu_R = \frac{(2)(21)(21)}{21 + 21} + 1 = \frac{882}{42} + 1 = 22 \tag{15.1}$$

$$\sigma_R = \sqrt{\frac{(2)(21)(21)[(2)(21)(21) - 21 - 21]}{(21 + 21)^2(21 + 21 - 1)}} \tag{15.2}$$

$$= \sqrt{10.24}$$

$$= 3.2$$

The value of z is

$$z = \frac{17 - 22}{3.2} = -1.56$$

The critical values of z for a two-tailed test at the 1% level of significance are ± 2.58; therefore, the hypothesis of randomness cannot be rejected. This test is illustrated in Figure 15.2.

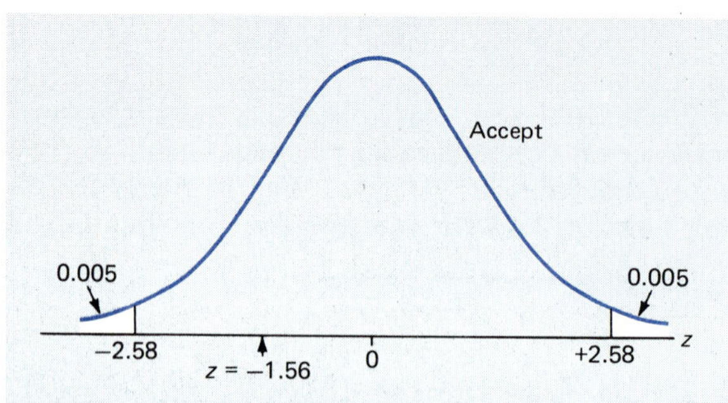

Figure 15.2
Runs test of randomness of sheet metal thickness data (Example 15.2).

15.2 The Sign Test

In Chapter 12, the t distribution was used to test the hypothesis that two population means are equal when the samples are small. However, in order to perform this test, both samples must come from normal populations with equal variances. If these requirements cannot be assumed, then this test cannot be used. In such a case, it is possible to use a nonparametric test, called the *sign test,* if the hypotheses can be modified slightly. Instead of testing the hypothesis that $\mu_1 = \mu_2$ against the alternate that $\mu_1 \neq \mu_2$, we test the hypothesis that the two populations have identical distributions against the alternate that they do not have identical distributions. Notice that these two sets of hypotheses are not exactly the same, since two populations may have equal means but different distributions. However, under the assumptions necessary for the parametric test, the two sets of hypotheses are practically equivalent.

Suppose that x and y are two random variables and we hypothesize that the distribution of the population of x is the same as the distribution of the population of y. In order to test this hypothesis, we draw two random samples of equal size, one from each population. The observations may be samples from two independent populations in which each x value is paired with a y value, or they may be matched pairs in which the x and y values may be "before" and "after" observations made on the same elementary unit. This latter situation is comparable to the test of data from two dependent samples illustrated in Example 12.6.

In that example, Metric Systems, Inc. tested the difference between assembly times using two different assembly procedures. The sign test is based on the signs, positive or negative, of the difference between each pair of x and y values. If the null hypothesis is true and both samples come from identical populations, we would expect that the differences between the pairs of x and y values, that is, $x - y$, will have the same number of pluses as minuses. This would be true for both dependent and independent samples. In other words, if the populations are identical, any x is just as likely to be greater than y as it is to be less than y. This is the same as saying that the sample proportion of pluses or minuses is expected to be 0.5. The procedure for the sign test is therefore similar to a parametric test of the hypothesis that the population proportion is 0.5.

The procedure for the test is to take the difference, $x - y$, of each pair of values and count the number of pluses. Any pair of values whose difference is zero is eliminated from the sample. Let $\bar{p}$ represent the proportion of pluses in the sample. The sampling distribution of $\bar{p}$ is binomial, but the normal approximation is generally appropriate since the null hypothesis of identical populations implies an expected proportion of 0.5 and, therefore, a symmetrical distribution.

If the null hypothesis is true, the mean and standard error of $\bar{p}$ are, respectively,

$$\mu_{\bar{p}} = p = 0.5$$

$$\sigma_{\bar{p}} = \sqrt{\frac{p(1-p)}{n}} = \sqrt{\frac{(0.5)(0.5)}{n}} = \sqrt{\frac{0.25}{n}} \qquad (9.16)$$

and a value of z can be computed from

$$z = \frac{\bar{p} - 0.5}{\sigma_{\bar{p}}}$$

The computed value of z is compared with critical values of z obtained from the tables of the standard normal distribution. Since the null hypothesis is rejected for too few pluses as well as too many, this is a two-tailed test.

Example 15.3 In Example 12.4 the production manager of the Quality Appliance Company tested the hypothesis that the average life of the television picture tubes produced by Lo-Voltage Electric was the same as the average life of tubes produced by Midwestern Electronics. Because the sample of tubes was small, it was necessary to make the assumption that both populations had normal distributions with equal variances in order to perform the test. Suppose, however, that this assumption cannot be justified. In that case, the parametric test could not be used. An alternative would be to use the nonparametric sign test. The null and alternate hypotheses would be modified to read:

H_0: The two samples come from identical populations.

H_1: The two samples do not come from identical populations.

Assume that random samples of 12 tubes made by each manufacturer are life tested. Each Lo-Voltage tube is randomly matched with a Midwestern Electronic tube. The results of the life tests are displayed in Table 15.1. The significance level for the test is 5%.

Since the null hypothesis specifies identical distributions, we would expect to observe the same number of pluses as minuses if H_0 is true. The expected proportion of pluses should be 0.5. With this implied value of $p = 0.5$ and $n = 12$, both np and $n(1 - p)$ are greater than 5, which meets the rule of thumb for using the normal approximation to the binomial distribution. Therefore, the sampling distribution of $\bar{p}$ is assumed to be normal with $\mu_{\bar{p}} = 0.5$ and $\sigma_{\bar{p}} = \sqrt{0.25/12} = 0.144$. The data in Table 15.1 indicate that there are seven pluses from the 12 pairs of observations; therefore, $\bar{p} = 7/12 = 0.583$.

TABLE 15.1
LIFE TEST DATA ON PICTURE TUBES (RANDOMLY PAIRED)
(EXAMPLE 15.3)

Pair	*Lo-Voltage* (x)	*Midwestern* (y)	*Difference* x − y	*Sign of Difference*
1	2320	2540	−220	−
2	2710	2700	+10	+
3	2640	2550	+90	+
4	2730	2710	+20	+
5	2480	2650	−170	−
6	2340	2740	−400	−
7	2680	2620	+60	+
8	2720	2680	+40	+
9	2550	2390	+160	+
10	2580	2610	−30	−
11	2600	2470	+130	+
12	2540	2660	−120	−

The computed value of z is

$$z = \frac{\bar{p} - 0.5}{\sigma_{\bar{p}}} = \frac{0.583 - 0.500}{0.144} = 0.576$$

The critical values of z for a two-tailed test at the 5% level of significance are ± 1.96. Since $-1.96 < 0.576 < 1.96$, the null hypothesis cannot be rejected and we can assume that both samples come from identical populations. This test is illustrated in Figure 15.3.

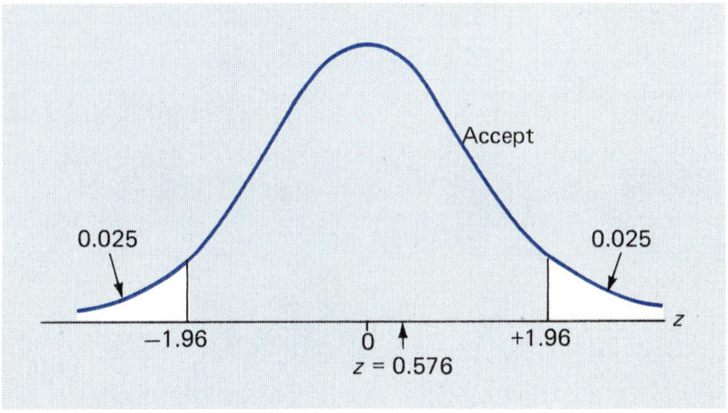

Figure 15.3
Sign test for identical distributions—life test data (Example 15.3).

15.3 The Rank Sum Test for Two Samples

A second nonparametric procedure to test the hypothesis that two samples come from identical populations is the two-sample rank sum test, also called the Mann–Whitney U test. The hypotheses tested are the same as in the sign test, that is,

H_0: The two samples come from identical populations.
H_1: The two samples do not come from identical populations.

However, while the sign test is applicable to any two samples whether independent or not, the rank sum test can *only* be used in tests of two independent random samples. On the other hand, while the sign test requires an equal number of observations in each sample, the rank sum test can also be used when there are unequal sample sizes.

This test is based on the sum of the ranks of the values of one sample in an array of the combined values of both samples. Since the rank sum test uses the ranks of the individual observations, it utilizes more of the sample information than the sign test.

The rationale for the rank sum test is: If the two samples come from identical populations, then the sums of the ranks of both samples should be approximately equal, allowing for differences in sample size. The statistic U is used as a measure of the difference between the ranked observations from the two samples.

$$U = n_1 n_2 + \frac{n_1(n_1 + 1)}{2} - R_1 \tag{15.3}$$

where n_1 = the size of sample number 1, n_2 = the size of sample number 2, and R_1 = the sum of the ranks of sample number 1.

If both n_1 and n_2 are greater than 10, the statistic U is approximately normally distributed, with

$$\mu_U = \frac{n_1 n_2}{2} \tag{15.4}$$

and

$$\sigma_U = \sqrt{\frac{n_1 n_2 (n_1 + n_2 + 1)}{12}} \tag{15.5}$$

As in the other tests discussed so far in this chapter, when the test statistic has a normal distribution, a value of z, the standard normal variate, can be determined from

$$z = \frac{U - \mu_U}{\sigma_U}$$

and compared with critical values of z for the significance level of the test. Since the null hypothesis should be rejected for large as well as small values of U, this is a two-tailed test.

Example 15.4 From the data provided in Example 15.3 and Table 15.1, test the hypothesis that both samples come from identical populations at the 5% level of significance.

1. The data from each sample from Table 15.1 are arranged in order of magnitude from smallest to largest. The rearranged data are shown in Table 15.2.

TABLE 15.2
LIFE TEST DATA ON PICTURE TUBES (ARRANGED IN ASCENDING ORDER) (EXAMPLE 15.4)

Sample #1 Lo-Voltage	*Sample #2 Midwestern*
2320	2390
2340	2470
2480	2540
2540	2550
2550	2610
2580	2620
2600	2650
2640	2660
2680	2680
2710	2700
2720	2710
2730	2740

2. The data from the two samples are merged in ascending order of magnitude and ranks are assigned to the combined data. The combined array is provided in Table 15.3 with the assigned ranks. Each observation is identified according to the sample to which it belongs. Notice that in the case of ties, each tied item is assigned the average rank for the tied observations. For example, both observations having the value 2540 are assigned the rank 6.5, which is the average of rank 6 and rank 7.

3. The sum of the ranks for sample number 1 is determined. Note that it makes no difference which sample is designated #1 and which #2. From Table 15.3,

$$\begin{aligned} R_1 &= 1 + 2 + 5 + 6.5 + 8.5 + 10 + 11 \\ &\quad + 14 + 17.5 + 20.5 + 22 + 23 \\ &= 141 \end{aligned}$$

4. U, μ_U, and σ_U are computed from the data and Equations 15.3, 15.4, and 15.5, respectively.

$$U = n_1 n_2 + \frac{n_1(n_1+1)}{2} - R_1 \tag{15.3}$$

$$= (12)(12) + \frac{(12)(13)}{2} - 141$$

$$= 144 + 78 - 141$$

$$= 81$$

$$\mu_U = \frac{n_1 n_2}{2} = \frac{144}{2} = 72 \tag{15.4}$$

TABLE 15.3
COMBINED ARRAY OF LIFE TEST DATA WITH RANKS (EXAMPLE 15.4)

Observation (hours)	*Rank*		*Sample*
2320	1		1
2340	2		1
2390	3		2
2470	4		2
2480	5		1
2540	6	6.5	1
2540	7		2
2550	8	8.5	2
2550	9		1
2580	10		1
2600	11		1
2610	12		2
2620	13		2
2640	14		1
2650	15		2
2660	16		2
2680	17	17.5	1
2680	18		2
2700	19		2
2710	20	20.5	2
2710	21		1
2720	22		1
2730	23		1
2740	24		2

$$\sigma_U = \sqrt{\frac{n_1 n_2(n_1 + n_2 + 1)}{12}} \tag{15.5}$$

$$= \sqrt{\frac{(12)(12)(12 + 12 + 1)}{12}}$$

$$= \sqrt{300}$$

$$= 17.3$$

5. The value of z is computed from the data and compared with the critical values of z for the significance level of the test.

$$z = \frac{U - \mu_U}{\sigma_U} = \frac{81 - 72}{17.3} = 0.52$$

The critical values of z for a two-tailed test at a 5% significance level are ± 1.96. Since $-1.96 < 0.52 < 1.96$, the null hypothesis cannot be rejected. Notice that this result is very similar to that which was obtained in Example 15.3 using the same data and the sign test. This test is illustrated in Figure 15.4.

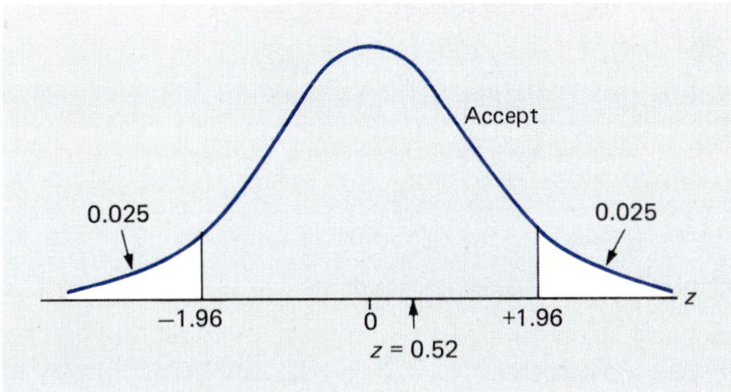

Figure 15.4
Rank sum test for two samples—life test data (Example 15.4).

15.4 The Rank Sum Test for Three or More Samples: The Kruskal–Wallace Test

This test is an extension of the Mann–Whitney U test to situations involving three or more samples. It is also the nonparametric equivalent of the analysis of variance technique discussed in Chapter 13.

In testing hypotheses about the differences among three or more population means using analysis of variance, it was necessary to assume that the populations were normally distributed with equal variances. When these assumptions are valid, we use the variance ratio and the F distribution to decide whether the observed differences among the sample means can be attributed to chance or are indicative of true differences among the populations. When these assumptions cannot be justified, the Kruskal–Wallace rank sum test provides a convenient alternative.

The null hypothesis for this test is that the various samples come from identical populations, while the alternate hypothesis is that they do not, that is,

H_0: All samples come from identical populations.

H_1: All samples do not come from identical populations.

Consider a problem in which we have k random samples of size $n_1, n_2, n_3, \ldots, n_k$. In order to perform this test, all the observations are combined, arranged in order of magnitude, and ranked from lowest to highest. The lowest value is ranked 1, the next lowest 2, and so on. The sum of the ranks for each sample is calculated and designated R_j for $j = 1, 2, 3, \ldots, k$. The test statistic, H, is then computed from:

$$H = \frac{12}{n(n+1)} \sum_{j=1}^{k} \frac{R_j^2}{n_j} - 3(n+1) \tag{15.6}$$

where $n = n_1 + n_2 + n_3 + \cdots + n_k$. The statistic, H, is distributed approximately as χ^2 with $k - 1$ degrees of freedom. As with the other tests involving the chi-square distribution, there must be a minimum of five observations in each of the k samples.

The value of H is compared to critical values of χ^2 for the appropriate degrees of freedom and for the significance level of the test. The null hypothesis is rejected if H is greater than or equal to the critical value of χ^2. Remember that only one tail of the chi-square distribution is used for the rejection region.

Example 15.5 Recall Example 13.4 in which a test was conducted to determine if there was any difference in the effects of three point-of-sale displays on the sales of Ego cosmetics. The data on the number of units sold in each of 10 department stores is reproduced in Table 15.4.

The first column in the table numbers the observations in each sample. In this problem, the sample sizes are all equal, with 10 in each sample. The "units sold" column under each display is the observed value of the response variable. The "Rank" column shows the rank assigned to that observation. Notice that ties receive the average value of the ranks held by the tied values. The method for assigning ranks is illustrated in Table 15.5. The significance level for this test is 1%.

The test statistic, H, is computed from Equation 15.6 as follows.

$$(15.6)\quad H = \frac{12}{n(n+1)} \sum_{j=1}^{k} \frac{R_j^2}{n_j} - 3(n+1)$$

$$= \frac{12}{(30)(31)}\left[\frac{(143.5)^2}{10} + \frac{(214.5)^2}{10} + \frac{(107.0)^2}{10}\right] - (3)(31)$$

$$= (0.013)(2059.2 + 4601.0 + 1144.9) - 93$$

$$= (0.013)(7805.1) - 93$$

$$= 101.47 - 93$$

$$= 8.47.$$

TABLE 15.4
DATA FOR RANK SUM TEST OF POINT-OF-SALE DISPLAYS (EXAMPLE 15.5)

	Display #1		*Display #2*		*Display #3*	
Observation	*Units Sold*	*Rank*	*Units Sold*	*Rank*	*Units Sold*	*Rank*
1	15	11.5	26	28	9	1
2	21	21	22	23.5	15	11.5
3	24	26	14	8	12	3.5
4	10	2	29	29	18	17.5
5	14	8	22	23.5	22	23.5
6	16	13.5	14	8	17	15.5
7	12	3.5	18	17.5	14	8
8	19	19	25	27	14	8
9	22	23.5	30	30	16	13.5
10	17	15.5	20	20	13	5
		$R_1 = 143.5$		$R_2 = 214.5$		$R_3 = 107.0$

TABLE 15.5
COMBINED ARRAY OF POINT-OF-SALE DATA WITH RANKS (EXAMPLE 15.5)

Observation (Units)	*Rank*		*Sample*
9	1		3
10	2		1
12	3	3.5	1
12	4		3
13	5		3
14	6		1
14	7		2
14	8	8	2
14	9		3
14	10		3
15	11	11.5	1
15	12		3
16	13	13.5	1
16	14		3
17	15	15.5	1
17	16		3
18	17	17.5	2
18	18		3
19	19		1
20	20		2
21	21		1
22	22		1
22	23	23.5	2
22	24		2
22	25		3
24	26		1
25	27		2
26	28		2
29	29		2
30	30		2

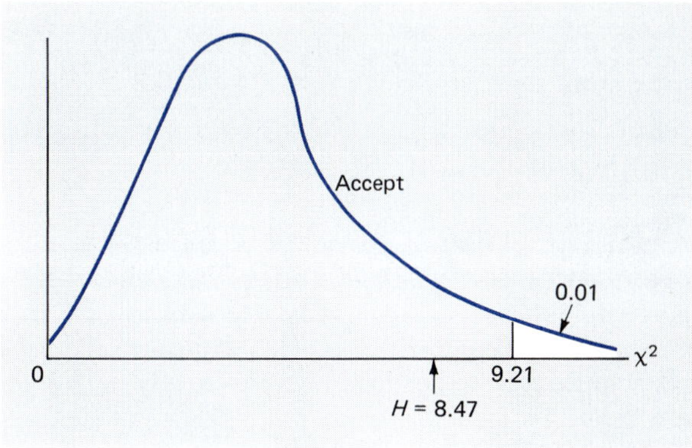

Figure 15.5
Rank sum test for three samples—point-of-sale data (Example 15.5).

The sampling distribution of H is approximately chi-square with $k - 1 = 2$ degrees of freedom. The critical value of χ^2 with two degrees of freedom for a significance level of 1% is 9.21. Since $8.47 < 9.21$, we cannot reject the null hypothesis. Notice that this result differs from the one obtained from the parametric analysis of variance in Example 13.4. This test is illustrated in Figure 15.5.

15.5 Rank Correlation

In Chapter 16 we will discuss a parametric measure of association between two variables called the coefficient of correlation. As with other parametric tests and measures, certain assumptions about the population are necessary if it is to be used. In situations where these assumptions cannot be made or where it is not possible to obtain numerical values for the variables, a nonparametric measure of association called Spearman's COEFFICIENT OF RANK CORRELATION may be applicable. The computation of the coefficient of rank correlation requires only that values of the two variables be assigned ranks. The coefficient of rank correlation, designated r_S, measures the degree of association between the two sets of ranks.

Any coefficient of correlation is a number between 0 and 1, where 0 represents no correlation at all and 1 represents perfect correlation. If the value of the coefficient is positive, then the correlation is said to be positive or direct; that is, large values of the one variable are associated with large values of the other, and small values are associated with small values. If the value of the coefficient is negative, then the correlation is said to be negative or indirect. In that case, large values of one variable are associated with small values of the other and vice versa.

In using the coefficient of rank correlation we deal with two sets of ranks assigned to the variables x and y. The original observations on the variables may be ranks or they may be numerical values ranked by magnitude. We will provide examples illustrating both cases.

The coefficient of rank correlation is

$$r_S = 1 - \frac{6\Sigma d^2}{n(n^2 - 1)} \tag{15.7}$$

where d is the difference between each pair of ranks and n is the number of pairs in the sample.

Example 15.6 A panel of men and a panel of women were asked by a consumer testing organization to rank six brands of beer according to taste. A rank of 1 was given to the best tasting beer and a rank of 6 to the worst. This is a case where the original observations are obtained in terms of ranks. The test was conducted to determine how closely men's and women's tastes in beer are related.

TABLE 15.6
BEER RANKINGS ACCORDING TO TASTE BY TWO PANELS (EXAMPLE 15.6)

Brand	Panel of Women (x)	Panel of Men (y)	Difference in Ranks $d = x - y$	d^2
A	6	6	0	0
B	1	2	−1	1
C	3	1	2	4
D	2	5	−3	9
E	4	3	1	1
F	5	4	1	1
				$\Sigma = 16$

Table 15.6 provides the rankings obtained and the computations necessary for r_S. Spearman's coefficient of rank correlation from these data and Equation 15.7 is

$$r_S = 1 - \frac{(6)(16)}{(6)(36 - 1)} \tag{15.7}$$
$$= 1 - \frac{96}{210}$$
$$= 1 - 0.457$$
$$= 0.543.$$

Example 15.7

An aptitude test was administered to a random sample of job applicants by the personnel department of Metric Systems, Inc. in order to determine the strength of the relationship between the number of years of schooling completed by an individual and his score on the test. The test score, x, and years of schooling, y, are shown in Table 15.7. This is a case where the original data are numerical but are converted to ranks.

The coefficient of rank correlation for these data is

$$r_S = 1 - \frac{(6)(33)}{(10)(100 - 1)} \tag{15.7}$$
$$= 1 - \frac{198}{990}$$
$$= 1 - 0.20$$
$$= 0.80.$$

The coefficient of correlation is somewhat difficult to interpret. Since it is a unitless number between 0 and 1, there is no simple answer to the question of what constitutes a high level of correlation and what does not. It is possible to test for the significance of the deviation of r_s from zero. That is, it is possible to determine if the calculated value is sufficiently greater than zero to indicate that a relationship does exist between the two

TABLE 15.7
DATA ON APTITUDE TEST SCORES AND YEARS OF SCHOOLING (EXAMPLE 15.7)

Applicant	*Test Score* (*x*)	*Rank* (*x*)	*Years of School* (*y*)	*Rank* (*y*)	*Difference in Ranks* (*d*)	d^2
1	80	4	13	2	2	4
2	50	8	10	6.5	1.5	2.25
3	75	5	11	5	0	0
4	65	7	9	8	−1	1
5	45	9	8	9	0	0
6	90	2	14	1	1	1
7	95	1	12	3.5	−2.5	6.25
8	40	10	7	10	0	0
9	85	3	10	6.5	−3.5	12.25
10	70	6	12	3.5	2.5	6.25
						$\Sigma = 33.00$

variables. These significance tests about the coefficient of correlation are covered in Chapter 16, so they will not be treated here. However, we will state that the test described in the next chapter is applicable to the coefficient of rank correlation.

Definitions

Coefficient of rank correlation: A measure of association between two variables based on ranks.

Distribution-free statistics: Statistical procedures that require no assumptions about the population distribution; this term is used interchangeably with "nonparametric statistics."

Nonparametric statistics: Statistical procedures that involve no hypotheses about population parameters.

Run: A sequence of observations preceded and followed by a different observation.

Formulas

Mean and standard error of R, the number of runs:

$$\mu_R = \frac{2n_1n_2}{n_1 + n_2} + 1 \tag{15.1}$$

$$\sigma_R = \sqrt{\frac{2n_1n_2(2n_1n_2 - n_1 - n_2)}{(n_1 + n_2)^2(n_1 + n_2 - 1)}} \tag{15.2}$$

Mann–Whitney's U:

$$U = n_1n_2 + \frac{n_1(n_1 + 1)}{2} - R_1 \tag{15.3}$$

$$\mu_U = \frac{n_1n_2}{2} \tag{15.4}$$

$$\sigma_U = \sqrt{\frac{n_1n_2(n_1 + n_2 + 1)}{12}} \tag{15.5}$$

Kruskal–Wallace H:

$$H = \frac{12}{n(n+1)} \sum_{j=1}^{k} \frac{R_j^2}{n_j} - 3(n+1) \tag{15.6}$$

Spearman's Coefficient of Rank Correlation:

$$r_S = 1 - \frac{6\Sigma d^2}{n(n^2 - 1)} \tag{15.7}$$

Problems

15.1 Test the following sequence of A's and B's for randomness at both the 1% and 5% level of significance.
BBBAABBBBAAAABBBBAAAABBBBAAABBAAAAABBAA

15.2 Test the following sequence of A's and B's for randomness at both the 1% and 5% level of significance.
BAABBABABBAABAABABBBABABABBABAABABABBABB

15.3 The production manager of the Advance Machine Company was displeased with the head of the purchasing department because a particular component used in manufacturing several different products was out of stock for the third time in one month. The purchasing manager claimed that he would be able to maintain better control of the inventory of that component if the production manager would give him reasonably accurate forecasts of the demand. The production manager claimed that accurate forecasts were impossible because the demand was random. The purchasing manager selected a random sample of 40 requisitions for the component, each representing one week's demand in hundreds of units. The data from the 40 requisitions appear below. Would you accept the production manager's claim of random demand at the 5% level of significance? Read the data down and then across.

5.9	5.4	6.3	7.2	6.8	5.9	6.7	7.3	6.6	4.6
6.3	6.8	7.1	5.4	6.1	5.6	4.9	6.5	7.0	4.8
4.7	4.9	6.2	4.5	7.3	6.4	6.8	5.8	6.2	5.7
5.4	5.9	4.9	6.8	5.3	5.8	6.2	5.2	6.1	5.2

15.4 In a recent criminal trial, the attorney for the defense made a motion for a mistrial on the grounds that the jurors in that jurisdiction were not selected in a random manner. A statistical consultant was called in to test this claim. He listed, by sex, the last 50 jurors in the order in which they were called. Test whether this sequence of male and female jurors is random at $\alpha = 0.05$.
FFFFMMMFFFFMMMMMMMMFFFFFFFFFFMMMMMFFFFFFFMMMFFMMMMM

15.5 An analysis on the return on investment in the common stocks of two companies listed on the New York Stock Exchange was to be conducted to determine if the average return was the same for both companies. Since there was no basis for the assumptions required for a parametric test, the sign test was used to test the hypothesis that these data come from identical distributions. Test this hypothesis at $\alpha = 0.05$ from the following data, using the sign test.

Year	Return on Investment (%)	
	Company A	*Company B*
1967	7.9	8.4
1968	8.5	8.1
1969	9.4	8.3
1970	7.8	8.0
1971	7.2	7.6
1972	5.6	6.1
1973	5.8	5.4
1974	5.9	6.0
1975	3.9	6.1
1976	5.8	5.9
1977	6.3	6.1
1978	5.9	6.3

15.6 In order to test whether or not background music had any effect on production efficiency, music was played in one production department for a period of 15 working days. The output on each of these days was compared with the production output on 15 similar days during which there was no music. Use the sign test to test the hypothesis that the distribution of output is identical under both conditions at $\alpha = 0.05$.

Day	*Output with Music* (x)	*Output without Music* (y)
1	870	880
2	940	910
3	920	910
4	850	880
5	860	850
6	840	820
7	900	870
8	830	850
9	840	850
10	860	840
11	800	820
12	850	830
13	940	860
14	890	880
15	880	880

15.7 The Advance Machine Company was considering the purchase of an automatic machine tool for precision cutting. Two manufacturers offered such a machine tool for sale; however, the price of machine A was $300 less than that of machine B. Advance wanted to test whether there was any difference in the precision of the two machines before making a purchase decision. A test was set up in which both machine tools were adjusted to exactly the same setting and a sample of 12 identical parts were cut on each. Each part was then measured with a micrometer and the absolute deviation of the measurement from the required value was recorded in thousandths of a centimeter. Use the sign test to test the hypothesis that both samples come from identical populations. In other words, there is no difference in the performance of the two machine tools at the 5% level of significance. The data follow.

Part	*Machine A*	*Machine B*
1	2	3
2	4	3
3	3	2
4	5	3
5	0	2
6	4	2
7	3	2
8	4	3
9	2	2
10	6	4
11	4	3
12	3	2

15.8 Use the data from problem 15.5 and the rank sum test to test the hypothesis that both samples come from identical populations at $\alpha = 0.05$.

15.9 Use the data from problem 15.7 and the rank sum test to test the hypothesis that both samples come from identical populations at $\alpha = 0.05$.

15.10 In a mileage test of their unleaded gasoline, Southwestern Oil Company drove 11 identical cars over their test track at their refinery in New Jersey. Another 14 identical cars were driven over a similar track at their California refinery. The miles driven by each car on a single gallon of gasoline are noted below. Use the rank sum test to test the hypothesis that the distribution of mileages from both samples are identical at $\alpha = 0.05$.

New Jersey Sample	*California Sample*
19.1	26.1
22.4	24.3
17.6	19.8
19.7	23.2
21.0	16.4
22.2	18.0
20.9	25.5
24.8	22.7
23.0	15.7
25.5	25.0
18.6	22.0
	19.4
	17.3
	16.0

15.11 An agency of the federal government administered a reading comprehension test to a random sample of 15 sixth-graders drawn from the student body of an urban elementary school. The same test was given to a random sample of 15 sixth-grade students drawn from the student body of a rural elementary school. The scores in both samples are listed below. Test the hypothesis that both sets of scores came from identical populations using the rank sum test and a significance level of 5%.

Urban Sample	*Rural Sample*
65	48
60	87
72	91
68	92
81	87
89	56
92	64
45	57
58	76
67	88
74	69
75	90
80	51
63	54
50	79

15.12 Use the Kruskal–Wallace rank sum test to test the hypothesis that the following three samples come from identical populations at $\alpha = 0.05$.

Sample #1	*Sample #2*	*Sample #3*
12.3	16.0	5.7
11.0	10.1	3.8
15.4	8.4	9.4
6.8	13.2	2.5
9.9	7.1	4.7

15.13 Use the Kruskall–Wallace rank sum test to test the hypothesis that the following four samples come from identical populations at $\alpha = 0.05$.

Sample #1	*Sample #2*	*Sample #3*	*Sample #4*
54.8	61.7	48.6	80.0
53.2	58.3	32.2	68.5
57.1	54.1	39.7	28.7
52.0	51.2	44.4	30.0
59.9	68.8	50.0	55.0
	63.5		71.4
			37.7

15.14 In problem 15.10, data from two mileage tests conducted by the Southwestern Oil Company were presented. After completing those tests, SWOCO conducted a third test using seven identical cars on a test track at their Louisiana refinery. The mileage data from the Louisiana test are presented below with the data from the two previous tests. Use the Kruskal–Wallace rank sum test to test the hypothesis, at the 5% level of significance, that all three samples came from identical populations.

Louisiana Sample	*New Jersey Sample*	*California Sample*
17.8	19.1	26.1
15.3	22.4	24.3
16.3	17.6	19.8
21.7	19.7	23.2
20.3	21.0	16.4
17.1	22.2	18.0
15.5	20.9	25.5
	24.8	22.7
	23.0	15.7
	25.5	25.0
	18.6	22.0
		19.4
		17.3
		16.0

15.15 In problem 11, reading comprehension tests were administered by a government agency to samples of sixth-graders from urban and rural elementary schools. The same test was given to a random sample of 10 sixth-graders drawn from the student body of a suburban elementary school. The scores from all three samples are listed below. Use the Kruskal–Wallace rank sum test to test the hypothesis, at $\alpha = 0.05$, that all three samples come from identical populations.

Urban Sample	*Rural Sample*	*Suburban Sample*
65	48	88
60	87	85
72	91	91
68	92	73
81	87	66
89	56	73
92	64	84
45	57	86
58	76	78
67	88	77
74	69	
75	90	
80	51	
63	54	
50	79	

15.16 The television critics of two New York newspapers ranked the five most popular television series on a scale of 1 to 5 according to their preferences. Compute the coefficient of rank correlation between the two sets of ranks from the data below.

	Critic A	*Critic B*
Mollie's Angels	4	5
LaVerne and Harry	5	2
Bad Times	2	4
Claude	1	3
Welcome Back Mother	3	1

15.17 One criterion used by the Reliance Insurance Company in hiring sales trainees was their class rank at the time of their graduation from college. Since most were graduates of large state universities, the class sizes were usually comparable. In order to determine if there is a relationship between class rank and future sales performance, the ten top sales representatives in the company were ranked according to their previous year's sales and also by class rank. Compute the coefficient of rank correlation from the data presented below.

Sales Representative	*Sales (in thousands of dollars)* (x)	*Class Rank* (y)
A	1100	14
B	989	21
C	863	87
D	851	9
E	802	31
F	798	45
G	754	28
H	750	94
I	740	73
J	739	60

15.18 The top executives of Midwestern Electric rank prospective managerial candidates on the basis of what they know about each candidate. In order to determine if there is any consistency in the rankings obtained in this manner, two vice-presidents were asked to rank the same ten candidates. Compute the coefficient of rank correlation from the following two sets of ranks.

Candidate	*V.P. #1*	*V.P. #2*
A	3	2
B	1	5
C	8	9
D	2	1
E	4	6
F	9	10
G	5	3
H	7	4
I	10	8
J	6	7

15.19 The management of Cinderella Clothes believed that it would be useful to determine the strength of any relationship between retail store frontage and number of customers. Eight branch stores in similar locations, but with different frontages, were selected. The number of shoppers entering each store during selected time periods of equal duration were noted. These data, together with the frontage of each store, are provided below. Compute the coefficient of rank correlation between these two variables.

Store	*Frontage in Feet* (x)	*Number of Shoppers* (y)
1	40	49
2	32	57
3	38	58
4	25	35
5	55	60
6	50	63
7	30	40
8	22	44

CHAPTER 16 REGRESSION ANALYSIS

16.1
Correlation

16.2
Inferences about the correlation coefficient

16.3
Simple linear regression

16.4
Confidence intervals for predictions

16.5
Multiple regression

Consider a problem in which it is necessary to make some sort of inference about a population parameter when it is impossible to observe any members of the population of interest. You might conclude that this is a very difficult, if not impossible, task. Nevertheless, this type of situation is common to problems in business involving prediction or forecasting. The term forecast usually implies the prediction of some future event. More precisely, a business forecast usually involves the estimation of some characteristic of a population that does not exist at the time the information is required. Business forecasts typically involve future sales or product demand. Most other forecasts made by a firm, such as requirements for raw materials or labor are based on sales or demand forecasts. However, if we need to predict next year's demand for some product, we have to be able to observe at least a sample of the population of next year's potential purchasers. This is impossible, since that particular population will not exist until next year.

Whenever it is necessary to make inferences about a population that cannot be observed directly, we must base these inferences on some related population that can be observed. For example, if we let y represent some variable characteristic of interest and x represent another variable related to y, then we may be able to use information obtained about x to make inferences about y. This is only possible, however, if we can be sure that such a relationship does indeed exist and if we can determine the nature of the relationship. Such a relationship between variables is called a function, as discussed in Section 2.6, and can be represented by the expression $y = f(x)$. In such a functional relationship, x is referred to as the independent variable and y is called the dependent variable. Whenever this functional relationship is of the form $y = a + bx$, it is said to be a linear relationship. That is, $y = a + bx$ is the equation of a straight line. The techniques involved in determining the relationship between a dependent variable and one or more independent variables and in using that relationship to make estimates are called REGRESSION ANALYSIS. A related technique, which measures the strength of the relationship between variables, is called CORRELATION ANALYSIS.

We will consider only linear regression, that is, regressions in which the relationship between variables can be expressed by a linear equation. Also, since we are primarily interested in regression, we will discuss correlation only as an adjunct to the regression analysis even though correlation is frequently considered as a topic by itself.

Technically, the distinction between correlation analysis and regression analysis relates to the nature of x, the independent variable. In regression, x is treated as a known value and the relationship between x and y is used to predict values of y for the given value of x. In correlation, x and y are both treated as random variables. This distinction has no practical effect on the regression analysis if certain conditions are satisfied. These conditions will be specified later.

16.1 Correlation

Correlation is a measure of the degree of association between two variables. Whenever we want to use this relationship to make inferences or predictions, it is certainly logical to determine first if there really is a relationship and, if so, to compute some measure of its strength.

In Chapter 1, we cited an example in which a small building contractor wanted to predict next year's demand for single-family dwelling units in Buckingham County. The demand for single-family homes next year is, of course, dependent on the population of next year's potential home buyers. Since that population cannot be observed this year, a prediction about that population must be based on some population that can be observed and which might logically be related to the population of interest. Mr. Wilson, the sole proprietor of Buckingham Builders, selected two factors which he believed might be related to the demand for housing. These factors were marriages, as measured by the number of marriage licenses issued during the previous year, and the unemployment rate during the previous year. In order to determine if either of these factors can be used as a predictor of housing demand as measured by building permits, he went to the Buckingham county court house and collected data on the following.

> *y*, the number of building permits issued for single-family dwellings in the years 1969–1978
>
> *x*, the number of marriage licenses issued in the years 1968 through 1977
>
> *v*, the unemployment rate in the county in the years 1968 through 1977

Each of the *y* values was paired with an *x* and a *v* value for the previous year; that is, he paired building permits in 1969 with marriage licenses and unemployment in 1968, building permits in 1970 with marriage licenses and unemployment in 1969, and so on. The data he obtained are summarized in Table 16.1.

A quick method for determining if there is an apparent relationship between two variables is to plot the points corresponding to each pair of values on a SCATTER DIAGRAM. A scatter diagram is simply a chart on which the horizontal axis is scaled in units

TABLE 16.1
DATA ON BUILDING PERMITS, MARRIAGE LICENSES, AND UNEMPLOYMENT FOR BUCKINGHAM COUNTY, 1968–1978 (EXAMPLE 16.1)

	Building Permits (in hundreds) (y)		*Marriage Licenses (in hundreds) (x)*		*Unemployment Rate (in %) (v)*	
Observation	*Year*	*Number*	*Year*	*Number*	*Year*	*Number*
1	1969	3.3	1968	21	1968	2.8
2	1970	3.9	1969	22	1969	1.8
3	1971	2.6	1970	18	1970	3.4
4	1972	1.6	1971	10	1971	4.0
5	1973	3.5	1972	23	1972	3.8
6	1974	1.8	1973	8	1973	4.5
7	1975	3.7	1974	24	1974	1.6
8	1976	2.2	1975	17	1975	4.5
9	1977	1.0	1976	6	1976	4.6
10	1978	2.0	1977	15	1977	2.8

corresponding to one of the variables—in this case, one of the independent variables—and the vertical axis is scaled in units corresponding to the dependent variable. Scatter diagrams for x and *y* and for *v* and *y* appear as Figures 16.1 and 16.2, respectively. By observing the pattern of points on these charts, it is possible to determine, at least roughly, whether or not there is a relationship between the two variables and, if so, to estimate the type of relationship.

For example, in Figure 16.1, an apparent relationship does exist between x and *y*. It appears that large values of x are associated

Figure 16.1
Scatter diagram of building permits versus marriage licenses.

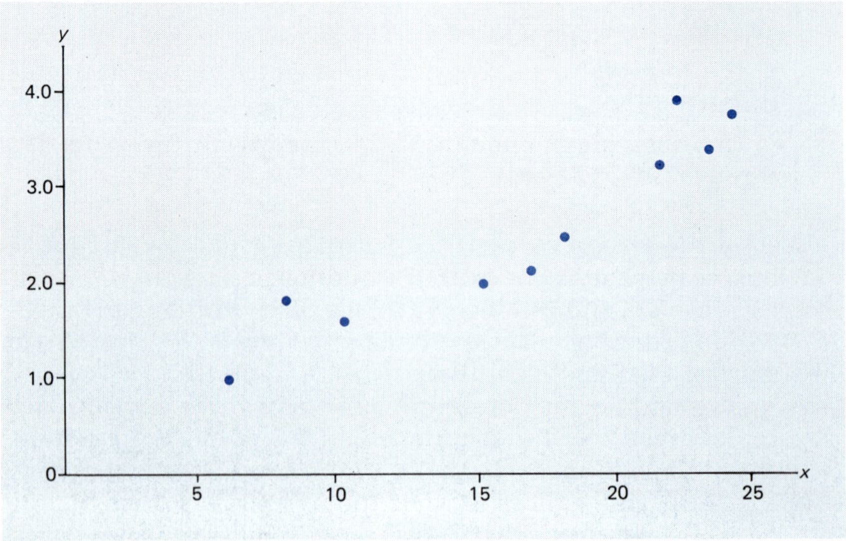

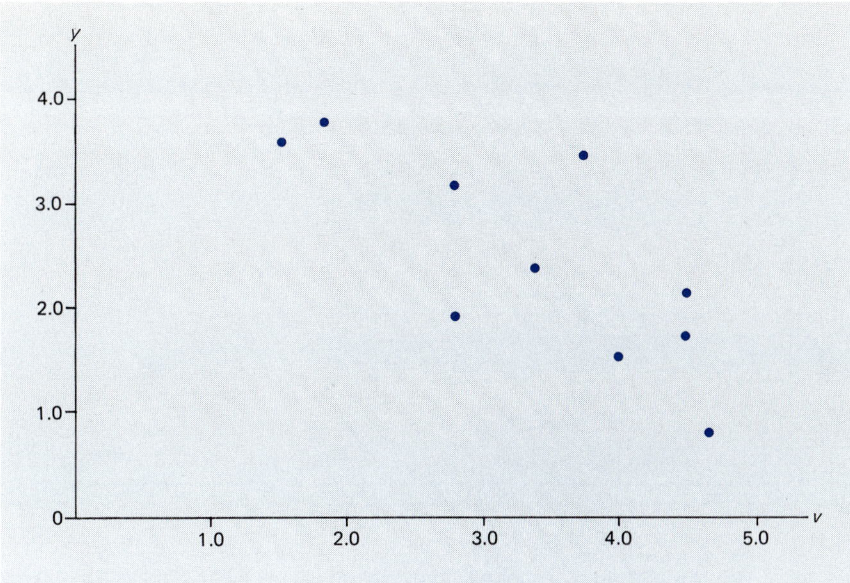

Figure 16.2
Scatter diagram of building permits versus unemployment.

with large values of y and small values of x with small values of y. The points also seem to be arranged along a straight line which slants upward from the lower left-hand corner of the chart to the upper right-hand corner. This pattern tends to indicate the existence of a *positive* linear correlation.

In Figure 16.2, there also appears to be a relationship between the two variables. However, in this case, large values of v appear to be associated with small values of y and vice versa. The points also seem to be arranged along a straight line, but this time it slants from upper left to lower right. This pattern tends to indicate a *negative* linear correlation. Notice, however, that the points in this diagram seem to be more widely scattered about an apparent line than the points in Figure 16.1. The amount of scatter or dispersion tends to indicate the strength of the association. We might conclude from observing these diagrams, at least tentatively, that x and y are more closely related than v and y.

In contrast, the diagram of Figure 16.3 illustrates a pattern that tends to indicate that there is no relationship between the variables plotted. The points seem to be scattered randomly across the chart. On the other hand, the diagram of Figure 16.4 does show a pattern, but one that does not appear to resemble a straight line. For the data of this figure, we might tentatively assume that there is an association between the two variables but that it is not linear.

Having studied the scatter diagrams of Figures 16.1 and 16.2, Mr. Wilson of Buckingham Builders tentatively assumed that both marriages and unemployment have an effect on the demand for

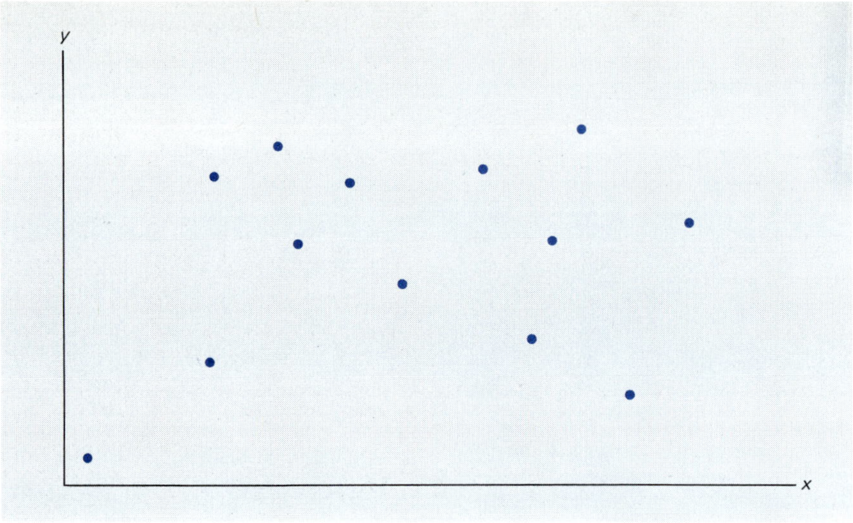

Figure 16.3
Scatter diagram showing no correlation.

housing in the following year. The scatter diagrams, however, should not be considered conclusive, but should be used simply as an indicator that a particular variable should be either eliminated for consideration as a predictor or investigated further. For example, a variable that produces a scatter diagram like Figure 16.3 does not warrant further consideration.

The next step is to compute a value called the COEFFICIENT OF CORRELATION between variables y and x and y and v from the sample data. The coefficient of correlation is a measure of the

Figure 16.4
Scatter diagram showing apparent nonlinear relationship.

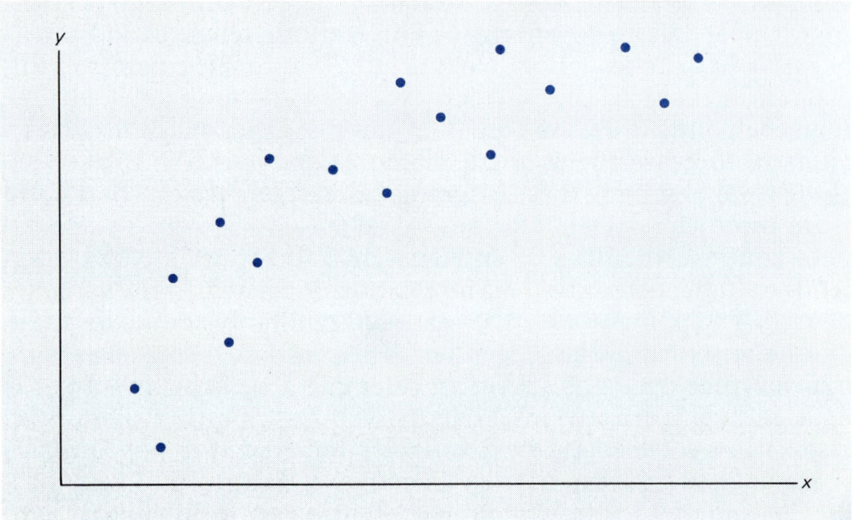

degree of linear association between the two variables. It is worth noting here that the value of the coefficient of correlation can be determined for any two variables without specifying one as being dependent and the other as independent. We have made that specification here because we are primarily interested in regression rather than correlation. The use of a regression equation for prediction requires that the variable to be predicted be designated as dependent and the other variable designated as independent.

There is a theoretical coefficient of correlation between the two populations represented by the variables x and y. This population coefficient of correlation is designated by the Greek letter ρ (rho), and it indicates the degree of association between the population of x's and the population of y's. The population coefficient of correlation is defined as

$$\rho = \frac{\Sigma(x - \mu_x)(y - \mu_y)}{N\sigma_x\sigma_y} \tag{16.1}$$

Since the businessman rarely, if ever, has access to entire populations, the population correlation coefficient must be estimated from sample data. The sample coefficient of correlation is defined as

$$r_{y.x} = \frac{\Sigma(x - \bar{x})(y - \bar{y})}{\sqrt{\Sigma(x - \bar{x})^2\Sigma(y - \bar{y})^2}} \tag{16.2}$$

where the subscript on the r designates the two variables involved. If a distinction is made between the dependent and the independent variable, the dependent variable appears first in the subscript.

A more convenient computational form for the sample correlation coefficient is

$$r_{y.x} = \frac{n\Sigma xy - \Sigma x\Sigma y}{\sqrt{[n\Sigma x^2 - (\Sigma x)^2][n\Sigma y^2 - (\Sigma y)^2]}} \tag{16.3}$$

The coefficient of correlation is always a number between -1 and $+1$ inclusive. A value of $r = +1$ represents perfect positive correlation; $r = -1$ represents perfect negative correlation; and $r = 0$ indicates no correlation whatsoever.

Example 16.1

In order to decide which of the two variables, x = marriage licenses and v = unemployment rate, is a better predictor of housing demand, Mr. Wilson, the building contractor, computes the sample correlation coefficients for each variable from the data in Tables 16.1, 16.2, and 16.3. The symbol $r_{y.x}$ represents the correlation coefficient between x and y, while $r_{y.v}$ represents the correlation coefficient between v and y. By Equation 16.3 and the data of Tables 16.1, 16.2, and 16.3, these values are

TABLE 16.2
CORRELATION AND REGRESSION ANALYSIS OF BUILDING PERMITS VS MARRIAGE LICENSES (DATA FOR COMPUTATION) (EXAMPLE 16.1)

x	y	x^2	y^2	xy
21	3.3	441	10.89	69.3
22	3.9	484	15.21	85.8
18	2.6	324	6.76	46.8
10	1.6	100	2.56	16.0
23	3.5	529	12.25	80.5
8	1.8	64	3.24	14.4
24	3.7	576	13.69	88.8
17	2.2	289	4.84	37.4
6	1.0	36	1.00	6.0
15	2.0	225	4.00	30.0
164	25.6	3068	74.44	475.0

TABLE 16.3
CORRELATION AND REGRESSION ANALYSIS OF BUILDING PERMITS VS UNEMPLOYMENT RATE (DATA FOR COMPUTATION) (EXAMPLE 16.1)

v	y	v^2	y^2	vy
2.8	3.3	7.84	10.89	9.24
1.8	3.9	3.24	15.21	7.02
3.4	2.6	11.56	6.76	8.84
4.0	1.6	16.00	2.56	6.40
3.8	3.5	14.44	12.25	13.30
4.5	1.8	20.25	3.24	8.10
1.6	3.7	2.56	13.69	5.92
4.5	2.2	20.25	4.84	9.90
4.6	1.0	21.16	1.00	4.60
2.8	2.0	7.84	4.00	5.60
33.8	25.6	125.14	74.44	78.92

$$(16.3) \qquad r_{y.x} = \frac{(10)(475) - (164)(25.6)}{\sqrt{[10(3068) - (164)^2][10(74.44) - (25.6)^2]}}$$

$$= \frac{4750 - 4198.4}{\sqrt{(3784)(89.04)}}$$

$$= \frac{551.6}{580.45}$$

$$= 0.95.$$

Substituting v's for x's in Equation 16.3, we get

$$(16.3) \qquad r_{y.v} = \frac{(10)(78.92) - (33.8)(25.6)}{\sqrt{[10(125.14) - (33.8)^2][10(74.44) - (25.6)^2]}}$$
$$= \frac{789.2 - 865.28}{\sqrt{(108.96)(89.04)}}$$
$$= \frac{-76.08}{98.50}$$
$$= -0.77$$

Notice that the coefficient of correlation between y and v is a *negative* 0.77. This indicates that an *inverse* correlation exists between these two variables. We could have anticipated the inverse relationship by virtue of the fact that the points on the scatter diagram of these two variables appear to slant downward from the upper left to the lower right. Although both x and v appear to be correlated with y, since only one independent variable is used in a simple regression Mr. Wilson chooses x, marriage licenses, as his predictor of the demand for housing. His choice is based on the fact that the absolute value of $r_{y.x}$ is considerably greater than the absolute value of $r_{y.v}$, indicating a higher degree of association.

Note that a high value of the coefficient of correlation does not necessarily mean that there is a cause and effect relationship between the two variables. A high sample value of r can occur by chance or because both variables are related to some other variable which has not even been considered in the analysis. As an example of the latter case, a correlation analysis indicated that there is a very strong relationship between the salaries of schoolteachers and an index of liquor sales. To conclude that changes in the sales of liquor are caused by changes in teachers' salaries is the same as attributing a large part of the liquor consumption in this country to schoolteachers. Although it is likely that teachers do consume their fair share of alcoholic beverages, this conclusion would have to be erroneous. This is a case in which fluctuations in both variables are no doubt related to changes in a third variable, that being general economic conditions.

16.2 Inferences about the Correlation Coefficient

Since r is a *sample* correlation coefficient, it is used to make inferences about ρ, the unknown population correlation coefficient. The most common inference about ρ is a test of the hypothesis that ρ is actually equal to zero, that is, that there is no relationship between the two variables. The null and alternate hypotheses to be tested are, respectively,

$$H_0\colon \rho = 0$$
$$H_1\colon \rho \neq 0$$

The statistic used for this test is r, the sample correlation coefficient computed from a sample of size n. You have learned in the chapters on hypothesis testing that in order to use any statistic for testing a hypothesis, the sampling distribution of that statistic must be known. In this case, r does not follow any of the commonly used and tabulated sampling distributions. Fortunately, a very simple transformation results in the following statistic, which has a t distribution with $n - 2$ degrees of freedom.

$$t = \frac{r\sqrt{n-2}}{\sqrt{1-r^2}} \tag{16.4}$$

The hypothesis about ρ can be tested with the statistic t computed from the sample data and compared with a critical value of t for the specified significance level.

Example 16.2 Test the hypothesis that $\rho = 0$ at $\alpha = 0.01$ for the data on building permits and marriage licenses provided in Example 16.1.

In Example 16.1, $r_{y.x} = 0.95$ is computed from a sample of $n = 10$. Therefore, from Equation 16.4,

$$\begin{aligned} t &= \frac{r\sqrt{n-2}}{\sqrt{1-r^2}} \qquad (16.4) \\ &= \frac{0.95\sqrt{10-2}}{\sqrt{1-(0.95)^2}} \\ &= \frac{0.95\sqrt{8}}{\sqrt{1-0.9025}} \\ &= 8.61 \end{aligned}$$

The critical value of t for eight degrees of freedom and $\alpha = 0.01$ is 2.896. Since the computed value of $t = 8.61$, which is greater than 2.896, the null hypothesis that $\rho = 0$ is rejected. Therefore, it can be assumed that some relationship does exist between the two variables.

16.3 Simple Linear Regression

Having obtained data on two variables, x and y, several procedures have indicated that a linear relationship exists between them. These procedures include the plotting of the data on a scatter diagram, the computation of the sample coefficient of correlation, and the testing of the hypothesis about the population coefficient of correlation. A linear relationship, that is, a straight line, can be expressed in the form $y = \alpha + \beta x$. In this equation, y represents a value of the dependent variable, x is the independent variable, and α and β are the parameters or coefficients of the regression line. The coefficient α is called the y intercept of the line. It is the value of y when $x = 0$. The coefficient β is simply the slope of the line. It represents the change in y for a given change in x. Figure 16.5 illustrates a typical regression line. From this figure, you can see that the slope of the line is the ratio of $y_1 - y_0$ to $x_1 - x_0$, that is,

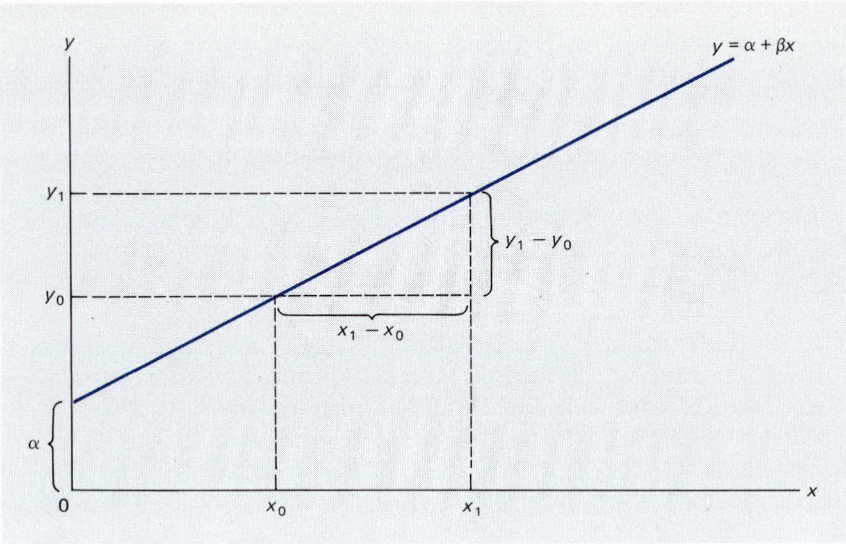

Figure 16.5
A regression line of y *on* x.

$$\beta = \frac{y_1 - y_0}{x_1 - x_0}$$

In the physical sciences, a regression relationship can frequently be exact; that is, given a value for x, the same value for y will always result. There are also some business situations where such an exact relationship exists. For example, a repair service may relate service charges to the amount of time spent on the job. The cost of a service call can be determined from the equation

$$y = \$10 + \$7.50x$$

where y represents the charge for the call and x represents the time required on the job in hours. In this equation, $\alpha = \$10$, a flat charge for the call regardless of the time spent, and $\beta = \$7.50$ per hour. It is, however, more likely that in business situations, instead of an exact relationship between x and y, there is a distribution of y values for each value of x. In that case, the regression line, $\alpha + \beta x$, represents an average value of y for the given x. The regression line is therefore a line of means. In this case we can state the relationship between the variables as

$$\mu_{y.x} = \alpha + \beta x \tag{16.5}$$

where $\mu_{y.x}$ represents the mean of a conditional distribution of y values for a given value of x.

When we estimate the population regression line from sample data, we are actually attempting to find the line that passes through the means of a series of conditional distributions of y. Figure 16.6 illustrates this situation. In order for Equation 16.5 to represent the regression of y on x, we must make the following assumptions.

1. The conditional distributions of y are normally distributed for all values of x.

2. The variances of all of the conditional distributions of y are equal.

3. The y values are independent of one another.

In estimating the regression line of means from sample data, the sample regression line is

$$\bar{y}_c = a + bx \tag{16.6}$$

The symbol $\bar{y}_c$ is the sample estimate of $\mu_{y.x}$ and represents the y values of points on the sample regression line. The subscript c on $\bar{y}$ is used to indicate that the value is *computed* from the regression equation. The a and b in Equation 16.6 are estimates of α and β, respectively. Once values are determined for a and b, the sample regression line is completely determined.

The Least Squares Regression Line. Look again at the scatter diagram of building permits versus marriage licenses in Figure 16.1. If you were asked to draw the straight line that best fits the points, you would probably be at a loss as to where to draw the line. You might, in fact, question what constitutes the criterion of the *best* fit.

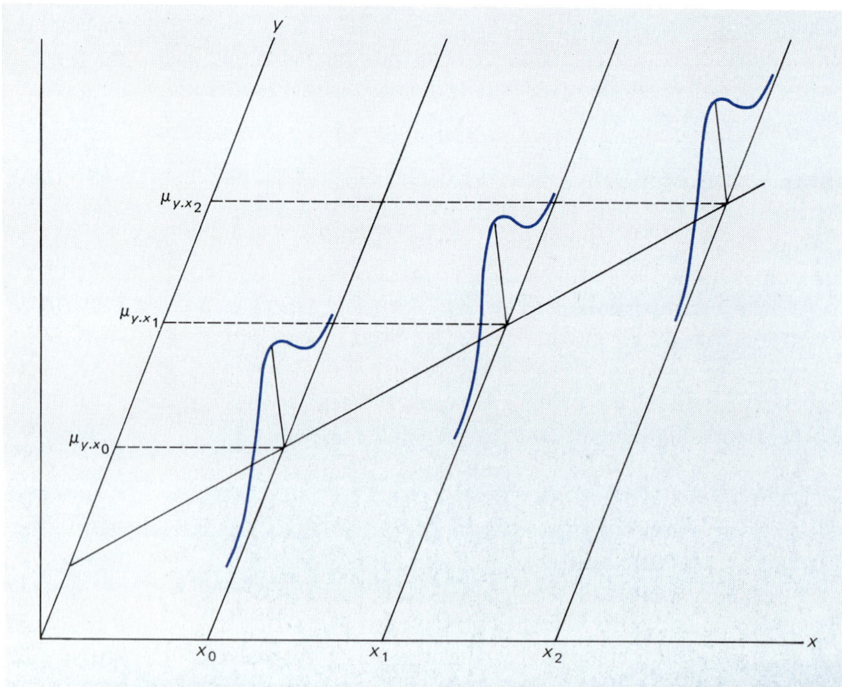

Figure 16.6
Regression line passing through the means of the conditional distributions of y.

It should be obvious that no straight line will pass through all of these points. On the other hand, any number of lines could be drawn that appear to approximate the arrangement of these points on the graph. In statistics, it is necessary to establish some criterion for the line of *best fit* so that for the same set of data only one line will meet the criterion. The criterion used in regression analysis is called the *least squares criterion*.

The least squares regression line has the property that it always passes through the point $(\bar{x}, \bar{y})$ and that the sum of the positive and negative deviations of the observed points about the line is zero. It also has the additional property that the sum of the squares of these deviations is less than for any other line. These two properties are summarized as follows.

$$\Sigma(y - \bar{y}_c) = 0$$

$$\Sigma(y - \bar{y}_c)^2 \quad \text{is a minimum}$$

For a given set of data, the estimated parameters of the least squares regression line are obtained by the simultaneous solution of a set of equations called the normal equations. In simple regression, the solution of these normal equations results in

$$a = \bar{y} - b\bar{x} \tag{16.7}$$

$$b = \frac{\Sigma xy - n\bar{x}\bar{y}}{\Sigma x^2 - n\bar{x}^2} \tag{16.8}$$

From the data in Table 16.2 we obtain $\bar{x}$ and $\bar{y}$ as follows.

$$\bar{x} = \frac{\Sigma x}{n} = \frac{164}{10} = 16.4$$

$$\bar{y} = \frac{\Sigma y}{n} = \frac{25.6}{10} = 2.56$$

Then, by substituting the appropriate values in Equations 16.7 and 16.8, we get b and then a as follows.

$$b = \frac{475 - (10)(16.4)(2.56)}{3068 - (10)(16.4)^2} = \frac{55.16}{378.4} \doteq 0.15 \tag{16.8}$$

$$a = 2.56 - (0.15)(16.4) = 2.56 - 2.46 = 0.10 \tag{16.7}$$

The regression equation of y on x is then

$$\bar{y}_c = 0.10 + 0.15x \tag{16.6}$$

This line is illustrated in Figure 16.7.

In terms of the original problem, $\bar{y}_c$ represents the predicted average number of building permits, in hundreds, for a specified value of x, the number of marriage licenses issued, in hundreds. If Mr. Wilson, the builder, finds that 2000, that is, 20 hundred, marriage licenses were issued in 1978, he can predict the number of

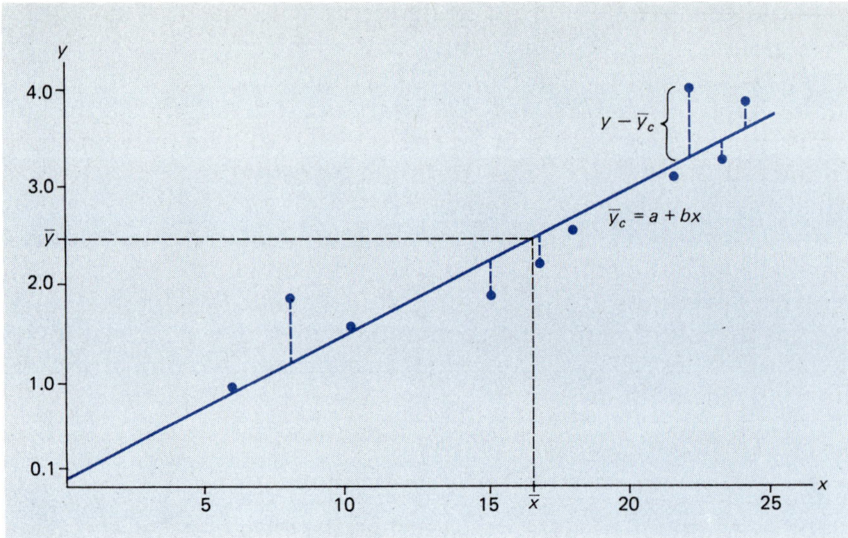

Figure 16.7
Regression line fitted to scatter diagram of building permits versus marriage licenses.

building permits for 1979 from the regression equation:

$$\bar{y}_c = 0.10 + (0.15)(20) = 0.10 + 3 = 3.1 \text{ hundreds} \tag{16.6}$$

that is, 310 building permits.

We interpret this result as follows. According to the regression equation, whenever 2000 marriage licenses are issued in Buckingham County in one year, the average number of building permits issued in the following year is estimated at 310.

Example 16.3 Using the data of Table 16.3, find the regression equation relating building permits, y, to unemployment rate, v, and predict the number of building permits for 1979 if the unemployment rate in 1978 is 3.1%.

1. Find b, the slope of the sample regression line, from Equation 16.8 after substituting v for x in that equation.

$$b = \frac{\Sigma vy - n\bar{v}\bar{y}}{\Sigma v^2 - n\bar{v}^2} = \frac{78.92 - (10)(3.38)(2.56)}{125.14 - (10)(3.38)^2} \tag{16.8}$$

$$= \frac{78.92 - 86.53}{125.14 - 114.24} = \frac{-7.61}{10.9} = -0.7$$

2. Find a, the y intercept from Equation 16.7, after substituting v for x in that equation.

$$\begin{aligned} a &= \bar{y} - b\bar{v} \\ &= 2.56 - (-0.7)(3.38) \\ &= 2.56 + 2.37 \\ &= 4.93 \end{aligned} \tag{16.7}$$

3. The regression equation is

$$\bar{y}_c = 4.93 - 0.7v \tag{16.6}$$

4. If $v = 3.1\%$, the predicted value from the regression equation is

$$\begin{aligned}(16.6) \qquad \bar{y}_c &= 4.93 - (0.7)(3.1)\\ &= 4.93 - 2.17\\ &= 2.76 \text{ hundreds} \quad \text{or} \quad 276 \text{ building permits.}\end{aligned}$$

The Standard Error of the Estimate. Any estimate based on data from a sample is subject to error due to the variation in the population sampled. This variation is usually measured by the variance and/or the standard deviation, which is the positive square root of the variance. In dealing with values of a single variable, say y, the population standard deviation is defined by Equation 5.3. This equation is reproduced here with the variable y substituted for x and the subscript y appended to the σ and the μ to indicate the variable being observed.

$$(5.3) \qquad \sigma_y = \sqrt{\frac{\Sigma(y - \mu_y)^2}{n}}$$

The corresponding formula for the sample standard deviation is given by Equation 9.4, which is reproduced below with the variable y substituted for x and the f, which designates frequency, omitted.

$$(9.4) \qquad s_y = \sqrt{\frac{\Sigma(y - \bar{y})^2}{(n - 1)}}$$

Notice that these measures of variation for a single variable measure deviations about the mean of the variable. Also remember that the numerator of the ratio under the square root sign is called the sum of squares, a value also used as a measure of variation.

In simple regression analysis, we are dealing with populations of two variables. The population regression line is a line of means of the distributions of the individual y values for different values of x. A measure of the scatter or dispersion of these individual y's about this line is provided by the STANDARD ERROR OF THE ESTIMATE, defined as

$$(16.9) \qquad \sigma_{y.x} = \sqrt{\frac{\Sigma(y - \mu_{y.x})^2}{N}}$$

The sample standard error of the estimate is defined as

$$(16.10) \qquad s_{y.x} = \sqrt{\frac{\Sigma(y - \bar{y}_c)^2}{n - 2}}$$

Notice that the degrees of freedom for $s_{y.x}$ are $n - 2$. The equation of the sample regression line requires estimates of the two parameters α and β and, therefore, two degrees of freedom are subtracted from n.

As with other formulas for measures of variation, an equivalent and much more convenient computational form for Equation 16.10 is

$$s_{y.x} = \sqrt{\frac{\Sigma y^2 - a\Sigma y - b\Sigma xy}{n-2}} \tag{16.11}$$

Keep in mind the distinction between the sample standard deviation of y, s_y, and the sample standard error of the estimate, $s_{y.x}$. The first measures the deviations of the individual y values around $\bar{y}$, while the second measures the deviations of these values around the regression line, $\bar{y}_c$.

Example 16.4 Using the data of Table 16.2 and the regression equation $\bar{y}_c = 0.1 + 0.15x$,

a) Find the standard error of the estimate for the regression line of building permits on marriage licenses;

b) Find the standard deviation of y and compare its value with $s_{y.x}$.

a)

$$\begin{aligned} s_{y.x} &= \sqrt{\frac{\Sigma y^2 - a\Sigma y - b\Sigma xy}{n-2}} \qquad (16.11) \\ &= \sqrt{\frac{74.44 - (0.1)(25.6) - (0.15)(475)}{10-2}} \\ &= \sqrt{\frac{0.63}{8}} = \sqrt{0.079} \\ &= 0.28 \end{aligned}$$

b)

$$\begin{aligned} s_y &= \sqrt{\frac{\Sigma(y-\bar{y})^2}{n-1}} = \sqrt{\frac{\Sigma y^2}{n-1} - \frac{(\Sigma y)^2}{n(n-1)}} \qquad (9.4) \\ &= \sqrt{\frac{74.44}{9} - \frac{(25.6)^2}{(10)(9)}} \\ &= \sqrt{8.271 - 7.282} = \sqrt{0.989} \\ &= 0.994 \end{aligned}$$

Comparing $s_y = 0.994$ with $s_{y.x} = 0.28$, one can see that the variation of the y values about the regression line is considerably less than their variation about $\bar{y}$. We might interpret this to mean that changes in the number of marriages account for a great deal of the variation in the demand for housing.

Example 16.5 Using the data of Table 16.3 and the regression equation $\bar{y}_c = 4.93 - 0.7v$,

a) Find the standard error of the estimate for the regression line of building permits on unemployment rate;

b) Compare the standard deviation of y determined in Example 16.4 with $s_{y.v}$.

a)

$$(16.11) \qquad s_{y.v} = \sqrt{\frac{\Sigma y^2 - a\Sigma y - b\Sigma vy}{n-2}}$$

$$= \sqrt{\frac{74.44 - (4.93)(25.6) - (-0.7)(78.92)}{10-2}}$$

$$= \sqrt{\frac{74.44 - 126.21 + 55.24}{8}} = \sqrt{\frac{3.47}{8}}$$

$$= \sqrt{0.434} = 0.66.$$

b) $s_y = 0.994$ from the previous example.
$s_{y.v} = 0.66.$

Comparing these two values, one can see that the variation of y about this regression line is less than the variation about $\bar{y}$. However, the difference is much less than that obtained in the previous example. We might conclude from this that changes in the rate of unemployment account for some of the variation in the demand for housing, but not as much as changes in marriages.

The Coefficient of Determination. In addition to its value as a predictor, the regression relationship is also useful for explaining the variation of the dependent variable. In the examples regarding the demand for housing in Buckingham County as measured by numbers of building permits, we observed that the number of permits issued varies substantially from year to year. In order to predict the demand for housing next year, we should also attempt to explain this variation. We have assumed, because of the high value of the correlation coefficient between building permits and marriage licenses, that marriages account for some of the variation in the demand for housing. This assumption has been reinforced by comparing the standard deviation of y with the standard error of the estimate in Example 16.4. It would be very useful, however, to have a measure that indicates exactly how much of the variation in the dependent variable is explained or accounted for by its relationship with the independent variable in the regression equation. In Example 16.4, we found that $s_{y.x}$ was considerably less than s_y, indicating that the variation of y about the regression line was considerably less than its variation about its own mean value. We could, and should, conclude from this that the variable x explains part of the variation in y. We would now like to know how much of the variation in y is explained by x.

The total amount of variation in y can be represented by the variance, s_y^2, which is simply the square of the standard deviation.

Another measure of variation that is related to the standard deviation and the variance is the sum of squares, which we introduced in Chapter 13, Analysis of Variance. Recall that any sample variance is simply a sum of squares divided by the number of degrees of freedom, that is, $s^2 = \mathrm{SS}/df$. Therefore,

$$s_y^2 = \frac{\mathrm{SS}_y}{n - 1}$$

and

$$s_{y.x}^2 = \frac{\mathrm{SS}_{y.x}}{n - 2}$$

where SS_y represents the sum of squares of y about $\bar{y}$ and $\mathrm{SS}_{y.x}$ represents the sum of squares of y about the regression line. We can obtain a sum of squares from a variance by simply multiplying the variance by the number of degrees of freedom, that is,

$$\mathrm{SS}_y = (n - 1)s_y^2 \tag{16.12}$$

and

$$\mathrm{SS}_{y.x} = (n - 2)s_{y.x}^2 \tag{16.13}$$

However, it is not necessary to calculate variances in order to obtain a sum of squares. The sums of squares in regression can be obtained directly from the following formulas.

$$\mathrm{SS}_y = \Sigma y^2 - n\bar{y}^2 \tag{16.14}$$

$$\mathrm{SS}_{y.x} = \Sigma y^2 - a\Sigma y - b\Sigma xy \tag{16.15}$$

If we use the sum of squares as our measure of variation, then SS_y represents the total variation in the dependent variable measured about its own mean. $\mathrm{SS}_{y.x}$ represents the variation of y about the regression line. If $\mathrm{SS}_{y.x}$ is less than SS_y, we can conclude that x has explained some of the variation in y. If x explained *all* the variation in y, then every y value would fall on the regression line and $\mathrm{SS}_{y.x}$ would be zero.

The deviation of the y values from the regression line represents variation not explained by x. This unexplained variation can be measured by $\mathrm{SS}_{y.x}$. Now, if SS_y is the total variation in y before using x, and if $\mathrm{SS}_{y.x}$ is the remaining or unexplained variation in y after using x, then $\mathrm{SS}_y - \mathrm{SS}_{y.x}$ represents the amount of variation explained by x. If we then divide the explained variation by the total variation, we get the *proportion* of the original variation in y that is explained by x, that is,

$$r_{y.x}^2 = \frac{\mathrm{SS}_y - \mathrm{SS}_{y.x}}{\mathrm{SS}_y} \tag{16.16}$$

This value, r^2, is called the COEFFICIENT OF DETERMINATION.

It is no accident that the symbol for the coefficient of determination is r^2 while the symbol for the coefficient of correlation is r.

The coefficient of correlation, which we obtained directly from Equation 16.3, can also be obtained by taking the square root of the coefficient of determination.

Example 16.6

From the data in Examples 16.4 and 16.5, compute the coefficient of determination between building permits and marriage licenses, $r^2_{y.x}$, and between building permits and unemployment rate, $r^2_{y.v}$, and specify the proportion of the variation in the dependent variable that can be explained by each of the independent variables. Also, compute the coefficient of correlation between y and each of the independent variables.

a) Building permits versus marriage licenses. Using the data in Example 16.4 and Equations 16.12 and 16.13,

$$SS_y = (n - 1)s^2_y = (9)(0.989) = 8.901 \tag{16.12}$$

$$SS_{y.x} = (n - 2)s^2_{y.x} = (8)(0.079) = 0.632 \tag{16.13}$$

From Equation 16.16,

$$\begin{aligned} r^2_{y.x} &= \frac{SS_y - SS_{y.x}}{SS_y} \\ &= \frac{8.901 - 0.632}{8.901} \\ &= 0.929 \end{aligned} \tag{16.16}$$

The value of $r^2_{y.x}$ indicates that 92.9% of the variation in numbers of building permits is explained by the number of marriages in the previous year. The coefficient of correlation between y and x is obtained from

$$r_{y.x} = \sqrt{0.929} = 0.964$$

The slight difference between this value and the one obtained from Equation 16.3 in Example 16.1 is due to rounding.

b) Building permits versus unemployment rate.

$$SS_y = 8.901$$

From Example 16.5 and Equation 16.13,

$$SS_{y.v} = (n - 2)s^2_{y.v} = (8)(0.434) = 3.47 \tag{16.13}$$

From Equation 16.16,

$$r^2_{y.v} = \frac{SS_y - SS_{y.v}}{SS_y} = \frac{8.901 - 3.47}{8.901} = 0.61 \tag{16.16}$$

The value of $r^2_{y.v}$ indicates that 61% of the variation in building permits can be attributed to the variation in unemployment rates. The coefficient of correlation between y and v is

$$r_{y.v} = \sqrt{0.61} = 0.78$$

This agrees with the result obtained in Example 16.1, except for the rounding error.

In commenting on the results of the previous example, it may appear to be inconsistent to conclude that marriages seem to account for 93% of the variation in building permits and at the same time that unemployment accounts for about 61%. This seeming inconsistency can be explained by noting that while both marriages and the unemployment rate are related to building permits, they are also probably related to each other as well. The effect of the independent variable in a simple regression also includes the effect of all other independent variables that are correlated with it.

Example 16.7 Several manufacturing employees of Metric Systems, Inc. are to be trained as inspectors. The basis for their selection was to be their scores on an aptitude test. However, in order to determine if the test scores are useful in predicting success on the job, 10 workers are chosen at random and given the test. They are then trained and put to work as inspectors. After a week, their proficiency in the job is measured by their output per shift. Table 16.4 shows the test score, x, and a measure of the average output per shift, y, for each of the 10 employees. From these data, the regression line of y on x is determined and the standard error of the estimate and coefficient of determination are computed as follows.

The regression coefficients are obtained from

$$(16.8) \qquad b = \frac{\Sigma xy - n\bar{x}\bar{y}}{\Sigma x^2 - n\bar{x}^2} = \frac{26{,}890 - 24{,}019}{42{,}875 - 36{,}603} = 0.46$$

$$(16.7) \qquad a = \bar{y} - b\bar{x} = 39.7 - (0.46)(60.5) = 11.9$$

and the equation of the regression line is

$$\bar{y}_c = 11.9 + 0.46x$$

The sample standard error of the estimate is found from the sample values by Equation 16.11.

TABLE 16.4
REGRESSION ANALYSIS OF APTITUDE TEST SCORES VERSUS PRODUCTION EFFICIENCY (DATA FOR COMPUTATION) (EXAMPLE 16.7)

x	y	x^2	y^2	xy
80	70	6400	4900	5600
80	60	6400	3600	4800
45	24	2025	576	1080
70	38	4900	1444	2660
95	45	9025	2025	4275
20	30	400	900	600
50	35	2500	1225	1750
90	50	8100	2500	4500
25	25	625	625	625
50	20	2500	400	1000
605	397	42,875	18,195	26,890

(16.11)
$$s_{y.x} = \sqrt{\frac{\Sigma y^2 - a\Sigma y - b\Sigma xy}{n-2}}$$
$$= \sqrt{\frac{18{,}195 - (11.9)(397) - (0.46)(26{,}890)}{10-2}}$$
$$= \sqrt{\frac{1101.3}{8}}$$
$$= \sqrt{138}$$
$$= 11.7$$

The sum of squares about the regression line is simply

$$SS_{y.x} \doteq 1101$$

In order to compute the coefficient of determination, we must find the sum of squares about the mean value of y; that is,

(16.14)
$$SS_y = \Sigma y^2 - n\bar{y}^2 = 2434$$

The coefficient of determination is simply

$$r^2 = \frac{2434 - 1101}{2434} = 0.55$$

which indicates that only 55% of the variation in performance can be accounted for by the aptitude measured in the test. The coefficient of correlation is

$$r = \sqrt{0.55} \doteq 0.74$$

This value of r could also be computed directly from Equation 16.3:

$$r = \frac{n\Sigma xy - \Sigma xy}{\sqrt{[n\Sigma x^2 - (\Sigma x)^2][n\Sigma y^2 - (\Sigma y)^2]}}$$
$$= \frac{(10)(26{,}890) - (605)(397)}{\sqrt{[(10)(42{,}875) - (605)^2][(10)(18{,}195) - (397)^2]}}$$
$$= \frac{28{,}715}{\sqrt{(62{,}725)(24{,}341)}}$$
$$= \frac{28{,}715}{39{,}074}$$
$$\doteq 0.74.$$

Note that what appears to be a relatively high value of the coefficient of correlation of 0.74 corresponds to a coefficient of determination of only 0.55.

16.4 Confidence Intervals for Predictions

The primary purpose of regression analysis is to make predictions about the dependent variable on the basis of information about the independent variable. The regression line $\bar{y}_c = a + bx$ is an estimated line of means of the conditional distributions of y. Since $\bar{y}_c$

is an estimate based on sample data, a different sample might result in a different regression line and a different value of $\overline{y}_c$. Consequently, predictions or estimates based on the sample regression line should be accompanied by statements regarding the likelihood that the estimated value will be in error by some specified amount. These statements are, in fact, confidence intervals as defined in Chapter 10, Estimation.

Any two-sided confidence interval is of the form

$$v \pm k_{(1-C)/2}\text{SE}(v),$$

where v is simply a point estimate, k is the confidence multiplier corresponding to the level of confidence, C is the confidence level specified, and SE(v) represents the standard error of the statistic used as the estimator.

Confidence intervals about the regression line can be constructed for two types of estimates. The regression line can be used to estimate the average value of the dependent variable for a given value of x. This is an estimate of $\mu_{y.x}$. For the various examples about building permits and marriage licenses, this estimate would answer the question, "If $x = x_0$, what is the average value of y?" On the other hand, we can also obtain a confidence interval estimate of an individual value of y for a given x. This would answer the question, "If $x = x_0$, what will y actually be?" The point estimate in both cases could be $\overline{y}_c$, the point on the regression line corresponding to the specified value of x. However, we know that individual values vary more widely than average values and, therefore, we should expect a confidence interval on the individual values to be wider than a confidence interval on the means. We discuss both types of intervals below.

Confidence Intervals on the Mean, $\mu_{y.x}$. The point estimate of the mean value of y for a given x would be $\overline{y}_c$. Therefore, a confidence interval on this conditional population mean, $\mu_{y.x}$, would be based on $\overline{y}_c$ plus and minus a confidence multiplier, either z or t, times the standard error of $\overline{y}_c$. The standard error of $\overline{y}_c$ is given by

$$s_{\overline{y}_c} = s_{y.x}\sqrt{\frac{1}{n} + \frac{(x - \overline{x})^2}{\Sigma(x - \overline{x})^2}} \tag{16.17}$$

Notice that $s_{y.x}$, the standard error of the estimate, is a part of $s_{\overline{y}_c}$. A more convenient computational form for $s_{\overline{y}_c}$ is

$$s_{\overline{y}_c} = s_{y.x}\sqrt{\frac{1}{n} + \frac{(x - \overline{x})^2}{\Sigma x^2 - n\overline{x}^2}} \tag{16.18}$$

An interval estimate on the average value of y, given x with confidence equal to C, is

$$\overline{y}_c \pm t_{n-2,\,(1-C)/2}s_{\overline{y}_c} \tag{16.19}$$

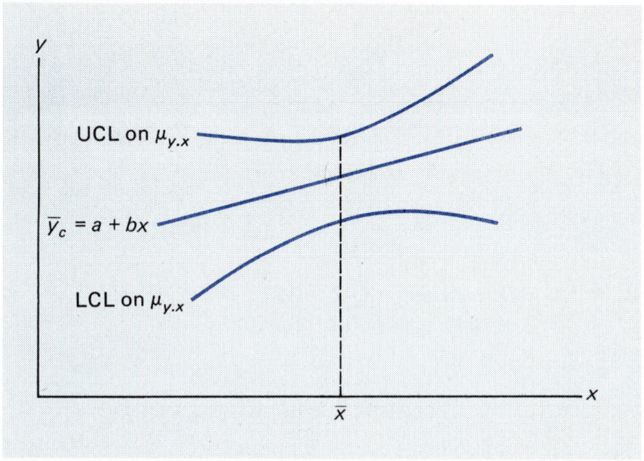

Figure 16.8
Confidence band on the conditional mean.

Of course, whenever the degrees of freedom in the regression are greater than 30, the corresponding z value can be used instead of t, as discussed in Chapters 9 and 10.

If you look closely at Equations 16.17 and 16.18, you will notice that $s_{\bar{y}_c}$ contains the factor $(x - \bar{x})$. The further the specified value of x deviates from $\bar{x}$ in either direction, the larger the value of $s_{\bar{y}_c}$. Therefore, if confidence intervals on the mean are computed for a series of different values of x, a confidence band about the regression line results. This confidence band consists of two curves which are closest at the point where $x = \bar{x}$ and diverge as x deviates from $\bar{x}$. When $x = \bar{x}$, the expression for $s_{\bar{y}_c}$ reduces to $s_{\bar{y}_c} = s_{y.x}/\sqrt{n}$, which is very similar to the standard error of the mean in one-variable statistics. A confidence band on $\mu_{y.x}$ is illustrated in Figure 16.8.

Example 16.8

Find the 95% confidence interval on the average number of building permits issued in Buckingham County in a given year when the number of marriage licenses issued in the previous year was 2000.

1. State the regression equation of building permits on marriage licenses from the section on the least squares regression line and determine the value of $\bar{y}_c$ when x is 20 hundreds.

$$\begin{aligned}\bar{y}_c &= 0.10 + 0.15x \\ &= 0.10 + (0.15)(20) \\ &= 3.1 \text{ hundreds}\end{aligned}$$

2. Compute the value of $s_{\bar{y}}$ from Equation 16.18 and the data provided in Table 16.2 and Example 16.4.

$$s_{\bar{y}_c} = s_{y.x}\sqrt{\frac{1}{n} + \frac{(x - \bar{x})^2}{\Sigma x^2 - n\bar{x}^2}} \quad (16.18)$$

$$s_{y.x} = 0.28, \qquad n = 10, \qquad \bar{x} = 16.4, \qquad \Sigma x^2 = 3068$$

$$(16.18) \quad s_{\bar{y}_c} = 0.28\sqrt{\frac{1}{10} + \frac{(20 - 16.4)^2}{3068 - (10)(16.4)^2}}$$
$$= 0.28\sqrt{0.1 + 0.034}$$
$$= 0.28\sqrt{0.134}$$
$$= (0.28)(0.366)$$
$$= 0.1025$$

3. Find the value of t for a two-sided 95% confidence interval and $n - 2 = 8$ degrees of freedom.

$$t_{8,0.05} = 2.306$$

4. Compute the upper and lower confidence limits on $\mu_{y.x}$ as follows.

$$\text{C.L.} = \bar{y}_c \pm t_{8,0.05}s_{\bar{y}_c}$$
$$= 3.1 \pm (2.306)(0.1025)$$
$$= 3.1 \pm 0.236$$
$$\text{L.C.L.} = 2.86 \text{ hundreds,} \qquad \text{U.C.L.} = 3.34 \text{ hundreds}$$

The 95% confidence interval on the average number of building permits when 2000 marriage licenses were issued in the previous year is 286 to 334.

Confidence Intervals on a Predicted y. If we want to construct a confidence interval on a predicted individual value of y for a given x, we still use $\bar{y}_c$ as the point estimate of this value, which we will designate y_c. Since individual values vary more than means, the confidence interval for y_c will be wider than the confidence interval for means. We construct this interval as follows.

$$(16.20) \qquad \bar{y}_c \pm t_{n-2,\,(1-C)/2}s_{y_c}$$

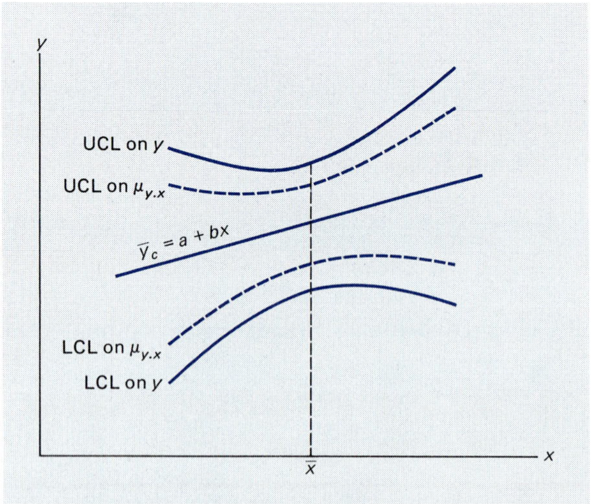

Figure 16.9
Confidence band on an individual value of y.

Notice that this is the same as Equation 16.19, except that s_{y_c} replaces $s_{\bar{y}_c}$. The standard error of the individual values is

$$s_{y_c} = s_{y.x}\sqrt{1 + \frac{1}{n} + \frac{(x - \bar{x})^2}{\Sigma(x - \bar{x})^2}} \tag{16.21}$$

The computational form of Equation 16.21 is

$$s_{y_c} = s_{y.x}\sqrt{1 + \frac{1}{n} + \frac{(x - \bar{x})^2}{\Sigma x^2 - n\bar{x}^2}} \tag{16.22}$$

Figure 16.9 shows the relationship of the confidence band on y to the confidence band on $\mu_{y.x}$.

Example 16.9

Find the 95% confidence interval on a predicted value of y, the number of building permits issued in Buckingham County in a given year, when the number of marriage licenses issued in the previous year was 2000.

1. Determine the value of $\bar{y}_c$ from the regression equation as in Example 16.8. Since the data for this example is the same as that used in the previous example, $\bar{y}_c = 3.1$ hundreds.
2. Compute the value of s_{y_c} from Equation 16.22.

$$s_{y_c} = s_{y.x}\sqrt{1 + \frac{1}{n} + \frac{(x - \bar{x})^2}{\Sigma x^2 - n\bar{x}^2}} \tag{16.22}$$

$$= 0.28\sqrt{1 + \frac{1}{10} + \frac{(20 - 16.4)^2}{3068 - (10)(16.4)^2}}$$

$$= 0.28\sqrt{1.1 + 0.034}$$

$$= 0.28\sqrt{1.134}$$

$$= (0.28)(1.065)$$

$$= 0.298$$

3. Find the value of t for a two-sided 95% confidence interval and $n - 2$ degrees of freedom.

$$t_{8,0.05} = 2.306$$

4. Compute the upper and lower confidence limits on y as follows.

$$\text{C.L.} = \bar{y}_c \pm t_{8,0.05}s_{y_c}$$

$$= 3.1 \pm (2.306)(0.298)$$

$$= 3.1 \pm 0.687$$

$$\text{L.C.L.} = 2.41 \text{ hundreds}, \qquad \text{U.C.L.} = 3.79 \text{ hundreds}$$

The 95% confidence interval on a predicted number of building permits issued when 2000 marriage licenses were issued in the previous year is 241 to 379.

16.5 Multiple Regression

So far we have been attempting to predict the value of one dependent variable, which we have labeled y, on the basis of a single independent variable, x. Referring to Example 16.7, y is some measure of job performance and x is the score on an aptitude test. After completing the simple regression analysis, we found that the coefficient of determination, r^2, is 0.55. By definition, the coefficient of determination indicates the proportion of the variation in the dependent variable that is accounted for by the regression relationship. In this case, we can conclude that only 55% of the variation in job performance is accounted for by the variable measured in the aptitude test. If we want to improve our prediction of performance, we might look for some other variable or variables that help to account for the remaining 46% of the variation in y, which is still unexplained.

When a dependent variable, y, is related in a regression equation to two or more independent variables, we are then performing a multiple regression analysis. We shall continue to limit ourselves to situations in which the relationship between the variables is linear; consequently, the multiple regression equation can be written

$$\bar{y}_c = a + b_1 x_1 + b_2 x_2 + \cdots + b_n x_n$$

We shall also continue to use y as the designation for the dependent variable; x_i will designate the ith independent variable; b_i is the regression coefficient corresponding to the x_i, and a is the value of $\bar{y}_c$ when all $x_i = 0$.

Most of the concepts introduced in multiple regression are simple extensions of those studied in simple regression. For example, in a three-variable multiple regression, one dependent and two independent variables, we will use the following symbols.

$s_{y.12}$ = the standard error of the estimate of y on variables 1 and 2.
$R^2_{y.12}$ = the coefficient of multiple determination.
$R_{y.12}$ = the coefficient of multiple correlation.

The meanings of these new terms and symbols are explained in the following paragraphs.

For illustration, let us continue with the problem of predicting job performance, which we encountered in Example 16.7. The variable y is still the measure of performance in terms of output per shift, and x is now designated x_1 and represents scores on an aptitude test. In an attempt to account for the variation in y not accounted for by the simple regression relationship between y and x_1, a third variable, x_2, will be considered, which represents years of previous experience. The data required for the multiple regression analysis are presented in Table 16.5.

TABLE 16.5
MULTIPLE REGRESSION ANALYSIS OF APTITUDE TEST SCORES AND YEARS OF EXPERIENCE VERSUS PRODUCTION EFFICIENCY (DATA FOR COMPUTATION) (EXAMPLE 16.10)

y	x_1	x_2	y^2	x_1^2	x_2^2	x_1y	x_2y	x_1x_2
70	80	5	4900	6400	25	5600	350	400
60	80	9	3600	6400	81	4800	540	720
24	45	1	576	2025	1	1080	24	45
38	70	0	1444	4900	0	2660	0	0
45	95	4	2025	9025	16	4275	180	380
30	20	2	900	400	4	600	60	40
35	50	3	1225	2500	9	1750	105	150
50	90	6	2500	8100	36	4500	300	540
25	25	2	625	625	4	625	50	50
20	50	2	400	2500	4	1000	40	100
397	605	34	18,195	42,875	180	26,890	1649	2425

Before proceeding with a complicated multiple regression analysis, it is desirable to determine whether there is any relationship at all between y and the new variable, x_2. The first step is to draw a scatter diagram for these two variables and, if a linear relationship is indicated, to use Equation 16.3 to compute the simple correlation coefficient between y and x_2. Assuming that the scatter diagram indicates that a linear relationship does exist between y and x_2, we then compute

$$r_{y.x_2} = \frac{n\Sigma x_2 y - \Sigma x_2 \Sigma y}{\sqrt{[n\Sigma x_2^2 - (\Sigma x_2)^2][n\Sigma y^2 - (\Sigma y)^2]}} \tag{16.3}$$

For the new data given,

$$\begin{aligned} r_{y.x_2} &= \frac{(10)(1649) - (34)(397)}{\sqrt{[10(180) - (34)^2][10(18{,}195) - (397)^2]}} \\ &= \frac{16{,}490 - 13{,}498}{\sqrt{(644)(24{,}341)}} \\ &= \frac{2992}{3959} \\ &\doteq 0.76 \end{aligned}$$

The previous coefficient of correlation between y and what is now designated x_1 was $r_{y.x} \doteq 0.74$.

It is apparent from the sample that x_2 is more closely related to y than x_1, and that using them together in a regression equation should provide a better prediction of y than that obtained by using either one by itself. Therefore, we would proceed with the multiple regression analysis.

The Multiple Regression Equation by Least Squares. The multiple regression equation is the least-squares line fitted to the observed y values. Consequently, the sum of the squares of the vertical distances from the point to the line, $\Sigma(y - \bar{y}_c)^2$, is to be minimized. The regression coefficients of the line are obtained from a set of three normal equations, when three variables are involved. These are

$$\Sigma y = na + b_1\Sigma x_1 + b_2\Sigma x_2 \tag{16.23}$$
$$\Sigma x_1 y = a\Sigma x_1 + b_1\Sigma x_1^2 + b_2\Sigma x_1 x_2$$
$$\Sigma x_2 y = a\Sigma x_2 + b_1\Sigma x_1 x_2 + b_2\Sigma x_2^2$$

There are several methods for solving sets of simultaneous equations. For three equations in three unknowns, an algebraic solution by hand is relatively simple. For larger sets of equations it is best to resort to the use of a computer.

Example 16.10 For the data given in Table 16.5, the normal equations are

$$397 = 10a + 605b_1 + 34b_2$$
$$26{,}890 = 605a + 42{,}875b_1 + 2425b_2$$
$$1649 = 34a + 2425b_1 + 180b_2$$

Step 1: Eliminate a from the three equations. Multiply the first equation by 17 and the third equation by 5, resulting in

$$6749 = 170a + 10{,}285b_1 + 578b_2$$
$$8245 = 170a + 12{,}125b_1 + 900b_2$$

Subtracting the top equation from the bottom one, we get

$$1496 = 1840b_1 + 322b_2$$

Multiplying the first equation by 121 and the second equation by 2 results in:

$$48{,}037 = 1210a + 73{,}205b_1 + 4114b_2$$
$$53{,}780 = 1210a + 85{,}750b_1 + 4850b_2$$

Subtracting the top equation from the bottom one, we get

$$5743 = 12{,}545b_1 + 736b_2$$

Step 2: Eliminate b_2 from the two resulting equations.

$$5743 = 12{,}545b_1 + 736b_2$$
$$1496 = 1840b_1 + 322b_2$$

Multiply the first equation by 14 and the second equation by 32.

$$80{,}402 = 175{,}630b_1 + 10{,}304b_2$$
$$47{,}872 = 58{,}880b_1 + 10{,}304b_2$$

Subtracting the bottom equation from the top one, we get

$$32{,}530 = 116{,}750b_1$$

or

$$b_1 = \frac{32{,}530}{116{,}750} \doteq 0.278$$

Step 3: Substitute the value for b_1 into one of the equations in b_1 and b_2 and solve for b_2.

$$\begin{aligned} 1496 &= 1840b_1 + 322b_2 \\ &= 1840(0.278) + 322b_2 \\ &= 511.52 + 322b_2 \\ 322b_2 &= 984.48 \\ b_2 &= 3.057 \end{aligned}$$

Step 4: Substitute the values for b_1 and b_2 into one of the equations in a, b_1, and b_2, and solve for a.

$$\begin{aligned} 397 &= 10a + 605b_1 + 34b_2 \\ &= 10a + 605(0.278) + 34(3.057) \\ 10a &= 397 - 168.19 - 103.938 \\ &= 124.872 \\ a &= 12.487 \end{aligned}$$

Step 5: Substitute the values for a, b_1, and b_2 into the regression equation.

$$\bar{y}_c = 12.487 + 0.278x_1 + 3.057x_2$$

The Standard Error of the Estimate—Multiple Regression. The standard error of the estimate in multiple regression measures the variation of the observed y values about the regression line just as in simple regression. If we could compute the regression line from the entire population of values, the regression equation would provide $\mu_{y.12} = \alpha + \beta_1 x_1 + \beta_2 x_2$, and the standard error of the estimate would be

$$\sigma_{y.12} = \sqrt{\frac{\Sigma(y - \bar{y}_c)^2}{N}}$$

As in the case of simple regression, however, the estimating equation is based on a sample from the population. Therefore, instead of having values for α, β_1, and β_2, estimates of these parameters, a, b_1, and b_2, are computed from the sample. Consequently, the sample standard error of the estimate is defined as

$$s_{y.12} = \sqrt{\frac{\Sigma(y - \bar{y}_c)^2}{n - 3}}$$

where one degree of freedom is subtracted from n in the denominator for each of the three parameters estimated from the sample values.

A more convenient computational form for the standard error of the estimate is

$$s_{y.12} = \sqrt{\frac{\Sigma y^2 - a\Sigma y - b_1\Sigma x_1 y - b_2\Sigma x_2 y}{n - 3}} \qquad (16.24)$$

For the data in Example 16.10,

$$s_{y.12} = \sqrt{\frac{18{,}195 - (12.487)(397) - (0.278)(26{,}890) - (3.057)(1649)}{7}}$$

$$= \sqrt{\frac{18{,}195 - 4957.339 - 7475.42 - 5040.993}{7}}$$

$$= \sqrt{\frac{721.248}{7}}$$

$$= \sqrt{103.035}$$

$$= 10.15$$

We can now compare the standard error of the multiple regression with the standard error of the simple regression, which was computed in Example 16.6:

$$s_{y.12} = 10.15 < 11.70 = s_{y.x_1}$$

It is apparent that by introducing the second independent variable, the variation in the observed y values has been reduced. This is even more obvious when we compare the sums of the squared deviations in the y values about $\bar{y}$, about the simple regression line, and about the multiple regression line. The variation of the y values can best be expressed in terms of the sum of squares.

$$SS_y = 2434$$
$$SS_{y.x_1} \doteq 1101$$
$$SS_{y.12} \doteq 721$$

where SS_y is the sum of squares about $\bar{y}$, $SS_{y.x_1}$ is the sum of squares about the simple regression line, and $SS_{y.12}$ is the sum of squares about the multiple regression line.

It is evident that the variation in y, as represented by the sum of squares, is much less about the simple regression line than it is about $\bar{y}$. There is even less variation about the multiple regression line than about the simple regression of y on x_1. However, as in the simple regression case, it is desirable to have some measure of the proportion of variation in the dependent random variable, which is explained by the multiple regression relationship. This measure is provided by the coefficient of multiple determination.

The Coefficients of Multiple Determination and Correlation. The COEFFICIENT OF MULTIPLE DETERMINATION is similar to the coefficient of determination in simple regression in that it measures the proportion of the variation in y, which is accounted for by the regression equation.

The sum of squares about $\bar{y}$, SS_y, still represents the total variation in y. The sum of squares about the regression line, $SS_{y.12}$, represents the variation still unaccounted for after utilizing the multiple regression relationship. The ratio

$$R^2_{y.12} = \frac{SS_y - SS_{y.12}}{SS_y} \tag{16.25}$$

is the coefficient of multiple determination, which is the proportion of the variation in y that is accounted for by the multiple regression relationship.

The square root of the coefficient of multiple determination is the coefficient of multiple correlation. Thus

$$R = \sqrt{R^2}$$

For the data in Example 16.10, $R^2 = (2434 - 721)/2434 = 0.704$, indicating that 70.4% of the variation in y has been accounted for by the relationship between y and x_1 and x_2 taken together in the regression equation. This should be compared with the coefficient of simple determination between y and x_1 computed in Example 16.7, which was $r^2 = 0.550$.

A considerable amount of the variation in y, which was still unaccounted for after the simple regression relationship was established, has now been explained. Even now, however, only 70.4% of the variation in y has been accounted for. This leaves 29.6% of the variation still unaccounted for, a situation which might lead us to look for still another independent variable related to y. However, we will not undertake that exercise here.

The coefficient of multiple correlation for this example is simply

$$R = \sqrt{R^2} = \sqrt{0.704} = 0.84$$

Definitions

Coefficient of correlation: A statistical measure of the degree or strength of the association between two variables.

Coefficient of determination: A statistical measure of the proportion of the variation in the dependent variable that is explained by the regression equation.

Coefficient of multiple correlation: A statistical measure of the degree of association between a dependent variable and two or more independent variables.

Coefficient of multiple determination: A statistical measure of the proportion of the variation in the dependent variable that is explained by the regression equation with two or more independent variables.

Correlation analysis: A statistical method for measuring the strength of association between any two random variables.

Regression analysis: A statistical method for determining the type of relationship that exists among two or more variables and for using that relationship to make estimates or predictions.

Scatter diagram: A chart in which pairs of values (x, y) are plotted to provide a visual display of the pattern of their relationship.

Standard error of the estimate: A measure of the variation of the observed values about the regression line.

Formulas

Population coefficient of correlation:

$$\rho = \frac{\Sigma(x - \mu_x)(y - \mu_y)}{N\sigma_x\sigma_y} \tag{16.1}$$

Sample coefficient of correlation:

$$r_{y.x} = \frac{\Sigma(x - \bar{x})(y - \bar{y})}{\sqrt{\Sigma(x - \bar{x})^2\Sigma(y - \bar{y})^2}} \tag{16.2}$$

Sample coefficient of correlation, computational form:

$$r_{y.x} = \frac{n\Sigma xy - \Sigma x\Sigma y}{\sqrt{[n\Sigma x^2 - (\Sigma x)^2][n\Sigma y^2 - (\Sigma y)^2]}} \tag{16.3}$$

Transformation of r to a t statistic:

$$t = \frac{r\sqrt{n - 2}}{\sqrt{1 - r^2}} \tag{16.4}$$

Population regression line:

$$\mu_{y.x} = \alpha + \beta x \tag{16.5}$$

Sample regression line:

$$\bar{y}_c = a + bx \tag{16.6}$$

Sample regression coefficients:

$$a = \bar{y} - b\bar{x} \tag{16.7}$$

$$b = \frac{\Sigma xy - n\bar{x}\bar{y}}{\Sigma x^2 - n\bar{x}^2} \tag{16.8}$$

Population standard error of the estimate:

$$\sigma_{y.x} = \sqrt{\frac{\Sigma(y - \mu_{y.x})^2}{N}} \tag{16.9}$$

Sample standard error of the estimate:

$$s_{y.x} = \sqrt{\frac{\Sigma(y - \bar{y}_c)^2}{n - 2}} \tag{16.10}$$

Sample standard error of the estimate, computational form:

$$s_{y.x} = \sqrt{\frac{\Sigma y^2 - a\Sigma y - b\Sigma xy}{n - 2}} \tag{16.11}$$

Sums of squares:

(16.12) $$SS_y = (n-1)s_y^2$$

(16.13) $$SS_{y.x} = (n-2)s_{y.x}^2$$

(16.14) $$SS_y = \Sigma y^2 - n\bar{y}^2$$

(16.15) $$SS_{y.x} = \Sigma y^2 - a\Sigma y - b\Sigma xy$$

Sample coefficient of determination:

(16.16) $$r_{y.x}^2 = \frac{SS_y - SS_{y.x}}{SS_y}$$

Standard error of $\bar{y}_c$:

(16.17) $$s_{\bar{y}_c} = s_{y.x}\sqrt{\frac{1}{n} + \frac{(x-\bar{x})^2}{\Sigma(x-\bar{x})^2}}$$

Standard error of $\bar{y}_c$, computational form:

(16.18) $$s_{\bar{y}_c} = s_{y.x}\sqrt{\frac{1}{n} + \frac{(x-\bar{x})^2}{\Sigma x^2 - n\bar{x}^2}}$$

Interval estimate on $\bar{y}_c$:

(16.19) $$\bar{y}_c \pm t_{n-2,\,(1-C)/2}s_{\bar{y}_c}$$

Interval estimate on y_c:

(16.20) $$\bar{y}_c \pm t_{n-2,\,(1-C)/2}s_{y_c}$$

Standard error of y_c:

(16.21) $$s_{y_c} = s_{y.x}\sqrt{1 + \frac{1}{n} + \frac{(x-\bar{x})^2}{\Sigma(x-\bar{x})^2}}$$

Standard error of y_c, computational form:

(16.22) $$s_{y_c} = s_{y.x}\sqrt{1 + \frac{1}{n} + \frac{(x-\bar{x})^2}{\Sigma x^2 - n\bar{x}^2}}$$

Normal equations for three-variable multiple regression:

(16.23) $$\begin{aligned} \Sigma y &= na + b_1\Sigma x_1 + b_2\Sigma x_2 \\ \Sigma x_1 y &= a\Sigma x_1 + b_1\Sigma x_1^2 + b_2\Sigma x_1 x_2 \\ \Sigma x_2 y &= a\Sigma x_2 + b_1\Sigma x_1 x_2 + b_2\Sigma x_2^2 \end{aligned}$$

Sample standard error of the estimate, three-variable multiple regression, computational form:

(16.24) $$s_{y.12} = \sqrt{\frac{\Sigma y^2 - a\Sigma y - b_1\Sigma x_1 y - b_2\Sigma x_2 y}{n-3}}$$

Coefficient of multiple determination:

(16.25) $$R_{y.12}^2 = \frac{SS_y - SS_{y.12}}{SS_y}$$

Problems *Note to the student:* Results in regression analysis are frequently sensitive to rounding errors. Therefore, all intermediate results should be carried out to at least three decimal places and carried as such from step to step. This also applies when data from one problem is used in some subsequent problem.

16.1 For the following pairs of data, plot a scatter diagram and decide whether or not a linear relationship exists between the variables.

Pair:	1	2	3	4	5	6	7	8	9	10
x:	8	7	9	1	11	1	4	8	2	6
y:	4	4	10	0	9	2	4	9	6	5

16.2 For the following pairs of data, plot a scatter diagram and decide whether or not a linear relationship exists between the variables.

Pair:	1	2	3	4	5	6	7	8	9	10
x:	7	10	6	9	4	5	0	1	1	3
y:	0	3	7	1	8	6	9	8	10	7

16.3 For the following pairs of data, plot a scatter diagram and decide whether or not a linear relationship exists between the variables.

Pair:	1	2	3	4	5	6	7	8	9	10
x:	4	3	0	8	1	10	4	9	7	6
y:	1	7	9	9	10	2	4	5	3	9

16.4 Using Equation 16.3, compute the coefficient of correlation between the variables in problem 16.1.

16.5 Using Equation 16.3, compute the coefficient of correlation between the variables in problem 16.2.

16.6 Using Equation 16.3, compute the coefficient of correlation between the variables in problem 16.3.

16.7 The Ego Cosmetics Company wants to determine if a relationship exists between advertising expenditures and sales. The following data were compiled for six major sales regions, where x represents advertising expenditures in thousands of dollars and y represents sales in millions of dollars. Compute the coefficient of correlation between these two variables.

Region	x	y
1	50	35
2	55	40
3	100	50
4	30	30
5	25	35
6	60	40

16.8 Metric Systems, Inc. wants to find a shortcut method for estimating fabrication costs of complex electronic systems. This shortcut could save the company thousands of dollars in preparing proposals for potential customers. The following data were obtained on nine previously manufactured systems. The variable y represents fabrication time in hundreds of hours, and x represents the number of semiconductors used in the system in hundreds. Compute the coefficient of correlation between these two variables.

System	x	*y*
1	1.80	2.79
2	2.70	5.50
3	1.44	3.55
4	1.60	3.57
5	1.44	3.64
6	1.80	3.42
7	2.00	3.01
8	4.29	5.31
9	7.80	9.44

16.9 For the data in problem 16.4, test the hypothesis that $\rho = 0$ at $\alpha = 0.05$.

16.10 For the data in problem 16.5, test the hypothesis that $\rho = 0$ at $\alpha = 0.05$.

16.11 For the data in problem 16.7, test the hypothesis that $\rho = 0$ at $\alpha = 0.01$.

16.12 For the data in problem 16.8, test the hypothesis that $\rho = 0$ at $\alpha = 0.01$.

16.13 **a)** Compute the coefficients of the regression line of y on x for the data in problem 16.1.
b) Using the regression equation, find $\bar{y}_c$ when $x = 3$.

16.14 **a)** Compute the coefficients of the regression line of y on x for the data of problem 16.2.
b) Using the regression equation, find $\bar{y}_c$ when $x = 8$.

16.15 Find the coefficients of the regression line of sales on advertising expenditures for the Ego Cosmetics Company data provided in problem 16.7. Predict sales if advertising expenditures are $40,000 and compute the standard error of the estimate, $s_{y.x}$.

16.16 Find the regression equation of fabrication hours on the number of semiconductors for the Metric Systems, Inc. data provided in problem 16.8. Estimate fabrication hours for a system containing 150 semiconductors and compute the standard error of the estimate, $s_{y.x}$.

16.17 From the data of problem 16.1 and your answers to problem 16.13, determine the following.
a) The standard error of the estimate of y on x;
b) the coefficient of determination of y on x using Equation 16.16.

16.18 From the data of problem 16.2 and your answers to problem 16.14, determine the following.
a) The standard error of the estimate of y on x;
b) the coefficient of determination of y on x using Equation 16.16.

16.19 Referring to problems 16.7 and 16.15 on the Ego Cosmetics Company, determine the following.
a) The coefficient of determination of sales on advertising expenditures using Equation 16.16.

b) Interpret the meaning of the coefficient of determination for this problem.

16.20 From the Ego Cosmetics Company data provided in problems 16.7 and 16.15:

a) Predict sales if advertising expenditures are \$35,000.
b) Find the 95% confidence interval on this predicted value of sales.
c) Find the 95% confidence interval on $\bar{y}_c$, the average sales when advertising expenditures are \$35,000.

16.21 From the Metric Systems, Inc. data on fabrication hours and numbers of semiconductors provided in problems 16.8 and 16.16, determine the following.

a) The coefficient of determination of y on x using Equation 16.16.
b) Interpret the meaning of the coefficient of determination in this problem.

16.22 Using the Metric Systems, Inc. data of problems 16.8 and 16.16:

a) Predict the number of fabrication hours required if the number of semiconductors in a system is 300.
b) Find a 99% confidence interval on the predicted value.
c) Find a 99% confidence interval on $\bar{y}_c$, the average number of fabrication hours when the number of semiconductors is 300.

16.23 The following data provide aptitude test scores and sales during one particular week for ten sales representatives of the Apex Specialty Company. If the dependent variable y is sales and the independent variable x is test score, determine the following.

a) The regression line of y on x;
b) The standard error of the estimate of y on x;
c) The coefficient of determination of y on x.
d) Interpret the meaning of the coefficient of determination in the context of this problem.
e) Predict sales, given an aptitude test score of 55.
f) Construct a 99% confidence interval on this prediction.
g) Construct a 99% confidence interval on $\bar{y}_c$, the average sales for a test score of 55.

Sales Representative	*Test Score*	*Sales (in thousands of dollars)*
1	80	4.0
2	60	3.0
3	60	4.5
4	40	3.5
5	70	6.0
6	50	2.0
7	90	6.0
8	50	2.5
9	50	4.0
10	40	4.5

16.24 For the following data determine:

a) The multiple regression equation of y on x_1 and x_2.

b) The standard error of the estimate, $s_{y.12}$.

c) The coefficient of multiple determination, $R^2_{y.12}$.

d) A predicted value of y if $x_1 = 2.5$ and $x_2 = 8$.

y	x_1	x_2
10	1	6
12	2	7
15	3	9
17	4	10
20	5	12

16.25 After conducting a simple regression analysis on sales and advertising expenditures, Ego Cosmetics Company decided to add another independent variable to the regression equation in order to account for variation in sales which was not explained by the variation in advertising expenditures. The new variable in the equation is the size of the population in the sales region. The data are provided below. The variable y still represents sales in millions of dollars. The variable x_1 represents advertising expenditures in thousands of dollars and x_2 represents population in millions.

a) Find the multiple regression equation of y on x_1 and x_2.

b) Find the standard error of the estimate, $s_{y.12}$.

c) Find the coefficient of multiple determination, $R^2_{y.12}$ and interpret its meaning in the context of this problem.

d) Predict sales if $x_1 = \$40{,}000$ and $x_2 = 2$ million.

y	x_1	x_2
35	50	5.0
40	55	4.0
50	100	8.0
30	30	1.5
35	25	3.5
40	60	1.0

16.26 Eastern Auto Stores, Inc. operates a chain of stores which sell automobile parts and accessories in small towns throughout the east and midwest. Corporate management believes that sales are affected by the number of automobile registrations in the marketing area served by each store, and also by the number of competitive stores in the area. Data were obtained from five towns in which the company had stores. These data are presented below. The variable y represents sales in hundred thousands of dollars, x_1 represents automobile registrations in ten thousands, and x_2 represents the number of competitors within a 30-mile radius of the Eastern Auto Store.

y	x_1	x_2
5	2	2
7	3	0
8	4	3
4	3	1
9	4	2

a) Determine the regression equation of y on x_1 and x_2.
b) Compute the standard error of the estimate, $s_{y.12}$.
c) Determine the proportion of the variation in sales which is explained by the two independent variables in the multiple regression equation.
d) Estimate sales when there are 25,000 automobile registrations and two competitors in the area.

NO SALE
SNOW SHOE
SALE

CHAPTER 17 TIME SERIES

17.1
The time series model

17.2
Trend

17.3
The cycle

17.4
The seasonal index

17.5
Forecasting

A TIME SERIES is a set of numerical observations taken at different points in time. These observations often vary from time period to time period. The basic problem in the study of time series is to analyze the nature of these variations and possibly to use the information obtained to predict future values in the series.

In Chapter 16 we studied regression analysis. In some ways, a time series resembles a regression relationship between two variables. In simple regression, one of the variables is called the independent variable and the other the dependent variable. In time series we are also interested in the relationship between two variables one of which, the independent variable, is time.

The basic assumption in time series analysis is that the values of the dependent variable are a function of time. This is certainly true for certain kinds of physical phenomena. For example, the growth of a living organism is certainly a function of time. The decay of a radioactive substance is also dependent upon time. However, we are primarily interested in the behavior of business and economic variables such as sales, production rates, and prices. There is no doubt that the values of these variables change with the passage of time, but whether a stable functional relationship between time and an economic variable can be determined and used for predicting future values is still debatable.

The ability to forecast future events with accuracy would be extremely valuable to a business executive or manager. Future sales, prices, interest rates, and the supply of labor and raw materials all have a bearing on the outcome of current decisions. Time series analysis provides an approach to forecasting such occurrences that

TABLE 17.1
ASSOCIATED DEPARTMENT STORES, INC.: SALES IN MILLIONS OF DOLLARS, 1964–1978 (EXAMPLE 17.1)

Year	*First Quarter*	*Second Quarter*	*Third Quarter*	*Fourth Quarter*	*Total for Year*
1964	2.97	4.95	3.96	7.92	19.8
1965	1.71	5.14	4.92	9.63	21.4
1966	4.75	6.50	7.00	6.75	25.0
1967	4.00	5.15	8.58	10.87	28.6
1968	3.68	6.12	4.90	9.80	24.5
1969	3.20	3.34	2.89	5.77	15.2
1970	2.26	5.07	4.88	6.59	18.8
1971	4.29	6.37	5.20	10.14	26.0
1972	5.83	9.25	7.43	15.09	37.6
1973	4.65	8.73	7.74	14.08	35.2
1974	6.30	10.62	8.65	16.73	42.3
1975	6.92	12.49	10.18	18.91	48.5
1976	8.70	13.06	10.80	21.44	54.0
1977	8.23	13.87	11.20	22.20	55.5
1978	9.05	13.50	12.10	23.35	58.0

can be useful, particularly in the context of short-term planning.

Time series analysis, however, is also of value in describing and explaining the past behavior of economic variables. For example, if a department store could determine the nature of seasonal fluctuations in sales it would be able to use this information to allocate its advertising budget over the course of the year in the most effective way.

We will use a single set of data to illustrate a time series analysis in this chapter. The data, provided in Table 17.1, represent gross sales of Associated Department Stores, Inc. over a period of 15 years. Associated Department Stores is a company that operates a chain of department stores throughout the United States. Sales data from all stores are aggregated and compiled quarterly in the New York office. Table 17.1 shows sales in millions of dollars for the years 1964–1978 by quarter and by year. We will refer to data from this table in the following examples.

17.1 The Time Series Model

The variations that occur in a time series are usually viewed as being the result of four factors or components: the SECULAR TREND, the SEASONAL VARIATION if the data are reported for time periods shorter than one year, the CYCLE if it exists, and RANDOM or IRREGULAR FLUCTUATIONS.

The secular trend, or simply the trend, refers to the long-term behavior of the dependent variable. The variation due to trend is the result of gradually changing factors such as population growth, technological advances, changes in consumer buying habits, and so on. Of the four components of variation in a time series, trend is probably the most important, particularly in forecasting.

Seasonal variations are patterns of variation that occur within the period of one year and recur with regularity year after year. The causes of seasonal variation are changes in the weather, the occurrence of holidays, and the conventions or customs of the population. Some examples of variables that exhibit seasonality are fuel oil consumption, which peaks during the cold winter months and then declines; electric power production, which increases in the cold and hot months of the year and decreases in the milder months of spring and fall; and retail sales, which tend to peak just before certain holidays such as Christmas and Mother's Day. Seasonal variation is second in practical importance only to the trend in time series analysis.

The cyclical component, or cycle, may or may not exist for a particular time series. When there is a cycle, it is characterized as a periodic growth followed by a decline in the values of the dependent variable, which occurs with regularity over a period of from two to 15 or more years. When evidence of a cycle is present in economic time series, it is seldom regular enough to be used in forecasting. Unless a regular cycle can be described and then projected forward to a specific point in time, it becomes more of a hindrance than a help in forecasting.

Irregular variation is variation in the dependent variable due to random causes. Unlike the other three types of variation, there is no fixed pattern to its behavior. Irregular variations are unpredictable. Wars, natural disasters such as earthquakes and floods, unseasonable weather patterns, and fads in fashion or behavior are all examples of unpredictable factors that can cause variation in a time series. The best that a time series analysis can do with respect to the irregular component is to attempt to eliminate it so that the remaining three components can be studied and described.

A graph of a typical time series with a trend line superimposed is illustrated in Figure 17.1. Figure 17.2 shows deviations of the cyclical and seasonal components about the trend line, which is drawn as a straight horizontal line. In this figure, the cycle is the smooth curve, while the seasonal component is represented by the jagged line. Notice that the seasonal pattern repeats itself every year, while it takes five years in this illustration to complete one cycle.

The most commonly used time series model is multiplicative, that is, the value of the dependent variable is considered to be the product of these four components. If we let y equal the value of the dependent variable, then

$$y = T \cdot S \cdot C \cdot I \tag{17.1}$$

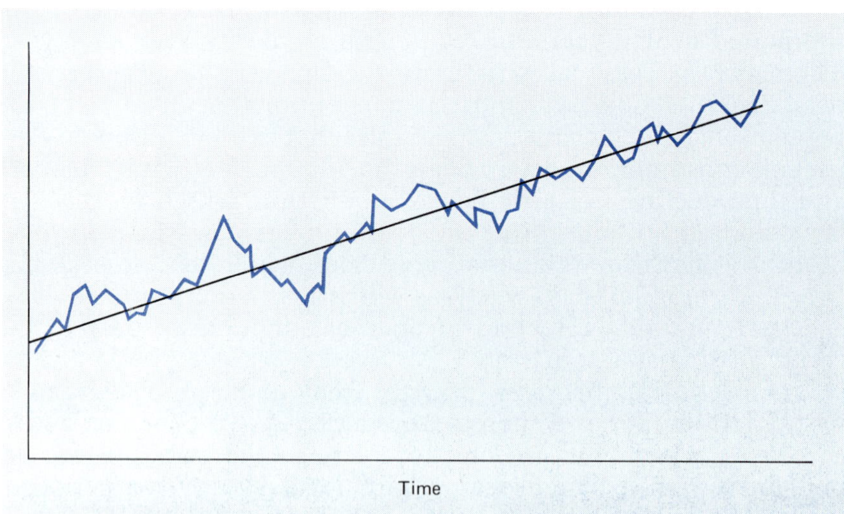

Figure 17.1
A time series with the trend superimposed.

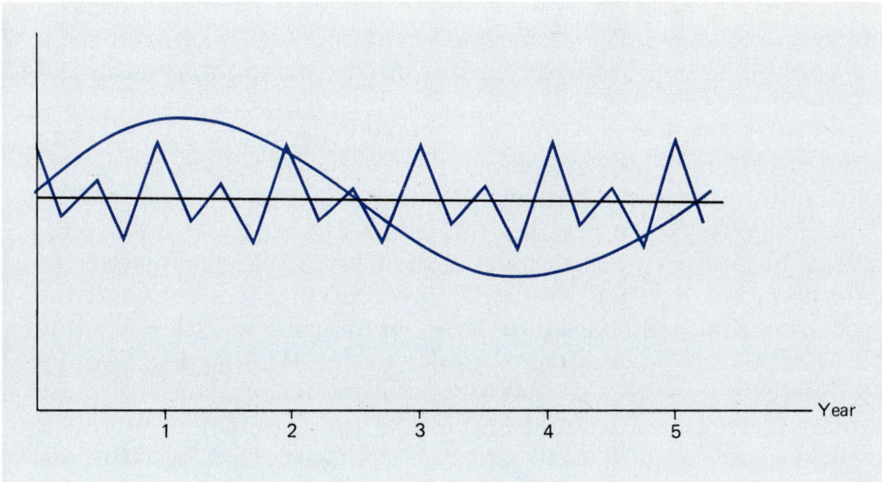

Figure 17.2
The seasonal and cyclical components of a time series shown as deviations from the trend.

where

- T = that part of the value of the dependent variable which is attributable to the trend. T is stated in the same units as y.
- S = that part of the value of the dependent variable which is attributable to the seasonal effect. S is stated as an index number.
- C = that part of the value of the dependent variable which is attributable to the cycle. C is stated as an index number.
- I = that part of the value of the dependent variable which is attributable to random causes. If I is quantified, it is stated as an index number.

For a series composed only of annual data, there is no seasonal component. In that case, the time series model becomes

$$y = T \cdot C \cdot I \tag{17.2}$$

Any component of the multiplicative time series model can be removed by the process of division. For example, if we obtain a value for the seasonal component, we can remove it from an observed value of the dependent variable as follows.

$$\frac{y}{S} = \frac{T \cdot S \cdot C \cdot I}{S} = T \cdot C \cdot I$$

Trend can be removed in the same way, that is,

$$\frac{T \cdot C \cdot I}{T} = C \cdot I$$

and so on for each component. The process of isolating each component of a time series is called DECOMPOSITION.

17.2 Trend

The trend refers to the general movement of a time series over a long period of time. For some time series, this general movement is well described by a straight line. For others, it should be described by one of several types of curves, including an exponential or power curve, some form of polynomial, or one of the several standardized trend curves such as the Gompertz curve or the logistic curve. We will consider only the straight line trend in this chapter.

In analyzing a time series for trend, the analyst must make several preliminary decisions. Included in these decisions are the type of curve that will be used to represent the trend and the procedure that will be used to define or describe it. The procedure for selecting a type of curve is similar to that described in Chapter 16, Regression Analysis; that is, a scatter diagram or a diagram similar to Figure 17.1 could be plotted and a judgment could be made by the analyst as to the type of curve that best fits the observations. We will use the method of least squares, which was explained in Chapter 16, to compute the equation of the trend line.

Least Squares Trend Line. A linear equation is an equation of the form $y = a + bx$, where a is the value of y when $x = 0$ and b is the slope of the line. In linear regression, we used the equation

$$\bar{y}_c = a + bx \quad (16.6)$$

where the subscript c indicated that this was a computed value of y and represented the estimated mean of a population of y values. We will use a similar expression for the trend line. The linear trend equation is

$$y_c = a + bx \quad (17.3)$$

The only difference between Equations 16.6 and 17.3 is that there is no bar over the y, since we do not assume that there is a population of y's at each value of x from which a sample is taken. In the trend equation, x represents the independent variable which, in this case, is time.

Given a set of observed values of the dependent variable y and the corresponding values of x, the time at which each of the y values was observed, the coefficients a and b of the linear trend line can be obtained from the following equations.

$$b = \frac{\Sigma xy - n\bar{x}\bar{y}}{\Sigma x^2 - n\bar{x}^2} \quad (16.8)$$

$$a = \bar{y} - b\bar{x} \quad (16.7)$$

Notice that these equations are identical to those used in Chapter 16.

In the equation for the trend, the variable x represents time. If the data is annual, for example, x could be 1971, 1972, 1973, and so on. Calculations can be simplified by using coded values for x, that is, we could assign the value 0 to the first year observed, the value 1 to the next, and so on. Even better, if we assign the value 0 to the middle year in the series, we can reduce the magnitude of the values used in calculating the coefficients of the trend line even further. We used this procedure to obtain shortcuts for computing means and standard deviations in Chapters 4 and 5. The linear trend equation for coded data is

$$y_c = a + bt \tag{17.4}$$

When the middle year in a time series is designated 0, then each year before the middle year is -1, -2, -3, etc., and each year after the middle year is 1, 2, 3, etc. Under this procedure, if we let t represent the coded value of time, then $\Sigma t = 0$.

Suppose that data from the years 1971 through 1977 are obtained. If x represents the year and t is its coded value, the relationship between t and x is as follows.

x:	1971	1972	1973	1974	1975	1976	1977
t:	-3	-2	-1	0	1	2	3

There is a slight complication when the series contains an even number of years. In that case, there is no single middle year and the zero point has to fall halfway between the two middle years. As an illustration, suppose that data for 1978 were added to the previous series. We would then have

x:	1971	1972	1973	1974	1975	1976	1977	1978
t:	$-3\frac{1}{2}$	$-2\frac{1}{2}$	$-1\frac{1}{2}$	$-\frac{1}{2}$	$\frac{1}{2}$	$1\frac{1}{2}$	$2\frac{1}{2}$	$3\frac{1}{2}$

In order to avoid having fractional values of t when there are an even number of years, it is common practice to state the deviations from 0 in units of one-half year. Therefore, for the eight years listed above, the coded values are

x:	1971	1972	1973	1974	1975	1976	1977	1978
t:	-7	-5	-3	-1	1	3	5	7

The computations of the constants a and b from the coded values are the same whether there is an odd or even number of years, except that the constant a now represents the computed value of trend at a point halfway between the two middle years for the series containing an even number of terms, and b is the change in the trend for each half-year period.

Since, for data coded in this manner, $\Sigma t = 0$ and, therefore, $\bar{t} = 0$, the equations for a and b reduce to

$$a = \bar{y} \tag{17.5}$$

$$b = \frac{\Sigma ty}{\Sigma t^2} \tag{17.6}$$

which are simplifications of Equations 16.7 and 16.8.

Example 17.1 From the Associated Department Stores annual sales data provided in column 6 of Table 17.1, find the least squares trend equation and compute y_c for the years 1964 through 1978.

The data for this problem and the computed values of y_c are provided in Table 17.2. Columns 1 and 2 simply list the years, x, and coded values of the years, t. Column 3 provides the annual sales data from Table 17.1. Columns 4 and 5 simply provide ty and t^2 for each year. From these data, we see that:

$$\Sigma t = 0 \qquad \text{and, therefore, } \bar{t} = \frac{\Sigma t}{n} = 0$$

$$\Sigma y = 510.4 \qquad \text{and, therefore, } \bar{y} = \frac{\Sigma y}{n} = \frac{510.4}{15} = 34.03$$

$$\Sigma ty = 809.8$$

$$\Sigma t^2 = 280$$

Then, from Equations 17.6 and 17.5,

$$b = \frac{\Sigma ty}{\Sigma t^2} = \frac{809.8}{280} = 2.89 \tag{17.6}$$

TABLE 17.2

ASSOCIATED DEPARTMENT STORES' ANNUAL SALES IN MILLIONS OF DOLLARS, 1964–1978 (DATA FOR COMPUTATION OF TREND) (EXAMPLE 17.1)

1 *Year* (x)	2 *Coded Year* (t)	3 *Sales* (y)	4 ty	5 t^2	6 *Trend Value* y_c
1964	−7	19.8	−138.6	49	13.80
1965	−6	21.4	−128.4	36	16.69
1966	−5	25.0	−125.0	25	19.58
1967	−4	28.6	−114.4	16	22.47
1968	−3	24.5	−73.5	9	25.36
1969	−2	15.2	−30.4	4	28.25
1970	−1	18.8	−18.8	1	31.14
1971	0	26.0	0	0	34.03
1972	1	37.6	37.6	1	36.92
1973	2	35.2	70.4	4	39.81
1974	3	42.3	126.9	9	42.70
1975	4	48.5	194.0	16	45.59
1976	5	54.0	270.0	25	48.48
1977	6	55.5	333.0	36	51.37
1978	7	58.0	406.0	49	54.26
Sums	0	510.4	809.8	280	

Trend equation $y_c = 34.03 + 2.89t$

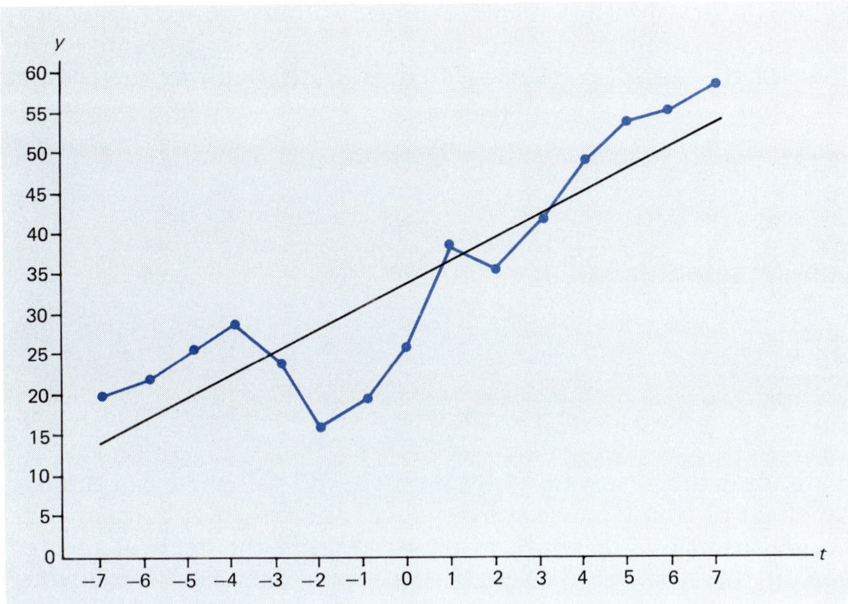

Figure 17.3
Associated Department Stores' data-sales versus time (Example 17.1).

$$a = \bar{y} = 34.03 \tag{17.5}$$

and the trend equation is

$$y_c = 34.03 + 2.89t$$

Column 6 in Table 17.2 lists the values of y_c computed from the trend equation for each of the 15 years. These values are obtained by substituting a numerical value of t into the trend equation and solving for y_c. For example, to obtain a value of y_c for the year 1964, when $t = -7$,

$$\begin{aligned} y_c &= 34.03 + (2.89)(-7) \\ &= 34.03 - 20.23 \\ &= 13.80. \end{aligned}$$

For the year 1973, $t = 2$ and

$$\begin{aligned} y_c &= 34.03 + (2.89)(2) \\ &= 34.03 + 5.78 \\ &= 39.81. \end{aligned}$$

The annual sales from column 3 of Table 17.2 and the trend values from column 6 are plotted in Figure 17.3. It appears that the trend line represents the long-term movement in sales of the Associated Department Stores Company quite well.

Adjusting Data for Trend. Consider the annual sales of Associated Department Stores as presented in column 3 of Table 17.2. Since these are annual figures, there is no seasonal component. Therefore, each value consists of a component due to trend, irregular variations, and possibly cycle. In order to determine if a cyclical component is present in the data, it is necessary to remove the effect of the trend in order to analyze the remaining or residual variations. Since, for annual data,

$$y = T \cdot C \cdot I \tag{17.2}$$

and

$$y_c \doteq T$$

then, by substitution,

$$y \doteq y_c \cdot C \cdot I$$

and

$$\frac{y}{y_c} = \frac{y_c \cdot C \cdot I}{y_c} = C \cdot I$$

By dividing the observed values by the trend value, we can remove the effect of trend, leaving the cyclical and irregular components. We will use this procedure to try to identify the cyclical component in the Associated Department Stores data in the following section.

17.3 The Cycle

The cycle is a periodic fluctuation in a time series that takes place with regularity over a period of from two to 15 or more years. It should be reemphasized that not all time series contain a cyclical component and, in fact, the presence of a regular cycle in most economic time series has yet to be established.

To illustrate the procedure for decomposing a time series, we use the Associated Department Stores' annual sales data and its trend component from Table 17.2 to isolate the cycle. As discussed in the previous section, the trend can be removed from the original observations by dividing each of these observations by y_c. The residual then consists of $C \cdot I$, that is,

$$\frac{y}{y_c} = C \cdot I$$

The result of this devision is that $C \cdot I$ is expressed as a proportion of the trend value. If we multiply this proportion by 100, we then have $C \cdot I$ expressed as a percent of trend, that is,

$$\frac{y}{y_c}(100) = C \cdot I\%$$

Table 17.3 provides the original data for y in column 3, the computed trend values y_c in column 4, and $C \cdot I$ as a percent of trend in column 5. Keep in mind that the values in column 5 of Table 17.3 represent both the cyclical and irregular components of the data. In

TABLE 17.3
DETRENDED ANNUAL SALES, ASSOCIATED DEPARTMENT STORES—1964–1978 (EXAMPLE 17.3)

1 Year	2 Coded Year	3 Sales (in millions of dollars)	4 Trend Value (in millions of dollars)	5 $C \cdot I$ % of Trend
1964	−7	19.8	13.80	143.5
1965	−6	21.4	16.69	128.2
1966	−5	25.0	19.58	127.7
1967	−4	28.6	22.47	127.3
1968	−3	24.5	25.36	96.6
1969	−2	15.2	28.25	53.8
1970	−1	18.8	31.14	60.4
1971	0	26.0	34.03	76.4
1972	1	37.6	36.92	101.8
1973	2	35.2	39.81	88.4
1974	3	42.3	42.70	99.1
1975	4	48.5	45.59	106.4
1976	5	54.0	48.48	111.4
1977	6	55.5	51.37	108.0
1978	7	58.0	54.26	106.9

order to isolate and study the cycle, it is still necessary to remove the irregular variations. This is most simply accomplished by the method of moving averages.

Moving Averages. As the name implies, a MOVING AVERAGE is a series of averages computed from the terms in a series by successively dropping one old term and adding one new term for each average. For example, in a three-term moving average, the averages are computed from three terms at a time. Each new average drops the first term that went into the previous average and adds the next term in succession in the original series.

Moving averages tend to smooth or eliminate erratic movements from the series. Consequently, taking moving averages is often referred to as a smoothing operation. The number of terms used in each average determines the amount of smoothing that occurs. Usually, the more terms used in the average, the smoother the results.

For a three-term moving average, three successive terms are added, the sum divided by three, and the resulting average placed to coincide with the middle term entering the average; that is,

$$\frac{x_0 + x_1 + x_2}{3} = \bar{x}_1 \qquad \text{centered at term } x_1$$

$$\frac{x_1 + x_2 + x_3}{3} = \bar{x}_2 \qquad \text{centered at term } x_2$$

$$\frac{x_2 + x_3 + x_4}{3} = \bar{x}_3 \qquad \text{centered at term } x_3$$

and so on.

For a moving average containing an even number of terms, each average is placed between the two middle terms entering the average. For example, for a four-term moving average,

$$\frac{x_0 + x_1 + x_2 + x_3}{4} = \bar{x}_1 \qquad \text{centered between } x_1 \text{ and } x_2$$

$$\frac{x_1 + x_2 + x_3 + x_4}{4} = \bar{x}_2 \qquad \text{centered between } x_2 \text{ and } x_3$$

$$\frac{x_2 + x_3 + x_4 + x_5}{4} = \bar{x}_3 \qquad \text{centered between } x_3 \text{ and } x_4$$

and so on.

Example 17.2 Construct a three-term moving average from the following sequence of values and center the average appropriately in relation to the original data.

5, 7, 6, 9, 3, 6, 2, 3, 9, 1, 4, 7

Step 1. Average the first three terms of the series.

$$\bar{x}_1 = \frac{5 + 7 + 6}{3} = \frac{18}{3} = 6.0$$

Step 2. Drop the first term in the previous average and add the next term from the series.

$$\bar{x}_2 = \frac{7 + 6 + 9}{3} = \frac{22}{3} = 7.3$$

Step 3. For each new average, drop the first term in the previous average and add the next term in the sequence until the series is exhausted.

Step 4. Arrange the data as follows.

x	*Sum of Three Terms*	*Average of Three Terms*
5		
7	18	6.0
6	22	7.3
9	18	6.0
3	18	6.0
6	11	3.7
2	11	3.7
3	14	4.7
9	13	4.3
1	14	4.7
4	12	4.0
7		

Notice that there is no average opposite the first and last terms in the original sequence. This is because each average corresponds to the middle value of all the terms included in the average. In a three-term moving average, neither the first nor the last term in the series can be a midpoint.

Example 17.3

Construct both a three-term moving average and a five-term moving average for the detrended Associated Department Stores' annual sales data provided in column 5 of Table 17.3. Plot the detrended data and both sets of moving averages on separate graphs and compare the results.

$C \cdot I$ as % of Trend (from Table 17.3)	3-Term Moving Total	3-Term Moving Average	5-Term Moving Total	5-Term Moving Average
143.5				
128.2	399.4	133.1		
127.7	383.2	127.7	623.3	124.7
127.3	351.6	117.2	533.6	106.7
96.6	277.7	92.6	465.8	93.2
53.8	210.8	70.3	414.5	82.9
60.4	190.6	63.5	389.0	77.8
76.4	238.6	79.5	380.8	76.2
101.8	266.6	88.9	426.1	85.2
88.4	289.3	96.4	472.1	94.4
99.1	293.9	98.0	507.1	101.4
106.4	316.9	105.6	513.3	102.7
111.4	325.8	108.6	531.8	106.4
108.0	326.3	108.8		
106.9				

Notice that for the five-term moving average, the first two and last two terms in the original sequence do not have an average associated with them.

The detrended annual data, $C \cdot I$ as a percent of trend, are plotted in Figure 17.4. The three-term and five-term moving averages are plotted in

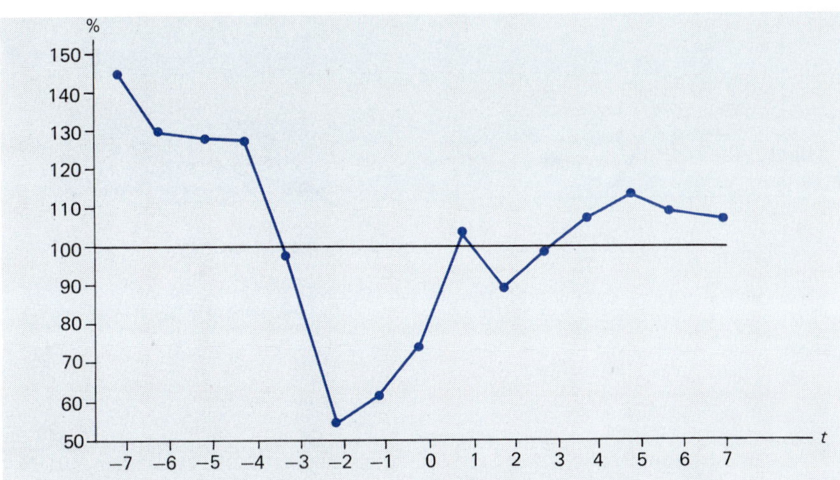

Figure 17.4 *Associated Department Stores' detrended annual sales data ($C \cdot I$ as percentage of trend) (Example 17.3).*

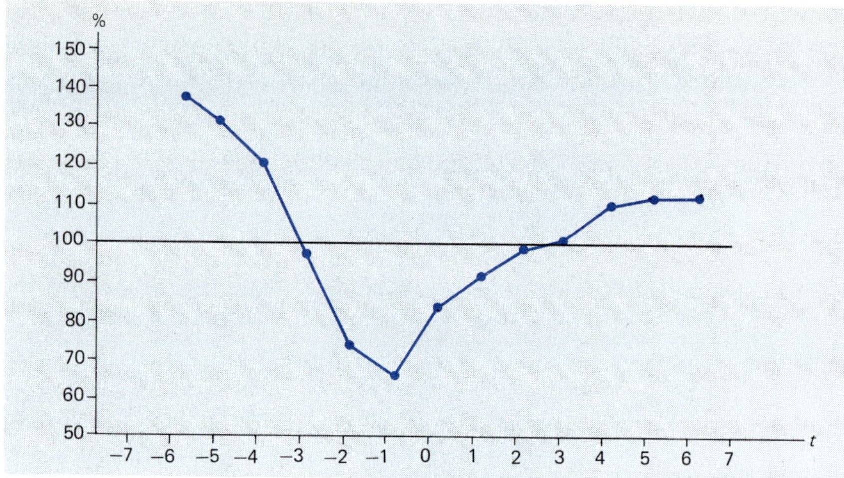

Figure 17.5
Associated Department Stores' three-year moving average of detrended annual sales data (Example 17.3).

Figures 17.5 and 17.6, respectively. Notice that the moving averages have a smoothing effect on the detrended data. The five-term average appears smoother than the three-term average. Figures 17.5 and 17.6 are graphs of what we assume to be the cyclical component of the original sales data, since the irregular fluctuations appear to have been substantially removed by the averaging process. In looking at these two graphs, they do appear to show a portion of a cycle. However, determining whether or not this apparent cyclical movement continues and repeats itself with any regularity requires many more years of data. If there is a cycle, we have apparently observed only about half of one.

Figure 17.6
Associated Department Stores' five-year moving average of detrended annual sales data (Example 17.3).

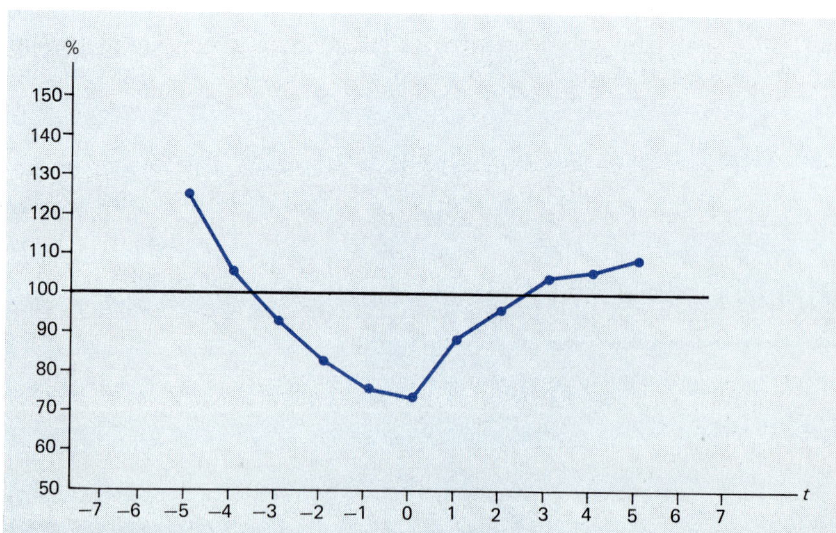

17.4 The Seasonal Index

Whenever observations on the dependent variable are made at periodic intervals within one year, the time series may contain seasonal variation. As mentioned previously, seasonal variation, or seasonality, can be attributed to holidays, weather patterns, customs and conventions, and so on.

Seasonal variation is generally expressed as an index number. Therefore, in order to measure seasonality over a year there must be a series of seasonal index numbers. Monthly data require twelve indexes to express the seasonality, while quarterly data require four.

In the previous sections, we worked only with annual data and, therefore, the data contained no seasonal component. However, if we must deal with monthly or quarterly data to determine the trend and the cycle, it is desirable to remove the seasonal variation first. Also, isolating the seasonal component allows us to study the effect of seasonality on the series. Consequently, whether the primary goal of the analysis is to determine the trend and the cycle or simply to study seasonality, we must first isolate the seasonal effect.

The simplest method for removing the seasonal component from the series is by the method of moving averages, discussed in the previous section. Since moving averages tend to smooth a series by eliminating the short-term and irregular fluctuations, a moving average applied to quarterly or monthly data tends to remove the seasonal and irregular variations, leaving the trend and the cycle; that is, the moving average (MA) is approximately equal to $T \cdot C$. Therefore, since our time-series model is multiplicative, if we divide the original data by the moving average, we should be left with $S \cdot I$. That is, since

$$y = T \cdot C \cdot S \cdot I \tag{17.1}$$

and

$$\text{MA} \doteq T \cdot C$$

then

$$\frac{y}{\text{MA}} = \frac{T \cdot C \cdot S \cdot I}{T \cdot C} = S \cdot I$$

Of course, having proceeded this far, we are still left with $S \cdot I$ and must still remove I. This is also done by averaging, but not with moving averages. The quarterly or monthly values of $S \cdot I$ are arranged by quarter or month and a MODIFIED MEAN is obtained for each of the periods. A *modified mean* is a simple arithmetic mean obtained after eliminating the highest and lowest value in the set of data from which the mean is calculated.

These modified means, multiplied by 100, constitute a set of unadjusted seasonal indexes. For monthly data, the sum of the twelve seasonal indexes should be 1200. For quarterly data, the sum of the four seasonal indexes should be 400. Frequently, the sums of the unadjusted indexes do not equal these values. When that happens, an adjustment must be made to force the sums of the index numbers to equal 1200 or 400 for monthly and quarterly indexes, respectively. The adjustment factor for quarterly indexes is 400 divided by the sum of the unadjusted indexes. For monthly indexes it is 1200 divided by the sum of the unadjusted indexes. When each unadjusted index number is multiplied by the adjustment factor, the result is the set of adjusted seasonal indexes.

This procedure is summarized as follows.

1. Isolate the trend and cyclical components of the data by computing a series of moving averages. These moving averages are approximately equal to $T \cdot C$.

2. Divide the original data by the moving average. This results in a series which contains only the seasonal and irregular components, $S \cdot I$.

3. Group the $S \cdot I$ data by period. For example, for quarterly data, group all the first-quarter values together, the second-quarter values together, and so on.

4. Compute a modified mean for each period by eliminating the highest and lowest values in each group and finding the arithmetic mean of the remaining values.

5. Multiply each of the modified means by 100. The result is the unadjusted seasonal index for each period.

6. Multiply each unadjusted index by the appropriate adjustment factor. The result is the adjusted seasonal index.

Example 17.4 Use the Associated Department Stores' quarterly data for the years 1973 through 1977 to compute seasonal indexes. Note that since we will be computing a four-term moving average from these data, use the last two quarters of 1972 and the first two quarters of 1978 in computing the moving averages. This will provide a moving average to correspond with every quarter of the years 1973 through 1977.

The solution is illustrated with reference to the data in Tables 17.4 and 17.5. Columns 1 and 2 in Table 17.4 list the original data obtained from Table 17.1 by quarter from the third quarter of 1972 through the second quarter of 1978. Column 1 identifies the years and quarters, while column 2 provides the observed values of sales in millions of dollars.

1. A four-quarter moving average must be computed from these data. We normally compute these averages by summing the first four values in the series and dividing by four. We then drop the first value, add the next and repeat the procedure. However, recall that a moving average is associated with the middle term of the values being averaged. When we

TABLE 17.4
ASSOCIATED DEPARTMENT STORES' QUARTERLY SALES IN MILLIONS OF DOLLARS—THIRD QUARTER 1972 THROUGH SECOND QUARTER 1978 (DATA FOR COMPUTATION OF SEASONAL INDEXES) (EXAMPLE 17.4)

1		2	3	4	5	6
Year and Quarter		*Sales*	*Four-Quarter Moving Total*	*Total of Two Four-Quarter Moving Totals*	*Moving Average* $MA = T \cdot C$	$\frac{Y}{MA} = S \cdot I$
1972	3rd	7.43				
	4th	15.09				
			35.90			
1973	1st	4.65		72.11	9.01	0.516
			36.21			
	2nd	8.73		71.41	8.93	0.978
			35.20			
	3rd	7.74		72.05	9.01	0.859
			36.85			
	4th	14.08		75.59	9.45	1.490
			38.74			
1974	1st	6.30		78.39	9.80	0.643
			39.65			
	2nd	10.62		81.95	10.24	1.037
			42.30			
	3rd	8.65		85.22	10.65	0.812
			42.92			
	4th	16.73		87.71	10.96	1.526
			44.79			
1975	1st	6.92		91.11	11.39	0.608
			46.32			
	2nd	12.49		94.82	11.85	1.054
			48.50			
	3rd	10.18		98.78	12.35	0.824
			50.28			
	4th	18.91		101.13	12.64	1.496
			50.85			
1976	1st	8.70		102.32	12.79	0.680
			51.47			
	2nd	13.06		105.47	13.18	0.991
			54.00			
	3rd	10.80		107.53	13.44	0.804
			53.53			
	4th	21.44		107.87	13.48	1.591
			54.34			
1977	1st	8.23		109.08	13.64	0.603
			54.74			
	2nd	13.87		110.24	13.78	1.007
			55.50			
	3rd	11.20		111.82	13.98	0.801
			56.32			
	4th	22.20		112.27	14.03	1.582
			55.95			
1978	1st	9.05				
	2nd	13.50				

average four terms at a time there is no single middle value. Therefore, the average is associated with a point between the two middle values while we require each average to correspond to one specific quarter. The positions of the four-quarter moving totals from column 3 of Table 17.4 illustrate this point. Each total falls in the space between two adjacent quarters. This difficulty can be remedied by averaging the moving averages two at a time and placing this new average at its appropriate position, which corresponds to one specific quarter. This is done by adding two adjacent moving totals from column 3 and placing their sum in column 4. The modified four-quarter average is obtained by dividing this last total by 8, since the sum of two moving totals contains data from eight quarters. The modified four-quarter moving average is shown in column 5. Notice that the sums in column 4 and the averages in column 5 both correspond to one specific quarter. The moving average in column 5 is approximately equal to $T \cdot C$.

2. When y is divided by the moving average, the result is equal to $S \cdot I$. These values appear in column 6 of Table 17.4.

TABLE 17.5
COMPUTATION OF QUARTERLY INDEXES FOR ASSOCIATED DEPARTMENT STORES' SALES (EXAMPLE 17.4)

	Quarter ($S \cdot I$)			
Year	*1*	*2*	*3*	*4*
1973	~~0.516~~	~~0.978~~	~~0.859~~	~~1.490~~
1974	0.643	1.037	0.812	1.526
1975	0.608	~~1.054~~	0.824	1.496
1976	~~0.680~~	0.991	0.804	~~1.591~~
1977	0.603	1.007	~~0.801~~	1.582
Sums	1.854	3.035	2.440	4.604
Modified means	0.618	1.012	0.813	1.535
Unadjusted seasonal index	61.8	101.2	81.3	153.5
Adjusted seasonal index	62.1	101.8	81.8	154.3

3. In order to remove the irregular fluctuations from these values, the $S \cdot I$ values are grouped by quarter in Table 17.5; that is, all the first-quarter values are placed in the first column, the second-quarter values in the second column, and so on.

4. The highest and lowest values in each column are eliminated and the arithmetic mean is calculated from the remaining values. The modified means of the columns are 0.618, 1.012, 0.813, and 1.535.

5. Each modified mean is multiplied by 100. The result is the set of unadjusted quarterly seasonal indexes: 61.8, 101.2, 81.3, and 153.5.

6. If the sum of the unadjusted quarterly indexes is equal to 400, no further adjustment is necessary. However, the sum of these indexes is

$$61.8 + 101.2 + 81.3 + 153.5 = 397.8$$

The adjustment factor is obtained by dividing the desired sum of 400 by the sum of the unadjusted indexes, which is 397.8, that is, the adjustment factor in this case is

$$\frac{400.0}{397.8} = 1.0055$$

Each of the unadjusted quarterly indexes is multiplied by the adjustment factor, resulting in the four adjusted quarterly indexes:

$$S_1 = (61.8)(1.0055) = 62.1$$
$$S_2 = (101.2)(1.0055) = 101.8$$
$$S_3 = (81.3)(1.0055) = 81.8$$
$$S_4 = (153.5)(1.0055) = 154.3$$

Notice that the sum of the adjusted seasonal indexes is now equal to 400.

If each of the adjusted seasonal indexes is divided by 400, we have the proportion or percent of annual sales which, on the average, will occur in each of the four quarters. For example,

$$\frac{S_1}{400} = \frac{62.1}{400} = 0.155 \quad \text{or} \quad 15.5\%$$

$$\frac{S_2}{400} = \frac{101.8}{400} = 0.255 \quad \text{or} \quad 25.5\%$$

$$\frac{S_3}{400} = \frac{81.8}{400} = 0.205 \quad \text{or} \quad 20.5\%$$

$$\frac{S_4}{400} = \frac{154.3}{400} = 0.386 \quad \text{or} \quad 38.6\%$$

Deseasonalizing Time Series Data. When interest centers on determining the pattern of seasonal variation, then the analysis of the time series data can stop with the computation of the seasonal indexes. However, if the purpose of the analysis is to estimate the trend and look for a cycle, then the seasonal component should be removed from the data first. The process of removing the seasonal component from a time series is called DESEASONALIZING. This is accomplished, for the multiplicative model, by dividing the original data by the appropriate seasonal index expressed as a decimal fraction. This is shown in Table 17.6, where column 1 lists the year and quarter, column 2 provides the original quarterly sales data, column 3 shows the seasonal index as a percentage, and column 4 shows the seasonal index expressed as a decimal fraction. Column 6 provides the deseasonalized sales data. The values in column 6 were obtained by dividing the sales data of column 2 by the index in column 5.

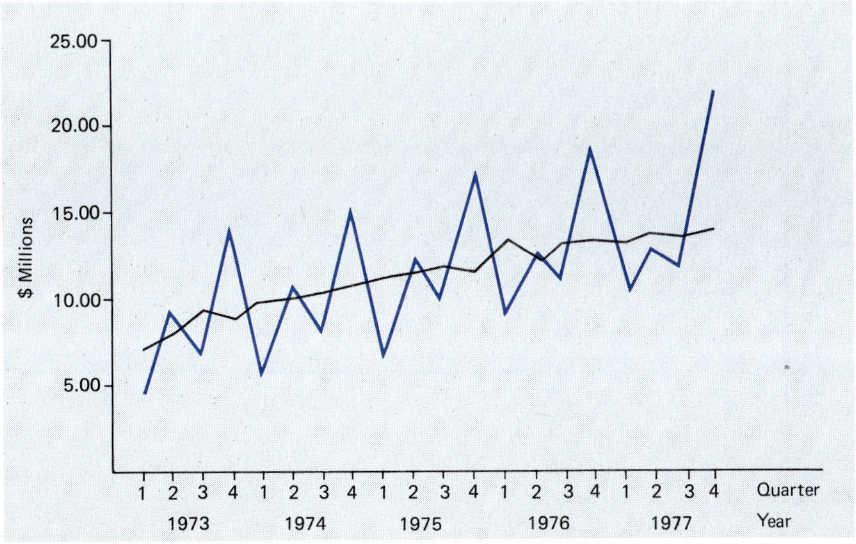

Figure 17.7
Associated Department Stores' quarterly sales, 1973–1977. The colored line represents original data; the black line, deseasonalized data.

TABLE 17.6

ASSOCIATED DEPARTMENT STORES' QUARTERLY SALES DATA, 1973–1977: ORIGINAL AND DESEASONALIZED DATA

Year and Quarter		*Sales (in millions of dollars)*	*Seasonal Index (as percentage)*	*Seasonal Index (as decimal)*	*Deseasonalized Sales (in millions of dollars)*
1973	1st	4.65	62.1	0.621	7.49
	2nd	8.73	101.8	1.018	8.58
	3rd	7.74	81.8	0.818	9.46
	4th	14.08	154.3	1.543	9.13
1974	1st	6.30	62.1	0.621	10.14
	2nd	10.62	101.8	1.018	10.43
	3rd	8.65	81.8	0.818	10.57
	4th	16.73	154.3	1.543	10.84
1975	1st	6.92	62.1	0.621	11.14
	2nd	12.49	101.8	1.018	12.27
	3rd	10.18	81.8	0.818	12.44
	4th	18.91	154.3	1.543	12.26
1976	1st	8.70	62.1	0.621	14.01
	2nd	13.06	101.8	1.018	12.83
	3rd	10.80	81.8	0.818	13.20
	4th	21.44	154.3	1.543	13.90
1977	1st	8.23	62.1	0.621	13.25
	2nd	13.87	101.8	1.018	13.62
	3rd	11.20	81.8	0.818	13.69
	4th	22.20	154.3	1.543	14.39

Figure 17.7 graphs the original data and the deseasonalized data on the same chart. The solid curve represents the deseasonalized sales. By removing the seasonal effect, the trend and the cycle, if any, become much more apparent.

17.5 Forecasting

The principal reason for analyzing time series in business is for forecasting. If future demand or sales can be predicted with accuracy, then major decisions can be made regarding purchases, plant expansions and other capital expenditures, hiring, advertising, and practically every other phase of business activity. For forecasting, the most important component of a time series is the trend. If the trend equation is determined, then a predicted value can be obtained by substituting a value of t which corresponds to the year for which the forecast is desired into the trend equation and solving for the value of the dependent variable. If a forecast is desired for a particular quarter or for all four quarters of that year, and if there is a seasonal component in the series, then the trend value must be adjusted by the appropriate seasonal index. Adjusting the trend to add seasonality is exactly opposite to the procedure for deseasonalizing data. In fact, since removing the seasonal effect is called deseasonalizing, adding the seasonal effect is called SEASONALIZING. A time series is seasonalized by multiplying each trend value by the appropriate seasonal index expressed as a decimal.

Since the irregular component cannot be predicted, it is simply ignored in forecasting. The cyclical component also creates difficulty. First, there must be enough data to show that a cycle does, in fact, exist. Second, for the cycle to be used in the forecast it must occur with regularity. If a regular cycle can be discerned, then it can be described and projected to the appropriate point in time. The problem in most business and economic forecasting, aside from the unpredictability of the irregular component, is that regular cycles seldom occur in economic time series. In most cases, the best that can be done is to base the forecast on the trend equation and the seasonal component.

Example 17.5

One of the important functions of the marketing research department of Associated Department Stores is to prepare sales forecasts which are used in both short- and long-range planning activities. A sales forecast for the year 1979 and for each of its quarters is required. The forecast is to be based on the trend equation developed in Section 17.2 and the seasonal indexes from Section 17.4.

The trend equation for annual sales is

$$y_c = 34.03 + 2.89t$$

Since t corresponding to 1979 would be 8, predicted sales for that year would be

$$\begin{aligned} y_{1979} &= 34.03 + (2.89)(8) \\ &= 34.03 + 23.12 \\ &= 57.15 \text{ millions of dollars} \end{aligned}$$

If there was no seasonal effect on department store sales, we would expect each quarter's sales to be approximately one-fourth of this amount. We could, in fact, predict quarterly sales from the trend equation by dividing the coefficients of that equation by four, that is, a trend equation for quarterly sales would be

$$\begin{aligned} y_{c_q} &= \tfrac{1}{4}(34.03 + 2.89t) \\ &= \frac{34.03}{4} + \frac{2.89}{4}t \\ &= 8.51 + 0.7225t \end{aligned}$$

Then quarterly sales in 1979 by the trend equation would be predicted to be

$$\begin{aligned} y_{c_q} &= 8.51 + (0.7225)(8) \\ &= 8.51 + 5.78 \\ &= 14.29 \text{ millions of dollars} \end{aligned}$$

This figure of 14.29 is based on trend only and does not account for seasonality. The quarterly forecast is obtained by multiplying the trend value by each seasonal index, expressed as a decimal, to obtain the four quarterly forecasts. This is done as follows.

Quarter	*Seasonal Index (Decimal Form)*	*Trend Value*	*Forecast*
1st	0.621	14.29	8.87
2nd	1.018	14.29	14.55
3rd	0.818	14.29	11.69
4th	1.543	14.29	22.05
			57.16

Except for a slight error due to rounding, the sum of the four quarterly forecasts is equal to the forecast resulting from the annual trend equation.

Definitions

Cycle: The periodic growth and decline in a variable that occurs with regularity over a period of from two to 15 or more years.

Decomposition: The process of isolating each component of a time series.

Deseasonalizing: The process of removing the seasonal component from a time series.

Modified mean: A simple arithmetic mean obtained from a set of data after eliminating the highest and lowest values in the set.

Moving averages: A series of averages computed from the terms in a series by successively dropping the oldest term and adding the next term in succession for each new average.

Seasonalizing: Adjusting the trend for the effect of the season.

Secular trend: Same as trend.

Time series: A set of numerical observations taken at different points in time.

Trend: The long-term behavior of a variable.

Formulas

The time series model:

(17.1) $$y = T \cdot S \cdot C \cdot I$$

The time series model for annual data:

(17.2) $$y = T \cdot C \cdot I$$

The linear trend equation:

(17.3) $$y_c = a + bx$$

The linear trend equation for coded data:

(17.4) $$y_c = a + bt$$

Coefficients of the least squares trend line with coded data:

(17.5) $$a = \bar{y}$$

(17.6) $$b = \frac{\Sigma ty}{\Sigma t^2}$$

Problems

17.1 Compute a series of five-term moving averages for the following set of data. Read the data across each row.

14	11	18	4	7	9	23	18	6	21
28	19	5	12	13	11	26	4	1	18
2	6	30	9	22	5	8	2	19	12

17.2 Compute a series of six-term moving averages from the following set of data. Read the data across each row.

18	37	45	22	14	48
32	12	27	31	29	19
44	46	16	27	15	35
23	45	19	42	17	28

17.3 Compute a series of three-term moving averages from the data given in problem 17.1. Read the data down each column.

17.4 Compute a series of four-term moving averages from the data in problem 17.2. Read the data down each column.

17.5 The following data represent annual sales in millions of dollars of Metric Systems, Inc. for the years 1964 through 1978. Find the least squares linear trend equation and predict sales for 1979 and 1980.

1964 – 20	1969 – 19	1974 – 31
1965 – 23	1970 – 20	1975 – 37
1966 – 25	1971 – 27	1976 – 40
1967 – 26	1972 – 25	1977 – 42
1968 – 22	1973 – 30	1978 – 44

17.6 The following data represent annual production of Portland cement by the Eastboro Portland Cement Company in millions of tons for the years 1961 through 1978. Find the least squares linear trend equation and predict production for 1979 and 1980.

1961 – 20.9	1967 – 27.8	1973 – 38.0
1962 – 21.2	1968 – 24.3	1974 – 37.5
1963 – 19.8	1969 – 26.2	1975 – 43.0
1964 – 24.4	1970 – 31.7	1976 – 39.8
1965 – 27.6	1971 – 39.4	1977 – 43.1
1966 – 28.1	1972 – 36.1	1978 – 42.1

17.7 For the Metric Systems sales data provided in problem 17.5, remove the trend component and show the cyclical and irregular components, $C \cdot I$, as a percentage of trend for each year.

17.8 For the Eastboro Portland Cement Company data provided in problem 17.6, remove the trend component and show the cyclical and irregular components, $C \cdot I$, as a percentage of trend for each year.

17.9 For the detrended Metric Systems sales data of problem 17.7, smooth the series using a three-term moving average to remove the irregular fluctuations and plot the smoothed series on a graph. Does there appear to be a cyclical component to these data?

17.10 For the detrended Eastboro Portland Cement data of problem 17.8, smooth the series using a three-term moving average to remove the irregular fluctuations and plot the smoothed series on a graph. Does there appear to be a cyclical component to these data?

17.11 Annual production figures of the Eastboro Portland Cement Company are broken down by quarter from mid-1971 through mid-1978. Compute the quarterly seasonal indexes from these data.

Year	*First Quarter*	*Second Quarter*	*Third Quarter*	*Fourth Quarter*
1971			15.8	5.9
1972	7.0	8.9	15.0	5.2
1973	7.6	9.2	15.4	5.8
1974	7.3	9.6	14.9	5.7
1975	8.5	10.7	17.5	6.3
1976	5.8	11.3	15.4	7.3
1977	7.4	10.9	17.7	7.1
1978	8.1	10.8		

17.12 The Middletown Power and Light Company compiled the following data on electric power consumption in their service area. Power consumption in millions of kilowatt hours by quarter from mid-1972 to mid-1978 is provided. Compute the quarterly seasonal indexes from these data.

Year	First Quarter	Second Quarter	Third Quarter	Fourth Quarter
1972			20.2	16.0
1973	24.4	13.0	21.5	17.1
1974	26.3	13.3	23.4	18.0
1975	26.5	15.0	23.2	19.3
1976	33.0	19.0	23.4	18.6
1977	30.7	23.7	25.3	19.3
1978	30.8	17.1		

17.13 Using the seasonal indexes from problem 17.12, deseasonalize the time series of power consumption in that problem and compute the equation of the least squares trend line for the deseasonalized quarterly data.

17.14 Use the trend equation of problem 17.13 and the seasonal indexes of problem 17.12 to predict power consumption in the Middletown area for each quarter of 1979.

17.15 Given the following trend equation for annual sales in thousands of dollars and the seasonal indexes of the Fortune Carpet Company, predict sales for each quarter of 1980. Let $t = 0$ in 1972.

$$y_c = 64.1 + 45.4t$$

$$S_1 = 117.4 \qquad S_2 = 91.8 \qquad S_3 = 82.7 \qquad S_4 = 108.1$$

17.16 Given the following trend equation for annual sales in thousands of dollars and the seasonal indexes of the Nationwide Lawn Care Company, predict sales for each quarter of 1980. Let $t = 0$ in 1973.

$$y_c = 101.7 + 350t$$

$$S_1 = 41.2 \qquad S_2 = 121.4 \qquad S_3 = 156.3 \qquad S_4 = 81.1$$

CHAPTER 18 INTRODUCTION TO STATISTICAL DECISION THEORY

18.1
Decision making under uncertainty

18.2
States of nature

18.3
Notation used in statistical decision theory

18.4
Payoff tables and functions

18.5
Prior distributions and Bayes' criterion

18.6
Opportunity loss

18.7
Decision trees

18.8
The expected value of perfect information

18.9
Sampling and revising the prior distribution

18.10
Preposterior analysis

18.11
The expected value of sample information

Business managers constantly use statistical data and statistical techniques to assist them in making decisions. This fact was emphasized in the very first chapter of this book. We pointed out in Chapter 1 that a great deal of information that is pertinent to business decision making is generated in the form of numerical or statistical data. These data are used in various ways in the decision-making process. For example, they may be used for describing certain conditions, or for making estimates or forecasts, or for testing hypotheses. In all these cases, the data provide information that is useful for decision making. However, none of these procedures can assure the decision maker that his decision will be best or optimal in an economic sense.

We now introduce a set of procedures by which statistical data can be used for making optimal decisions. These procedures are generally referred to as statistical decision theory or Bayesian decision theory. How we define an optimal decision will be explained as we proceed.

Most of the techniques of statistical inference that we have discussed so far have involved hypothesis testing. Hypothesis testing is a procedure for making a decision. In fact, hypothesis testing is sometimes called classical decision theory. In hypothesis testing, the statistical decision is to either accept or reject the null hypothesis. In practical applications of the technique, such as in business, the statistical decision should lead to some practical decision related to some specific business problem. If the statistical test leads to acceptance of the null hypothesis, the decision maker will act in some appropriate manner. If the null hypothesis is rejected, the decision maker will act in a different manner. Keep in mind that taking no action at all, that is, maintaining some status quo, is an act in the sense of decision making.

There are three major deficiencies in this classical decision-making procedure as it applies to practical situations. First, it provides for only two possible actions. These correspond to acceptance or rejection of the null hypothesis. Also, it allows only two states of the parameter being tested, that is, those parameter values that make the null hypothesis true and those values for which the null hypothesis is false. In most practical decision problems, a decision maker must make a choice from among several different acts subject to a wide variety of states or conditions over which he or she has no control.

Second, in practical business decision problems there are real economic consequences that result from making a wrong decision. Although economic consequences of a wrong decision might be considered by the decision maker using the classical procedure in selecting the significance level for the test, these considerations are never an explicit part of the decision model or procedure.

The third major deficiency in the classical method is that classical statistics does not recognize the validity of any information pertaining to the decision that does not exist in the form of empirical data that result from a process of sampling.

The Bayesian approach to decision making attempts to remedy these deficiencies in the following ways. First, it provides a model for decision making in situations that involve multiple acts and multiple states of the parameter, or *nature*, which is the term commonly used in decision theory. Second, it incorporates the economic consequences of wrong decisions directly into the decision-making model. Finally, it allows the use of information pertinent to the problem which exists prior to any sampling or experimentation, whether this information is in the form of empirical data or is a subjective assessment of the decision maker.

18.1 Decision Making under Uncertainty

The process of decision making involves the selection of a single act from among some set of alternative acts. Unless there are alternatives from which to choose, there really is no decision problem. A rational decision maker should attempt to select that act, from the set of possible acts, which will produce a best result or consequence in the context of his decision problem.

However, the consequence of the selection of any act is dependent upon factors that are beyond the control of the decision maker. These factors are usually referred to as the STATES OF NATURE. Since the decision maker does not know which of the possible states of nature will occur, his decision must be made under a condition of uncertainty.

Because the consequence of any act selected by the decision maker will depend on the state of nature, information about the state of nature would be valuable. The terms "risk" and "uncertainty" refer to two different types of information available to the decision maker. When the state of nature is unknown, but objective or empirical data is available so that he can use these data to assign probabilities to the various states of nature, the procedure is generally referred to as decision making under *risk*. When the state of nature is unknown and there is no objective information on which probabilities can be based, the procedure is referred to as decision making under *uncertainty*. However, even when no objective information is available, the decision maker may, in Bayesian decision theory, assign subjective probabilities to the states of nature to assist him in making a decision. Once probabilities are assigned, regardless of the manner in which they were obtained, the decision procedure that follows is exactly the same. Therefore, for practical purposes, risk and uncertainty are essentially the same and decision making under both circumstances will be referred to here as decision making under uncertainty.

18.2 States of Nature

We previously explained that the consequences of any action on the part of a decision maker are dependent upon certain factors beyond his control. These factors are referred to as the STATES OF NATURE. The term "state of nature" does not necessarily mean nature in the ordinary sense of the word. "States of nature" is a general term which is used to encompass all those factors beyond the control of the decision maker that affect the outcome of his decision. In addition to actual states of nature such as floods, storms, droughts, etc., it could include such factors as the prices of raw materials, the availability and cost of labor, the cost of money, that is, interest rates, as well as the conscious and unconscious acts of other individuals who may or may not be competitors.

For example, the consequences of a farmer's decision to plant a particular crop would be affected by the actual state of nature over the next growing season, that is, the amount of rainfall. This decision could also be affected by the acts of many individuals in expressing their preferences for different types of foodstuffs, which in turn will affect the demand for, and the price at which he can sell, his product. In this case, the factors affecting this farmer's decision consist of an actual state of nature, rainfall, and a sort of pseudo state of nature, the demand for the product.

A manufacturer's decision whether or not to market a new product may be affected by the decision of his competitor with respect to the marketing of a similar product. The competitor may deliberately attempt to make a decision that will have an adverse effect on this particular manufacturer. His goal may be to force this manufacturer out of business or to cause him to lose money in order to create a market situation more beneficial to himself. In this case, the state of nature bearing on the manufacturer's decision would be the conscious acts of a competitor whose goals are in direct opposition to his own.

The state of nature in the manufacturer's decision problem would be the occurrence or lack of occurrence of a particular event. We could say that the state of nature affecting that particular decision is dichotomous. On the other hand, for the farmer's decision, the state of nature could assume numerous values. Rainfall over the growing season could range, let us say, between 18 and 36 inches. Each increment of rainfall would have a varying effect on the harvest. The demand for the product as reflected in the market price per bushel could also vary, and just as widely. For the farmer's decision problem there would be many states of nature.

In many practical decision situations, the total number of possible states of nature that might have some effect on the consequences of a decision could be so great as to make analysis of the problem an almost insurmountable task; it is usually necessary, therefore, to reduce the states of nature to a reasonable number.

This is generally done by eliminating those factors or states whose effect on the decision consequences is determined to be minimal and retaining only those factors that would appear to have a significant effect.

18.3 Notation Used in Statistical Decision Theory

Statistical decision theory differs in several respects from classical statistics. Several of these differences have already been pointed out and others will become apparent as the discussion proceeds. Because of these differences, some special symbols are needed to describe and use these new techniques. For example, we have defined a concept called the states of nature which includes various kinds of factors. What type of symbol should we use to represent these factors?

In some decision problems, the state of nature is a true random variable. The unknown demand for some product would be a good example of this. However, not all states of nature are random variables. Sometimes the states of nature represent possible values of some population parameter such as a mean or a proportion. We know that a parameter is not variable, and yet because of the decision maker's uncertainty about the true value of the parameter, he treats it as though it were. In classical statistics, Greek letters are usually used to represent parameters. We will adopt the Greek letter θ, theta, as a general symbol for all the states of nature. However, whenever we treat it as though it were a variable, we will place a $\sim$, tilde, over it. In other words, $\tilde{\theta}$ represents the states of nature that are or should be treated as a variable quantity or condition. The new symbols used in this chapter and their meanings are summarized below.

- $\tilde{\theta}$: the states of nature.
- θ_i: a particular state of nature.
- $P(\tilde{\theta})$: a probability distribution on the state of nature.
- A: the set of acts in the decision problem.
- a_j: a particular act.
- Q: the set of consequences or payoffs in the decision problem; $Q = f(\tilde{\theta}, a)$.
- q_{ij}: a particular payoff associated with θ_i and a_j.

18.4 Payoff Tables and Functions

In any decision problem, the choice of a specific act from the set of alternative acts should be based on an evaluation of the possible consequences. As a first step in this evaluation, it is convenient to display the set of acts, the set of states of nature, and the conse-

TABLE 18.1
GENERAL FORMAT OF A PAYOFF TABLE

States of Nature: $\tilde{\theta}$	Acts: A a_0	a_1	a_2	. . .	a_j
θ_0	q_{00}	q_{01}	q_{02}		q_{0j}
θ_1	q_{10}	q_{11}	q_{12}		q_{1j}
θ_2	q_{20}	q_{21}	q_{22}		q_{2j}
⋮					
θ_i	q_{i0}	q_{i1}	q_{i2}		q_{ij}

quence associated with each act and each state in the form of a table called a "payoff table."

Table 18.1 illustrates the general form of a payoff table. Generally, the set of acts is listed across the top of the table, while the set of states of nature constitutes the extreme left-hand column. The construction of payoff tables is illustrated in the following examples.

Example 18.1 Executives of Metric Systems, Inc. must decide whether or not to submit a bid for a particular government contract. Before the bid can be submitted, a detailed proposal must be written providing extensive information about the proposed system and its performance. This information could only be obtained as a result of a study which would be conducted by a team of the company's top engineers and scientists. It is estimated that the cost of this study and the proposal preparation would be $80,000. If MSI is awarded the contract, their profit will be $225,000. However, if they fail to get the contract, the $80,000 spent on the study and the proposal will be a total loss. If they decide not to submit a bid, their payoff will be zero.

This is a simple two-action problem with only two states of nature. The acts are

a_1: Bid on contract.

a_2: Do not bid.

The states of nature are

θ_1: MSI is awarded the contract.

θ_2: MSI is not awarded the contract.

The payoff table for this problem is illustrated in Table 18.2. The payoffs are

q_{11}: θ_1 and a_1, \$225,000

q_{12}: θ_1 and a_2, \$0

q_{21}: θ_2 and a_1, −\$80,000

q_{22}: θ_2 and a_2, \$0

Notice that neither act is better for both states of nature. If θ_1 is the state which occurs, then a_1 is the better act, since a $250,000 profit is preferable

TABLE 18.2
PAYOFF TABLE FOR BID DECISION (EXAMPLE 18.1)

	Acts: A	
States: $\tilde{\theta}$	a_1	a_2
θ_1	\$225,000	\$0
θ_2	−\$80,000	\$0

to a zero profit. However, if θ_2 occurs, then a_2 is preferable to a_1, since zero profit is better than a loss of \$80,000. In order to decide which act is optimal under the conditions of uncertainty about the state of nature, some criterion of optimality and an assessment of the probability of each state of nature is required. Without such a criterion and a probability distribution, the payoff table serves principally as a device for displaying the states, acts, and consequences in a concise and clear manner.

For a simple two-action problem with two states of nature, determining the payoff for each act and each state is relatively simple. However, for problems involving several acts and states of nature, a payoff function is a useful shortcut. We will illustrate a somewhat more complicated decision problem in the following example.

Example 18.2

Consider the problem of Mr. Wilson, the sole proprietor of Buckingham Builders. He has been approached by the president of the Universal Development Company, a developer of exclusive residential communities. Universal owns 100 acres of prime land in Buckingham County, which is zoned for single-family residences on two-acre lots. Universal plans to subdivide this acreage into 50 residential lots, to put in streets, sewers, and utilities, and to sell the lots to builders who would construct houses which would range in price from \$85,000 to \$125,000. Universal offers Mr. Wilson the opportunity to purchase up to five of these lots at \$12,000 each. Mr. Wilson's decision involves a choice among six alternate acts. These are: purchase no lots, purchase one lot, purchase two lots, and so on, up to a maximum of five lots. We designate these acts by a_0 through a_5, where the subscript corresponds to the number of lots purchased.

Mr. Wilson's choice among these acts should be influenced by the payoff to him. He has plans for a house that he could price to sell for \$89,000. His construction cost would be \$60,000. This, plus the price of the lot, would bring his total cost up to \$72,000, providing him with a profit of \$17,000 on each house he can sell.

The state of nature in this problem is the demand for Mr. Wilson's houses at a price of \$89,000. There are six possible states of nature corresponding to the demand for no houses, the demand for one house, etc., up to a demand for five houses. We designate each of the states with θ_i, where i ranges from 0 to 5 and represents the number of houses demanded. For this problem, there are 36 consequences or payoffs corresponding to each act and each state of nature. The simplest way to calculate these payoffs is to determine the function that relates payoff, Q, to Mr. Wilson's choice of an act and to the demand for his houses.

For each house that Mr. Wilson builds, his investment is \$72,000. His total investment depends on the number of lots he buys and the number of houses he builds. Therefore, his total investment is \$72,000a_j. He will receive \$89,000 for each house that he can sell. His payoff is then the difference between his receipts and his investment, that is, $Q = \$89{,}000\theta_i - \$72{,}000a_j$. However, we must realize that Mr. Wilson can never sell more houses than he has built. Therefore, even if there is a demand for five houses, if he has purchased only three lots and constructed three houses, he can only sell three. Consequently, we must have two payoff functions, one for those conditions where demand is less than the number of houses built by Mr. Wilson, that is, $\tilde{\theta}_i < a_j$, and one for those conditions where demand is greater than or equal to the supply, that is, $\tilde{\theta}_i \geq a_j$. These payoff functions are

$$Q(\theta_i < a_j) = \$89{,}000\theta_i - \$72{,}000a_j$$
$$Q(\theta_i \geq a_j) = \$89{,}000a_j - \$72{,}000a_j = \$17{,}000a_j$$

As an example, suppose Mr. Wilson purchases four lots and builds four houses, that is, he selects act a_4. Suppose the demand is only for two houses, that is, $\tilde{\theta} = \theta_2$. The payoff would be

$$\begin{aligned} Q(2, 4) &= (\$89{,}000)(2) - (\$72{,}000)(4) \\ &= \$178{,}000 - \$288{,}000 \\ &= -\$110{,}000, \end{aligned}$$

or a loss of \$110,000.

Suppose, on the other hand, that Mr. Wilson built only two houses but the demand was for four. It should be obvious that Mr. Wilson cannot sell more than two houses. His payoff would then be \$17,000 for each house built and sold, or (\$17,000)(2) = \$34,000.

These two payoff functions were used to fill in all the payoffs in Table 18.3. All entries in that table are in thousands of dollars, and the minus signs indicate losses. These values are actual profits and losses associated with each act and each state. They are referred to as *conditional payoffs*, because they are conditional upon the state of nature.

TABLE 18.3
PAYOFF TABLE—LOT PURCHASE DECISION (PAYOFF IN THOUSANDS OF DOLLARS) (EXAMPLE 18.2)

	Acts: A					
States: $\tilde{\theta}$	a_0	a_1	a_2	a_3	a_4	a_5
$\tilde{\theta} = 0$	0	−72	−144	−216	−288	−360
$\tilde{\theta} = 1$	0	17	−55	−127	−199	−271
$\tilde{\theta} = 2$	0	17	34	−38	−110	−182
$\tilde{\theta} = 3$	0	17	34	51	−21	−93
$\tilde{\theta} = 4$	0	17	34	51	68	−4
$\tilde{\theta} = 5$	0	17	34	51	68	85

Payoff tables can also be constructed in terms of losses or costs. When the payoffs are in terms of losses or costs, then negative values represent profits or gains.

The payoff table indicates that there is no single act which is best for all the states of nature. Therefore, as in the previous example, in order to make an optimal decision, some criterion and additional information are necessary.

18.5 Prior Distributions and Bayes' Criterion

There are various criteria for selecting an act that is optimal in some sense from among the set of acts. We will discuss only one of these, called the Bayes' criterion. The Bayes' criterion is an expected value criterion. You should remember that in order to compute an expected value, a set of probabilities is necessary.

In any decision problem, the decision maker is concerned with a set of states of nature that affect the consequences of his decision. In a somewhat narrower sense, he is concerned with the value of some unknown population parameter. If he knew the value of this particular parameter, he would also know the corresponding state of nature, and the decision would no longer have to be made under conditions of uncertainty.

Since the state of nature, or the value of the pertinent parameter, is unknown, the decision maker should express his uncertainty about that value in terms of a probability distribution. Remember that probability has previously been defined as the relative likelihood of occurrence of an uncertain event. Once the decision maker has assigned probabilities to each possible state of nature, we have a probability distribution on the states of nature or on the possible values of the unknown parameter. Of course, the probabilities assigned must conform to the requirements of the rules of probability. That is, they must all be nonnegative numbers between zero and one, inclusive, and the sum of the probabilities over all the possible states of nature must equal one.

We have already mentioned that, in a strict sense, there is no such thing as a probability distribution on a parameter. Probability distributions apply to quantities that are variable, not to parameters that are constants. However, *estimates* of a parameter derived from samples do have distributions, and statisticians have adopted various conventions for making probability statements about these estimates. Regardless of the terms used, statements that express the degree of uncertainty inherent in an estimate of some unknown parameter are expressed, or at least interpreted, in terms of probability. Therefore, whether the state of nature is a true variable or represents possible values of an unknown parameter, we will use the concept of a probability distribution.

The probability distribution used at the beginning of the decision analysis is generally referred to as the PRIOR DISTRIBUTION. In many instances, it is possible to develop the prior distribution from empirical information or data that are already available. For example, historical data taken from order forms, invoices, or inspection records regarding orders received, merchandise shipped, or numbers of nonconforming items per lot manufactured could be used to provide frequency distributions from which probabilities could be estimated. In some instances, it might be reasonable to assume that the phenomenon of interest follows a particular theoretical distribution and that the parameters of that distribution can be estimated from available data, such as that mentioned above.

In all cases similar to those mentioned, where there is some objective basis for assigning prior probabilities to the states of nature, this prior distribution is said to be an *objective* prior distribution.

When no empirical information is available, the prior distribution must be determined subjectively. The subjective estimates leading to the prior distribution should not be guesses. They should result from a careful study and analysis of the decision situation and an assessment by the decision maker of the probabilities or odds that a particular state of nature will occur. It is likely that a decision maker has at his disposal considerable expertise and experience relating to the decision problem. The fact that this information does not exist in some objective, quantitative form does not mean that it should be ignored.

Bayes' criterion states that the expected payoff should be computed for each act in the set of acts, and that act should be chosen for which the expected payoff is the best. "Best" would refer to the highest value if the payoffs are expressed in terms of profit or gain, and the lowest value when payoffs are expressed in terms of cost or loss.

Example 18.3 In Example 18.2, all the data related to the acts, states, and economic consequences in Mr. Wilson's decision problem were arranged in the form of a payoff table. Although the table displays these data clearly and concisely, it does not provide sufficient information for making an optimal decision, at least not by Bayes' criterion. In order to apply that criterion, it is necessary to have a probability distribution on the states of nature. In this case, the states of nature are the various levels of demand for Mr. Wilson's houses priced at $89,000.

Consider the fact that Mr. Wilson is a builder with considerable experience in the buying and selling of property. To assume that he is without any information regarding the demand for a particular type of property would be a foolish assumption. It is quite likely that he is capable of estimating the probability of the demand for properties of different types at various prices, if not exactly, then within reasonable limits.

Utilizing his experience, his information about the market for real estate, and his knowledge about the value of these lots in comparison with other property in the area, he should be able to assign weights to each level of demand which are related to his feelings regarding the likelihood of the demand assuming each level. The only rule he must follow is that the sum of all the weights or probabilities must equal one, and that each probability must be a nonnegative number between zero and one inclusive.

Suppose Mr. Wilson believes that the most likely number of his houses that will be demanded is four, that is, $\tilde{\theta} = 4$. He feels that it is possible, but very unlikely, that there will be no demand for his houses at all, that is, that $\tilde{\theta} = 0$. He also believes that the probability that $\tilde{\theta} = 3$ is only slightly less than the probability that $\tilde{\theta} = 4$, and that the probability that $\tilde{\theta} = 2$ and the probability that $\tilde{\theta} = 5$ are equal. Translating these subjective beliefs into a probability distribution that conforms to the rules of probability, Mr. Wilson arrived at the following distribution.

Demand ($\tilde{\theta}$)	$P(\tilde{\theta})$
0	0.05
1	0.10
2	0.15
3	0.25
4	0.30
5	0.15
	1.00

Since the Bayes' criterion is an expected value criterion, it is now necessary to compute an expected value for each act. These calculations are the same as for any expected value as defined by Equation 8.1. We repeat Equation 8.1 for the expected value of a random variable below.

$$E(x) = \sum_{x} xf(x) \tag{8.1}$$

The expected value of an act is simply a weighted average taken over all values of the states of nature where the weights are the probabilities. Since the payoffs in this and in most business problems are expressed in terms of dollars or monetary value, we will call the expected value of an act the *expected monetary value* and abbreviate it EMV. The expected monetary value of an act is

$$\text{EMV}(a_j) = \sum_{i} q_{ij} f(\theta_i) \tag{18.1}$$

Equation 18.1 simply means that in order to obtain the EMV of an act, multiply each conditional payoff of that act for a given state of nature by the probability of that state, and sum all these products over all the states.

Table 18.4 repeats the data provided in Table 18.3 and adds the probability distribution on the states of nature. The expected monetary value of

TABLE 18.4
PAYOFF TABLE WITH PROBABILITIES, LOT PURCHASE DECISION (PAYOFF IN THOUSANDS OF DOLLARS) (EXAMPLE 18.3)

			Acts: A					
States: $\tilde{\theta}$	$P(\tilde{\theta})$		a_0	a_1	a_2	a_3	a_4	a_5
$\tilde{\theta} = 0$	0.05		0	−72	−144	−216	−288	−360
$\tilde{\theta} = 1$	0.10		0	17	−55	−127	−199	−271
$\tilde{\theta} = 2$	0.15		0	17	34	−38	−110	−182
$\tilde{\theta} = 3$	0.25		0	17	34	51	−21	−93
$\tilde{\theta} = 4$	0.30		0	17	34	51	68	−4
$\tilde{\theta} = 5$	0.15		0	17	34	51	68	85
	1.00	EMV	0	12.55	16.20	6.50	−25.45	−84.10

each act is also provided. The expected monetary value of a_3 was calculated as follows.

$$\begin{aligned}\text{EMV}(a_3) &= (-216)(0.05) + (-127)(0.10) + (-38)(0.15)\\ &\quad + (51)(0.25) + (51)(0.30) + (51)(0.15)\\ &= 6.50\end{aligned}$$

All the other EMV's were computed in exactly the same way. Remember, all the payoffs in the table are expressed in terms of thousands of dollars, therefore, each EMV is also in thousands of dollars.

According to Bayes' criterion, the optimal act is a_2, purchase two lots, since $\text{EMV}(a_2) = \$16{,}200$, which is higher than the EMV of any other act.

18.6 Opportunity Loss

It is common in decision theory to specify the consequences of an act for any given state of nature in terms of OPPORTUNITY LOSS instead of payoff. Opportunity loss is simply the absolute value of the difference between the payoff actually realized for the act selected and the payoff which could have been obtained had the best or optimal act been selected. The opportunity loss, which is associated with the optimal act for a given state of nature, would always be 0, that is, no opportunity loss. For any other act, the opportunity loss is always a positive value. If all the consequences are specified in terms of opportunity loss, then the optimal act is always the act with the lowest expected opportunity loss, abbreviated EOL.

A table in which all the consequences are shown in terms of opportunity loss is called an *opportunity loss table*. Any payoff table can be converted to an opportunity loss table in the following way. For each state of nature find the optimal act, that is, the act with the maximum payoff. The opportunity loss for that act and the given state of nature is 0. For all other acts and the given state, the opportunity loss is simply the absolute difference between the conditional value of that act and the conditional value of the optimal act. The opportunity loss table corresponding to Table 18.4 is shown in Table 18.5.

Let's see how some of the opportunity losses in Table 18.5 were calculated. For example, consider the state of nature $\tilde{\theta} = 1$. If there is demand for only one house, then the best possible act that Mr. Wilson could select is a_1, purchase one lot. His payoff for that act and state of nature would be \$17,000, as shown in Table 18.4. The opportunity loss corresponding to the best act is zero, that is, no opportunity loss. If Mr. Wilson had selected a_0, that is, purchased no lots and the demand was for one, his payoff would be zero. However, his opportunity loss would be the absolute difference between that zero payoff and the \$17,000 payoff he would have received had he selected a_1. In other words, the opportunity loss corresponding to a_0 and θ_1 is \$17,000. The opportunity loss for a_2 is the absolute difference between the payoff of −\$55,000 for that act and the \$17,000 payoff for the best act, or \$72,000. The opportunity loss for a_3 when θ_1 is the state of nature is the absolute difference between −\$127,000 and \$17,000, or \$144,000. This procedure is followed for all the other acts for each state of nature. The expected opportunity loss for each act, or EOL, is calculated in the same way as the EMV. Each opportunity loss in a column is multiplied by the probability of the corresponding state of nature and summed over all the states. The EOLs for each act appear in the last line of the table.

The best act is that act with the *lowest* expected opportunity loss. From Table 18.5, that act is a_2, with an EOL of 36.50 or \$36,500. This is the act which also has the highest EMV from Table 18.4. The optimal act will be the same whether the choice is based on EMV or EOL.

Notice in Table 18.5, that the opportunity loss increases by 17,000 for each unit for which a is less than $\tilde{\theta}$. In other words, for

TABLE 18.5
OPPORTUNITY LOSS TABLE WITH PROBABILITIES, LOT PURCHASE DECISION (OPPORTUNITY LOSS IN THOUSANDS OF DOLLARS)

			Acts: A					
States:	$P(\tilde{\theta})$		a_0	a_1	a_2	a_3	a_4	a_5
$\tilde{\theta} = 0$	0.05		0	72	144	216	288	360
$\tilde{\theta} = 1$	0.10		17	0	72	144	216	288
$\tilde{\theta} = 2$	0.15		34	17	0	72	144	216
$\tilde{\theta} = 3$	0.25		51	34	17	0	72	144
$\tilde{\theta} = 4$	0.30		68	51	34	17	0	72
$\tilde{\theta} = 5$	0.15		85	68	51	34	17	0
	1.00							
		EOL	52.70	40.15	36.50	46.20	78.15	136.80

every house demanded that Mr. Wilson cannot sell, he forgoes the \$17,000 profit that he could have made had he purchased that number of lots. This is the cost of underestimating the demand or understocking. Also, for each unit for which a is greater than $\tilde{\theta}$, his opportunity loss increases by 72,000. This is his investment in each house that he cannot recover if he can't sell the house and is the cost of overstocking.

Example 18.4 From the payoff table for the Metric Systems, Inc. bidding problem in Table 18.2, and the information that MSI management estimates that the probability of getting the contract if they submit a proposal and bid is 0.25,

a) find the expected payoff, EMV, of each act and select the optimal act;

b) construct an opportunity loss table and compute EOL for each act.

a)
$$\begin{aligned}\text{EMV}(a_1) &= (0.25)(\$225{,}000) + (0.75)(-\$80{,}000)\\ &= \$56{,}250 - \$60{,}000\\ &= -\$3750.\end{aligned}$$

$$\text{EMV}(a_2) = (0.25)(0) + (0.75)(0) = 0$$

Since $\text{EMV}(a_2) = 0 > \text{EMV}(a_1) = -\$3750$; therefore, a_2 is optimal.

b)

Opportunity Loss Table

State $\tilde{\theta}$	$P(\tilde{\theta})$		a_1	a_2
θ_1: Contract	0.25		0	\$225,000
θ_2: No contract	0.75		\$80,000	0
	1.00			
		EOL	\$60,000	\$56,250

$$\text{EOL}(a_1) = (0.25)(0) + (0.75)(\$80{,}000) = \$60{,}000$$
$$\text{EOL}(a_2) = (0.25)(\$225{,}000) + (0.75)(0) = \$56{,}250$$

Example 18.5 Advance Machine Company ships a particular product to customers in lots of 500 units. The proportion of defective pieces in each lot varies from lot to lot but is generally either 0.01, 0.05, or 0.10. From past experience, they estimate the probability that a lot contains a defective proportion equal to θ as follows.

$\tilde{\theta}$	$P(\tilde{\theta})$
0.01	0.45
0.05	0.25
0.10	0.30

The production manager has a choice to make between two possible courses of action every time a lot is ready for shipment to the customer. He could inspect each lot entirely before shipping and thereby eliminate all defective items from the lot. This involves a cost. The cost of inspection is 15¢ per article inspected, and therefore the cost of inspecting the entire lot would be \$75. We will let the act "inspect 100%" he designated a_1.

The other course of action, a_2, is to ship each lot without inspecting. However, each defective discovered by the customer must be replaced by Advance Machine Company at a cost of \$2.50 per defective item. Obviously, the number of defectives in any lot, and therefore the cost to Advance, of each act is a function of $\tilde{\theta}$. Let the act "ship without inspecting" be designated a_2.

The payoff function for each act in terms of *cost* is: For act a_1, inspect all items, it is

$$Q(a_1) = \$75$$

For act a_2, it is

$$Q(a_2) = (\$2.50)(500)\tilde{\theta} = \$1250\tilde{\theta}$$

The states of nature, prior probabilities, payoff and opportunity loss tables are provided in Table 18.6. The expected payoffs and expected opportunity losses for both acts are shown in the table. Since payoffs are provided in terms of cost, the better act is the act with the lower EMV and, of course, the lower EOL. That act is a_2, "ship as is."

TABLE 18.6
PAYOFF AND OPPORTUNITY LOSS TABLE WITH PRIOR PROBABILITIES, INSPECTION DECISION (PAYOFF AND OPPORTUNITY LOSS IN DOLLARS) (EXAMPLE 18.5)

			Payoff (Cost)			*Opportunity Loss*	
$\tilde{\theta}$	$P(\tilde{\theta})$		a_1	a_2		a_1	a_2
$\tilde{\theta} = 0.01$	0.45		75.00	12.50		62.50	0
$\tilde{\theta} = 0.05$	0.25		75.00	62.50		12.50	0
$\tilde{\theta} = 0.10$	0.30		75.00	125.00		0	50.00
	1.00						
		EMV	75.00	58.75	EOL	31.25	15.00

18.7 Decision Trees

Decision problems, such as those illustrated in the previous examples, can also be solved using *decision tree* analysis, providing the number of acts and states in the problem are not so numerous as to make this approach cumbersome and time consuming. A decision tree is simply a graphic device for displaying the information about the acts, the states, and the consequences. In this respect, it is similar to the tabular displays that we have been using, such as Table 18.6.

In a decision tree, lines that issue from branching points, called NODES, represent occurrences, either the selection of an act by the decision maker or the selection of a state by nature. Lines that

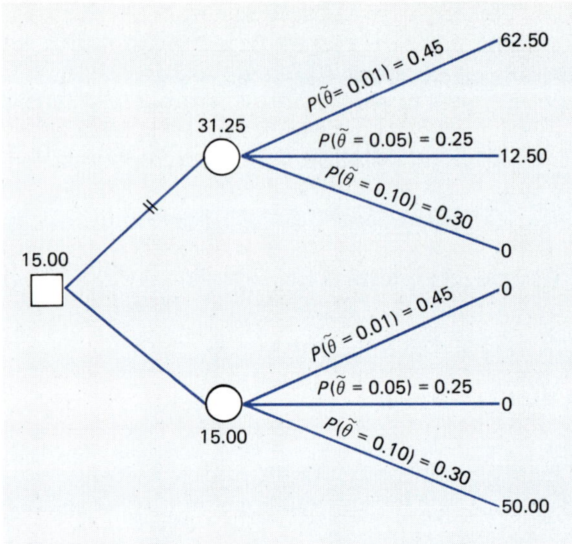

Figure 18.1
Decision tree for inspection problem (Example 18.5).

represent the selection of an act by the decision maker issue from *decision nodes,* designated by small squares. Lines that represent the selection of a state by nature issue from *chance nodes,* designated by small circles. The decision tree corresponding to the problem of Example 18.5 and Table 18.6 is illustrated in Figure 18.1.

The first node in the decision tree is a decision node. The decision maker begins by selecting an act from the set of acts, that is, either a_1 or a_2. The lines issuing from the decision node lead to a chance node where nature selects one of its possible states. In this example, the states of nature are the possible proportions of defectives in the lot, that is 0.01, 0.05, or 0.10. Therefore, three paths issue from each chance node, one for each possible state. At the end of each of these paths, the payoff or opportunity loss resulting from the act and the state of nature leading to that point is noted.

In analyzing a decision problem under uncertainty, it is the prevailing state of nature, or the state that *will* prevail, that the decision maker is uncertain about. The information at his disposal regarding the likelihood of the occurrence of any specific state of nature is in terms of probabilities. Consequently, on each path leading from a chance node, we have noted the probability that chance or nature will select that path or state. This probability is a *prior* probability.

By using the opportunity losses that have been listed at the end of each path on the decision tree and the probabilities associated with each chance path, it is possible, by working backward, to assign a value at each chance node. That value represents the expected value to the decision maker of proceeding to that particular node but as yet with no knowledge of the path that nature will pick except as a probability. This expected value is computed in exactly the same way that expected payoffs and expected opportunity losses were computed from the tabular arrangement of data in Table 18.5. Each opportunity loss is multiplied by its probability of occurrence, and the products are summed. These values, \$31.25 and \$15.00, appear at the appropriate chance nodes and represent expected opportunity loss.

Now let us place the decision maker at the origin of the problem, that is, at the decision node. Before nature makes its selection of a state, or what is essentially the same thing to the decision maker, before nature's selection becomes apparent to him, the decision maker sees an expected opportunity loss of \$31.25 along path a_1 and an expected opportunity loss of \$15.00 along path a_2. Seeking to minimize loss, his logical choice should be act a_2. Having made this decision, path a_1 is blocked off with a double line. This leaves act a_2 open, which is the better act.

18.8 The Expected Value of Perfect Information

We can see that a decision maker in a typical decision situation is faced with the problem of making a choice between alternate acts. However, before actually selecting his course of action, the decision maker usually has another decision to make. This decision is whether to make the final or *terminal decision* immediately, thereby stopping the analysis after using only the information already available to him, or to postpone this terminal decision until he can obtain additional information about the state of nature.

The process of obtaining additional information has a cost associated with it. The decision maker should be concerned about the cost of the additional information and also with its potential value. It would be poor business practice to spend more for anything, including information, than it can possibly be worth. Consequently, the decision maker should have some way of estimating the value of any additional information about the state of nature that he could obtain. For the time being, we will determine the upper limit on this value, that is, the maximum value that any amount of additional information about the state of nature could possibly have. The maximum value of any additional information about the state of nature should coincide with perfect information. In other words, perfect information or absolute certainty about the state of nature is the maximum amount of information possible in any decision problem. In an attempt to evaluate the worth of perfect information we will arrive at a measure which we will call the *expected value of perfect information,* or simply the EVPI.

If a decision maker had perfect knowledge about the state of nature, then he would always be able to choose that act which represents the best or optimal decision under the circumstances. The difference between his payoff with perfect information, that is, under conditions of certainty, and his expected payoff under uncertainty represents the *cost of the uncertainty*. We have already found a measure of the cost of uncertainty. This measure is the expected opportunity loss of the *optimal* act under the Bayes' criterion. Opportunity loss would always be 0 with perfect information because under those conditions the right decision would always be made. Therefore, the expected opportunity loss of the best act under uncertainty represents the amount by which the decision maker could improve the consequences of his decision making if he had perfect information. We will define the EVPI as being exactly equal to the expected opportunity loss of the best act (by the Bayes' criterion) under uncertainty.

The EVPI can also be defined as the absolute difference between expected payoff under conditions of certainty and the expected payoff of the best act under uncertainty. Either way, it represents the maximum improvement that could be expected given perfect information. Since it is never possible to obtain perfect information through sampling, the value of additional information obtained by the decision maker by sampling must always be *less* than the EVPI. This provides a guide for the decision maker as to the maximum amount he should spend in order to obtain additional information about the state of nature. The *maximum* expenditure for this purpose should always be *less* than the EVPI.

Example 18.6 Refer to Example 18.5 regarding the production manager's problem in deciding whether or not to inspect each lot of his product. The decision facing him was to choose between the acts "inspect each lot 100%" or "ship each lot without inspecting." The cost or payoff of the act chosen depends on the state of nature which, in this case, was the true proportion of defectives in each lot. Based on his prior probability distribution on this parameter, the expected opportunity loss was computed for each act. For act a_1, "inspect 100%," $\text{EOL}(a_1) = \$31.25$. For act a_2, "ship without inspecting," $\text{EOL}(a_2) = \$15.00$.

If a terminal decision were to be made immediately, based only on his prior information, the manufacturer should select act a_2, which in the long run and in the absence of additional information will incur an average opportunity loss for each lot shipped of only \$15.00 as opposed to \$31.25 for the other act. If the exact proportion of defectives in each lot was known before shipping, the manufacturer could *always* select the better act and his opportunity loss for each lot shipped would be 0. Therefore, the amount \$15.00 is the maximum improvement he could expect per lot if he had perfect information about the proportion of defective product in each lot.

Since some information about the state of nature could be obtained at a cost by sampling, and since *perfect* information can never be obtained from a sample because sampling error is always present, then the value of perfect information should never be spent in obtaining information from a sample. In this case, the manufacturer should never spend as much as $15.00 to sample from a lot before shipment.

Let us assume that when sampling is undertaken it involves a fixed cost of $10.00 plus 20¢ per sampled article. In that case, a sample of size 25 would cost $10.00 + $5.00 = $15.00, which is exactly equal to EVPI. Since the expenditure for sampling should never be equal to EVPI, the maximum size sample that should be taken would be a sample of size 24, which would cost $14.80.

18.9 Sampling and Revising the Prior Distribution

Suppose that, facing a choice among alternate acts, the decision maker concludes that it would be worthwhile to defer his decision until he can obtain additional information about the state of nature. This information would be obtained by experimentation or sampling. However, having collected data from a sample, the decision maker is now faced with the problem of how to use it.

The information available to the decision maker before a sample is taken is called the prior information. The prior information is expressed in the form of a probability distribution called the *prior distribution*. Information obtained from a sample is called the sampling information. The sampling information should be combined with the prior information so that all the information about the state of nature available to the decision maker can be used in making the decision. The method for combining the prior and sampling information will result in a new probability distribution called the POSTERIOR DISTRIBUTION. We will designate the prior distribution on the states of nature as $P_0(\tilde{\theta})$ and the posterior distribution as $P_1(\tilde{\theta})$. The posterior distribution is used for selecting the optimal act after the sample is taken in exactly the same way as the prior distribution is used before the sample is taken.

When a sample is taken, we observe some random variable which is related to the unknown states of nature and compute a value called a statistic. Recall that a sample statistic has a distribution called the sampling distribution. It is knowledge about the characteristics of the sampling distribution which enabled the statistician to make confidence statements about a population parameter, or to devise decision rules for hypothesis tests about that parameter in the chapters on statistical inference. In this chapter, we refer to the statistic computed from the sample values in general as x, and the sampling distribution of x provides $P(x = x_i | \theta)$. We will designate the sampling distribution simply as $P(x | \theta)$. When x is the result of a particular sampling process, the probability of having obtained that particular result for a given value of $\tilde{\theta}$ is called the *likelihood*.

The reason for combining sampling information with prior information is to obtain an estimate of the likelihood of occurrence of each possible state of nature that uses all the data available to the decision maker after or *posterior* to sampling. Consequently, the result of combining the prior information and the sampling information is the new probability distribution on the states of nature, called the POSTERIOR DISTRIBUTION. This process of obtaining the posterior distribution involves a revision of the prior distribution based on data obtained from the sample. Since the posterior distribution provides the probability that $\tilde{\theta}$ takes on a particular value or state given the sampling results, this probability can also be expressed as $P_1(\tilde{\theta} = \theta_i \,|\, x)$, or simply as $P_1(\tilde{\theta} \,|\, x)$.

The method for combining the prior and sampling information into a new posterior distribution is provided by Bayes' Rule, which states

$$P(\theta_i \,|\, x) = \frac{P(x \,|\, \theta_i)P(\theta_i)}{\sum_i P(x \,|\, \theta_i)P(\theta_i)} \tag{18.2}$$

Notice that Equation 18.2 is the same as Equation 7.15 for Bayes' rule, which was used in the chapter on probability and is reproduced here.

$$P(A_i \,|\, B) = \frac{P(B \,|\, A_i)P(A_i)}{\sum_i P(B \,|\, A_i)P(A_i)} \tag{7.15}$$

Only the symbols are different in Equation 18.2 in order to be consistent with the notation of this chapter.

Example 18.7 Refer once more to the problem of the production manager of the Advance Machine Company in Example 18.6, who must make a decision between the acts "inspect each lot 100%," or "ship each lot without inspecting." The payoff function for each act was given as

$$Q(a_1) = \$75$$
$$Q(a_2) = \$1250\tilde{\theta}$$

The unknown state of nature or parameter θ is the proportion of defective product in each lot. From past experience, the manufacturer assigned the following probabilities to the three values of θ that he considered possible.

$\tilde{\theta}$	$P_0(\tilde{\theta})$
0.01	0.45
0.15	0.25
0.10	0.30

Using these prior probabilities, the production manager computed the expected opportunity loss of each act with the following results

$$\text{EOL}(a_1) = \$31.25$$
$$\text{EOL}(a_2) = \$15.00$$

leading to a decision to ship each lot without inspecting.

The EVPI, the expected value of perfect information, which is equal to the expected opportunity loss of the better act, is \$15.00. The manager could reasonably spend something less than \$15.00 to obtain additional information about the proportion of defectives in each lot. Since obtaining information involves sampling, and the cost of sampling is \$10.00 plus 20¢ per sampled article, he could conceivably take a sample of up to 24 items at a cost ranging up to \$14.80. Suppose, however, that he decided to take a sample of only two items and observe the number of defectives. A sample is actually taken from the next lot ready to be shipped and no defectives are found in the sample. How should this affect his decision?

It is reasonable to assume that a sample containing 0 defectives is more likely to have come from a lot with a small proportion of defectives than from one containing a large proportion of defectives. Consequently, more weight should be given to a low value of θ than to a high one. Exactly what weights should be assigned are determined by Bayes' rule. First, however, the conditional probabilities or likelihoods must be determined. In order to determine the likelihoods, something must be known about the sampling distribution of the statistic obtained.

When sampling from a large, dichotomous population and observing the number of elements possessing or failing to possess some attribute, the sampling distribution of the "number of defectives" is binomial with parameters n and θ. Here, θ is used to represent the parameter of a binomial distribution instead of p simply to be consistent with the use of θ in this chapter to represent any parameter or state of nature. For any specific value of n and θ, the probability of getting exactly X defectives is easily determined either directly from the binomial probability function or from the binomial tables in Appendix Table A.1.

In this example, the probabilities of finding no defectives in a sample of two items for each possible value of θ turn out to be

$$P(X = 0 | \theta = 0.01) = 0.98$$
$$P(X = 0 | \theta = 0.05) = 0.90$$
$$P(X = 0 | \theta = 0.10) = 0.81$$

These conditional probabilities are the likelihoods of observing the sample outcome, $X = 0$, for each of the possible parameter values or states of nature affecting the problem. Bayes' rule provides a method of revising the prior probabilities assigned to the states of nature, using the information obtained from the sample. The prior probabilities were

$$P_0(\theta = 0.01) = 0.45$$
$$P_0(\theta = 0.05) = 0.25$$
$$P_0(\theta = 0.10) = 0.30$$

Bayes' rule, Equation 18.2, specifies that

$$P_1(\theta_i \mid X) = \frac{P(X \mid \theta_i) P_0(\theta_i)}{\sum_i P(X \mid \theta_i) P_0(\theta_i)}$$

By substituting in Equation 18.2 we get

$$P_1(\theta = 0.01 \mid X = 0) = \frac{(0.98)(0.45)}{(0.98)(0.45) + (0.90)(0.25) + (0.81)(0.30)} = \frac{0.441}{0.909} = 0.485$$

$$P_1(\theta = 0.05 \mid X = 0) = \frac{(0.90)(0.25)}{(0.98)(0.45) + (0.90)(0.25) + (0.81)(0.30)} = \frac{0.225}{0.909} = 0.248$$

$$P(\theta = 0.10 \mid X = 0) = \frac{(0.81)(0.30)}{(0.98)(0.45) + (0.90)(0.25) + (0.81)(0.30)} = \frac{0.243}{0.909} = 0.267$$

Having observed no defectives in a sample of size 2, the prior probabilities that were assigned to the three possible states of nature have been revised as follows.

Probability That θ Equals	*Prior to Sample*	*Posterior to Sample*
0.01	0.450	0.485
0.05	0.250	0.248
0.10	0.300	0.267

Since the sample was very small, the revised probabilities are not greatly different from the prior probabilities; however, somewhat greater weight is now given to the low value of θ and less weight to the high value.

Using the new posterior probabilities, the decision maker can now perform the same analysis that was previously performed with the prior probabilities. This is done in Table 18.7.

TABLE 18.7
PAYOFF AND OPPORTUNITY LOSS TABLE AFTER SAMPLE WITH ZERO DEFECTIVES (EXAMPLE 18.7)

			Payoff (Cost)			*Opportunity Loss*	
$\tilde{\theta}$	$P_1(\tilde{\theta})$		a_1	a_2		a_1	a_2
$\tilde{\theta} = 0.01$	0.485		75.00	12.50		62.50	0
$\tilde{\theta} = 0.05$	0.248		75.00	62.50		12.50	0
$\tilde{\theta} = 0.10$	0.267		75.00	125.00		0	50.00
		EMV	75.00	54.94	EOL	33.41	13.35

After the sample results in no defectives out of two items sampled, we can see that the better act is still a_2, since it has the lower expected cost and the lower expected opportunity loss. The EOL of a_2 after the sample is \$13.35, as compared to an EOL of a_2 prior to the sample of \$15.00. Since the EOL of the better act is the EVPI, and is also called the cost of uncertainty, we can see that after observing the sample result, the cost of uncertainty has been reduced by \$1.65.

It might be interesting to determine what the effect would be on the decision to select a_1 or a_2 if the sample had produced one or two defectives. We could conduct exactly the same analysis as the one just concluded; however, this can be done more simply if the data are arranged in tabular form.

Suppose, for example, that instead of finding no defectives in the sample of size 2, we observe one defective. The data can be arranged as shown in Table 18.8.

Column 1 lists the states of nature and column 2 provides the prior probabilities. The conditional probabilities of the sample result, in this case one defective, are given in column 3. These conditional probabilities or likelihoods can be taken directly from the binomial tables. The joint probabilities, column 4, result from multiplying the entries in column 3 by the entries in column 2. The sum of the joint probabilities in column 4 is the marginal probability of the event "one defective."

The posterior probabilities in column 5 are obtained by dividing each of the joint probabilities in column 4 by the marginal probability of that sample outcome, which is the sum of the entries in that column. Columns 6 and 7 simply reproduce the opportunity loss table of the problem. The expected opportunity loss of each act is computed using the new set of posterior probabilities instead of the prior probabilities.

We can see that if one defective is observed in a sample of size 2, much more weight is given to the high value of θ and much less to the low value. When these posterior probabilities are applied to the opportunity loss table we can see that a_1 is now the preferred act, since it has the smaller expected opportunity loss.

TABLE 18.8
CALCULATION OF POSTERIOR PROBABILITIES AFTER OBSERVING ONE DEFECTIVE (EXAMPLE 18.7)

1	2	3	4	5	6	7
State of Nature $\tilde{\theta}$	*Prior Probability* $P_0(\tilde{\theta})$	*Likelihood* $P(X = 1 \mid \tilde{\theta})$	*Joint Probability* $P(X \cap \tilde{\theta})$	*Posterior Probability* $P_1(\tilde{\theta})$	*Opportunity Loss* a_1	a_2
0.01	0.45	0.020	0.009	$\frac{0.009}{0.081} = 0.11$	62.50	0
0.05	0.25	0.095	0.024	$\frac{0.024}{0.081} = 0.30$	12.50	0
0.10	0.30	0.160	0.048	$\frac{0.048}{0.081} = 0.59$	0	50.00
Total	1.00		0.081	1.00		
				EOL	10.63	29.50

Also note that EOL(a_1) is \$10.63, which is the cost of uncertainty after a sample of size 2 is taken, resulting in one defective. This can be compared to the EOL of the better act prior to the sample, which was \$15.00. The information from the sample with one defective reduced the cost of uncertainty by \$4.37.

18.10 Preposterior Analysis

The previous example has illustrated how data obtained from a sample can be used to revise a set of prior probabilities on the states of nature. A decision rule that specifies the better or best act corresponding to each possible sample outcome can be determined by this analysis even before the sample is actually taken.

In the previous section, when considering the decision to inspect each lot of parts or to ship without inspection, we were able to determine the better act if the sample of size 2 produced no defectives or one defective. It is just one more step to determine the better act for the one remaining sample outcome that could occur, that is, two defectives. Table 18.9 shows the data leading to a new posterior distribution and expected opportunity losses of the two acts if a sample of size 2 results in two defectives.

TABLE 18.9
POSTERIOR PROBABILITIES AFTER OBSERVING TWO DEFECTIVES

1	2	3	4	5		6	7
						Opportunity Loss	
$\tilde{\theta}$	$P_0(\tilde{\theta})$	$P(X = 2\|\theta)$	$P(X \cap \theta)$	$P_1(\tilde{\theta})$		a_1	a_2
0.01	0.45	0.0001	0.000045	0.01		62.50	0
0.05	0.25	0.0025	0.000625	0.17		12.50	0
0.10	0.30	0.0100	0.003000	0.82		0	50.00
Total	1.00		0.003670	1.00			
					EOL	2.75	41.00

For a_1, EOL is \$2.75, while for a_2, EOL is \$41.00. If a sample of size 2 results in two defectives, the better act is to inspect 100%.

An analysis of the type performed in this and the preceding section is called a PREPOSTERIOR ANALYSIS. It consists of determining the posterior distribution that would result for every possible outcome from a sample of a specified size. Each resulting posterior distribution is used to compute the expected opportunity loss of all the acts available and to select that act for which the EOL is the smallest.

The decision rule resulting from the analysis just concluded can be shown as follows.

Take a sample of size 2 and choose:

act a_2 if there are 0 defectives;
act a_1 if there is 1 defective;
act a_1 if there are 2 defectives.

18.11 The Expected Value of Sample Information

The value of information in the context of a particular decision problem was discussed previously. The maximum value of information to the decision maker was considered to be equal to the cost he would incur due to his uncertainty or lack of knowledge about the state of nature. This maximum value was called EVPI, and is equal to the difference between the economic consequences the decision maker would realize under conditions of certainty and the consequences he would incur under the conditions of uncertainty existing at the time the decision was made.

Because a decision maker can never obtain better results than would be realized under conditions of certainty, that is, with perfect information about the state of nature, the expected value of perfect information, or EVPI, represents the maximum that should ever be spent for obtaining such information. However, as long as information about the state of nature must be obtained by means of sampling, perfect information is never possible to achieve. Therefore, an amount equal to the EVPI should never be spent on the information-gathering process. At the same time, the information obtained from the sample should be worth more to the decision maker than it costs him to obtain it. It would be desirable to be able to determine the value of information that can be obtained from a sample of size n before actually taking the sample. Although it is not possible to determine the exact value of sample information before taking the sample, it is possible to compute an expected value, which is referred to as the *expected value of sample information,* or simply the EVSI.

One method of measuring the value of the information that could be obtained from a sample is to determine the difference between the expected consequence of making a decision without the sample information and the expected consequence of the decision made after a sample of size n is taken. Obviously, these expected consequences must be determined and compared before the actual sampling is done, since the reason for making this comparison is to determine whether or not it would be worthwhile to take the sample.

Prior to sampling, the measure of the expected consequence of making an immediate terminal decision is simply the expected opportunity loss of the best or better act. This value is called the prior EVPI. If a sample of size n were to be taken, it would be possible to compute a posterior expected opportunity loss for each act and for all of the possible sample outcomes. This procedure has been illustrated in the example provided in the previous sections. The posterior expected opportunity loss for a given sample outcome is the expected opportunity loss of the best act available, given that specific sample outcome. This value will, of course, differ depending on the sample outcome and, since prior to sampling we do not know what the outcome will be, the question arises as to how to compute a single value which can be used to represent posterior expected opportunity loss for all possible sample outcomes.

Although we do not know which sample outcome will result from the sampling experiment that we may conduct, we have a way of determining the probability of each possible sample outcome occurring. The probability of any sample outcome occurring is simply the marginal probability of that outcome, which is the sum of the joint probabilities for that sample result. This can be written as

$$P(\mathrm{x} = \mathrm{x}_i) = \sum_{\tilde{\theta}} P(\mathrm{x}_i \cap \tilde{\theta}) = \sum_{\tilde{\theta}} P(\mathrm{x}_i|\tilde{\theta})P(\tilde{\theta})$$

We can, therefore, find the *expected value of the posterior* EOL by taking the posterior expected opportunity loss of the best act for *each* sample outcome, weighting each of these values by the marginal probability of that sample outcome occurring, and then summing these weighted values over all possible sample outcomes. The value obtained is the *expected risk of the decision rule* and is also called the *Bayes' risk of the decision rule*. The expected value of sample information or EVSI is simply the difference between the prior EVPI and the Bayes' risk of the decision rule.

Example 18.8 To continue with the previous example, the EVSI for a sample of size 2 is obtained as follows.

Prior to sampling, the EVPI was equal to the EOL of act a_2, which was \$15.00. After conducting a preposterior analysis for a sample of size 2, the following results were obtained.

If the Number of Defectives	*The Better Act*	*EOL of the Better Act*	*Marginal Probability of the Sample Outcome*	*Product of EOL and Marginal Probability*
$X = 0$	a_2	\$13.35	0.909	\$12.14
$X = 1$	a_1	\$10.63	0.081	.86
$X = 2$	a_1	\$ 2.75	0.004	.01
			0.994	\$13.01

The sum of the marginal probabilities of all sample outcomes should equal 1. A minor discrepancy due to rounding errors, as in this case, is relatively unimportant.

The Bayes' risk of this decision rule is \$13.01. The EVSI for this sample of size 2 is computed as follows.

$$\text{EVSI} = \text{Prior EVPI} - \text{Bayes' risk} = \$15.00 - \$13.01 = \$1.99 \qquad (18.3)$$

We have determined that the expected value of the information from a sample of size 2 is \$1.99. Would it be worthwhile, therefore, to take a sample of size 2 under these circumstances? That would depend on the cost of taking the sample. If, as previously specified, there was a fixed cost of sampling of \$10.00 plus a variable cost of 20¢ per item sampled, then a sample of size 2 would cost \$10.40. Since the cost of the sample is greater than its expected value, the sample should not be taken. There is a value called the *expected net gain from sampling*, abbreviated ENGS, which is equal to the EVSI minus the cost of the sample, that is,

$$\text{ENGS} = \text{EVSI} - C(n) \qquad (18.4)$$

where $C(n)$ represents the cost of a sample of size n. A sample of size n is worthwhile only if ENGS > 0. In this example, ENGS $= \$1.99 - \$10.40 = -\$8.41$. Obviously it would not be worthwhile to take a sample of size two in this decision problem.

Definitions

Decision making under risk: Decision problems in which a probability distribution on the states of nature can be based on empirical data.

Decision making under uncertainty: Decision problems in which the probability distribution on the states of nature is based on subjective information.

Decision tree: A graphical device for displaying the factors in a decision problem in the form of a tree diagram.

ENGS: Expected net gain from sampling.

EVPI: Expected value of perfect information.

EVSI: Expected value of sampling information.

Node: A branch point on a decision tree.

Opportunity loss: The absolute value of the difference between the payoff actually realized for the act selected and the payoff which could have been obtained had the optimal act been selected.

Opportunity loss table: A table displaying the acts, states of nature, and consequences in a decision problem in terms of opportunity loss.

Payoff table: A table displaying the acts, states of nature, and consequences in a decision problem in terms of actual costs, losses, profits, or gains.

Posterior distribution: A probability distribution on the states of nature which results from combining the prior distribution with information obtained from a sample.

Pre-posterior analysis: A procedure for determining the optimal decision rule associated with a sample of size n before the sample is actually taken.

Prior distribution: A probability distribution on the states of nature obtained prior to any sampling or experimentation.

States of nature: Factors in a decision problem which affect the consequences of a decision and are beyond the control of the decision maker.

Formulas

Expected monetary value:

(18.1) $$\text{EMV}(a_j) = \sum_{\tilde{\theta}} q_{ij} f(\theta_i)$$

Bayes' rule:

(18.2) $$P(\theta_i | \text{x}) = \frac{P(\text{x}|\theta_i)P(\theta_i)}{\sum_i P(\text{x}|\theta_i)P(\theta_i)}$$

EVSI:

(18.3) $$\text{EVSI} = \text{Prior EVPI} - \text{Bayes' risk.}$$

ENGS:

(18.4) $$\text{ENGS} = \text{EVSI} - C(n)$$

Problems

18.1 Satellite Systems Corporation received a government contract for the design, development, and construction of a new communications satellite. Satellite Systems in turn asked companies A, B, and C to submit bids for one of the major subsystems of the satellite. SSC plans to award the subcontract to the lowest bidder among the three. List the acts and states of nature for SSC's decision problem.

18.2 Mr. Granger owns 20 acres of land near the route of a proposed expressway. If the expressway route is approved, he will be paid \$600,000 by the state highway department for the land. If the route is not approved, he intends to proceed with his previous plan to subdivide the land into one-acre building lots. He believes that he could sell each lot for \$15,000; however, he would incur subdividing and sales costs of \$1500 per lot. Mr. Granger has just received an offer of \$400,000 for the land from a large insurance company which would use it for the construction of a new home office building. He must either accept or reject this offer within 48 hours. List the acts and states of nature for this decision problem and construct the payoff table.

18.3 The owner of the Riverside Men's Clothing Outlet has been offered a lot of 100 men's suits at a price of \$1500 from a clothing manufacturer who was forced into bankruptcy when his factory was flooded during a recent storm. If the suits are undamaged by the water, Riverside can sell them for \$40 each. However, damaged suits would be a total loss.

a) List the acts in this decision problem.

b) How many states of nature are there in this problem?
c) Describe the states of nature.

18.4 Harry's Discount Appliance Store has the opportunity to purchase up to five color television sets at $100 each from a manufacturer who is discontinuing the line. The sets are of good quality and were designed to sell for $400. If Harry buys any of the sets, he would price them at $299.
a) What are the acts in this problem?
b) What are the states of nature in this problem?
c) What are the payoff functions in this problem?
d) Construct the payoff table.

18.5 Suppose that the bankrupt manufacturer in problem 18.3 had examined the suits and in his letter offering them to Riverside had specified the number of damaged suits in the lot. However, the first digit of the number was smudged, which made it difficult to read. The owner of Riverside believed that the number of damaged suits indicated in the letter was either 30, 50, or 80, but he wasn't sure which number was correct. The bankrupt manufacturer could not be contacted and a response to the offer was due the next day. Construct the payoff table for this problem.

18.6 The owner of a small neighborhood bakery bakes "homemade" French bread each night to sell in the bakery the next day. The bread costs him 45¢ per loaf to bake and he sells it for $1.00 per loaf. Any loaves that are unsold at the end of the day are purchased by a nearby supermarket at 20¢ a loaf. Each night the baker must decide how many loaves to bake for the next day. Assume that in the past he has never sold fewer than 20 loaves or more than 25. Construct the payoff table for this problem. What are the payoff functions?

18.7 Refer to problem 18.2. Suppose that Mr. Granger estimated that the probability that the expressway route is approved is 0.30. Find the EMV of each act in his decision problem and select the optimal act by Bayes' criterion.

18.8 Suppose that in problem 18.7 Mr. Granger estimated the probability that the expressway route is approved at 0.40. Find the EMV of each act and select the optimal act by Bayes' criterion.

18.9 Refer to problem 18.4. Harry, of Harry's Discount Appliance store, believes that there is one chance in 10 that he will be able to sell none of the discontinued television models and exactly the same probability that he will sell four. He feels that the probability that he can sell either two or three is 40%, and on further consideration concludes that the probability of selling two is equal to the probability of selling three. He also believes that the probability of selling only one is about 35%. List the values of the prior probability distribution on the demand for these sets and compute the EMV for each act. How many of the sets should Harry buy?

18.10 What is the optimal act in problem 18.9 if the prior distribution on the demand is as follows?

$\tilde{\theta}$	$P(\tilde{\theta})$
0	0.05
1	0.20
2	0.30
3	0.35
4	0.05
5	0.05

18.11 Refer to problems 18.3 and 18.5. The owner of the Riverside Men's Clothing Outlet examined the bankrupt manufacturer's letter very closely and finally concluded that the following probabilities reflected his uncertainty about the value of the smudged number.

$\tilde{\theta}$	$P(\tilde{\theta})$
30	0.20
50	0.30
80	0.50

Find the EMV of the acts in the problem and select the optimal act by Bayes' criterion.

18.12 Suppose that in problem 18.11 the owner of Riverside assigned equal probabilities to each of the possible values of the states of nature. Find the EMV for each act and select the optimal act by Bayes' criterion.

18.13 Suppose that, in the baker's decision problem of problem 18.6, he assigned the following probabilities to the demand for the French bread.

$\tilde{\theta}$	$P(\tilde{\theta})$
20	0.05
21	0.20
22	0.25
23	0.20
24	0.20
25	0.10

Find the EMV of all acts and select the optimal act.

18.14 Satellite Systems Corporation has a government contract to produce 5000 sophisticated communications devices, each of which requires a critical component that must be produced to very tight tolerances. Advance Machine Company can supply these components and guarantee that they will meet the specifications at a price of \$19.50 each. SSC could manufacture these components themselves at a fixed tooling and setup cost of \$27,500 plus \$10.50 per unit produced. However, the proportion of defectives could be as high as 40% of the numbers produced. For every defective part made by SSC, the cost of detection, rework, and replacement is $7.50.

Although SSC's manufacturing engineers cannot predict the exact proportion of defectives that would occur, they have provided the following probability distribution on this parameter.

$\tilde{\theta}$	$P(\tilde{\theta})$
0	0.10
0.10	0.35
0.20	0.25
0.30	0.20
0.40	0.10

a) Construct the payoff and opportunity loss table for this problem.
b) Determine the EMV and the EOL for each act.
c) Should SSC make or buy this component?
d) What is the EVPI?

18.15 Use the data provided in problems 18.3, 18.5, and 18.11 to construct the opportunity loss table for the Riverside Men's Clothing Outlet problem. Determine the EOL for each act and specify the EVPI.

18.16 Use the data provided in problems 18.6 and 18.13 to construct the opportunity loss table for the baker's problem. Determine the EOL for each act and specify the EVPI.

18.17 Use the data provided in problems 18.2 and 18.7 to construct the opportunity loss table for the decision problem. Compute the EOL for each act and specify the EVPI.

18.18 Construct a decision tree in terms of opportunity loss for problem 18.17.

18.19 Construct a decision tree in terms of opportunity loss for the Riverside Men's Clothing Outlet problem. Refer to problems 18.3, 18.5, 18.11, and 18.15 for data.

18.20 From the data and your solution to problem 18.17, determine the maximum amount that Mr. Granger should pay for information regarding the approval of the proposed expressway route.

18.21 Refer to problem 18.15. Suppose that the owner of the Riverside Men's Clothing Outlet was offered the opportunity to take a random sample of suits from the lot offered for sale and inspect the sample for damage. If the cost of the sample was $20 per unit sampled, what is the maximum size sample he should consider taking?

18.22 Before deciding whether or not to subcontract the manufacture of the critical component described in problem 18.14, the production manager of Satellite Systems Corporation considered the possibility of producing a small pilot sample of components in order to better estimate the proportion defective. If the cost of producing a pilot sample is \$1000 plus \$20 per unit produced, what is the largest sample he should consider?

18.23 Refer to the data and your solutions to problems 18.11, 18.15, and 18.21. Suppose that the owner of Riverside Men's Clothing Outlet takes a sample of two suits from the lot of suits offered for sale and finds that neither suit is damaged.

a) Find the posterior distribution on the number of damaged suits in the lot.

b) What is the optimal decision after the sample is taken?

c) What is the posterior EVPI, after the sample?

Hint: The three possible values of the number of damaged suits should be converted to proportions on the basis of 100 suits in the lot. Use the binomial distribution to find the conditional probabilities or likelihoods.

18.24 Suppose that in problem 18.23, one of the two suits sampled was damaged.

a) Find the posterior distribution on the number of damaged suits in the lot.

b) What is the optimal decision after considering this sample result?

c) What is the posterior EVPI after this sample result?

18.25 In the Satellite Systems Corporation make-or-buy decision, problems 18.14 and 18.22, suppose a pilot sample of three components resulted in one defective.

a) Find the posterior distribution on the proportion defective.

b) What is the optimal decision after this sample result?

c) What is the posterior EVPI after this sample result?

Hint: Use the binomial distribution to obtain the likelihoods for each possible value of the proportion defective.

18.26 Use preposterior analysis and the results of problems 18.23 and 18.24 to determine the decision rule that should be used by the Riverside Men's Clothing Outlet for all possible samples of size 2.

18.27 Use preposterior analysis and the data from problems 18.11, 18.15, and 18.21 to determine the decision rule that should be used by the Riverside Men's Clothing Outlet for all possible outcomes of a sample of size 3.

18.28 Use preposterior analysis and the data from problems 18.14 and 18.22 to determine the decision rule that should be used by Satellite Systems Corporation in making their make-or-buy decision based on the outcome of a sample of size 3.

18.29 What are the Bayes' risk, EVSI, and ENGS of a sample of size 2 in the Riverside Men's Clothing Outlet decision problem when the cost of sampling is \$20 per suit sampled. Use the data of problems 18.23, 18.24, and 18.26.

18.30 What are the Bayes' risk, EVSI, and ENGS of a sample of size 3 in the Riverside Men's Clothing Outlet decision problem when the cost of sampling is \$20 per suit sampled. Use the data of problems 18.11 and 18.15.

18.31 What are the Bayes' risk, EVSI, and ENGS of a sample of size 2 in the Satellite Systems Corporation decision problem when the cost of sampling is \$1000 plus \$20 per component. Use the data of problem 18.14.

18.32 What are the Bayes' risk, EVSI, and ENGS of a sample of size 3 in the Satellite Systems Corporation decision problem when the cost of sampling is \$1000 plus \$20 per component. Use the data of problem 18.28.

18.33 Refer to Examples 18.5, 18.6, 18.7, and 18.8 in the text. Find EVSI and ENGS of a sample of size 3 in the Advance Machine Company inspection decision problem.

GLOSSARY

Acceptance region: That portion of the area under the sampling distribution of a test statistic that corresponds to acceptance of the null hypothesis.

Aggregative index number: A composite index number which is constructed by taking the ratio of two sets of aggregated numbers, either weighted or unweighted.

Alpha (α): The probability of making a type I error. The significance level of a test.

ANOVA: Abbreviation for analysis of variance.

Arithmetic mean: The arithmetic mean is defined best by Equations 4.3 and 4.6.

Array: A set of numerical data arranged in order of magnitude.

Attribute: A characteristic of an elementary unit that can only be observed as to its presence or absence.

Average of relatives index number: A composite index number constructed by first converting pairs of values in the base and nonbase periods to relatives and then computing the arithmetic mean of the relatives.

Base period: The time period used as the base or basis of comparison when constructing an index number.

Bernoulli population: A population divided into two mutually exclusive subsets according to the presence or absense of an attribute of interest.

Beta (β): The probability of making a type II error.

Census: A survey in which observations are made on the entire population of elementary units.

Class boundaries: The highest and lowest actual values of observations that belong to a specific class.

Class interval: The range of actual values that belong to a given class.

Class limits: The highest and lowest recorded values that can go into a specified class.

Class mark: The class midpoint. A representative value for the class.

Classification variable: Any variable characteristic of data which is used to divide individual observations into categories or cells.

Cluster sampling: A sampling method in which heterogeneous groups of elements are randomly selected from a population of such groups.

Coefficient of correlation: A statistical measure of the degree or strength of the association between two variables.

Coefficient of determination: A statistical measure of the proportion of the variation in the dependent variable that is explained by the regression equation.

Coefficient of multiple correlation: A statistical measure of the degree of association between a dependent variable and two or more independent variables.

Coefficient of multiple determination: A statistical measure of the proportion of the variation in the dependent variable that is explained by the regression equation with two or more independent variables.

Coefficient of rank correlation: A measure of association between two variables based on ranks.

Coefficient of variation: A measure of relative variation. See Equation 5.12.

Composite index number: A relative which summarizes the relationship between two sets of prices, quantities, or other values.

Confidence: The probability that a specified interval contains the population parameter of interest.

Confidence interval estimate: A range of values which includes the population parameter with a stated probability.

Confidence limit: An endpoint of a confidence interval.

Contingency table: A table in which data are classified according to two classification variables.

Continuous random variable: A numerical random variable that can assume any value in some range of values.

Continuous variable: A variable that can assume any value within some range of values. Observations on a continuous variable are the result of measurements.

Correlation analysis: A statistical method for measuring the strength of association between any two random variables.

Critical region: That portion of the area under the sampling distribution of a test statistic that corresponds to rejection of the null hypothesis.

Critical value: The value of a test statistic that divides the acceptance region from the critical region.

Crude mode: The class mark of the class in a frequency distribution that has the greatest number of observations.

Cumulative frequency distribution: A frequency distribution that indicates the number of observations that are "less than" or "more than" certain specified values.

Cumulative relative frequency distribution: A frequency distribution that indicates the proportion of observations that are "less than" or "more than" certain specified values.

Cycle: The periodic growth and decline in a variable that occurs with regularity over a period of from two to 15 or more years.

Deciles: Those values that divide a set of data into ten equal parts.

Decision making under risk: Decision problems in which a probability distribution on the states of nature can be based on empirical data.

Decision making under uncertainty: Decision problems in which the probability distribution on the states of nature is based on subjective information.

Decision rule: A rule that specifies the action to be taken with respect to the null hypothesis for all possible values of the test statistic.

Decision tree: A graphical device for displaying the factors in a decision problem in the form of a tree diagram.

Decomposition: The process of isolating each component of a time series.

Descriptive statistics: The branch of statistics which is concerned only with describing and summarizing sets of numerical data. In general, we assume that descriptive statistics deals with statistical populations.

Deseasonalizing: The process of removing the seasonal component from a time series.

Discrete random variable: A random variable that is limited to certain values.

Discrete variable: A variable that is restricted to certain values within some range of values. Generally these values are integers. Observations of a discrete variable are counts.

Distribution-free statistics: Statistical procedures that require no assumptions about the population distribution; this term is used interchangeably with "nonparametric statistics."

Elementary unit: The physical entity on which an observation is made.

ENGS: Expected net gain from sampling.

Estimator: A statistic used for estimating the value of a population parameter.

Event: Any set of outcomes of an experiment; a subset of the sample space.

Event space—event set: The subset of outcomes corresponding to some designated event.

EVPI: Expected value of perfect information.

EVSI: Expected value of sampling information.

Expected value: The expected value of a random variable is equal to the mean of its probability distribution.

Experiment: Any well-defined action that leads to a single well-defined result.

Factor: A variable of interest that is controlled in an analysis of variance.

Factor level: A value at which the factor is controlled; same as a treatment.

Frequency: The number of observations that fall into a specified class.

Frequency distribution: A tabular arrangement or grouping of numerical data into classes and a count of the number of observations that fall into each class.

Frequency polygon: A graphical representation of a frequency distribution in which frequencies are plotted by placing a dot above the class mark and connecting these dots with straight lines.

Function: A relationship between two sets, called the domain and the range, which assigns a single element of the range to each element in the domain.

Geometric mean: The geometric mean is defined best by Equation 4.10.

Goodness-of-fit test: A test of the hypothesis that data observed from a sample belongs to a stated population distribution.

Independent trials: Trials of an experiment in which the probability of a specific outcome on any trial is unaffected by the outcome of any other trial.

Judgment sampling: A sampling method in which the decision to include any population element in the sample is based on human judgment.

Laspeyres method: A weighting system for the construction of index numbers which uses base period values as weights.

Least significant difference: The smallest difference between any two sample means in an analysis of variance that can be considered to be significant at the level selected for the test.

Mean square: The sum of squares divided by the number of degrees of freedom; a sample variance.

Median: The value in a data set that divides the set into two equal parts.

Mode: The value in a set of data that occurs with the greatest frequency.

Modified mean: A simple arithmetic mean obtained from a set of data after eliminating the highest and lowest values in the set.

Moving averages: A series of averages computed from the terms in a series by successively dropping the oldest term and adding the next term in succession for each new average.

Node: A branch point on a decision tree.

Nonparametric statistics: Statistical procedures that involve no hypotheses about population parameters.

Numerical random variable: A random variable whose values are numbers.

Ogive: A frequency polygon of a cumulative frequency distribution.

One-tailed test: A hypothesis test in which the entire critical or rejection region is located in only one tail of the sampling distribution of the test statistic.

Operating characteristic curve: A graph of the probability of accepting the null hypothesis for alternate values of the parameter being tested.

Opportunity loss: The absolute value of the difference between the payoff actually realized for the act selected and the payoff which could have been obtained had the optimal act been selected.

Opportunity loss table: A table displaying the acts, states of nature, and consequences in a decision problem in terms of opportunity loss.

Outcome: Any single, possible result of an experiment.

Paasche method: A weighting system for the construction of index numbers which uses current period values as weights.

Parameter: Some constant characteristic of a population.

Parameter of a probability distribution: A constant value that characterizes the distribution.

Payoff table: A table displaying the acts, states of nature, and consequences in a decision problem in terms of actual costs, losses, profits, or gains.

Percentiles: Those values that divide a set of data into 100 equal parts.

Point estimate: The numerical value of the estimator computed from the sample.

Pooled estimate: A parameter estimate obtained from two or more samples by pooling or combining all of the sample data.

Population or **statistical population:** The entire set of possible observations that may be made on the statistical universe. Although the term universe typically refers to the elementary units themselves while the term population refers to the observations made on the units, the two terms are frequently used interchangeably. In this book we use the term population to refer to both the elementary units and the observations.

Posterior distribution: A probability distribution on the states of nature which results from combining the prior distribution with information obtained from a sample.

Power: The probability that a test will reject the null hypothesis when it is false.

Power curve: A graph of the probability of rejecting the null hypothesis for alternate values of the parameter being tested.

Pre-posterior analysis: A procedure for determining the optimal decision rule associated with a sample of size n before the sample is actually taken.

Prior distribution: A probability distribution on the states of nature obtained prior to any sampling or experimentation.

Probability:

General definition: A numerical measure of the likelihood of occurrence of an uncertain event.

Classical definition: The probability of an event E is the ratio of the number of outcomes in the event space of E to the total number of outcomes in the sample space of the experiment, provided that all outcomes are equally likely to occur.

Relative frequency definition: The probability of an event E is the limit, as n approaches infinity, of the ratio of the number of times that E is observed to occur in n trials of an experiment.

Probability distribution: A function that assigns probabilities to values of a random variable.

Probability function: Same as a probability distribution.

Probability sampling: A sampling method in which every element in the population has a known and nonzero probability of being included in the sample.

Quartiles: Those values that divide a set of data into four equal parts.

Random sampling: Same as probability sampling.

Random variable: A variable whose values are determined by the outcome of an experiment.

Range: A measure of variation determined by subtracting the lowest value in a set from the highest.

Raw data: The statistical data in its original form, before any statistical techniques are used to refine, process, or summarize.

Regression analysis: A statistical method for determining the type of relationship that exists among two or more variables and for using that relationship to make estimates or predictions.

Rejection region: Same as critical region.

Relative: The ratio of two numbers usually expressed as a percent.

Relative frequency distribution: A frequency distribution which indicates the proportion of observations that fall into each class.

Run: A sequence of observations preceded and followed by a different observation.

Sample: Any portion of the statistical population. Generally, a sample consists of fewer elementary units or observations than are contained in the population.

Sample number: The number of successes in a sample.

Sample proportion: The proportion of successes in a sample.

Sample space: The set of all possible outcomes of an experiment.

Sample survey: A survey in which observations are made on a sample of elementary units drawn from the population of interest.

Sampling: The process of selecting samples from a population.

Sampling fraction: The ratio of the sample size to the population size.

Scatter diagram: A chart in which pairs of values (x, y) are plotted to provide a visual display of the pattern of their relationship.

Seasonalizing: Adjusting the trend for the effect of the season.

Secular trend: Same as trend.

Set: A collection of elements having some describable similarity.

Significance level: The probability level at which an observed deviation can no longer be attributed to chance.

Simple index number: A percentage which expresses the relationship between two numbers, one of which is called the base.

Simple random sampling: A sampling method which provides that every possible combination of n elements in the population has the same probability of being selected. A method of sampling so that on each selection every member of the remaining population has an equal probability of being selected.

Skewness: Departure from symmetry.

Standard deviation: The standard deviation is best defined by Formula 5.3.

Standard error of the estimate: A measure of the variation of the observed values about the regression line.

States of nature: Factors in a decision problem which affect the consequences of a decision and are beyond the control of the decision maker.

Statistic: A value computed from a sample.

Statistical independence: Two events are statistically independent if the occurrence or nonoccurrence of one has no effect on the probability of the occurrence of the other.

Statistical inference: The branch of statistics which is concerned with drawing conclusions about populations on the basis of incomplete data, that is, on the basis of data obtained from a sample.

Statistics: The branch of applied mathematics which is concerned with the collection and analysis of numerical data. Also a body or set of numerical data.

Stratified random sampling: A sampling method in which the population is divided into homogeneous subsets called strata, from which simple random samples are drawn and then combined into the overall sample.

Sum of squares: The sum of the squared deviations of sample values about the sample mean.

Survey or **statistical survey:** A planned process of collecting statistical data by making observations on elementary units.

Systematic sampling: A sampling method in which every kth element of the population is selected after the starting point is selected randomly.

Test of independence: A test of the hypothesis that membership in a particular category of one classification variable has no effect on membership in a particular category of another classification variable.

Test statistic: The statistic used to test a hypothesis about a population parameter.

Time series: A set of numerical observations taken at different points in time.

Treatment: A value of a controlled factor; see factor level.

Treatment combination: A combination of one level of one factor with one level of another factor in a two-factor analysis of variance.

Treatment population: The theoretical population of elementary units that could potentially be exposed to a treatment.

Trend: The long-term behavior of a variable.

Two-tailed test: A hypothesis test in which the critical or rejection region for the test is divided between the two tails of the sampling distribution of the test statistic.

Type I error: Rejection of a true null hypothesis.

Type II error: Acceptance of a false null hypothesis.

Unbiased estimator: An estimator whose expected value is equal to the value of the parameter to be estimated.

Universe or **statistical universe:** The entire body of elementary units that are of interest and are the subject of a statistical investigation.

Variable: An observable quantitative characteristic of an elementary unit that may vary from unit to unit.

Variance: The variance is best defined by Formula 5.4.

Variance ratio: The ratio of two sample variances.

Weighted arithmetic mean: A modified arithmetic mean that weights each value in a set according to its relative importance.

ANSWERS TO EVEN-NUMBERED PROBLEMS

Chapter 2

2.2 **a)** 712 **b)** 496 **c)** 558 **d)** 272

2.4 **a)** 2136 **b)** 7120 **c)** 2236 **d)** 7150

2.6 **a)** 4 **b)** 1 **c)** 2 or 3 **d)** 4 **e)** 2 **f)** 3

2.8 **a)** 229 **b)** 475 **c)** 224 **d)** 970 **e)** 40.8 **f)** 9.3

2.10 **a)** antilog 2.68727 is approximately 487 **b)** antilog 0.69075 is approximately 4.91 **c)** antilog 2.49102 is approximately 310 **d)** antilog 1.2546 is approximately 18.0

2.12 **a)** antilog 2.62818 is approximately 425 **b)** antilog 1.25890 is approximately 18.2 **c)** antilog 3.05372 is approximately 1130 **d)** antilog 0.96198 is approximately 9.16

2.14 **a)** $\{\varnothing\}$ **b)** $\{3,4,5,6,7\}$ **c)** $\{3\}$ **d)** $\{4,5,6\}$ **e)** $\{7\}$ **f)** $\{1,2,3,4,5,6,7,8,9,10\}$ **g)** $\{8,9,10\}$

2.16 **a)** $n(S) - n(A \cap B \cap C)' = 0$
b) $n(A) + n(C) - n(A \cap C) = 7$
c) $n(B) + n(C) - n(B \cap C) = 5$
d) $n(A) + n(C') - n(A \cap C') = 6$
e) $n(S) - n(A \cup C) = 3$ **f)** $n(A') + n(B) - n(A' \cap B) = 8$
g) $n(A) + n(B) + n(C) - n(A \cap B) - n(A \cap C) - n(B \cap C) + n(A \cap B \cap C) = 7$

2.18

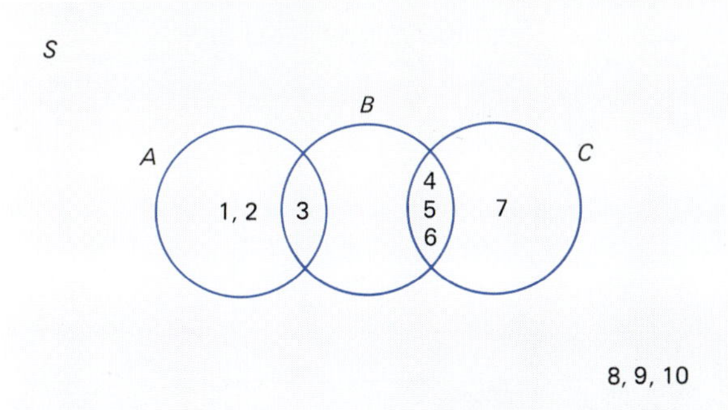

2.20 **a)** $A = \{x,y,z\}$ **b)** $B = \{x,w\}$ **c)** $C = \{v,y\}$

2.22 **a)** $(A \cap B) \cup (A \cup B)'$
b) $(A \cap B \cap C') \cup (A \cap B' \cap C) \cup (A' \cap B \cap C)$
c) $(A \cap B) \cup (B \cap C)$

2.24 **a)**

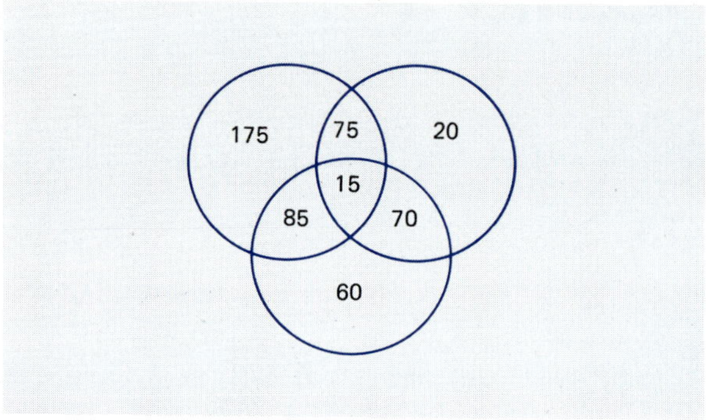

b) 500

2.26 **a)** ${}_5P_5 = 120$ **b)** ${}_8P_5 = 6720$

2.28 $\binom{20}{5} = 15{,}504$

2.30 ${}_{12}P_4 = 11{,}880$

2.32 \$1500; \$3000; \$4500; \$6000; \$7500

2.34 \$37,500

Chapter 3 **3.2** The number of classes should be six, since $n = 60$. $2^5 = 32$; $2^6 = 64$

Classes	Class Mark x	Frequency f	Boundaries
\$5 to \$24	\$14.50	17	4.50– 24.50
\$25 to \$44	\$34.50	18	24.50– 44.50
\$45 to \$64	\$54.50	9	44.50– 64.50
\$65 to \$84	\$74.50	10	64.50– 84.50
\$85 to \$104	\$94.50	4	84.50–104.50
\$105 to \$124	\$114.50	2	104.50–124.50
		60	

The class interval is \$20.00.

3.4 "LESS THAN" CUMULATIVE DISTRIBUTION

Class	*Cumulative Frequency*
less than 20	0
less than 30	1
less than 40	5
less than 50	19
less than 60	34
less than 70	61
less than 80	93
less than 90	100

"MORE THAN" CUMULATIVE DISTRIBUTION

Class	*Cumulative Frequency*
20 and over	100
30 and over	99
40 and over	95
50 and over	81
60 and over	66
70 and over	39
80 and over	7
90 and over	0

3.6

Class	*Cumulative Relative Frequency*
3.9 or less	0
5.9 or less	0.025
7.9 or less	0.325
9.9 or less	0.700
11.9 or less	0.950
13.9 or less	1.000

3.8

Class Boundaries	*Class Mark*
65.5–68.5	67
68.5–71.5	70
71.5–74.5	73
74.5–77.5	76
77.5–80.5	79
80.5–83.5	82

3.10

Class	*Cumulative Frequency*
less than $2.30	0
less than $2.80	50
less than $3.30	120
less than $3.80	222
less than $4.30	332
less than $4.80	428
less than $5.30	469
less than $5.80	483
less than $6.30	492
less than $6.80	500

a) 428 **b)** 492 **c)** 8 **d)** cannot be determined

3.12 **a)** 0.300 **b)** 0.267 **c)** 0.733

3.14 **a)** categorical **b)** Since "brand" is an attribute, there are no numerical class boundaries or limits.

c)

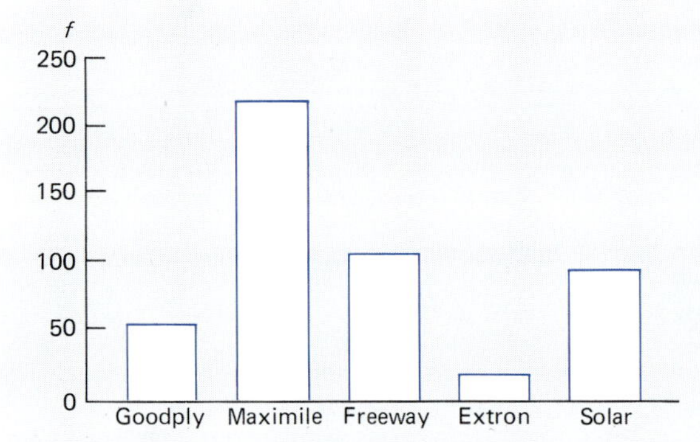

3.16

Class	*Cumulative Frequency*
36.0 and over	50
37.8 and over	44
39.6 and over	37
41.4 and over	13
43.2 and over	6
45.0 and over	4
46.8 and over	0

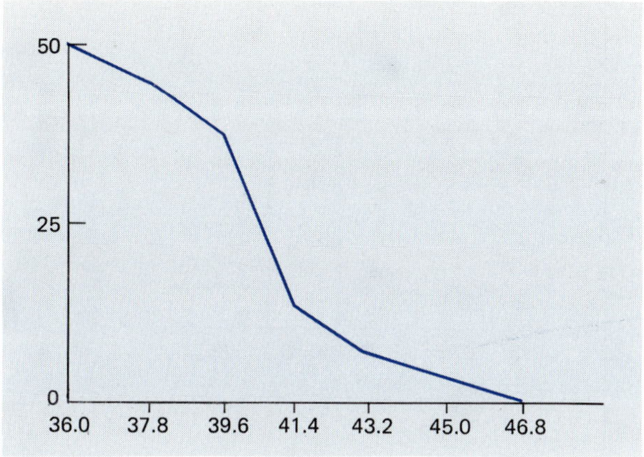
50
25
0
36.0
37.8
39.6
41.4
43.2
45.0
46.8

3.18

a)

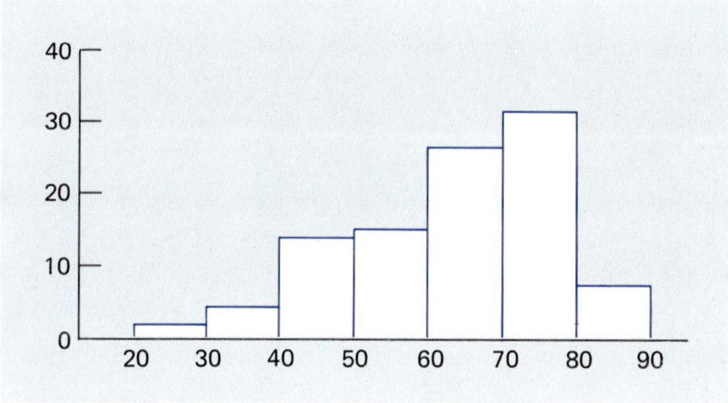
40
30
20
10
0
20
30
40
50
60
70
80
90

b)

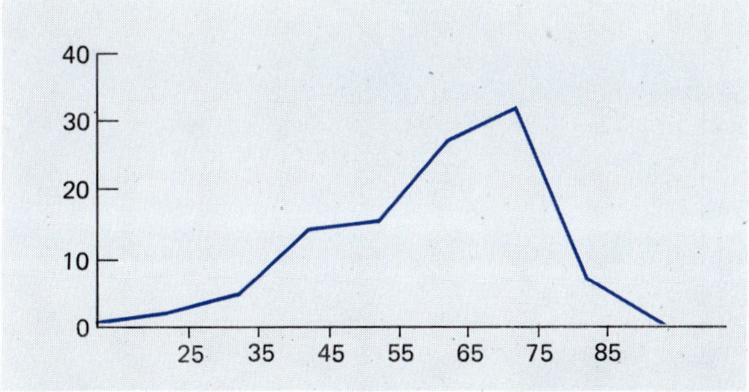
40
30
20
10
0
25
35
45
55
65
75
85

3.20

a)

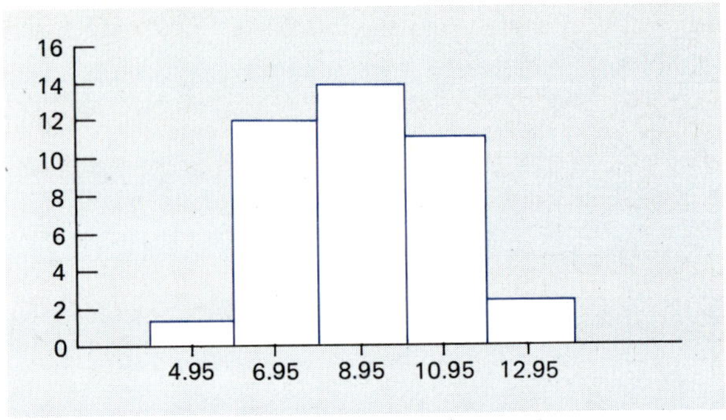
16
14
12
10
8
6
4
2
0
4.95
6.95
8.95
10.95
12.95

b)

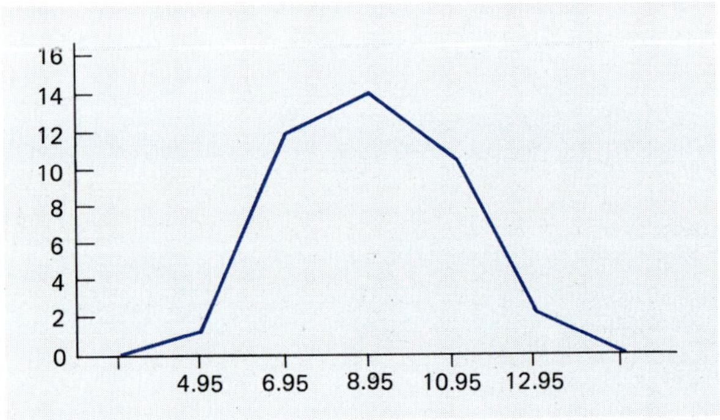
16
14
12
10
8
6
4
2
0
4.95
6.95
8.95
10.95
12.95

3.22

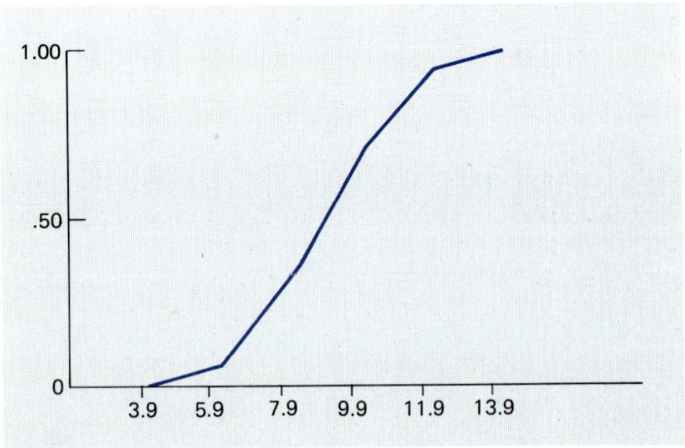
1.00
.50
0
3.9
5.9
7.9
9.9
11.9
13.9

3.24 The frequency polygons provide a rough indication of the shape of each distribution. However, since, in a properly constructed histogram, the area in each bar is proportional to the frequency or relative frequency in that class, the histogram provides a better picture of the distribution and, therefore, more information.

Chapter 4

4.2 **a)** $\mu = 49.63$ or 50; Md = 52; mode = 27 **b)** The mean or the median.

4.4 $\mu = \$1.33$

4.6 $\mu = 8.58$

4.8 Md = 8.3; mode = 7

4.10 Md = 65.13; mode = 52

4.12 **a)** Mode = 225.5; Md = 210.5 **d)** The mode is the most common life of the bulbs. However, half of the bulbs will fail prior to the median life of 210.5 hours, while half will last longer than the median life.

4.14 The median price is \$41,429. Half of the houses sold in 1978 cost less than \$41,429, while half cost more.

4.16 $\mu = 28.05$ or 28; Md = 27.18 or 27

4.18 $D_2 = \$32{,}500$; twenty percent of the houses sold in 1978 cost less than \$32,500, while 80% cost more.

4.20 **a)** $F_3 = L = \left(\frac{0.6N - F}{f}\right)c$ **b)** $F_3 = 72.625$ or 73

4.22 $G = 130.8\%$

Chapter 5

5.2 $R = 48$; $\sigma^2 = 254.484$; $\sigma = 15.953$

5.4 $R = 15$

5.6 $\sigma^2 = 22.16$; $\sigma = 4.71$

5.8 $R = 104$; $\sigma^2 = 658.69$; $\sigma = 25.66$

5.10 $\sigma = 11.489$ thousands or \$11,489

5.12 $\sigma^2 = 42.58$; $\sigma = 6.53$

5.14 $V = 0.264$

5.16 V (accidents) = 0.51; V (diameters) = 0.054. The distribution of accidents is about ten times more variable than the distribution of diameters.

5.18 $\text{Sk}_p = -0.48$

5.20 $\text{Sk}_p = -0.24$

5.22 Sk_p (diameters) = 0.04; Sk_p (accounts) = 0.48. The distribution of past due accounts has the greater degree of skewness and is therefore more asymmetrical than the distribution of diameters.

Chapter 6

6.2 42.9, 57.1, 71.4, 85.7, 100.0, 114.3, 128.6

6.4 Philadelphia = 100.0, San Francisco = 106.1, Boston = 111.4

6.6 1975 = 96.5, 1976 = 100.00, 1977 = 105.1

6.8 1975 = 98.6, 1976 = 100.0, 1977 = 103.6, 1978 = 108.5

6.10 $P_{77} = 109.3$

6.12 $P_{76} = 104.2$, $P_{77} = 109.4$

6.14 1975 = 95.1, 1976 = 96.7, 1977 = 100.0, 1978 = 104.5

6.16 $Q_{78} = 121.6$

6.18 $Q_{78} = 130.5$

6.20 $P_{77} = 107.5$

6.22 AFTER-TAX PROFITS IN THOUSANDS IN CONSTANT 1967 DOLLARS

1975	\$70.2
1976	\$74.4
1977	\$73.7
1978	\$75.0

Chapter 7

7.2 a) $\frac{4}{13}$ **b)** $\frac{8}{13}$ **c)** $\frac{3}{26}$ **d)** $\frac{1}{13}$

7.4

Coin 1	*Coin 2*	*Coin 3*
H	H	H
H	H	T
H	T	H
T	H	H
H	T	T
T	H	T
T	T	H
T	T	T

7.6 a) $\frac{1}{2}$ **b)** $\frac{3}{8}$ **c)** $\frac{1}{2}$ **d)** $\frac{1}{2}$

7.8 a) $\frac{1}{4}$ **b)** $\frac{3}{4}$ **c)** $\frac{2}{3}$ **d)** $\frac{1}{2}$

7.10 a) $\frac{1}{6}$ **b)** $\frac{1}{6}$ **c)** $\frac{1}{36}$ **d)** $\frac{1}{9}$

7.12 a) $\frac{1}{2}$ **b)** $\frac{1}{4}$ **c)** $\frac{13}{102}$ **d)** $\frac{13}{102}$

7.14 a) 0 **b)** 0.115 **c)** 0.25 **d)** 0.441

7.16 a) 0.06 **b)** 0.60 **c)** 0.301 **d)** 0.84

7.18 a) 0.927 **b)** grade and line are not independent

7.20 a) $\frac{1}{2}$ **b)** 0 **c)** $\frac{3}{4}$ **d)** $\frac{1}{4}$

7.22 a) $\frac{1}{6}$ **b)** $\frac{1}{3}$ **c)** $\frac{1}{6}$ **d)** 0

7.24 0.09

7.26 0.078; 0.9865

7.28 $P(A|D) = 0.244$; $P(B|D) = 0.585$; $P(C|D) = 0.171$

7.30 0.000006

7.32 **a)** 0.85 **b)** 0.135 **c)** 0.64 **d)** 0.36

7.34 $0.028675 \doteq 0.03$

Chapter 8

8.2 **a)** x is time until failure, a continuous random variable **b)** x is number of children, a discrete random variable **c)** x is age, a continuous random variable

8.4

x	$f(x)$
0	$\frac{1}{8}$
1	$\frac{3}{8}$
2	$\frac{3}{8}$
3	$\frac{1}{8}$

8.6

y	$f(y)$
0	$\frac{4}{9}$
1	$\frac{4}{9}$
2	$\frac{1}{9}$

8.8 $E(x) = \frac{3}{2}$

8.10 $E(y) = \frac{2}{3}$

8.12 **a)** $E(x + c) = 3\frac{1}{3}$ **b)** $E(cx) = 1$

8.14 $E(y) = 5.4$

8.16 $\text{Var}(x) = \frac{5}{18}$

8.18 $\text{Var}(x) = \frac{9}{16}$

8.20 **a)** 0.602 **b)** 0.398 **c)** 0.164 **d)** binomial **e)** $\mu = 12.0$; $\sigma^2 = 2.4$

8.22 $P(x \leq 8) = 0.274$; $E(x) = np = 10$

8.24 0.41

8.26 **a)** $\frac{21}{45}$ **b)** $\frac{21}{45}$ **c)** $\frac{3}{45}$

8.28 **a)** 0.154 **b)** 0.440 **c)** 0.846

8.30 $E(x) = 0.15$; $\text{Var}(x) = 0.13$

8.32 **a)** 0.367 **b)** 0.265 **c)** 0.919

8.34 **a)** 0.184 **b)** 0.916 **c)** 0.185

8.36 **a)** 0.409 **b)** 0.410

8.38 **a)** 0.606 **b)** 0.303 **c)** 0.091

8.40 **a)** 0.3085 **b)** 0.5328 **c)** 0.1587 **d)** 0.6247 **e)** 0.4332

8.42 **a)** 0.0104 **b)** 0.0446 **c)** 0.9450

Chapter 9

9.2 The population variation is of greater importance than the population size. The greater the variation in the population, the larger a sample must be in order to be representative. If there were no variation in the population, a sample of size 1 would be representative. The size of the population has little influence on the sample size required in order to be representative.

9.4 $\sigma^2 = 12.49$; $\sigma = 3.53$

9.6 **a)** No, because not all customers will have charge accounts. **b)** Yes. This is called systematic sampling and does result in a random sample.

9.8 $\sigma_{\overline{x}} = 5$

9.10 $E(\overline{x}) = 75$; $\sigma^2_{\overline{x}} = 9$

9.12 0.3859

9.14 0.022

9.16 **a)** 0.403 **b)** 0.403

9.18 $E(\overline{p}) = 0.4$; $\sigma_{\overline{p}} = 0.1265$; $\sigma_x = 1.897$

9.20 $\sigma_p = 0.1219$; $\sigma_x = 1.829$

9.22 0.0384

9.24 **a)** 0.1075 **b)** 0.7805

9.26 0.1271

9.28 0.0102

9.30 0.121

Chapter 10

10.2 LCL = 227.1; UCL = 252.9

10.4 **a)** LCL = \$38,952.64; UCL = \$39,247.36 **b)** LCL = \$38,969.71

10.6 **a)** LCL = \$38,854.54; UCL = \$40,945.56 **b)** UCL = \$40,766.66

10.8 **a)** LCL = 171.7; UCL = 178.3 **b)** LCL = 171.1

10.10 **a)** \$76.68 **b)** \$76.14 **c)** \$77.21 **d)** \$72.02

10.12 LCL = 32.7; UCL = 37.3

10.14 LCL = 0.138; UCL = 0.362

10.16 **a)** LCL = 0.211; UCL = 0.289 **b)** LCL = 0.217

10.18 **a)** LCL = 0.114 **b)** UCL = 0.193 **c)** LCL = 0.114; UCL = 0.206

10.20 **a)** LCL = 0.200 **b)** LCL = 0.194; UCL = 0.306

10.22 **a)** UCL = 0.309 **b)** LCL = 0.297; UCL = 0.403

10.24 UCL = 56.61

10.26 LCL = 10.70; UCL = 13.66

10.28 $n = 2172$

10.30 $n = 57$

10.32 $n = 167$

10.34 **a)** $n = 1068$ **b)** $n = 385$

10.36 $n = 521$

Chapter 11

11.2 H_0: $p = 0.10$; H_1: $p \neq 0.10$; two-tailed test

11.4 H_0: $p \leq 0.50$; H_1: $p > 0.50$; one-tailed test

11.6 Type II error

11.8 The Type I error would be the error of concluding that more people in Middletown drink Green Label beer than all other brands when this is not true. The Type II error would be the error of concluding that 50% or less drink Green Label beer when, in fact, more than 50% actually do.

11.10 H_0: $\mu \geq 25$; H_1: $\mu < 25$. If $\bar{x}$ is less than or equal to 19.62, reject H_0; otherwise, accept H_0. The lack of information about the population makes no difference in this case because the sample size is large enough for the Central Limit Theorem to apply.

11.12 H_0: $\mu = 25$; H_1: $\mu \neq 25$. If $\bar{x}$ is less than or equal to 8.22 or greater than or equal to 41.78, reject H_0; otherwise, accept H_0.

11.14 0.287

11.16 **a)** If $\bar{x}$ is greater than or equal to 50.59, reject H_0; otherwise, accept H_0. **b)** Since $\bar{x} = 46.3$, H_0 is accepted and the company's claim is rejected.

11.18 If $\bar{p}$ is greater than or equal to 0.5392, reject H_0; otherwise, accept H_0.

11.20 **a)** yes **b)** no

11.22 Yes, conclude a change had occurred at $\alpha = 0.05$. The critical value of $\bar{x}$ for this test was 1127.32. Since $\bar{x} = 1150$ is greater than $\bar{x}_c = 1127.32$, the null hypothesis is rejected.

11.24 The null hypothesis is accepted at both the 5% and 1% levels of significance.

11.26 **a)** H_0: $p \geq 0.60$; H_1: $p < 0.60$ **b)** If $\bar{p}$ is less than or equal to 0.56, reject H_0; otherwise, accept H_0. **c)** If X is less than or equal to 141, reject H_0; otherwise, accept H_0.

11.28 **a)** H_0: $p = 0.04$; H_1: $p \neq 0.04$ **b)** If $\bar{p}$ is less than or equal to 0.025 or greater than or equal to 0.055, reject H_0; otherwise, accept H_0. **c)** Since $\bar{p} = 0.057$, reject H_0.

11.30 $\alpha \doteq 0.04$

11.32 $\beta = 0.1151$; power $= 0.8849$

11.34 $\beta = 0.8413$; power $= 0.1587$

11.36 If z is less than or equal to -2.58 or greater than or equal to $+2.58$, reject H_0; otherwise, accept H_0.

11.38 If t is greater than or equal to 2.32, reject H_0; otherwise, accept H_0.

11.40 If z is greater than or equal to 2.58, reject H_0; otherwise, accept H_0.

Chapter 12

12.2 $d_c^* = \pm 34.5$; $\bar{x}_1 - \bar{x}_2 = -29.5$. Since $-34.5 < -29.5 < +34.5$, accept H_0.

12.4 $d_c^* = \pm 2.35$; $\bar{x}_1 - \bar{x}_2 = 2.70$. Since $2.70 > 2.35$, reject H_0.

12.6 $d_c^* = \pm 84.6$; $\bar{x}_1 - \bar{x}_2 = -29.5$. Since $-84.6 < -29.5 < +84.6$, accept H_0.

12.8 $d_c^* = \pm 4.13$; $\bar{x}_1 - \bar{x}_2 = -1.75$. Since $-4.13 < -1.75 < +4.13$, accept H_0. Assumption: populations are normal, with equal variances.

12.10 $d_c^* = \pm 0.069$; $\bar{p}_1 - \bar{p}_2 = -0.09$. Since $-0.09 < -0.069$, reject H_0.

12.12 $d_c^* = \pm 0.045$; $\bar{p}_1 - \bar{p}_2 = 0.0265$. Since $-0.045 < 0.0265 < +0.045$, accept H_0.

12.14 $d_c^* = \pm 0.063$; $\bar{p}_1 - \bar{p}_2 = 0.088$. Since $0.088 > 0.063$, reject H_0. It cannot be concluded that the proportions are the same.

12.16 $\bar{d}_c = \pm 0.77$; $\bar{d} = 1.25 > 0.77$, therefore, reject H_0.

12.18 H_0: $\sigma_2^2 = \sigma_1^2$; H_1: $\sigma_2^2 > \sigma_1^2$. Since $F = 1.653 < F_c = 3.51$, accept H_0.

12.20 H_0: $\sigma_1^2 = \sigma_2^2$; H_1: $\sigma_1^2 > \sigma_2^2$. Since $F = 3.02 > F_c = 1.90$, reject H_0.

Chapter 13

13.2 Treatment effect is significant at $\alpha = 0.05$.

13.4 Since $VR = 3$, we conclude that the treatment effect is significant at the 5% level but it is not significant at the 1% level.

13.6 Since $VR = 14.8$, we conclude that there is a significant difference in the means at $\alpha = 0.01$.

13.8 Since $VR = 0.04$, we accept the hypothesis of equal means.

13.10 Since $VR = 4.24$, we reject the null hypothesis and conclude that there is a difference in tire mileage at $\alpha = 0.05$.

13.12 LSD = 4.33; $|\bar{x}_{.1} - \bar{x}_{.2}| = 1 < 4.33$, not significant; $|\bar{x}_{.1} - \bar{x}_{.3}| = 6 > 4.33$, significant; $|\bar{x}_{.1} - \bar{x}_{.4}| = 5 > 4.33$, significant; $|\bar{x}_{.2} - \bar{x}_{.3}| = 7 > 4.33$, significant; $|\bar{x}_{.2} - \bar{x}_{.4}| = 6 > 4.33$, significant; $|\bar{x}_{.3} - \bar{x}_{.4}| = 1 < 4.33$, not significant

13.14 LSD = 5.36; $|\bar{x}_{.1} - \bar{x}_{.2}| = 6 > 5.36$, significant; $|\bar{x}_{.1} - \bar{x}_{.3}| = 1 < 5.36$, not significant; $|\bar{x}_{.1} - \bar{x}_{.4}| = 7.25 > 5.36$, significant; $|\bar{x}_{.2} - \bar{x}_{.3}| = 5 < 5.36$, not significant;

$|\bar{x}_{.2} - \bar{x}_{.4}| = 1.25 < 5.36$, not significant;
$|\bar{x}_{.3} - \bar{x}_{.4}| = 6.25 > 5.36$, significant

13.16 There is a significant difference among both the column and row means at $\alpha = 0.05$.

13.18 There is no significant column factor effect at $\alpha = 0.01$. The row factor effect is significant at $\alpha = 0.01$.

13.20 There is a significant row factor effect at both the 5% and 1% levels of significance.

13.22 Both the row and column factors have significant effects at $\alpha = 0.05$.

13.24 LSD = 8.71; $|\bar{x}_{1.} - \bar{x}_{2.}| = 6.3 < 8.71$, not significant;
$|\bar{x}_{1.} - \bar{x}_{3.}| = 7.3 < 8.71$, not significant;
$|\bar{x}_{2.} - \bar{x}_{3.}| = 13.67 > 8.71$, significant

Chapter 14

14.2 $\chi^2 = 16.0$

14.4 Reject H_0.

14.6 Accept H_0.

14.8 Computed $\chi^2 = 10.14$; critical value of $\chi^2 = 16.81$; H_0 cannot be rejected at $\alpha = 0.01$.

14.10 Computed $\chi^2 \doteq 0.01$; critical value of $\chi^2 = 3.84$; H_0 cannot be rejected at $\alpha = 0.05$.

14.12 Reject H_0.

14.14 The critical value of χ^2 with six degrees of freedom at $\alpha = 0.01$ is 16.81. Since the calculated χ^2 of 18.94 exceeds the critical value, the hypothesis of independence is rejected. In the example α was 0.05, resulting in the same decision. However, in testing at $\alpha = 0.05$, there is a 5% probability of rejecting the null hypothesis if it is true. With α at 0.01, this probability is only 1%.

14.16 Computed $\chi^2 = 107.1$; critical value of $\chi^2 = 13.28$; reject H_0.

14.18 Computed $\chi^2 \doteq 0.17$; critical value of $\chi^2 = 9.49$; accept H_0.

14.20 Computed χ^2 with Yates correction factor is approximately 2.7; critical value of $\chi^2 = 3.84$; accept H_0.

Chapter 15

15.2 $z = 2.65$; reject H_0.

15.4 $z = -4.26$; reject H_0.

15.6 $z = 1.04$; accept H_0.

15.8 $z = 0.72$; accept H_0.

15.10 $z = -0.30$; accept H_0.

15.12 $H = 7.26$; reject H_0.

15.14 $H = 6.45$; accept H_0.

15.16 $r_s = -0.10$

15.18 $r_s = 0.745$

Chapter 16

16.2 A linear relationship does appear to exist.

16.4 $r = 0.77$

16.6 $r = -0.44$

16.8 $r = 0.95$

16.10 Computed value of $t \doteq -3.5$. The critical values of t are ± 2.306, therefore, reject H_0.

16.12 Computed value of $t \doteq 8$. The critical values of t are ± 3.499, therefore, reject H_0.

16.14 **a)** $a = 9.77$; $b = -0.84$ **b)** $\bar{y}_c \doteq 3.05$

16.16 $\bar{y}_c = 1.84 + 0.95x$ if $x = 150 \bar{y}_c = 3.27$ or 327 hours; $s_{y.x} = 0.694$ or 69.4 hours

16.18 **a)** $s_{y.x} = 1.92$ **b)** $r^2_{y.x} = 0.72$

16.20 **a)** \$33.9 million **b)** LCL = \$24.7 million; UCL = \$43.1 million **c)** LCL = \$29.7 million; UCL = \$38.1 million

16.22 **a)** 469 hours **b)** LCL = 213 hours; UCL = 725 hours **c)** LCL = 387 hours; UCL = 551 hours

16.24 **a)** $\bar{y}_c = 3 + x_1 + x_2$ **b)** $s_{y.12} = 0$ **c)** $R^2_{y.12} = 1.0$ **d)** $\bar{y}_c = 13.5$

16.26 **a)** $\bar{y}_c = 0.429 + 1.873x_1 + 0.111x_2$ **b)** $s_{y.12} = 1.834$ **c)** $R^2_{y.12} = 0.61$ **d)** \$533,350

Chapter 17

17.2 The series of moving averages is

30.7 33.0 28.8 25.8 27.3 29.8 25.0 27.0 32.7 30.8 30.2
27.8 30.5 27.0 26.8 27.3 29.8 30.2 29.0

17.4 The series of moving averages is

29.3 34.0 29.0 29.5 35.0 37.0 40.8 33.3 26.8 21.0 22.0
24.8 30.5 28.5 28.0 25.0 18.8 27.3 24.8 29.8 32.5

17.6 $y_c = 31.72 + 0.72t$; 1979 45.40 million tons; 1980 46.84 million tons

17.8

Year	*CI as % of Trend*	*Year*	*CI as % of Trend*
1961	107.3	1970	97.7
1962	101.3	1971	116.3
1963	88.6	1972	102.2
1964	102.5	1973	103.4
1965	109.4	1974	98.2
1966	105.3	1975	108.5
1967	98.9	1976	96.9
1968	82.2	1977	94.9
1969	84.5	1978	95.8

17.10

Year	Three-term Moving Average
1962	99.1
1963	97.5
1964	100.2
1965	105.7
1966	104.5
1967	95.5
1968	88.5
1969	88.1
1970	99.5
1971	105.4
1972	107.3
1973	101.3
1974	103.4
1975	101.2
1976	100.1
1977	95.9

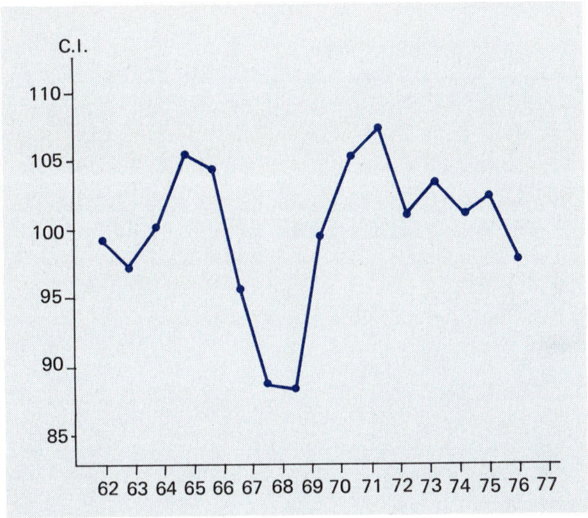

For the period represented by the data there does not appear to be a cycle. Additional data may show otherwise.

17.12 $S_1 = 132.3$; $S_2 = 74.7$; $S_3 = 108.1$; $S_4 = 84.9$

17.14

1979	1st quarter	36.1 millions
1979	2nd quarter	20.7 millions
1979	3rd quarter	30.3 millions
1979	4th quarter	24.1 millions

17.16

1980	1st quarter	\$262.8 thousands
1980	2nd quarter	\$774.4 thousands
1980	3rd quarter	\$997.1 thousands
1980	4th quarter	\$517.4 thousands

Chapter 18 **18.2** a_1: Sell land to insurance company; a_2: Don't sell land to insurance company; θ_1: Expressway route is approved; θ_2: Expressway route is not approved

PAYOFF TABLE

	a_1	a_2
θ_1	\$400,000	\$600,000
θ_2	\$400,000	\$270,000

18.4 **a)** a_0: Buy no sets; a_1: Buy one set; a_2: Buy two sets; a_3: Buy three sets; a_4: Buy four sets; a_5: Buy five sets **b)** θ_0: Demand for no sets; θ_1: Demand for one set; θ_2: Demand for two sets; θ_3: Demand for three sets; θ_4: Demand for four sets; θ_5: Demand for five sets
c) $Q(\theta > a) = \$299a - \$100a = \$199a$

d)

PAYOFF TABLE

	a_0	a_1	a_2	a_3	a_4	a_5
θ_0	0	−100	−200	−300	−400	−500
θ_1	0	199	99	−1	−101	−201
θ_2	0	199	398	298	198	98
θ_3	0	199	398	597	497	397
θ_4	0	199	398	597	796	696
θ_5	0	199	398	597	796	995

18.6

PAYOFF TABLE

States	Acts					
	Bake 20	*Bake 21*	*Bake 22*	*Bake 23*	*Bake 24*	*Bake 25*
θ	a_0	a_1	a_2	a_3	a_4	a_5
Demand 20	11.00	10.75	10.50	10.25	10.00	9.75
Demand 21	11.00	11.55	11.30	11.05	10.80	10.55
Demand 22	11.00	11.55	12.10	11.85	11.60	11.35
Demand 23	11.00	11.55	12.10	12.65	12.40	12.15
Demand 24	11.00	11.55	12.10	12.65	13.20	12.95
Demand 25	11.00	11.55	12.10	12.65	13.20	13.75

$Q(\theta > a) = \$1.00a - \$0.45a = \$0.55a$
$Q(\theta \leq a) = \$1.00 - \$0.45a + (a - \theta)(\$0.20)$

18.8 $\text{EMV}(a_1) = \$400{,}000$; $\text{EMV}(a_2) = \$402{,}000$; select a_2

18.10 a_3 is optimal

18.12 $\text{EMV}(a_1) = \$366.67$; $\text{EMV}(a_2) = 0$; select a_1

18.14 a)

		Payoff		*Opportunity Loss*	
		Subcontract	*Make*	*Subcontract*	*Make*
$\tilde{\theta}$	$P(\tilde{\theta})$	a_1	a_2	a_1	a_2
0	0.10	\$87,500	\$80,000	\$7500	0
0.1	0.35	\$87,500	\$83,750	\$3750	0
0.2	0.25	\$87,500	\$87,500	0	0
0.3	0.20	\$87,500	\$91,250	0	\$3750
0.4	0.10	\$87,500	\$95.000	0	\$7500

b) $\text{EMV}(a_1) = \$87{,}500$; $\text{EMV}(a_2) = \$86{,}937.50$; $\text{EOL}(a_1) = \$2062.50$; $\text{EOL}(a_2) = \$1500$ **c)** Make **d)** EVPI = \$1500

18.16

OPPORTUNITY LOSS TABLE

States $\tilde{\theta}$	$P(\tilde{\theta})$		*Acts* a_0	a_1	a_2	a_3	a_4	a_5
$\tilde{\theta} = 20$	0.05		0	0.25	0.50	0.75	1.00	1.25
$\tilde{\theta} = 21$	0.20		0.55	0	0.25	0.50	0.75	1.00
$\tilde{\theta} = 22$	0.25		1.10	0.55	0	0.25	0.50	0.75
$\tilde{\theta} = 23$	0.20		1.65	1.10	0.55	0	0.25	0.50
$\tilde{\theta} = 24$	0.20		2.20	1.65	1.10	0.55	0	0.25
$\tilde{\theta} = 25$	0.10		2.75	2.20	1.65	1.10	0.55	0
		EOL	1.43	0.92	0.57	0.42	0.43	0.60

EVPI = 0.42

18.18

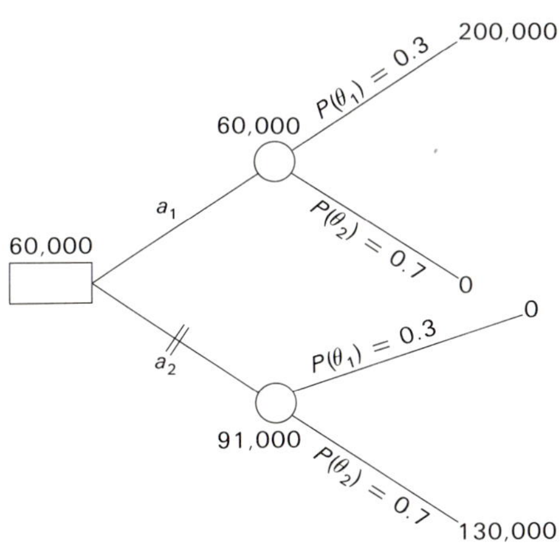

18.20 \$60,000

18.22 24

18.24 **a)**

$\tilde{\theta}$	$P_1(\tilde{\theta})$
30	0.213
50	0.381
80	0.406

b) a_1 is optimal **c)** \$284.20

18.26 If x = 0 or 1, select a_1. If x = 2, select a_2.

18.28 If x = 0, select a_2. If x = 1, 2, or 3, select a_1.

18.30 Bayes' risk = \$167.31; EVSI = \$182.69; ENGS = \$122.69

18.32 Bayes' risk = \$769.20; EVSI = \$730.80; ENGS = −\$329.20

APPENDIX A

TABLE A.1

CUMULATIVE BINOMIAL PROBABILITIES $P[X \leqslant c] = \sum_{x=0}^{c} \binom{n}{x} p^x (1-p)^{n-x}$

	c	.01	.05	.10	.20	.30	.40	p .50	.60	.70	.80	.90	.95
$n=1$	0	.990	.950	.900	.800	.700	.600	.500	.400	.300	.200	.100	.050
	1	1.000	1.000	1.000	1.000	1.000	1.000	1.000	1.000	1.000	1.000	1.000	1.000
$n=2$	0	.980	.902	.810	.640	.490	.360	.250	.160	.090	.040	.010	.002
	1	1.000	.997	.990	.960	.910	.840	.750	.640	.510	.360	.190	.097
	2	1.000	1.000	1.000	1.000	1.000	1.000	1.000	1.000	1.000	1.000	1.000	1.000
$n=3$	0	.970	.857	.729	.512	.343	.216	.125	.064	.027	.008	.001	.000
	1	1.000	.993	.972	.896	.784	.648	.500	.352	.216	.104	.028	.007
	2	1.000	1.000	.999	.992	.973	.936	.875	.784	.657	.488	.271	.143
	3	1.000	1.000	1.000	1.000	1.000	1.000	1.000	1.000	1.000	1.000	1.000	1.000
$n=4$	0	.961	.815	.656	.410	.240	.130	.063	.026	.008	.002	.000	.000
	1	.999	.986	.948	.819	.652	.475	.313	.179	.084	.027	.004	.000
	2	1.000	1.000	.996	.973	.916	.821	.688	.525	.348	.181	.052	.014
	3	1.000	1.000	1.000	.998	.992	.974	.938	.870	.760	.590	.344	.185
	4	1.000	1.000	1.000	1.000	1.000	1.000	1.000	1.000	1.000	1.000	1.000	1.000
$n=5$	0	.951	.774	.590	.328	.168	.078	.031	.010	.002	.000	.000	.000
	1	.999	.977	.919	.737	.528	.337	.188	.087	.031	.007	.000	.000
	2	1.000	.999	.991	.942	.837	.683	.500	.317	.163	.058	.009	.001
	3	1.000	1.000	1.000	.993	.969	.913	.813	.663	.472	.263	.081	.023

Example: The probability that x is less than or equal to 2 given that $n=6$ and $p=0.40$ is 0.544.

TABLE A.1 (CONTINUED)

	c	.01	.05	.10	.20	.30	.40	p .50	.60	.70	.80	.90	.95
	4	1.000	1.000	1.000	1.000	.998	.990	.969	.922	.832	.672	.410	.226
	5	1.000	1.000	1.000	1.000	1.000	1.000	1.000	1.000	1.000	1.000	1.000	1.000
$n=6$	0	.941	.735	.531	.262	.118	.047	.016	.004	.001	.000	.000	.000
	1	.999	.967	.886	.655	.420	.233	.109	.041	.011	.002	.000	.000
	2	1.000	.998	.984	.901	.744	.544	.344	.179	.070	.017	.001	.000
	3	1.000	1.000	.999	.983	.930	.821	.656	.456	.256	.099	.016	.002
	4	1.000	1.000	1.000	.998	.989	.959	.891	.767	.580	.345	.114	.033
	5	1.000	1.000	1.000	1.000	.999	.996	.984	.953	.882	.738	.469	.265
	6	1.000	1.000	1.000	1.000	1.000	1.000	1.000	1.000	1.000	1.000	1.000	1.000
$n=7$	0	.932	.698	.478	.210	.082	.028	.008	.002	.000	.000	.000	.000
	1	.998	.956	.850	.577	.329	.159	.063	.019	.004	.000	.000	.000
	2	1.000	.996	.974	.852	.647	.420	.227	.096	.029	.005	.000	.000
	3	1.000	1.000	.997	.967	.874	.710	.500	.290	.126	.033	.003	.000
	4	1.000	1.000	1.000	.995	.971	.904	.773	.580	.353	.148	.026	.004
	5	1.000	1.000	1.000	1.000	.996	.981	.938	.841	.671	.423	.150	.044
	6	1.000	1.000	1.000	1.000	1.000	.998	.992	.972	.918	.790	.522	.302
	7	1.000	1.000	1.000	1.000	1.000	1.000	1.000	1.000	1.000	1.000	1.000	1.000
$n=8$	0	.923	.663	.430	.168	.058	.017	.004	.001	.000	.000	.000	.000
	1	.997	.943	.813	.503	.255	.106	.035	.009	.001	.000	.000	.000
	2	1.000	.994	.962	.797	.552	.315	.145	.050	.011	.001	.000	.000
	3	1.000	1.000	.995	.944	.806	.594	.363	.174	.058	.010	.000	.000
	4	1.000	1.000	1.000	.990	.942	.826	.637	.406	.194	.056	.005	.000
	5	1.000	1.000	1.000	.999	.989	.950	.855	.685	.448	.203	.038	.006
	6	1.000	1.000	1.000	1.000	.999	.991	.965	.894	.745	.497	.187	.057
	7	1.000	1.000	1.000	1.000	1.000	.999	.996	.983	.942	.832	.570	.337
	8	1.000	1.000	1.000	1.000	1.000	1.000	1.000	1.000	1.000	1.000	1.000	1.000
$n=9$	0	.914	.630	.387	.134	.040	.010	.002	.000	.000	.000	.000	.000
	1	.997	.929	.775	.436	.196	.071	.020	.004	.000	.000	.000	.000
	2	1.000	.992	.947	.738	.463	.232	.090	.025	.004	.000	.000	.000
	3	1.000	.999	.992	.914	.730	.483	.254	.099	.025	.003	.000	.000
	4	1.000	1.000	.999	.980	.901	.733	.500	.267	.099	.020	.001	.000
	5	1.000	1.000	1.000	.997	.975	.901	.746	.517	.270	.086	.008	.001
	6	1.000	1.000	1.000	1.000	.996	.975	.910	.768	.537	.262	.053	.008
	7	1.000	1.000	1.000	1.000	1.000	.996	.980	.929	.804	.564	.225	.071
	8	1.000	1.000	1.000	1.000	1.000	1.000	.998	.990	.960	.866	.613	.370
	9	1.000	1.000	1.000	1.000	1.000	1.000	1.000	1.000	1.000	1.000	1.000	1.000
$n=10$	0	.904	.599	.349	.107	.028	.006	.001	.000	.000	.000	.000	.000
	1	.996	.914	.736	.376	.149	.046	.011	.002	.000	.000	.000	.000
	2	1.000	.988	.930	.678	.383	.167	.055	.012	.002	.000	.000	.000
	3	1.000	.999	.987	.879	.650	.382	.172	.055	.011	.001	.000	.000
	4	1.000	1.000	.998	.967	.850	.633	.377	.166	.047	.006	.000	.000
	5	1.000	1.000	1.000	.994	.953	.834	.623	.367	.150	.033	.002	.000

TABLE A.1

	c	.01	.05	.10	.20	.30	.40	p .50	.60	.70	.80	.90	.95
	6	1.000	1.000	1.000	.999	.989	.945	.828	.618	.350	.121	.013	.001
	7	1.000	1.000	1.000	1.000	.998	.988	.945	.833	.617	.322	.070	.012
	8	1.000	1.000	1.000	1.000	1.000	.998	.989	.954	.851	.624	.264	.086
	9	1.000	1.000	1.000	1.000	1.000	1.000	.999	.994	.972	.893	.651	.401
	10	1.000	1.000	1.000	1.000	1.000	1.000	1.000	1.000	1.000	1.000	1.000	1.000
$n=11$	0	.895	.569	.314	.086	.020	.004	.000	.000	.000	.000	.000	.000
	1	.995	.898	.697	.322	.113	.030	.006	.001	.000	.000	.000	.000
	2	1.000	.985	.910	.617	.313	.119	.033	.006	.001	.000	.000	.000
	3	1.000	.998	.981	.839	.570	.296	.113	.029	.004	.000	.000	.000
	4	1.000	1.000	.997	.950	.790	.533	.274	.099	.022	.022	.000	.000
	5	1.000	1.000	1.000	.988	.922	.753	.500	.247	.078	.012	.000	.000
	6	1.000	1.000	1.000	.998	.978	.901	.726	.467	.210	.050	.003	.000
	7	1.000	1.000	1.000	1.000	.996	.971	.887	.704	.430	.161	.019	.002
	8	1.000	1.000	1.000	1.000	.999	.994	.967	.881	.687	.383	.090	.015
	9	1.000	1.000	1.000	1.000	1.000	.999	.994	.970	.887	.678	.303	.102
	10	1.000	1.000	1.000	1.000	1.000	1.000	1.000	.996	.980	.914	.686	.431
	11	1.000	1.000	1.000	1.000	1.000	1.000	1.000	1.000	1.000	1.000	1.000	1.000
$n=12$	0	.886	.540	.282	.069	.014	.002	.000	.000	.000	.000	.000	.000
	1	.994	.882	.659	.275	.085	.020	.003	.000	.000	.000	.000	.000
	2	1.000	.980	.889	.558	.253	.083	.019	.003	.000	.000	.000	.000
	3	1.000	.998	.974	.795	.493	.225	.073	.015	.002	.000	.000	.000
	4	1.000	1.000	.996	.927	.724	.438	.194	.057	.009	.001	.000	.000
	5	1.000	1.000	.999	.981	.882	.665	.387	.158	.039	.004	.000	.000
	6	1.000	1.000	1.000	.996	.961	.842	.613	.335	.118	.019	.001	.000
	7	1.000	1.000	1.000	.999	.991	.943	.806	.562	.276	.073	.004	.000
	8	1.000	1.000	1.000	1.000	.998	.985	.927	.775	.507	.205	.026	.002
	9	1.000	1.000	1.000	1.000	1.000	.997	.981	.917	.747	.442	.111	.020
	10	1.000	1.000	1.000	1.000	1.000	1.000	.997	.980	.915	.725	.341	.118
	11	1.000	1.000	1.000	1.000	1.000	1.000	1.000	.998	.986	.931	.718	.460
	12	1.000	1.000	1.000	1.000	1.000	1.000	1.000	1.000	1.000	1.000	1.000	1.000
$n=13$	0	.878	.513	.254	.055	.010	.001	.000	.000	.000	.000	.000	.000
	1	.878	.865	.621	.234	.064	.013	.002	.000	.000	.000	.000	.000
	2	.993	.975	.866	.502	.202	.058	.011	.001	.000	.000	.000	.000
	3	1.000	.997	.966	.747	.421	.169	.046	.008	.001	.000	.000	.000
	4	1.000	1.000	.994	.901	.654	.353	.133	.032	.004	.000	.000	.000
	5	1.000	1.000	.999	.970	.835	.574	.291	.098	.018	.001	.000	.000
	6	1.000	1.000	1.000	.993	.938	.771	.500	.229	.062	.007	.000	.000
	7	1.000	1.000	1.000	.999	.982	.902	.709	.426	.165	.030	.001	.000
	8	1.000	1.000	1.000	1.000	.996	.968	.867	.647	.346	.099	.006	.000
	9	1.000	1.000	1.000	1.000	.999	.992	.954	.831	.579	.253	.034	.003
	10	1.000	1.000	1.000	1.000	1.000	.999	.989	.942	.798	.498	.134	.025
	11	1.000	1.000	1.000	1.000	1.000	1.000	.998	.987	.936	.766	.379	.135
	12	1.000	1.000	1.000	1.000	1.000	1.000	1.000	.999	.990	.945	.746	.487
	13	1.000	1.000	1.000	1.000	1.000	1.000	1.000	1.000	1.000	1.000	1.000	1.000

(CONTINUED)

TABLE A.1 (CONTINUED)

	c	.01	.05	.10	.20	.30	.40	p .50	.60	.70	.80	.90	.95
$n=14$	0	.869	.488	.229	.044	.007	.001	.000	.000	.000	.000	.000	.000
	1	.992	.847	.585	.198	.047	.008	.001	.000	.000	.000	.000	.000
	2	1.000	.970	.842	.448	.161	.040	.006	.001	.000	.000	.000	.000
	3	1.000	.996	.956	.698	.355	.124	.029	.004	.000	.000	.000	.000
	4	1.000	1.000	.991	.870	.584	.279	.090	.018	.002	.000	.000	.000
	5	1.000	1.000	.999	.956	.781	.486	.212	.058	.008	.000	.000	.000
	6	1.000	1.000	1.000	.988	.907	.692	.395	.150	.031	.002	.000	.000
	7	1.000	1.000	1.000	.998	.969	.850	.605	.308	.093	.012	.000	.000
	8	1.000	1.000	1.000	1.000	.992	.942	.788	.514	.219	.044	.001	.000
	9	1.000	1.000	1.000	1.000	.998	.982	.910	.721	.416	.130	.009	.000
	10	1.000	1.000	1.000	1.000	1.000	.996	.971	.876	.645	.302	.044	.004
	11	1.000	1.000	1.000	1.000	1.000	.999	.994	.960	.839	.552	.158	.030
	12	1.000	1.000	1.000	1.000	1.000	1.000	.999	.992	.953	.802	.415	.153
	13	1.000	1.000	1.000	1.000	1.000	1.000	1.000	.999	.993	.956	.771	.512
	14	1.000	1.000	1.000	1.000	1.000	1.000	1.000	1.000	1.000	1.000	1.000	1.000
$n=15$	0	.860	.463	.206	.035	.005	.000	.000	.000	.000	.000	.000	.000
	1	.990	.829	.549	.167	.035	.005	.000	.000	.000	.000	.000	.000
	2	1.000	.964	.816	.398	.127	.027	.004	.000	.000	.000	.000	.000
	3	1.000	.995	.944	.648	.297	.091	.018	.002	.000	.000	.000	.000
	4	1.000	.999	.987	.836	.515	.217	.059	.009	.001	.000	.000	.000
	5	1.000	1.000	.998	.939	.722	.403	.151	.034	.004	.000	.000	.000
	6	1.000	1.000	1.000	.982	.869	.610	.304	.095	.015	.001	.000	.000
	7	1.000	1.000	1.000	.996	.950	.787	.500	.213	.050	.004	.000	.000
	8	1.000	1.000	1.000	.999	.985	.905	.696	.390	.131	.018	.000	.000
	9	1.000	1.000	1.000	1.000	.996	.966	.849	.597	.278	.061	.002	.000
	10	1.000	1.000	1.000	1.000	.999	.991	.941	.783	.485	.164	.013	.001
	11	1.000	1.000	1.000	1.000	1.000	.998	.982	.909	.703	.352	.056	.005
	12	1.000	1.000	1.000	1.000	1.000	1.000	.996	.973	.873	.602	.184	.036
	13	1.000	1.000	1.000	1.000	1.000	1.000	1.000	.995	.965	.833	.451	.171
	14	1.000	1.000	1.000	1.000	1.000	1.000	1.000	1.000	.995	.965	.794	.537
	15	1.000	1.000	1.000	1.000	1.000	1.000	1.000	1.000	1.000	1.000	1.000	1.000
$n=16$	0	.851	.440	.185	.028	.003	.000	.000	.000	.000	.000	.000	.000
	1	.989	.811	.515	.141	.026	.003	.000	.000	.000	.000	.000	.000
	2	.999	.957	.789	.352	.099	.018	.002	.000	.000	.000	.000	.000
	3	1.000	.993	.932	.598	.246	.065	.011	.001	.000	.000	.000	.000
	4	1.000	.999	.983	.798	.450	.167	.038	.005	.000	.000	.000	.000
	5	1.000	1.000	.997	.918	.660	.329	.105	.019	.002	.000	.000	.000
	6	1.000	1.000	.999	.973	.825	.527	.227	.058	.007	.000	.000	.000
	7	1.000	1.000	1.000	.993	.926	.716	.402	.142	.026	.001	.000	.000
	8	1.000	1.000	1.000	.999	.974	.858	.598	.284	.074	.007	.000	.000
	9	1.000	1.000	1.000	1.000	.993	.942	.773	.473	.175	.027	.001	.000
	10	1.000	1.000	1.000	1.000	.998	.981	.895	.671	.340	.082	.003	.000
	11	1.000	1.000	1.000	1.000	1.000	.995	.962	.833	.550	.202	.017	.001
	12	1.000	1.000	1.000	1.000	1.000	.999	.989	.935	.754	.402	.068	.007
	13	1.000	1.000	1.000	1.000	1.000	1.000	.998	.982	.901	.648	.211	.043
	14	1.000	1.000	1.000	1.000	1.000	1.000	1.000	.997	.974	.859	.485	.189

TABLE A.1

	c	.01	.05	.10	.20	.30	.40	p .50	.60	.70	.80	.90	.95
	15	1.000	1.000	1.000	1.000	1.000	1.000	1.000	1.000	.997	.972	.815	.560
	16	1.000	1.000	1.000	1.000	1.000	1.000	1.000	1.000	1.000	1.000	1.000	1.000
$n=17$	0	.843	.418	.167	.023	.002	.000	.000	.000	.000	.000	.000	.000
	1	.988	.792	.482	.118	.019	.002	.000	.000	.000	.000	.000	.000
	2	.999	.950	.762	.310	.077	.012	.001	.000	.000	.000	.000	.000
	3	1.000	.991	.917	.549	.202	.046	.006	.000	.000	.000	.000	.000
	4	1.000	.999	.978	.758	.389	.126	.025	.003	.000	.000	.000	.000
	5	1.000	1.000	.995	.894	.597	.264	.072	.011	.001	.000	.000	.000
	6	1.000	1.000	.999	.962	.775	.448	.166	.035	.003	.000	.000	.000
	7	1.000	1.000	1.000	.989	.895	.641	.315	.092	.013	.000	.000	.000
	8	1.000	1.000	1.000	.997	.960	.801	.500	.199	.040	.003	.000	.000
	9	1.000	1.000	1.000	1.000	.987	.908	.685	.359	.105	.011	.000	.000
	10	1.000	1.000	1.000	1.000	.997	.965	.834	.552	.225	.038	.001	.000
	11	1.000	1.000	1.000	1.000	.999	.989	.928	.736	.403	.106	.005	.000
	12	1.000	1.000	1.000	1.000	1.000	.997	.975	.874	.611	.242	.022	.001
	13	1.000	1.000	1.000	1.000	1.000	1.000	.994	.954	.798	.451	.083	.009
	14	1.000	1.000	1.000	1.000	1.000	1.000	.999	.988	.923	.690	.238	.050
	15	1.000	1.000	1.000	1.000	1.000	1.000	1.000	.998	.981	.882	.518	.208
	16	1.000	1.000	1.000	1.000	1.000	1.000	1.000	1.000	.998	.977	.833	.582
	17	1.000	1.000	1.000	1.000	1.000	1.000	1.000	1.000	1.000	1.000	1.000	1.000
$n=18$	0	.835	.397	.150	.018	.002	.000	.000	.000	.000	.000	.000	.000
	1	.986	.774	.450	.099	.014	.001	.000	.000	.000	.000	.000	.000
	2	.999	.942	.734	.271	.060	.008	.001	.000	.000	.000	.000	.000
	3	1.000	.989	.902	.501	.165	.033	.004	.000	.000	.000	.000	.000
	4	1.000	.998	.972	.716	.333	.094	.015	.001	.000	.000	.000	.000
	5	1.000	1.000	.994	.867	.534	.209	.048	.006	.000	.000	.000	.000
	6	1.000	1.000	.999	.949	.722	.374	.119	.020	.001	.000	.000	.000
	7	1.000	1.000	1.000	.984	.859	.563	.240	.058	.006	.000	.000	.000
	8	1.000	1.000	1.000	.996	.940	.737	.407	.135	.021	.001	.000	.000
	9	1.000	1.000	1.000	.999	.979	.865	.593	.263	.060	.004	.000	.000
	10	1.000	1.000	1.000	1.000	.994	.942	.760	.437	.141	.016	.000	.000
	11	1.000	1.000	1.000	1.000	.999	.980	.881	.626	.278	.051	.001	.000
	12	1.000	1.000	1.000	1.000	1.000	.994	.952	.791	.466	.133	.006	.000
	13	1.000	1.000	1.000	1.000	1.000	.999	.985	.906	.667	.284	.028	.002
	14	1.000	1.000	1.000	1.000	1.000	1.000	.996	.967	.835	.499	.098	.011
	15	1.000	1.000	1.000	1.000	1.000	1.000	.999	.992	.940	.729	.266	.058
	16	1.000	1.000	1.000	1.000	1.000	1.000	1.000	.999	.986	.901	.550	.226
	17	1.000	1.000	1.000	1.000	1.000	1.000	1.000	1.000	.998	.982	.850	.603
	18	1.000	1.000	1.000	1.000	1.000	1.000	1.000	1.000	1.000	1.000	1.000	1.000
$n=19$	0	.826	.377	.135	.014	.001	.000	.000	.000	.000	.000	.000	.000
	1	.985	.755	.420	.083	.010	.001	.000	.000	.000	.000	.000	.000
	2	.999	.933	.705	.237	.046	.005	.000	.000	.000	.000	.000	.000
	3	1.000	.987	.885	.455	.133	.023	.002	.000	.000	.000	.000	.000
	4	1.000	.998	.965	.673	.282	.070	.010	.001	.000	.000	.000	.000
	5	1.000	1.000	.991	.837	.474	.163	.032	.003	.000	.000	.000	.000

(CONTINUED)

TABLE A.1 (CONTINUED)

	c	.01	.05	.10	.20	.30	.40	p .50	.60	.70	.80	.90	.95
	6	1.000	1.000	.998	.932	.666	.308	.084	.012	.001	.000	.000	.000
	7	1.000	1.000	1.000	.977	.818	.488	.180	.035	.003	.000	.000	.000
	8	1.000	1.000	1.000	.993	.916	.667	.324	.088	.011	.000	.000	.000
	9	1.000	1.000	1.000	.998	.967	.814	.500	.186	.033	.002	.000	.000
	10	1.000	1.000	1.000	1.000	.989	.912	.676	.333	.084	.007	.000	.000
	11	1.000	1.000	1.000	1.000	.997	.965	.820	.512	.182	.023	.000	.000
	12	1.000	1.000	1.000	1.000	.999	.988	.916	.692	.334	.068	.002	.000
	13	1.000	1.000	1.000	1.000	1.000	.997	.968	.837	.526	.163	.009	.000
	14	1.000	1.000	1.000	1.000	1.000	.999	.990	.930	.718	.327	.035	.002
	15	1.000	1.000	1.000	1.000	1.000	1.000	.998	.977	.867	.545	.115	.013
	16	1.000	1.000	1.000	1.000	1.000	1.000	1.000	.995	.954	.763	.295	.067
	17	1.000	1.000	1.000	1.000	1.000	1.000	1.000	.999	.990	.917	.580	.245
	18	1.000	1.000	1.000	1.000	1.000	1.000	1.000	1.000	.999	.986	.865	.623
	19	1.000	1.000	1.000	1.000	1.000	1.000	1.000	1.000	1.000	1.000	1.000	1.000
$n=20$	0	.818	.358	.122	.012	.001	.000	.000	.000	.000	.000	.000	.000
	1	.983	.736	.392	.069	.008	.001	.000	.000	.000	.000	.000	.000
	2	.999	.925	.677	.206	.035	.004	.000	.000	.000	.000	.000	.000
	3	1.000	.984	.867	.411	.107	.016	.001	.000	.000	.000	.000	.000
	4	1.000	.997	.957	.630	.238	.051	.006	.000	.000	.000	.000	.000
	5	1.000	1.000	.989	.804	.416	.126	.021	.002	.000	.000	.000	.000
	6	1.000	1.000	.998	.913	.608	.250	.058	.006	.000	.000	.000	.000
	7	1.000	1.000	1.000	.968	.772	.416	.132	.021	.001	.000	.000	.000
	8	1.000	1.000	1.000	.990	.887	.596	.252	.057	.005	.000	.000	.000
	9	1.000	1.000	1.000	.997	.952	.755	.412	.128	.017	.001	.000	.000
	10	1.000	1.000	1.000	.999	.983	.872	.588	.245	.048	.003	.000	.000
	11	1.000	1.000	1.000	1.000	.995	.943	.748	.404	.113	.010	.000	.000
	12	1.000	1.000	1.000	1.000	.999	.979	.868	.584	.228	.032	.000	.000
	13	1.000	1.000	1.000	1.000	1.000	.994	.942	.750	.392	.087	.002	.000
	14	1.000	1.000	1.000	1.000	1.000	.998	.979	.874	.584	.196	.011	.000
	15	1.000	1.000	1.000	1.000	1.000	1.000	.994	.949	.762	.370	.043	.003
	16	1.000	1.000	1.000	1.000	1.000	1.000	.999	.984	.893	.589	.133	.016
	17	1.000	1.000	1.000	1.000	1.000	1.000	1.000	.996	.965	.794	.323	.075
	18	1.000	1.000	1.000	1.000	1.000	1.000	1.000	.999	.992	.931	.608	.264
	19	1.000	1.000	1.000	1.000	1.000	1.000	1.000	1.000	.999	.988	.878	.642
	20	1.000	1.000	1.000	1.000	1.000	1.000	1.000	1.000	1.000	1.000	1.000	1.000
$n=25$	0	.778	.277	.072	.004	.000	.000	.000	.000	.000	.000	.000	.000
	1	.974	.642	.271	.027	.002	.000	.000	.000	.000	.000	.000	.000
	2	.998	.873	.537	.098	.009	.000	.000	.000	.000	.000	.000	.000
	3	1.000	.966	.764	.234	.033	.002	.000	.000	.000	.000	.000	.000
	4	1.000	.993	.902	.421	.090	.009	.000	.000	.000	.000	.000	.000
	5	1.000	.999	.967	.617	.193	.029	.002	.000	.000	.000	.000	.000
	6	1.000	1.000	.991	.780	.341	.074	.007	.000	.000	.000	.000	.000
	7	1.000	1.000	.998	.891	.512	.154	.022	.001	.000	.000	.000	.000
	8	1.000	1.000	1.000	.953	.677	.274	.054	.004	.000	.000	.000	.000
	9	1.000	1.000	1.000	.983	.811	.425	.115	.013	.000	.000	.000	.000

TABLE A.1

c	.01	.05	.10	.20	.30	.40	p .50	.60	.70	.80	.90	.95
10	1.000	1.000	1.000	.994	.902	.586	.212	.034	.002	.000	.000	.000
11	1.000	1.000	1.000	.998	.956	.732	.345	.078	.006	.000	.000	.000
12	1.000	1.000	1.000	1.000	.983	.846	.500	.154	.017	.000	.000	.000
13	1.000	1.000	1.000	1.000	.994	.922	.655	.268	.044	.002	.000	.000
14	1.000	1.000	1.000	1.000	.998	.966	.788	.414	.098	.006	.000	.000
15	1.000	1.000	1.000	1.000	1.000	.987	.885	.575	.189	.017	.000	.000
16	1.000	1.000	1.000	1.000	1.000	.996	.946	.726	.323	.047	.000	.000
17	1.000	1.000	1.000	1.000	1.000	.999	.978	.846	.488	.109	.002	.000
18	1.000	1.000	1.000	1.000	1.000	1.000	.993	.926	.659	.220	.009	.000
19	1.000	1.000	1.000	1.000	1.000	1.000	.998	.971	.807	.383	.033	.001
20	1.000	1.000	1.000	1.000	1.000	1.000	1.000	.991	.910	.579	.098	.007
21	1.000	1.000	1.000	1.000	1.000	1.000	1.000	.998	.967	.766	.236	.034
22	1.000	1.000	1.000	1.000	1.000	1.000	1.000	1.000	.991	.902	.463	.127
23	1.000	1.000	1.000	1.000	1.000	1.000	1.000	1.000	.998	.973	.729	.358
24	1.000	1.000	1.000	1.000	1.000	1.000	1.000	1.000	1.000	.996	.928	.723
25	1.000	1.000	1.000	1.000	1.000	1.000	1.000	1.000	1.000	1.000	1.000	1.000

TABLE A.2

TABLE OF THE CUMULATIVE POISSON DISTRIBUTION $P(x \leqslant c) = \sum_{x=0}^{c} \frac{e^{-\mu}\mu^{x}}{x!}$

c \ μ	0.01	0.05	0.10	0.20	0.30	0.40	0.50	0.60	0.70	0.80	0.90	1.00	1.10	1.20	1.30	1.40	1.50	1.60
0	0.990	0.951	0.904	0.818	0.740	0.670	0.606	0.548	0.496	0.449	0.406	0.367	0.332	0.301	0.272	0.246	0.223	0.201
1	0.999	0.998	0.995	0.982	0.963	0.938	0.909	0.878	0.844	0.808	0.772	0.735	0.699	0.662	0.626	0.591	0.557	0.524
2		0.999	0.999	0.998	0.996	0.992	0.985	0.976	0.965	0.952	0.937	0.919	0.900	0.897	0.857	0.833	0.808	0.783
3				0.999	0.999	0.999	0.998	0.996	0.994	0.990	0.986	0.981	0.974	0.966	0.956	0.946	0.934	0.921
4					0.999	0.999	0.999	0.999	0.999	0.998	0.997	0.996	0.994	0.992	0.989	0.985	0.981	0.976
5							0.999	0.999	0.999	0.999	0.999	0.999	0.999	0.998	0.997	0.996	0.995	0.993
6										0.999	0.999	0.999	0.999	0.999	0.999	0.999	0.999	0.998
7												0.999	0.999	0.999	0.999	0.999	0.999	0.999
8															0.999	0.999	0.999	0.999

c \ μ	1.70	1.80	1.90	2.00	2.10	2.20	2.30	2.40	2.50	2.60	2.70	2.80	2.90	3.00	3.1	3.2	3.3
0	0.182	0.165	0.149	0.135	0.122	0.110	0.100	0.090	0.082	0.074	0.067	0.060	0.055	0.049	0.045	0.041	0.037
1	0.493	0.462	0.433	0.406	0.379	0.354	0.330	0.308	0.287	0.267	0.248	0.231	0.214	0.199	0.185	0.171	0.159
2	0.757	0.730	0.703	0.676	0.649	0.622	0.596	0.569	0.543	0.518	0.493	0.469	0.445	0.423	0.401	0.380	0.359
3	0.906	0.891	0.874	0.857	0.838	0.819	0.799	0.778	0.757	0.736	0.714	0.691	0.669	0.647	0.625	0.603	0.580
4	0.970	0.963	0.955	0.947	0.937	0.927	0.916	0.904	0.891	0.877	0.862	0.847	0.831	0.815	0.798	0.781	0.763
5	0.992	0.989	0.986	0.983	0.979	0.975	0.970	0.964	0.957	0.950	0.943	0.934	0.925	0.916	0.906	0.895	0.883
6	0.998	0.997	0.996	0.995	0.994	0.992	0.990	0.988	0.985	0.982	0.979	0.975	0.971	0.966	0.961	0.955	0.949
7	0.999	0.999	0.999	0.998	0.998	0.998	0.997	0.996	0.995	0.994	0.993	0.991	0.990	0.988	0.986	0.983	0.980
8	0.999	0.999	0.999	0.999	0.999	0.999	0.999	0.999	0.998	0.998	0.998	0.997	0.996	0.996	0.995	0.994	0.993
9	0.999	0.999	0.999	0.999	0.999	0.999	0.999	0.999	0.999	0.999	0.999	0.999	0.999	0.998	0.999	0.998	0.998
10					0.999	0.999	0.999	0.999	0.999	0.999	0.999	0.999	0.999	0.999	0.999	0.999	0.999
11									0.999	0.999	0.999	0.999	0.999	0.999	0.999	0.999	0.999
12													0.999	0.999	0.999	0.999	0.999

Example: The probability that x is less than or equal to 2 when $\mu = 1.50$ is 0.808.

	μ																
c	3.4	3.50	4.00	4.50	5.00	5.50	6.00	6.50	7.00	7.50	8.00	8.50	9.00	9.50	10.0	15.0	20.0
0	0.033	0.030	0.018	0.011	0.006	0.004	0.002	0.001	0.000	0.000	0.000	0.000	0.000	0.000	0.000	0.000	0.000
1	0.147	0.135	0.091	0.061	0.040	0.026	0.017	0.011	0.007	0.004	0.003	0.001	0.001	0.000	0.000	0.000	0.000
2	0.340	0.320	0.238	0.173	0.124	0.088	0.061	0.043	0.029	0.020	0.013	0.009	0.006	0.004	0.002	0.000	0.000
3	0.558	0.536	0.433	0.342	0.265	0.201	0.151	0.111	0.081	0.059	0.042	0.030	0.021	0.014	0.010	0.000	0.000
4	0.744	0.725	0.628	0.532	0.440	0.357	0.285	0.223	0.172	0.132	0.099	0.074	0.054	0.040	0.029	0.000	0.000
5	0.871	0.857	0.785	0.702	0.615	0.528	0.445	0.369	0.300	0.241	0.191	0.149	0.115	0.088	0.067	0.002	0.000
6	0.942	0.934	0.889	0.831	0.762	0.686	0.606	0.526	0.449	0.378	0.313	0.256	0.206	0.164	0.130	0.007	0.000
7	0.977	0.973	0.948	0.913	0.866	0.809	0.743	0.672	0.598	0.524	0.452	0.385	0.323	0.268	0.220	0.018	0.000
8	0.992	0.990	0.978	0.959	0.931	0.894	0.847	0.791	0.729	0.661	0.592	0.523	0.455	0.391	0.332	0.037	0.002
9	0.997	0.996	0.991	0.982	0.968	0.946	0.916	0.877	0.830	0.776	0.716	0.652	0.587	0.521	0.457	0.069	0.005
10	0.999	0.998	0.997	0.993	0.986	0.974	0.957	0.933	0.901	0.862	0.815	0.763	0.705	0.645	0.583	0.118	0.010
11	0.999	0.999	0.999	0.997	0.994	0.989	0.979	0.966	0.946	0.920	0.888	0.848	0.803	0.751	0.696	0.184	0.021
12	0.999	0.999	0.999	0.999	0.997	0.995	0.991	0.983	0.973	0.957	0.936	0.909	0.875	0.836	0.791	0.267	0.039
13		0.999	0.999	0.999	0.999	0.998	0.996	0.992	0.987	0.978	0.965	0.948	0.926	0.898	0.864	0.363	0.066
14			0.999	0.999	0.999	0.999	0.998	0.997	0.994	0.989	0.982	0.972	0.958	0.940	0.916	0.465	0.104
15				0.999	0.999	0.999	0.999	0.998	0.997	0.995	0.991	0.986	0.977	0.966	0.951	0.568	0.156
16					0.999	0.999	0.999	0.999	0.999	0.998	0.996	0.993	0.988	0.982	0.972	0.664	0.221
17						0.999	0.999	0.999	0.999	0.999	0.998	0.997	0.994	0.991	0.985	0.748	0.297
18							0.999	0.999	0.999	0.999	0.999	0.998	0.997	0.995	0.992	0.819	0.381
19								0.999	0.999	0.999	0.999	0.999	0.998	0.998	0.996	0.875	0.470
20									0.999	0.999	0.999	0.999	0.999	0.999	0.998	0.917	0.559
21										0.999	0.999	0.999	0.999	0.999	0.999	0.946	0.643
22											0.999	0.999	0.999	0.999	0.999	0.967	0.720
23												0.999	0.999	0.999	0.999	0.980	0.787
24														0.999	0.999	0.988	0.843
25															0.999	0.993	0.887
26																0.996	0.922
27																0.998	0.947
28																0.999	0.965
29																0.999	0.978
30																0.999	0.986
31																0.999	0.991
32																0.999	0.995
33																0.999	0.997
34																	0.998

TABLE A.3

TABLE OF AREAS FOR THE STANDARD NORMAL PROBABILITY DISTRIBUTION

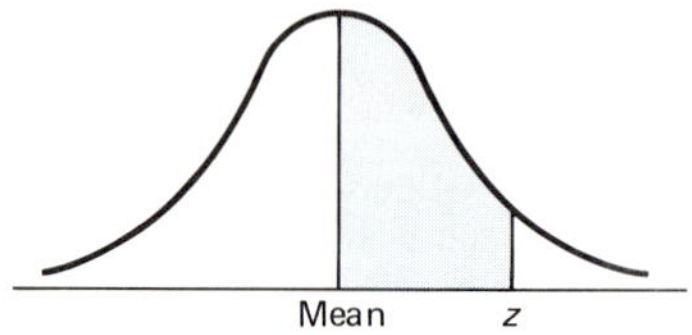

z	.00	.01	.02	.03	.04	.05	.06	.07	.08	.09
0.0	.0000	.0040	.0080	.0120	.0160	.0199	.0239	.0279	.0319	.0359
0.1	.0398	.0438	.0478	.0517	.0557	.0596	.0636	.0675	.0714	.0753
0.2	.0793	.0832	.0871	.0910	.0948	.0987	.1026	.1064	.1103	.1141
0.3	.1179	.1217	.1255	.1293	.1331	.1368	.1406	.1443	.1480	.1517
0.4	.1554	.1591	.1628	.1664	.1700	.1736	.1772	.1808	.1844	.1879
0.5	.1915	.1950	.1985	.2019	.2054	.2088	.2123	.2157	.2190	.2224
0.6	.2257	.2291	.2324	.2357	.2389	.2422	.2454	.2486	.2518	.2549
0.7	.2580	.2612	.2642	.2673	.2704	.2734	.2764	.2794	.2823	.2852
0.8	.2881	.2910	.2939	.2967	.2995	.3023	.3051	.3078	.3106	.3133
0.9	.3159	.3186	.3212	.3238	.3264	.3289	.3315	.3340	.3365	.3389
1.0	.3413	.3438	.3461	.3485	.3508	.3531	.3554	.3577	.3599	.3621
1.1	.3643	.3665	.3686	.3708	.3729	.3749	.3770	.3790	.3810	.3830
1.2	.3849	.3869	.3888	.3907	.3925	.3944	.3962	.3980	.3997	.4015
1.3	.4032	.4049	.4066	.4082	.4099	.4115	.4131	.4147	.4162	.4177
1.4	.4192	.4207	.4222	.4236	.4251	.4265	.4279	.4292	.4306	.4319
1.5	.4332	.4345	.4357	.4370	.4382	.4394	.4406	.4418	.4429	.4441
1.6	.4452	.4463	.4474	.4484	.4495	.4505	.4515	.4525	.4535	.4545
1.7	.4554	.4564	.4573	.4582	.4591	.4599	.4608	.4616	.4625	.4633
1.8	.4641	.4649	.4656	.4664	.4671	.4678	.4686	.4693	.4699	.4706
1.9	.4713	.4719	.4726	.4732	.4738	.4744	.4750	.4756	.4761	.4767
2.0	.4772	.4778	.4783	.4788	.4793	.4798	.4803	.4808	.4812	.4817
2.1	.4821	.4826	.4830	.4834	.4838	.4842	.4846	.4850	.4854	.4857
2.2	.4861	.4864	.4868	.4871	.4875	.4878	.4881	.4884	.4887	.4890
2.3	.4893	.4896	.4898	.4901	.4904	.4906	.4909	.4911	.4913	.4916
2.4	.4918	.4920	.4922	.4925	.4927	.4929	.4931	.4932	.4934	.4936
2.5	.4938	.4940	.4941	.4943	.4945	.4946	.4948	.4949	.4951	.4952
2.6	.4953	.4955	.4956	.4957	.4959	.4960	.4961	.4962	.4963	.4964
2.7	.4965	.4966	.4967	.4968	.4969	.4970	.4971	.4972	.4973	.4974
2.8	.4974	.4975	.4976	.4977	.4977	.4978	.4979	.4979	.4980	.4981
2.9	.4981	.4982	.4982	.4983	.4984	.4984	.4985	.4985	.4986	.4986
3.0	.49865	.4987	.4987	.4988	.4988	.4989	.4989	.4989	.4990	.4990
4.0	.4999683									

Example: For $z = 1.24$, shaded area is 0.3925 out of the total area of 1.

TABLE A.4
CRITICAL VALUES OF THE t DISTRIBUTION

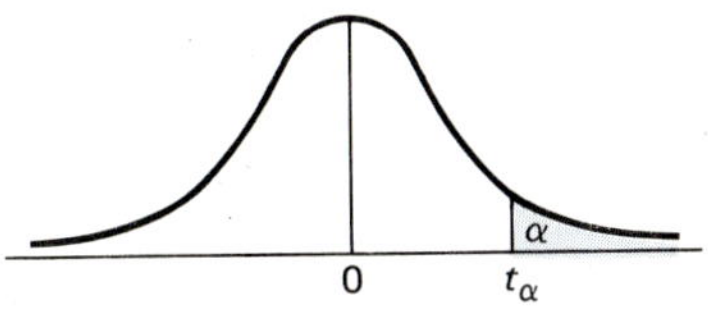

df	$t_{.100}$	$t_{.050}$	$t_{.025}$	$t_{.010}$	$t_{.005}$
1	3.078	6.314	12.706	31.821	63.657
2	1.886	2.920	4.303	6.965	9.925
3	1.638	2.353	3.182	4.541	5.841
4	1.533	2.132	2.776	3.747	4.604
5	1.476	2.015	2.571	3.365	4.032
6	1.440	1.943	2.447	3.143	3.707
7	1.415	1.895	2.365	2.998	3.499
8	1.397	1.860	2.306	2.896	3.355
9	1.383	1.833	2.262	2.821	3.250
10	1.372	1.812	2.228	2.764	3.169
11	1.363	1.796	2.201	2.718	3.106
12	1.356	1.782	2.179	2.681	3.055
13	1.350	1.771	2.160	2.650	3.012
14	1.345	1.761	2.145	2.624	2.977
15	1.341	1.753	2.131	2.602	2.947
16	1.337	1.746	2.120	2.583	2.921
17	1.333	1.740	2.110	2.567	2.898
18	1.330	1.734	2.101	2.552	2.878
19	1.328	1.729	2.093	2.539	2.861
20	1.325	1.725	2.086	2.528	2.845
21	1.323	1.721	2.080	2.518	2.831
22	1.321	1.717	2.074	2.508	2.819
23	1.319	1.714	2.069	2.500	2.807
24	1.318	1.711	2.064	2.492	2.797
25	1.316	1.708	2.060	2.485	2.787
26	1.315	1.706	2.056	2.479	2.779
27	1.314	1.703	2.052	2.473	2.771
28	1.313	1.701	2.048	2.467	2.763
29	1.311	1.699	2.045	2.462	2.756
30	1.310	1.697	2.042	2.457	2.750
∞	1.282	1.645	1.960	2.326	2.576

Example: The value of t with 12 degrees of freedom for which 1% of the area is in the tail is 2.681.

TABLE A.5

CRITICAL VALUES OF THE CHI-SQUARE DISTRIBUTION

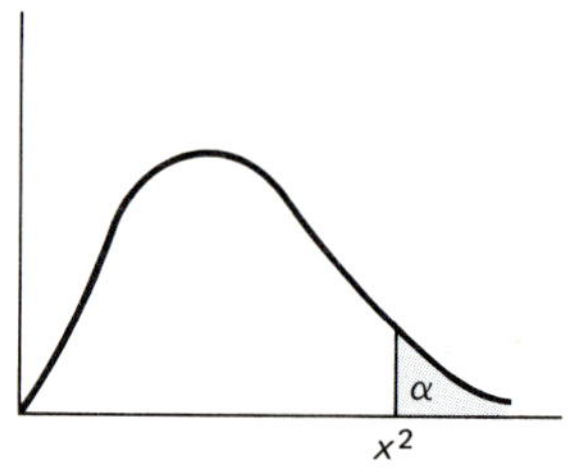

Proportion of area

	Proportion of Area										
df	*0.995*	*0.990*	*0.975*	*0.950*	*0.900*	*0.500*	*0.100*	*0.050*	*0.025*	*0.010*	*0.005*
1	0.00004	0.00016	0.00098	0.00393	0.0158	0.455	2.71	3.84	5.02	6.63	7.88
2	0.0100	0.0201	0.0506	0.103	0.211	1.386	4.61	5.99	7.38	9.21	10.60
3	0.072	0.115	0.216	0.352	0.584	2.366	6.25	7.81	9.35	11.34	12.84
4	0.207	0.297	0.484	0.711	1.064	3.357	7.78	9.49	11.14	13.28	14.86
5	0.412	0.554	0.831	1.145	1.61	4.251	9.24	11.07	12.83	15.09	16.75
6	0.676	0.872	1.24	1.64	2.20	5.35	10.64	12.50	14.45	16.81	18.55
7	0.989	1.24	1.69	2.17	2.83	6.35	12.02	14.07	16.01	18.48	20.23
8	1.34	1.65	2.18	2.73	3.49	7.34	13.36	15.51	17.53	20.09	21.96
9	1.73	2.09	2.70	3.33	4.17	8.34	14.68	16.92	19.02	21.67	23.59
10	2.16	2.56	3.25	3.94	4.87	9.34	15.99	18.31	20.48	23.21	25.19
11	2.60	3.05	3.82	4.57	5.58	10.34	17.28	19.63	21.92	24.72	26.76
12	3.07	3.57	4.40	5.23	6.30	11.34	18.55	21.03	23.34	26.22	28.30
13	3.57	4.11	5.01	5.89	7.04	12.34	19.81	22.36	24.74	27.69	29.82
14	4.07	4.66	5.63	6.57	7.79	13.34	21.06	23.68	26.12	29.14	31.32
15	4.60	5.23	6.26	7.26	8.55	14.34	22.31	25.00	27.49	30.58	32.80
16	5.14	5.81	6.91	7.96	9.31	15.34	23.54	26.30	28.85	32.00	34.27
17	5.70	6.41	7.56	8.67	10.09	16.34	24.77	27.59	30.19	33.41	35.72
18	6.26	7.01	8.23	9.39	10.86	17.34	25.99	28.87	31.53	34.81	37.16
19	6.84	7.63	8.91	10.12	11.65	18.34	27.20	30.14	32.85	36.19	38.58

Example: The value of χ^2 with 15 degrees of freedom for which 5% of the area is in the tail is 25.00.

TABLE A.5

	Proportion of Area										
df	0.995	0.990	0.975	0.950	0.900	0.500	0.100	0.050	0.025	0.010	0.005
20	7.43	8.26	9.59	10.85	12.44	19.34	28.41	31.41	34.17	37.57	40.00
21	8.03	8.90	10.28	11.59	13.24	20.34	29.62	32.67	35.48	38.93	41.40
22	8.64	9.54	10.98	12.34	14.04	21.34	30.81	33.92	36.78	40.29	42.80
23	9.26	10.20	11.69	13.09	14.85	22.34	32.01	35.17	38.08	41.64	44.13
24	9.89	10.36	12.40	13.85	15.66	23.34	33.20	36.42	39.36	42.98	45.56
25	10.52	11.52	13.12	14.61	16.47	24.34	34.38	37.65	40.65	44.31	46.93
26	11.16	12.20	13.84	15.38	17.29	25.34	35.56	38.89	41.92	45.64	48.29
27	11.81	12.83	14.57	16.15	18.11	26.34	36.74	40.11	43.19	46.96	49.64
28	12.46	13.56	15.31	16.93	18.94	27.34	37.92	41.34	44.46	38.28	50.99
29	13.12	14.26	16.05	17.71	19.77	28.34	39.09	42.56	45.72	49.59	52.34
30	13.79	14.95	16.79	18.49	20.60	29.34	40.26	43.77	46.98	50.89	53.67
40	20.71	22.16	24.43	26.51	29.05	39.34	51.80	55.76	59.34	63.69	66.77
50	27.99	29.71	32.36	34.76	37.69	49.33	63.17	67.50	71.42	76.15	79.49
60	35.53	37.43	40.48	43.19	46.46	59.33	74.40	79.08	83.30	83.38	91.95
70	43.28	45.44	48.76	51.74	55.33	69.33	85.53	90.53	95.02	100.4	104.22
80	57.17	53.54	51.17	60.39	64.28	79.33	98.58	101.9	106.6	112.3	116.32
90	59.20	61.75	65.65	69.13	73.29	89.33	107.6	113.1	118.1	124.1	123.3
100	67.33	70.06	74.22	77.93	82.36	99.33	113.5	124.3	129.6	135.3	140.2

TABLE A.6

CRITICAL VALUES OF THE *F* DISTRIBUTION

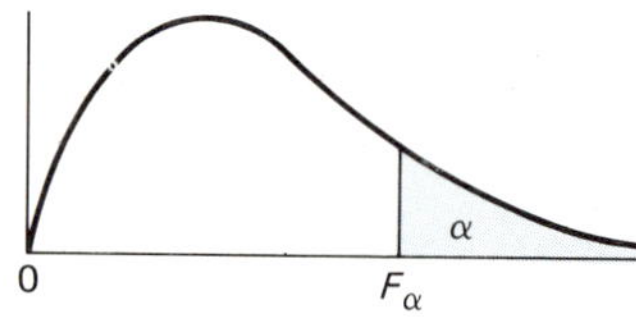

Degrees of Freedom ($\alpha = .05$)

df	1	2	3	4	5	6	7	8	9
1	161.4	199.5	215.7	224.6	230.2	234.0	236.8	238.9	240.5
2	18.51	19.00	19.16	19.25	19.30	19.33	19.35	19.37	19.38
3	10.13	9.55	9.28	9.12	9.01	8.94	8.89	8.85	8.81
4	7.71	6.94	6.59	6.39	6.26	6.16	6.09	6.04	6.00
5	6.61	5.79	5.41	5.19	5.05	4.95	4.88	4.82	4.77
6	5.99	5.14	4.76	4.53	4.39	4.28	4.21	4.15	4.10
7	5.59	4.74	4.35	4.12	3.97	3.87	3.79	3.73	3.68
8	5.32	4.46	4.07	3.84	3.69	3.58	3.50	3.44	3.39
9	5.12	4.26	3.86	3.63	3.48	3.37	3.29	3.23	3.18
10	4.96	4.10	3.71	3.48	3.33	3.22	3.14	3.07	3.02
11	4.84	3.98	3.59	3.36	3.20	3.09	3.01	2.95	2.90
12	4.75	3.89	3.49	3.26	3.11	3.00	2.91	2.85	2.80
13	4.67	3.81	3.41	3.18	3.03	2.92	2.83	2.77	2.71
14	4.60	3.74	3.34	3.11	2.96	2.85	2.76	2.70	2.65
15	4.54	3.68	3.29	3.06	2.90	2.79	2.71	2.64	2.59
16	4.49	3.63	3.24	3.01	2.85	2.74	2.66	2.59	2.54
17	4.45	3.59	3.20	2.96	2.81	2.70	2.61	2.55	2.49
18	4.41	3.55	3.16	2.93	2.77	2.66	2.58	2.51	2.46
19	4.38	3.52	3.13	2.90	2.74	2.63	2.54	2.48	2.42
20	4.35	3.49	3.10	2.87	2.71	2.60	2.51	2.45	2.39
21	4.32	3.47	3.07	2.84	2.68	2.57	2.49	2.42	2.37
22	4.30	3.44	3.05	2.82	2.66	2.55	2.46	2.40	2.34
23	4.28	3.42	3.03	2.80	2.64	2.53	2.44	2.37	2.32
24	4.26	3.40	3.01	2.78	2.62	2.51	2.42	2.36	2.30
25	4.24	3.39	2.99	2.76	2.60	2.49	2.40	2.34	2.28
26	4.23	3.37	2.98	2.74	2.59	2.47	2.39	2.32	2.27
27	4.21	3.35	2.96	2.73	2.57	2.46	2.37	2.31	2.25
28	4.20	3.34	2.95	2.71	2.56	2.45	2.36	2.29	2.24
29	4.18	3.33	2.93	2.70	2.55	2.43	2.35	2.28	2.22
30	4.17	3.32	2.92	2.69	2.53	2.42	2.33	2.27	2.21
40	4.08	3.23	2.84	2.61	2.45	2.34	2.25	2.18	2.12
60	4.00	3.15	2.76	2.53	2.37	2.25	2.17	2.10	2.04
120	3.92	3.07	2.68	2.45	2.29	2.17	2.09	2.02	1.96
∞	3.84	3.00	2.60	2.37	2.21	2.10	2.01	1.94	1.88

Example: The value of *F* with 4 degrees of freedom in the numerator and 20 degrees of freedom in the denominator for which 5% of the area is in the tail is 2.87.

TABLE A.6

	Degrees of Freedom				(α = .05)					
df	10	12	15	20	24	30	40	60	120	∞
1	241.9	243.9	245.9	248.0	249.1	250.1	251.1	252.2	253.3	254.3
2	19.40	19.41	19.43	19.45	19.45	19.46	19.47	19.48	19.49	19.50
3	8.79	8.74	8.70	8.66	8.64	8.62	8.59	8.57	8.55	8.53
4	5.96	5.91	5.86	5.80	5.77	5.75	5.72	5.69	5.66	5.63
5	4.74	4.68	4.62	4.56	4.53	4.50	4.46	4.43	4.40	4.36
6	4.06	4.00	3.94	3.87	3.84	3.81	3.77	3.74	3.70	3.67
7	3.64	3.57	3.51	3.44	3.41	3.38	3.34	3.30	3.27	3.23
8	3.35	3.28	3.22	3.15	3.12	3.08	3.04	3.01	2.97	2.93
9	3.14	3.07	3.01	2.94	2.90	2.86	2.83	2.79	2.75	2.71
10	2.98	2.91	2.85	2.77	2.74	2.70	2.66	2.62	2.58	2.54
11	2.85	2.79	2.72	2.65	2.61	2.57	2.53	2.49	2.45	2.40
12	2.75	2.69	2.62	2.54	2.51	2.47	2.43	2.38	2.34	2.30
13	2.67	2.60	2.53	2.46	2.42	2.38	2.34	2.30	2.25	2.21
14	2.60	2.53	2.46	2.39	2.35	2.31	2.27	2.22	2.18	2.13
15	2.54	2.48	2.40	2.33	2.29	2.25	2.20	2.16	2.11	2.07
16	2.49	2.42	2.35	2.28	2.24	2.19	2.15	2.11	2.06	2.01
17	2.45	2.38	2.31	2.23	2.19	2.15	2.10	2.06	2.01	1.96
18	2.41	2.34	2.27	2.19	2.15	2.11	2.06	2.02	1.97	1.92
19	2.38	2.31	2.23	2.16	2.11	2.07	2.03	1.98	1.93	1.88
20	2.35	2.28	2.20	2.12	2.08	2.04	1.99	1.95	1.90	1.84
21	2.32	2.25	2.18	2.10	2.05	2.01	1.96	1.92	1.87	1.81
22	2.30	2.23	2.15	2.07	2.03	1.98	1.94	1.89	1.84	1.78
23	2.27	2.20	2.13	2.05	2.01	1.96	1.91	1.86	1.81	1.76
24	2.25	2.18	2.11	2.03	1.98	1.94	1.89	1.84	1.79	1.73
25	2.24	2.16	2.09	2.01	1.96	1.92	1.87	1.82	1.77	1.71
26	2.22	2.15	2.07	1.99	1.95	1.90	1.85	1.80	1.75	1.69
27	2.20	2.13	2.06	1.97	1.93	1.88	1.84	1.79	1.73	1.67
28	2.19	2.12	2.04	1.96	1.91	1.87	1.82	1.77	1.71	1.65
29	2.18	2.10	2.03	1.94	1.90	1.85	1.81	1.75	1.70	1.64
30	2.16	2.09	2.01	1.93	1.89	1.84	1.79	1.74	1.68	1.62
40	2.08	2.00	1.92	1.84	1.79	1.74	1.69	1.64	1.58	1.51
60	1.99	1.92	1.84	1.75	1.70	1.65	1.59	1.53	1.47	1.39
120	1.91	1.83	1.75	1.66	1.61	1.55	1.50	1.43	1.35	1.25
∞	1.83	1.75	1.67	1.57	1.52	1.46	1.39	1.32	1.22	1.00

(CONTINUED)

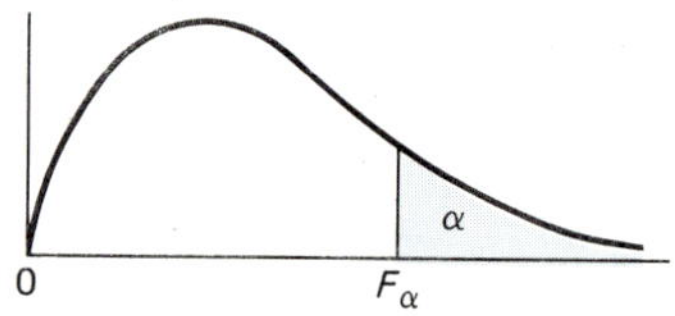

TABLE A.6 (CONTINUED)

	Degrees of Freedom				($\alpha = .01$)				
df	1	2	3	4	5	6	7	8	9
1	4052.	4999.5	5403.	5625.	5764.	5859.	5928.	5982.	6022.
2	98.50	99.00	99.17	99.25	99.30	99.33	99.36	99.37	99.39
3	34.12	30.82	29.46	28.71	28.24	27.91	27.67	27.49	27.35
4	21.20	18.00	16.69	15.98	15.52	15.21	14.98	14.80	14.66
5	16.26	13.27	12.06	11.39	10.97	10.67	10.46	10.29	10.16
6	13.75	10.92	9.78	9.15	8.75	8.47	8.26	8.10	7.98
7	12.25	9.55	8.45	7.85	7.46	7.19	6.99	6.84	6.72
8	11.26	8.65	7.59	7.01	6.63	6.37	6.18	6.03	5.91
9	10.56	8.02	6.99	6.42	6.06	5.80	5.61	5.47	5.35
10	10.04	7.56	6.55	5.99	5.64	5.39	5.20	5.06	4.94
11	9.65	7.21	6.22	5.67	5.32	5.07	4.89	4.74	4.63
12	9.33	6.93	5.95	5.41	5.06	4.82	4.64	4.50	4.39
13	9.07	6.70	5.74	5.21	4.86	4.62	4.44	4.30	4.19
14	8.86	6.51	5.56	5.04	4.69	4.46	4.28	4.14	4.03
15	8.68	6.36	5.42	4.89	4.56	4.32	4.14	4.00	3.89
16	8.53	6.23	5.29	4.77	4.44	4.20	4.03	3.89	3.78
17	8.40	6.11	5.18	4.67	4.34	4.10	3.93	3.79	3.68
18	8.29	6.01	5.09	4.58	4.25	4.01	3.84	3.71	3.60
19	8.18	5.93	5.01	4.50	4.17	3.94	3.77	3.63	3.52
20	8.10	5.85	4.94	4.43	4.10	3.87	3.70	3.56	3.46
21	8.02	5.78	4.87	4.37	4.04	3.81	3.64	3.51	3.40
22	7.95	5.72	4.82	4.31	3.99	3.76	3.59	3.45	3.35
23	7.88	5.66	4.76	4.26	3.94	3.71	3.54	3.41	3.30
24	7.82	5.61	4.72	4.22	3.90	3.67	3.50	3.36	3.26
25	7.77	5.57	4.68	4.18	3.85	3.63	3.46	3.32	3.22
26	7.72	5.53	4.64	4.14	3.82	3.59	3.42	3.29	3.18
27	7.68	5.49	4.60	4.11	3.78	3.56	3.39	3.26	3.15
28	7.64	5.45	4.57	4.07	3.75	3.53	3.36	3.23	3.12
29	7.60	5.42	4.54	4.04	3.73	3.50	3.33	3.20	3.09
30	7.56	5.39	4.51	4.02	3.70	3.47	3.30	3.17	3.07
40	7.31	5.18	4.31	3.83	3.51	3.29	3.12	2.99	2.89
60	7.08	4.98	4.13	3.65	3.34	3.12	2.95	2.82	2.72
120	6.85	4.79	3.95	3.48	3.17	2.96	2.79	2.66	2.56
∞	6.63	4.61	3.78	3.32	3.02	2.80	2.64	2.51	2.41

Example: The value of F with 6 degrees of freedom in the numerator and 15 degrees of freedom in the denominator for which 1% of the area is ain the tail is 4.32.

TABLE A.6

	Degrees of Freedom ($\alpha = .01$)									
f	10	12	15	20	24	30	40	60	120	∞
1	6056.	6106.	6157.	6209.	6235.	6261.	6287.	6313.	6339.	6366.
2	99.40	99.42	99.43	99.45	99.46	99.47	99.47	99.48	99.49	99.50
3	27.23	27.05	26.87	26.69	26.60	26.50	26.41	26.32	26.22	26.13
4	14.55	14.37	14.20	14.02	13.93	13.84	13.75	13.65	13.56	13.46
5	10.05	9.89	9.72	9.55	9.47	9.38	9.29	9.20	9.11	9.02
6	7.87	7.72	7.56	7.40	7.31	7.23	7.14	7.06	6.97	6.88
7	6.62	6.47	6.31	6.16	6.07	5.99	5.91	5.82	5.74	5.65
8	5.81	5.67	5.52	5.36	5.28	5.20	5.12	5.03	4.95	4.86
9	5.26	5.11	4.96	4.81	4.73	4.65	4.57	4.48	4.40	4.31
10	4.85	4.71	4.56	4.41	4.33	4.25	4.17	4.08	4.00	3.91
11	4.54	4.40	4.25	4.10	4.02	3.94	3.86	3.78	3.69	3.60
12	4.30	4.16	4.01	3.86	3.78	3.70	3.62	3.54	3.45	3.36
13	4.10	3.96	3.82	3.66	3.59	3.51	3.43	3.34	3.25	3.17
14	3.94	3.80	3.66	3.51	3.43	3.35	3.27	3.18	3.09	3.00
15	3.80	3.67	3.52	3.37	3.29	3.21	3.13	3.05	2.96	2.87
16	3.69	3.55	3.41	3.26	3.18	3.10	3.02	2.93	2.84	2.75
17	3.59	3.46	3.31	3.16	3.08	3.00	2.92	2.83	2.75	2.65
18	3.51	3.37	3.23	3.08	3.00	2.92	2.84	2.75	2.66	2.57
19	3.43	3.30	3.15	3.00	2.92	2.84	2.76	2.67	2.58	2.49
20	3.37	3.23	3.09	2.94	2.86	2.78	2.69	2.61	2.52	2.42
21	3.31	3.17	3.03	2.88	2.80	2.72	2.64	2.55	2.46	2.36
22	3.26	3.12	2.98	2.83	2.75	2.67	2.58	2.50	2.40	2.31
23	3.21	3.07	2.93	2.78	2.70	2.62	2.54	2.45	2.35	2.26
24	3.17	3.03	2.89	2.74	2.66	2.58	2.49	2.40	2.31	2.21
25	3.13	2.99	2.85	2.70	2.62	2.54	2.45	2.36	2.27	2.17
26	3.09	2.96	2.81	2.66	2.58	2.50	2.42	2.33	2.23	2.13
27	3.06	2.93	2.78	2.63	2.55	2.47	2.38	2.29	2.20	2.10
28	3.03	2.90	2.75	2.60	2.52	2.44	2.35	2.26	2.17	2.06
29	3.00	2.87	2.73	2.57	2.49	2.41	2.33	2.23	2.14	2.03
30	2.98	2.84	2.70	2.55	2.47	2.39	2.30	2.21	2.11	2.01
40	2.80	2.66	2.52	2.37	2.29	2.20	2.11	2.02	1.92	1.80
60	2.63	2.50	2.35	2.20	2.12	2.03	1.94	1.84	1.73	1.60
120	2.47	2.34	2.19	2.03	1.95	1.86	1.76	1.66	1.53	1.38
∞	2.32	2.18	2.04	1.88	1.79	1.70	1.59	1.47	1.32	1.00

TABLE A.7

TABLE OF COMMON LOGARITHMS

N	*0*	*1*	*2*	*3*	*4*	*5*	*6*	*7*	*8*	*9*
0	$-\infty$	00000	30103	47712	60206	69897	77815	84510	90309	95424
10	00000	00432	00860	01284	01703	02119	02531	02938	03342	03743
11	04139	04532	04922	05308	05690	06070	06446	06819	07188	07555
12	07918	08279	08636	08991	09342	09691	10037	10380	10721	11059
13	11394	11727	12057	12385	12710	13033	13354	13672	13988	14301
14	14613	14922	15229	15534	15836	16137	16435	16732	17026	17319
15	17609	17898	18184	18469	18752	19033	19312	19590	19866	20140
16	20412	20683	20952	21219	21484	21748	22011	22272	22531	22789
17	23045	23300	23533	23805	24055	24304	24551	24797	25042	25285
18	25527	25768	26007	26245	26482	26717	26951	27184	27416	27646
19	27875	28103	28330	28556	28780	29003	29226	29447	29667	29885
20	30103	30320	30535	30750	30963	31175	31387	31597	31806	32015
21	32222	32428	32634	32838	33041	33244	33445	33646	33846	34044
22	34242	34439	34635	34830	35025	35218	35411	35603	35793	35984
23	36173	36361	36549	36736	36922	37107	37291	37475	37658	37840
24	38021	38202	38382	38561	38739	38917	39094	39270	39445	39620
25	39794	39967	40140	40312	40483	40654	40824	40993	41162	41330
26	41497	41664	41830	41996	42160	42325	42488	42651	42813	42975
27	43136	43297	43457	43616	43775	43933	44091	44248	44404	44560
28	44716	44871	45025	45179	45332	45484	45637	45788	45939	46090
29	46240	46389	46538	46687	46835	46982	47129	47246	47422	47567
30	47712	47857	48001	48144	48287	48430	48572	48714	48855	48996
31	49136	49276	49415	49554	49693	49831	49969	50106	50243	50379
32	50515	50651	50786	50920	51055	51188	51322	51455	51587	51720
33	51851	51983	52114	52244	52375	52504	52634	52763	52892	53020
34	53148	53275	53403	53529	53656	53782	53908	54033	54158	54283
35	54407	54531	54654	54777	54900	55023	55145	55267	55388	55509
36	55630	55751	55871	55991	56110	56229	56348	56467	56585	56703
37	56820	56937	57054	57171	57287	57403	57519	57634	57749	57864
38	57978	58092	58206	58320	58433	58546	58659	58771	58883	58995
39	59106	59218	59329	59439	59550	59660	59770	59879	59988	60097
40	60206	60314	60423	60531	60638	60746	60853	60959	61066	61172
41	61278	61384	61490	61595	61700	61805	61909	62014	62118	62221
42	62325	62428	62531	62634	62737	62839	62941	63043	63144	63246
43	63347	63448	63548	63649	63749	63849	63949	64048	64147	64246
44	64345	64444	64542	64640	64738	64836	64933	65031	65128	65225
45	65321	65418	65514	65610	65706	65801	65896	65992	66087	66181
46	66276	66370	66464	66558	66652	66745	66839	66932	67025	67117
47	67210	67302	67394	67486	67578	67669	67761	67852	67943	68034
48	68124	68215	68305	68395	68485	68574	68664	68753	68842	68931
49	69020	69108	69197	69285	69373	69461	69548	69636	69723	69810

TABLE A.7

N	*0*	*1*	*2*	*3*	*4*	*5*	*6*	*7*	*8*	*9*
50	69897	36684	70070	70157	70243	70329	70415	70501	70586	70672
51	70757	70842	70927	71012	71096	71181	71265	71349	71433	71517
52	71600	71684	71767	71850	71933	72016	72099	72181	72263	72346
53	72428	72509	72591	72673	72754	72835	72916	72997	73078	73159
54	73239	73320	73400	73480	73560	73640	73719	73799	73878	73957
55	74036	74115	74194	74273	74351	74429	74507	74586	74663	74741
56	74819	74896	74974	75051	75128	75205	75282	75358	75435	75511
57	75587	75664	75740	75815	75891	75967	76042	76118	76193	76268
58	76343	76418	76492	76567	76641	76716	76790	76864	76938	77012
59	77085	77159	77232	77305	77379	77452	77525	77597	77670	77743
60	77815	77887	77960	78032	78104	78176	78247	78319	78390	78462
61	78533	78604	78675	78746	78817	78888	78958	79029	79099	79169
62	79239	79309	79379	79449	79518	79588	79657	79727	70796	79865
63	79934	80003	80072	80140	80209	80277	80346	80414	80482	80550
64	80618	80686	80754	80821	80889	80956	81023	81090	81158	81224
65	81291	81358	81425	81491	81558	81624	81690	81757	81823	81889
66	81954	82020	82086	82151	82217	82282	82347	82413	82478	82543
67	82607	82672	82737	82802	82866	82930	82995	83059	83123	83187
68	83251	83315	83378	83442	83506	83569	83632	83696	83759	83822
69	83885	83948	84011	84073	84136	84198	84261	84323	84386	84448
70	84510	84572	84634	84696	84757	84819	84880	84942	85003	85065
71	85126	85187	85248	85309	85370	85431	85491	85552	85612	85673
72	85733	85794	85854	85914	85974	86034	86094	86153	86213	86273
73	86332	86392	86451	86510	86570	86629	86688	86747	86806	86864
74	86923	86982	87040	87099	87157	87216	87274	87332	87390	87448
75	87506	87564	87622	87679	87737	87795	87852	87910	87967	88024
76	88081	88138	88195	88252	88309	88366	88423	88480	88536	88593
77	88649	88705	88762	88818	88874	88930	88986	89042	89098	89154
78	89209	89265	89321	89376	89432	89487	89542	89597	89653	89708
79	89763	89818	89873	89927	89982	90037	90091	90146	90200	90255
80	90309	90363	90417	90472	90526	90580	90634	90687	90741	90795
81	90849	90902	90956	91009	91062	91116	91169	91222	91275	91328
82	91381	91434	91487	91540	91593	91645	91698	91751	91803	91855
83	91908	91960	92012	92065	92117	92169	92221	92273	92324	92376
84	92428	92480	92531	92583	92634	92686	92737	92788	92840	92891
85	92942	92993	93044	93095	93146	93197	93247	93289	93349	93399
86	93450	93500	93551	93601	93651	93702	93752	93802	93852	93902
87	93952	94002	94052	94101	94151	94201	94250	94300	94349	94399
88	94448	94498	94547	94596	94645	94694	94743	94792	94841	94890
89	94939	94988	95036	95085	95134	95182	95231	95279	95328	95376
90	95424	95472	95521	95569	95617	95665	95713	95761	95809	95856

(CONTINUED)

TABLE A.7 (CONTINUED)

N	0	1	2	3	4	5	6	7	8	9
91	95904	95952	95999	96047	96095	96142	96190	96237	96284	96332
92	96379	96426	96473	96520	96567	96614	96661	96708	96755	96802
93	96848	96895	96942	96988	97035	97081	97128	97174	97220	97267
94	97313	97359	97405	97451	97497	97543	97589	97635	97681	97727
95	97772	97818	97864	97909	97955	98000	98046	98091	98137	98182
96	98227	98272	98318	98363	98408	98453	98498	98543	98588	98632
97	98677	98722	98767	98811	98856	98900	98945	98989	99034	99078
98	99123	99167	99211	99255	99300	99344	99388	99432	99476	99520
99	99564	99607	99651	99695	99739	99782	99826	99870	99913	99957

TABLE A.8
SQUARES, SQUARE ROOTS, AND RECIPROCALS 1–1000

N	N^2	$\sqrt{N}$	$\sqrt{10N}$	$1/N$	N	N^2	$\sqrt{N}$	$\sqrt{10N}$	$1/N$ .0
					50	2 500	7.071 068	22.36068	2000000
1	1	1.000 000	3.162 278	1.0000000	51	2 601	7.141 428	22.58318	1960784
2	4	1.414 214	4.472 136	.5000000	52	2 704	7.211 103	22.80351	1923077
3	9	1.732 051	5.477 226	.3333333	53	2 809	7.280 110	23.02173	1886792
4	16	2.000 000	6.324 555	.2500000	54	2 916	7.348 469	23.23790	1851852
5	25	2.236 068	7.071 068	.2000000	55	3 025	7.416 198	23.45208	1818182
6	36	2.449 490	7.745 967	.1666667	56	3 136	7.483 315	23.66432	1785714
7	49	2.645 751	8.366 600	.1428571	57	3 249	7.549 834	23.87467	1754386
8	64	2.828 427	8.944 272	.1250000	58	3 364	7.615 773	24.08319	1724138
9	81	3.000 000	9.486 833	.1111111	59	3 481	7.681 146	24.28992	1694915
10	100	3.162 278	10.00000	.1000000	60	3 600	7.745 967	24.49490	1666667
11	121	3.316 625	10.48809	.09090909	61	3 721	7.810 250	24.69818	1639344
12	144	3.464 102	10.95445	.08333333	62	3 844	7.874 008	24.89980	1612903
13	169	3.605 551	11.40175	.07692308	63	3 969	7.937 254	25.09980	1587302
14	196	3.741 657	11.83216	.07142857	64	4 096	8.000 000	25.29822	1562500
15	225	3.872 983	12.24745	.06666667	65	4 225	8.062 258	25.49510	1538462
16	256	4.000 000	12.64911	.06250000	66	4 356	8.124 038	25.69047	1515152
17	289	4.123 106	13.03840	.05882353	67	4 489	8.185 353	25.88436	1492537
18	324	4.242 641	13.41641	.05555556	68	4 624	8.246 211	26.07681	1470588
19	361	4.358 899	13.78405	.05263158	69	4 761	8.306 624	26.26785	1449275
20	400	4.472 136	14.14214	.05000000	70	4 900	8.366 600	26.45751	1428571
21	441	4.582 576	14.49138	.04761905	71	5 041	8.426 150	26.64583	1408451
22	484	4.690 416	14.83240	.04545455	72	5 184	8.485 281	26.83282	1388889
23	529	4.795 832	15.16575	.04347826	73	5 329	8.544 004	27.01851	1369863
24	576	4.898 979	15.49193	.04166667	74	5 476	8.602 325	27.20294	1351351
25	625	5.000 000	15.81139	.04000000	75	5 625	8.660 254	27.38613	1333333
26	676	5.099 020	16.12452	.03846154	76	5 776	8.717 798	27.56810	1315789
27	729	5.196 152	16.43168	.03703704	77	5 929	8.774 964	27.74887	1298701
28	784	5.291 503	16.73320	.03571429	78	6 084	8.831 761	27.92848	1282051
29	841	5.385 165	17.02939	.03448276	79	6 241	8.888 194	28.10694	1265823
30	900	5.477 226	17.32051	.03333333	80	6 400	8.944 272	28.28427	1250000
31	961	5.567 764	17.60682	.03225806	81	6 561	9.000 000	28.46050	1234568
32	1 024	5.656 854	17.88854	.03125000	82	6 724	9.055 385	28.63564	1219512
33	1 089	5.744 563	18.16590	.03030303	83	6 889	9.110 434	28.80972	1204819
34	1 156	5.830 952	18.43909	.02941176	84	7 056	9.165 151	28.98275	1190476
35	1 225	5.916 080	18.70829	.02857143	85	7 225	9.219 544	29.15476	1176471
36	1 296	6.000 000	18.97367	.02777778	86	7 396	9.273 618	29.32576	1162791
37	1 369	6.082 763	19.23538	.02702703	87	7 569	9.327 379	29.49576	1149425
38	1 444	6.164 414	19.49359	.02631579	88	7 744	9.380 832	29.66479	1136364
39	1 521	6.244 998	19.74842	.02564103	89	7 921	9.433 981	29.83287	1123596
40	1 600	6.324 555	20.00000	.02500000	90	8 100	9.486 833	30.00000	1111111
41	1 681	6.403 124	20.24846	.02439024	91	8 281	9.539 392	30.16621	1098901
42	1 764	6.480 741	20.49390	.02380952	92	8 464	9.591 663	30.33150	1086957
43	1 849	6.557 439	20.73644	.02325581	93	8 649	9.643 651	30.49590	1075269
44	1 936	6.633 250	20.97618	.02272727	94	8 836	9.695 360	30.65942	1063830
45	2 025	6.708 204	21.21320	.02222222	95	9 025	9.746 794	30.82207	1052632
46	2 116	6.782 330	21.44761	.02173913	96	9 216	9.797 959	30.98387	1041667
47	2 209	6.855 655	21.67948	.02127660	97	9 409	9.848 858	31.14482	1030928
48	2 304	6.928 203	21.90890	.02083333	98	9 604	9.899 495	31.30495	1020408
49	2 401	7.000 000	22.13594	.02040816	99	9 801	9.949 874	31.46427	1010101
50	2 500	7.071 068	22.36068	.02000000	100	10 000	10.00000	31.62278	1000000

(CONTINUED)

TABLE A.8 (CONTINUED)

N	N^2	$\sqrt{N}$	$\sqrt{10N}$	$1/N$ .0	N	N^2	$\sqrt{N}$	$\sqrt{10N}$	$1/N$ .00
100	10 000	10.00000	31.62278	10000000	150	22 500	12.24745	38.72983	6666667
101	10 201	10.04988	31.78050	09900990	151	22 801	12.28821	38.85872	6622517
102	10 404	10.09950	31.93744	09803922	152	23 104	12.32883	38.98718	6578947
103	10 609	10.14889	32.09361	09708738	153	23 409	12.36932	39.11521	6535948
104	10 816	10.19804	32.24903	09615385	154	23 176	12.40967	39.24283	6493506
105	11 025	10.24695	32.40370	09523810	155	24 025	12.44990	39.37004	6451613
106	11 236	10.29563	32.55764	09433962	156	24 336	12.49000	39.49684	6410256
107	11 449	10.34408	32.71085	09345794	157	24 649	12.52996	39.62323	6369427
108	11 664	10.39230	32.86335	09259259	158	24 964	12.56981	39.74921	6329114
109	11 881	10.44031	33.01515	09174312	159	25 281	12.60952	39.87480	6289308
110	12 100	10.48809	33.16625	09090909	160	25 600	12.64911	40.00000	6250000
111	12 321	10.53565	33.31666	09009009	161	25 921	12.68858	40.12481	6211180
112	12 544	10.58301	33.46640	08928571	162	26 244	12.72792	40.24922	6172840
113	12 769	10.63015	33.61547	08849558	163	26 569	12.76715	40.37326	6134969
114	12 996	10.67708	33.76389	08771930	164	26 896	12.80625	40.49691	6097561
115	13 225	10.72381	33.91165	08695652	165	27 225	12.84523	40.62019	6060606
116	13 456	10.77033	34.05877	08620690	166	27 556	12.88410	40.74310	6024096
117	13 689	10.81665	34.20526	08547009	167	27 889	12.92285	40.86563	5988024
118	13 924	10.86278	34.35113	08474576	168	28 224	12.96148	40.98780	5952381
119	14 161	10.90871	34.49638	08403361	169	28 561	13.00000	41.10961	5917160
120	14 400	10.95445	34.64102	08333333	170	28 900	13.03840	41.23106	5882353
121	14 641	11.00000	34.78505	08264463	171	29 241	13.07670	41.35215	5847953
122	14 884	11.04536	34.92850	08196721	172	29 584	13.11488	41.47288	5813953
123	15 129	11.09054	35.07136	08130081	173	29 929	13.15295	41.59327	5780347
124	15 376	11.13553	35.21363	08064516	174	30 276	13.19091	41.71331	5747126
125	15 625	11.18034	35.35534	08000000	175	30 625	13.22876	41.83300	5714286
126	15 876	11.22497	35.49648	07936508	176	30 976	13.26650	41.95235	5681818
127	16 129	11.26943	35.63706	07874016	177	31 329	13.30413	42.07137	5649718
128	16 384	11.31371	35.77709	07812500	178	31 684	13.34166	42.19005	5617978
129	16 641	11.35782	35.91657	07751938	179	32 041	13.37909	42.30839	5586592
130	16 900	11.40175	36.05551	07692308	180	32 400	13.41641	42.42641	5555556
131	17 161	11.44552	36.19392	07633588	181	32 761	13.45362	42.54409	5524862
132	17 424	11.48913	36.33180	07575758	182	33 124	13.49074	42.66146	5494505
133	17 689	11.53256	36.46917	07518797	183	33 489	13.52775	42.77850	5464481
134	17 956	11.57584	36.60601	07462687	184	33 856	13.56466	42.89522	5434783
135	18 225	11.61895	36.74235	07407407	185	34 225	13.60147	43.01163	5405405
136	18 496	11.66190	36.87818	07352941	186	34 596	13.63818	43.12772	5376344
137	18 769	11.70470	37.01351	07299270	187	34 969	13.67479	43.24350	5347594
138	19 044	11.74734	37.14835	07246377	188	35 344	13.71131	43.35897	5319149
139	19 321	11.78983	37.28270	07194245	189	35 721	13.74773	43.47413	5291005
140	19 600	11.83216	37.41657	07142857	190	36 100	13.78405	43.58899	5263158
141	19 881	11.87434	37.54997	07092199	191	36 481	13.82027	43.70355	5235602
142	20 164	11.91638	37.68289	07042254	192	36 864	13.85641	43.81780	5208333
143	20 449	11.95826	37.81534	06993007	193	37 249	13.89244	43.93177	5181347
144	20 736	12.00000	37.94733	06944444	194	37 636	13.92839	44.04543	5154639
145	21 025	12.04159	38.07887	06896552	195	38 025	13.96424	44.15880	5128205
146	21 316	12.08305	38.20995	06849315	196	38 416	14.00000	44.27189	5102041
147	21 609	12.12436	38.34058	06802721	197	38 809	14.03567	44.38468	5076142
148	21 904	12.16553	38.47077	06756757	198	39 204	14.07125	44.49719	5050505
149	22 201	12.20656	38.60052	06711409	199	39 601	14.10674	44.60942	5025126
150	22 500	12.24745	38.72983	06666667	200	40 000	14.14214	44.72136	5000000

TABLE A.8

N	N^2	$\sqrt{N}$	$\sqrt{10N}$	$1/N$ .00	N	N^2	$\sqrt{N}$	$\sqrt{10N}$	$1/N$ .00
200	40 000	14.14214	44.72136	5000000	250	62 500	15.81139	50.00000	4000000
201	40 401	14.17745	44.83302	4975124	251	63 001	15.84298	50.09990	3984064
202	40 804	14.21267	44.94441	4950495	252	63 504	15.87451	50.19960	3968254
203	41 209	14.24781	45.05552	4926108	253	64 009	15.90597	50.29911	3952569
204	41 616	14.28286	45.16636	4901961	254	64 516	15.93738	50.39841	3937008
205	42 025	14.31782	45.27693	4878049	255	65 025	15.96872	50.49752	3921569
206	42 436	14.35270	45.38722	4854369	256	65 536	16.00000	50.59644	3906250
207	42 849	14.38749	45.49725	4830918	257	66 049	16.03122	50.69517	3891051
208	43 264	14.42221	45.60702	4807692	258	66 564	16.06238	50.79370	3875969
209	43 681	14.45683	45.71652	4784689	259	67 081	16.09348	50.89204	3861004
210	44 100	14.49138	45.82576	4761905	260	67 600	16.12452	50.99020	3846154
211	44 521	14.52584	45.93474	4739336	261	68 121	16.15549	51.08816	3831418
212	44 944	14.56022	46.04346	4716981	262	68 644	16.18641	51.18594	3816794
213	45 369	14.59452	46.15192	4694836	263	69 169	16.21727	51.28353	3802281
214	45 796	14.62874	46.26013	4672897	264	69 696	16.24808	51.38093	3787879
215	46 225	14.66288	46.36809	4651163	265	70 225	16.27882	51.47815	3773585
216	46 656	14.69694	46.47580	4629630	266	70 756	16.30951	51.57519	3759398
217	47 089	14.73092	46.58326	4608295	267	71 289	16.34013	51.67204	3745318
218	47 524	14.76482	46.69047	4587156	268	71 824	16.37071	51.76872	3731343
219	47 961	14.79865	46.79744	4566210	269	72 361	16.40122	51.86521	3717472
220	48 400	14.83240	46.90416	4545455	270	72 900	16.43168	51.96152	3703704
221	48 841	14.86607	47.01064	4524887	271	73 441	16.46208	52.05766	3690037
222	49 284	14.89966	47.11688	4504505	272	73 984	16.49242	52.15362	3676471
223	49 729	14.93318	47.22288	4484305	273	74 529	16.52271	52.24940	3663004
224	50 176	14.96663	47.32864	4464286	274	75 076	16.55295	52.34501	3649635
225	50 625	15.00000	47.43416	4444444	275	75 625	16.58312	52.44044	3636364
226	51 076	15.03330	47.53946	4424779	276	76 176	16.61325	52.53570	3623188
227	51 529	15.06652	47.64452	4405286	277	76 729	16.64332	52.63079	3610108
228	51 984	15.09967	47.74935	4385965	278	77 284	16.67333	52.72571	3597122
229	52 441	15.13275	47.85394	4366812	279	77 841	16.70329	52.82045	3584229
230	52 900	15.16575	47.95832	4347826	280	78 400	16.73320	52.91503	3571429
231	53 361	15.19868	48.06246	4329004	281	78 961	16.76305	53.00943	3558719
232	53 824	15.23155	48.16638	4310345	282	79 524	16.79286	53.10367	3546099
233	54 289	15.26434	48.27007	4291845	283	80 089	16.82260	53.19774	3533569
234	54 756	15.29706	48.37355	4273504	284	80 656	16.85230	53.29165	3521127
235	55 225	15.32971	48.47680	4255319	285	81 225	16.88194	53.38539	3508772
236	55 696	15.36229	48.57983	4237288	286	81 796	16.91153	53.47897	3496503
237	56 169	15.39480	48.68265	4219409	287	82 369	16.94107	53.57238	3484321
238	56 644	15.42725	48.78524	4201681	288	82 944	16.97056	53.66563	3472222
239	57 121	15.45962	48.88763	4184100	289	83 521	17.00000	53.75872	3460208
240	57 600	15.49193	48.98979	4166667	290	84 100	17.02939	53.85165	3448276
241	58 081	15.52417	49.09175	4149378	291	84 681	17.05872	53.94442	3436426
242	58 564	15.55635	49.19350	4132231	292	85 264	17.08801	54.03702	3424658
243	59 049	15.58846	49.29503	4115226	293	85 849	17.11724	54.12947	3412969
244	59 536	15.62050	49.39636	4098361	294	86 436	17.14643	54.22177	3401361
245	60 025	15.65248	49.49747	4081633	295	87 025	17.17556	54.31390	3389831
246	60 516	15.68439	49.59839	4065041	296	87 616	17.20465	54.40588	3378378
247	61 009	15.71623	49.69909	4048583	297	88 209	17.23369	54.49771	3367003
248	61 504	15.74802	49.79960	4032258	298	88 804	17.26268	54.58938	3355705
249	62 001	15.77973	49.89990	4016064	299	89 401	17.29162	54.68089	3344482
250	62 500	15.81139	50.00000	4000000	300	90 000	17.32051	54.77226	3333333

(CONTINUED)

TABLE A.8 (CONTINUED)

N	N^2	$\sqrt{N}$	$\sqrt{10N}$	$1/N$ .00	N	N^2	$\sqrt{N}$	$\sqrt{10N}$	$1/N$ .00
300	90 000	17.32051	54.77226	3333333	350	122 500	18.70829	59.16080	2857143
301	90 601	17.34935	54.86347	3322259	351	123 201	18.73499	59.24525	2849003
302	91 204	17.37815	54.95453	3311258	352	123 904	18.76166	59.32959	2840909
303	91 809	17.40690	55.04544	3300330	353	124 609	18.78829	59.41380	2832861
304	92 416	17.43560	55.13620	3289474	354	125 316	18.81489	59.49790	2824859
305	93 025	17.46425	55.22681	3278689	355	126 025	18.84144	59.58188	2816901
306	93 636	17.49286	55.31727	3267974	356	126 736	18.86796	59.66574	2808989
307	94 249	17.52142	55.40758	3257329	357	127 449	18.89444	59.74948	2801120
308	94 864	17.54993	55.49775	3246753	358	128 164	18.92089	59.83310	2793296
309	95 481	17.57840	55.58777	3236246	359	128 881	18.94730	59.91661	2785515
310	96 100	17.60682	55.67764	3225806	360	129 600	18.97367	60.00000	2777778
311	96 721	17.63519	55.76737	3215434	361	130 321	19.00000	60.08328	2770083
312	97 344	17.66352	55.85696	3205128	362	131 044	19.02630	60.16644	2762431
313	97 969	17.69181	55.94640	3194888	363	131 769	19.05256	60.24948	2754821
314	98 596	17.72005	56.03570	3184713	364	132 496	19.07878	60.33241	2747253
315	99 225	17.74824	56.12486	3174603	365	133 225	19.10497	60.41523	2739726
316	99 856	17.77639	56.21388	3164557	366	133 956	19.13113	60.49793	2732240
317	100 489	17.80449	56.30275	3154574	367	134 689	19.15724	60.58052	2724796
318	101 124	17.83255	56.39149	3144654	368	135 424	19.18333	60.66300	2717391
319	101 761	17.86057	56.48008	3134796	369	136 161	19.20937	60.74537	2710027
320	102 400	17.88854	56.56854	3125000	370	136 900	19.23538	60.82763	2702703
321	103 041	17.91647	56.65686	3115265	371	137 641	19.26136	60.90977	2695418
322	103 684	17.94436	56.74504	3105590	372	138 384	19.28730	60.99180	2688172
323	104 329	17.97220	56.83309	3095975	373	139 129	19.31321	61.07373	2680965
324	104 976	18.00000	56.92100	3086420	374	139 876	19.33908	61.15554	2673797
325	105 625	18.02776	57.00877	3076923	375	140 625	19.36492	61.23724	2666667
326	106 276	18.05547	57.09641	3067485	376	141 376	19.39072	61.31884	2659574
327	106 929	18.08314	57.18391	3058104	377	142 129	19.41649	61.40033	2652520
328	107 584	18.11077	57.27128	3048780	378	142 884	19.44222	61.48170	2645503
329	108 241	18.13836	57.35852	3039514	379	143 641	19.46792	61.56298	2638522
330	108 900	18.16590	57.44563	3030303	380	144 400	19.49359	61.64414	2631579
331	109 561	18.19341	57.53260	3021148	381	145 161	19.51922	61.72520	2624672
332	110 224	18.22087	57.61944	3012048	382	145 924	19.54483	61.80615	2617801
333	110 889	18.24829	57.70615	3003003	383	146 689	19.57039	61.88699	2610966
334	111 556	18.27567	57.79273	2994012	384	147 456	19.59592	61.96773	2604167
335	112 225	18.30301	57.87918	2985075	385	148 225	19.62142	62.04837	2597403
336	112 896	18.33030	57.96551	2976190	386	148 996	19.64688	62.12890	2590674
337	113 569	18.35756	58.05170	2967359	387	149 769	19.67232	62.20932	2583979
338	114 244	18.38478	58.13777	2958580	388	150 544	19.69772	62.28965	2577320
339	114 921	18.41195	58.22371	2949853	389	151 321	19.72308	62.36986	2570694
340	115 600	18.43909	58.30952	2941176	390	152 100	19.74842	62.44998	2564103
341	116 281	18.46619	58.39521	2932551	391	152 881	19.77372	62.52999	2557545
342	116 964	18.49324	58.48077	2923977	392	153 664	19.79899	62.60990	2551020
343	117 649	18.52026	58.56620	2915452	393	154 449	19.82423	62.68971	2544529
344	118 336	18.54724	58.65151	2906977	394	155 236	19.84943	62.76942	2538071
345	119 025	18.57418	58.73670	2898551	395	156 025	19.87461	62.84903	2531646
346	119 716	18.60108	58.82176	2890173	396	156 816	19.89975	62.92853	2525253
347	120 409	18.62794	58.90671	2881844	397	157 609	19.92486	63.00794	2518892
348	121 104	18.65476	58.99152	2873563	398	158 404	19.94994	63.08724	2512563
349	121 801	18.68154	59.07622	2865330	399	159 201	19.97498	63.16645	2506266
350	122 500	18.70829	59.16080	2857143	400	160 000	20.00000	63.24555	2500000

TABLE A.8

N	N^2	$\sqrt{N}$	$\sqrt{10N}$	$1/N$ .00	N	N^2	$\sqrt{N}$	$\sqrt{10N}$	$1/N$ .00
400	160 000	20.00000	63.24555	2500000	450	202 500	21.21320	67.08204	2222222
401	160 801	20.02498	63.32456	2493766	451	203 401	21.23676	67.15653	2217295
402	161 604	20.04994	63.40347	2487562	452	204 304	21.26029	67.23095	2212389
403	162 409	20.07486	63.48228	2481390	453	205 209	21.28380	67.30527	2207506
404	163 216	20.09975	63.56099	2475248	454	206 116	21.30728	67.37952	2202643
405	164 025	20.12461	63.63961	2469136	455	207 025	21.33073	67.45369	2197802
406	164 836	20.14944	63.71813	2463054	456	207 936	21.35416	67.52777	2192982
407	165 649	20.17424	63.79655	2457002	457	208 849	21.37756	67.60178	2188184
408	166 464	20.19901	63.87488	2450980	458	209 764	21.40093	67.67570	2183406
409	167 281	20.22375	63.95311	2444988	459	210 681	21.42429	67.74954	2178649
410	168 100	20.24846	64.03124	2439024	460	211 600	21.44761	67.82330	2173913
411	168 921	20.27313	64.10928	2433090	461	212 521	21.47091	67.89698	2169197
412	169 744	20.29778	64.18723	2427184	462	213 444	21.49419	67.97058	2164502
413	170 569	20.32240	64.26508	2421308	463	214 369	21.51743	68.04410	2159827
414	171 396	20.34699	64.34283	2415459	464	215 296	21.54066	68.11755	2155172
415	172 225	20.37155	64.42049	2409639	465	216 225	21.56386	68.19091	2150538
416	173 056	20.39608	64.49806	2403846	466	217 156	21.58703	68.26419	2145923
417	173 889	20.42058	64.57554	2398082	467	218 089	21.61018	68.33740	2141328
418	174 724	20.44505	64.65292	2392344	468	219 024	21.63331	68.41053	2136752
419	175 561	20.46949	64.73021	2386635	469	219 961	21.65641	68.48357	2132196
420	176 400	20.49390	64.80741	2380952	470	220 900	21.67948	68.55655	2127660
421	177 241	20.51828	64.88451	2375297	471	221 841	21.70253	68.62944	2123142
422	178 084	20.54264	64.96153	2369668	472	222 784	21.72556	68.70226	2118644
423	178 929	20.56696	65.03845	2364066	473	223 729	21.74856	68.77500	2114165
424	179 776	20.59126	65.11528	2358491	474	224 676	21.77154	68.84766	2109705
425	180 625	20.61553	65.19202	2352941	475	225 625	21.79449	68.92024	2105263
426	181 476	20.63977	65.26868	2347418	476	226 576	21.81742	68.99275	2100840
427	182 329	20.66398	65.34524	2341920	477	227 529	21.84033	69.06519	2096436
428	183 184	20.68816	65.42171	2336449	478	228 484	21.86321	69.13754	2092050
429	184 041	20.71232	65.49809	2331002	479	229 441	21.88607	69.20983	2087683
430	184 900	20.73644	65.57439	2325581	480	230 400	21.90800	69.28203	2083333
431	185 761	20.76054	65.65059	2320186	481	231 361	21.93171	69.35416	2079002
432	186 624	20.78461	65.72671	2314815	482	232 324	21.95450	69.42622	2074689
433	187 489	20.80865	65.80274	2309469	483	233 289	21.97726	69.49820	2070393
434	188 356	20.83267	65.87868	2304147	484	234 256	22.00000	69.57011	2066116
435	189 225	20.85665	65.95453	2298851	485	235 225	22.02272	69.64194	2061856
436	190 096	20.88061	66.03030	2293578	486	236 196	22.04541	69.71370	2057613
437	190 969	20.90454	66.10598	2288330	487	237 169	22.06808	69.78539	2053388
438	191 844	20.92845	66.18157	2283105	488	238 144	22.09072	69.85700	2049180
439	192 721	20.95233	66.25708	2277904	489	239 121	22.11334	69.92853	2044990
440	193 600	20.97618	66.33250	2272727	490	240 100	22.13594	70.00000	2040816
441	194 481	21.00000	66.40783	2267574	491	241 081	22.15852	70.07139	2036660
442	195 364	21.02380	66.48308	2262443	492	242 064	22.18107	70.14271	2032520
443	196 249	21.04757	66.55825	2257336	493	243 049	22.20360	70.21396	2028398
444	197 136	21.07131	66.63332	2252252	494	244 036	22.22611	70.28513	2024291
445	198 025	21.09502	66.70832	2247191	495	245 025	22.24860	70.35624	2020202
446	198 916	21.11871	66.78323	2242152	496	246 016	22.27106	70.42727	2016129
447	199 809	21.14237	66.85806	2237136	497	247 009	22.29350	70.49823	2012072
448	200 704	21.16601	66.93280	2232143	498	248 004	22.31591	70.56912	2008032
449	201 601	21.18962	67.00746	2227171	499	249 001	22.33831	70.63993	2004008
450	202 500	21.21320	67.08204	2222222	500	250 000	22.36068	70.71068	2000000

(CONTINUED)

TABLE A.8 (CONTINUED)

N	N^2	$\sqrt{N}$	$\sqrt{10N}$	$1/N$ .00	N	N^2	$\sqrt{N}$	$\sqrt{10N}$	$1/N$ .00
500	250 000	22.36068	70.71068	2000000	550	302 500	23.45208	74.16198	1818182
501	251 001	22.38303	70.78135	1996008	551	303 601	23.47339	74.22937	1814882
502	252 004	22.40536	70.85196	1992032	552	304 704	23.49468	74.29670	1811594
503	253 009	22.42766	70.92249	1988072	553	305 809	23.51595	74.36397	1808318
504	254 016	22.44994	70.99296	1984127	554	306 916	23.53720	74.43118	1805054
505	255 025	22.47221	71.06335	1980198	555	308 025	23.55844	74.49832	1801802
506	256 036	22.49444	71.13368	1976285	556	309 136	23.57965	74.56541	1798561
507	257 049	22.51666	71.20393	1972387	557	310 249	23.60085	74.63243	1795332
508	258 064	22.53886	71.27412	1968504	558	311 364	23.62202	74.69940	1792115
509	259 081	22.56103	71.34424	1964637	559	312 481	23.64318	74.76630	1788909
510	260 100	22.58318	71.41428	1960784	560	313 600	23.66432	74.83315	1785714
511	261 121	22.60531	71.48426	1956947	561	314 721	23.68544	74.89993	1782531
512	262 144	22.62742	71.55418	1953125	562	315 844	23.70654	74.96666	1779359
513	263 169	22.64950	71.62402	1949318	563	316 969	23.72762	75.03333	1776199
514	264 196	22.67157	71.69379	1945525	564	318 096	23.74868	75.09993	1773050
515	265 225	22.69361	71.76350	1941748	565	319 225	23.76973	75.16648	1769912
516	266 256	22.71563	71.83314	1937984	566	320 356	23.79075	75.23297	1766784
517	267 289	22.73763	71.90271	1934236	567	321 489	23.81176	75.29940	1763668
518	268 324	22.75961	71.97222	1930502	568	322 624	23.83275	75.36577	1760563
519	269 361	22.78157	72.04165	1926782	569	323 761	23.85372	75.43209	1757469
520	270 400	22.80351	72.11103	1923077	570	324 900	23.87467	75.49834	1754386
521	271 441	22.82542	72.18033	1919386	571	326 041	23.89561	75.56454	1751313
522	272 484	22.84732	72.24957	1915709	572	327 184	23.91652	75.63068	1748252
523	273 529	22.86919	72.31874	1912046	573	328 329	23.93742	75.69676	1745201
524	274 576	22.89105	72.38784	1908397	574	329 476	23.95830	75.76279	1742160
525	275 625	22.91288	72.45688	1904762	575	330 625	23.97916	75.82875	1739130
526	276 676	22.93469	72.52586	1901141	576	331 776	24.00000	75.89466	1736111
527	277 729	22.95648	72.59477	1897533	577	332 929	24.02082	75.96052	1733102
528	278 784	22.97825	72.66361	1893939	578	334 084	24.04163	76.02631	1730104
529	279 841	23.00000	72.73239	1890359	579	335 241	24.06242	76.09205	1727116
530	280 900	23.02173	72.80110	1886792	580	336 400	24.08319	76.15773	1724138
531	281 961	23.04344	72.86975	1883239	581	337 561	24.10394	76.22336	1721170
532	283 024	23.06513	72.93833	1879699	582	338 724	24.12468	76.28892	1718213
533	284 089	23.08679	73.00685	1876173	583	339 889	24.14539	76.35444	1715266
534	285 156	23.10844	73.07530	1872659	584	341 056	24.16609	76.41989	1712329
535	286 225	23.13007	73.14369	1869159	585	342 225	24.18677	76.48529	1709402
536	287 296	23.15167	73.21202	1865672	586	343 396	24.20744	76.55064	1706485
537	288 369	23.17326	73.28028	1862197	587	344 569	24.22808	76.61593	1703578
538	289 444	23.19483	73.34848	1858736	588	345 744	24.24871	76.68116	1700680
539	290 521	23.21637	73.41662	1855288	589	346 921	24.26932	76.74634	1697793
540	291 600	23.23790	73.48469	1851852	590	348 100	24.28992	76.81146	1694915
541	292 681	23.25941	73.55270	1848429	591	349 281	24.31049	76.87652	1692047
542	293 764	23.28089	73.62065	1845018	592	350 464	24.33105	76.94154	1689189
543	294 849	23.30236	73.68853	1841621	593	351 649	24.35159	77.00649	1686341
544	295 936	23.32381	73.75636	1838235	594	352 836	24.37212	77.07140	1683502
545	297 025	23.34524	73.82412	1834862	595	354 025	24.39262	77.13624	1680672
546	298 116	23.36664	73.89181	1831502	596	355 216	24.41311	77.20104	1677852
547	299 209	23.38803	73.95945	1828154	597	356 409	24.43358	77.26578	1675042
548	300 304	23.40940	74.02702	1824818	598	357 604	24.45404	77.33046	1672241
549	301 401	23.43075	74.09453	1821494	599	358 801	24.47448	77.39509	1669449
550	302 500	23.45208	74.16198	1818182	600	360 000	24.49490	77.45967	1666667

TABLE A.8

N	N^2	$\sqrt{N}$	$\sqrt{10N}$	$1/N$ .00	N	N^2	$\sqrt{N}$	$\sqrt{10N}$	$1/N$ .00
600	360 000	24.49490	77.45967	1666667	650	422 500	25.49510	80.62258	1538462
601	361 201	24.51530	77.52419	1663894	651	423 801	25.51470	80.68457	1536098
602	362 404	24.53569	77.58866	1661130	652	425 104	25.53429	80.74652	1533742
603	363 609	24.55606	77.65307	1658375	653	426 409	25.55386	80.80842	1531394
604	364 816	24.57641	77.71744	1655629	654	427 716	25.57342	80.87027	1529052
605	366 025	24.59675	77.78175	1652893	655	429 025	25.59297	80.93207	1526718
606	367 236	24.61707	77.84600	1650165	656	430 336	25.61250	80.99383	1524390
607	368 449	24.63737	77.91020	1647446	657	431 649	25.63201	81.05554	1522070
608	369 664	24.65766	77.97435	1644737	658	432 964	25.65151	81.11720	1519757
609	370 881	24.67793	78.03845	1642036	659	434 281	25.67100	81.17881	1517451
610	372 100	24.69818	78.10250	1639344	660	435 600	25.69047	81.24038	1515152
611	373 321	24.71841	78.16649	1636661	661	436 921	25.70992	81.30191	1512859
612	374 544	24.73863	78.23043	1633987	662	438 244	25.72936	81.36338	1510574
613	375 769	24.75884	78.29432	1631321	663	439 569	25.74879	81.42481	1508296
614	376 996	24.77902	78.35815	1628664	664	440 896	25.76820	81.48620	1506024
615	378 225	24.79919	78.42194	1626016	665	442 225	25.78759	81.54753	1503759
616	379 456	24.81935	78.48567	1623377	666	443 556	25.80698	81.60882	1501502
617	380 689	24.83948	78.54935	1620746	667	444 889	25.82634	81.67007	1499250
618	381 924	24.85961	78.61298	1618123	668	446 224	25.84570	81.73127	1497006
619	383 161	24.87971	78.67655	1615509	669	447 561	25.86503	81.79242	1494768
620	384 400	24.89980	78.74008	1612903	670	448 900	25.88436	81.85353	1492537
621	385 641	24.91987	78.80355	1610306	671	450 241	25.90367	81.91459	1490313
622	386 884	24.93993	78.86698	1607717	672	451 584	25.92296	81.97561	1488095
623	388 129	24.95997	78.93035	1605136	673	452 929	25.94224	82.03658	1485884
624	389 376	24.97999	78.99367	1602564	674	454 276	25.96151	82.09750	1483680
625	390 625	25.00000	79.05694	1600000	675	455 625	25.98076	82.15838	1481481
626	391 876	25.01999	79.12016	1597444	676	456 976	26.00000	82.21922	1479290
627	393 129	25.03997	79.18333	1594896	677	458 329	26.01922	82.28001	1477105
628	394 384	25.05993	79.24645	1592357	678	459 684	26.03843	82.34076	1474926
629	395 641	25.07987	79.30952	1589825	679	461 041	26.05763	82.40146	1472754
630	396 900	25.09980	79.37254	1587302	680	462 400	26.07681	82.46211	1470588
631	398 161	25.11971	79.43551	1584786	681	463 761	26.09598	82.42272	1468429
632	399 424	25.13961	79.49843	1582278	682	465 124	26.11513	82.58329	1466276
633	400 689	25.15949	79.56130	1579779	683	466 489	26.13427	82.64381	1464129
634	401 956	25.17936	79.62412	1577287	684	467 856	26.15339	82.70429	1461988
635	403 225	25.19921	79.68689	1574803	685	469 225	26.17250	82.76473	1459854
636	404 496	25.21904	79.74961	1572327	686	470 596	26.19160	82.82512	1457726
637	405 769	25.23886	79.81228	1569859	687	471 969	26.21068	82.88546	1455604
638	407 044	25.25866	79.87490	1567398	688	473 344	26.22975	82.94577	1453488
639	408 321	25.27845	79.93748	1564945	689	474 721	26.24881	83.00602	1451379
640	409 600	25.29822	80.00000	1562500	690	476 100	26.26785	83.06624	1449275
641	410 881	25.31798	80.06248	1560062	691	477 481	26.28688	83.12641	1447178
642	412 164	25.33772	80.12490	1557632	692	478 864	26.30589	83.18654	1445087
643	413 449	25.35744	80.18728	1555210	693	480 249	26.32489	83.24662	1443001
644	414 736	25.37716	80.24961	1552795	694	481 636	26.34388	83.30666	1440922
645	416 025	25.39685	80.31189	1550388	695	483 025	26.36285	83.36666	1438849
646	417 316	25.41653	80.37413	1547988	696	484 416	26.38181	83.42661	1436782
647	418 609	25.43619	80.43631	1545595	697	485 809	26.40076	83.48653	1434720
648	419 904	25.45584	80.49845	1543210	698	487 204	26.41969	83.54639	1432665
649	421 201	25.47548	80.56054	1540832	699	488 601	26.43861	83.60622	1430615
650	422 500	25.49510	80.62258	1538462	700	490 000	26.45751	83.66600	1428571

(CONTINUED)

TABLE A.8 (CONTINUED)

N	N^2	$\sqrt{N}$	$\sqrt{10N}$	$1/N$ .00	N	N^2	$\sqrt{N}$	$\sqrt{10N}$	$1/N$ .00
700	490 000	26.45751	83.66600	1428571	750	562 500	27.38613	86.60254	1333333
701	491 401	26.47640	83.72574	1426534	751	564 001	27.40438	86.66026	1331558
702	492 804	26.49528	83.78544	1424501	752	565 504	27.42262	86.71793	1329787
703	494 209	26.51415	83.84510	1422475	753	567 009	27.44085	86.77557	1328021
704	495 616	26.53300	83.90471	1420455	754	568 516	27.45906	86.83317	1326260
705	497 025	26.55184	83.96428	1418440	755	570 025	27.47726	86.89074	1324503
706	498 436	26.57066	84.02381	1416431	756	571 536	27.49545	86.94826	1322751
707	499 849	26.58947	84.08329	1414427	757	573 049	27.51363	87.00575	1321004
708	501 264	26.60827	84.14274	1412429	758	574 564	27.53180	87.06320	1319261
709	502 681	26.62705	84.20214	1410437	759	576 081	27.54995	87.12061	1317523
710	504 100	26.64583	84.26150	1408451	760	577 600	27.56810	87.17798	1315789
711	505 521	26.66458	84.32082	1406470	761	579 121	27.58623	87.23531	1314060
712	506 944	26.68333	84.38009	1404494	762	580 644	27.60435	87.29261	1312336
713	508 369	26.70206	84.43933	1402525	763	582 169	27.62245	87.34987	1310616
714	509 796	26.72078	84.49852	1400560	764	583 696	27.64055	87.40709	1308901
715	511 225	26.73948	84.55767	1398601	765	585 225	27.65863	87.46428	1307190
716	512 656	26.75818	84.61678	1396648	766	586 756	27.67671	87.52143	1305483
717	514 089	26.77686	84.67585	1394700	767	588 289	27.69476	87.57854	1303781
718	515 524	26.79552	84.73488	1392758	768	589 824	27.71281	87.63561	1302083
719	516 961	26.81418	84.79387	1390821	769	591 361	27.73085	87.69265	1300390
720	518 400	26.83282	84.85281	1388889	770	592 900	27.74887	87.74964	1298701
721	519 841	26.85144	84.91172	1386963	771	594 441	27.76689	87.80661	1297017
722	521 284	26.87006	84.97058	1385042	772	595 984	27.78489	87.86353	1295337
723	522 729	26.88866	85.02941	1383126	773	597 529	27.80288	87.92042	1293661
724	524 176	26.90725	85.08819	1381215	774	599 076	27.82086	87.97727	1291990
725	525 625	26.92582	85.14693	1379310	775	600 625	27.83882	88.03408	1290323
726	527 076	26.94439	85.20563	1377410	776	602 176	27.85678	88.09086	1288660
727	528 529	26.96294	85.26429	1375516	777	603 729	27.87472	88.14760	1287001
728	529 984	26.98148	85.32292	1373626	778	605 284	27.89265	88.20431	1285347
729	531 411	27.00000	85.38150	1371742	779	606 841	27.91057	88.26098	1283697
730	532 900	27.01851	85.44004	1369863	780	608 400	27.92848	88.31761	1282051
731	534 361	27.03701	85.49854	1367989	781	609 961	27.94638	88.37420	1280410
732	535 824	27.05550	85.55700	1366120	782	611 524	27.96426	88.43076	1278772
733	537 289	27.07397	85.61542	1364256	783	613 089	27.98214	88.48729	1277139
734	538 756	27.09243	85.67380	1362398	784	614 656	28.00000	88.54377	1275510
735	540 225	27.11088	85.73214	1360544	785	616 225	28.01785	88.60023	1273885
736	541 696	27.12932	85.79044	1358696	786	617 796	28.03569	88.65664	1272265
737	543 169	27.14774	85.84870	1356852	787	619 369	28.05352	88.71302	1270648
738	544 644	27.16616	85.90693	1355014	788	620 944	28.07134	88.76936	1269036
739	546 121	27.18455	85.96511	1353180	789	622 521	28.08914	88.82567	1267427
740	547 600	27.20294	86.02325	1351351	790	624 100	28.10694	88.88194	1265823
741	549 081	27.22132	86.08136	1349528	791	625 681	28.12472	88.93818	1264223
742	550 564	27.23968	86.13942	1347709	792	627 264	28.14249	88.99438	1262626
743	552 049	27.25803	86.19745	1345895	793	628 849	28.16026	89.05055	1261034
744	553 536	27.27636	86.25543	1344086	794	630 436	28.17801	89.10668	1259446
745	555 025	27.29469	86.31338	1342282	795	632 025	28.19574	89.16277	1257862
746	556 516	27.31300	86.37129	1340483	796	633 616	28.21347	89.21883	1256281
747	558 009	27.33130	86.42916	1338688	797	635 209	28.23119	89.27486	1254705
748	559 504	27.34959	86.48699	1336898	798	636 804	28.24889	89.33085	1253133
749	561 001	27.36786	86.54479	1335113	799	638 401	28.26659	89.38680	1251564
750	562 500	27.38613	86.60254	1333333	800	640 000	28.28427	89.44272	1250000

TABLE A.8

N	N^2	$\sqrt{N}$	$\sqrt{10N}$	$1/N$ .00	N	N^2	$\sqrt{N}$	$\sqrt{10N}$	$1/N$ .00
800	640 000	28.28427	89.44272	1250000	850	722 500	29.15476	92.19544	1176471
801	641 601	28.30194	89.49860	1248439	851	724 201	29.17190	92.24966	1175088
802	643 204	28.31960	89.55445	1246883	852	725 904	29.18904	92.30385	1173709
803	644 809	28.33725	89.61027	1245330	853	727 609	29.20616	92.35800	1172333
804	646 416	28.35489	89.66605	1243781	854	729 316	29.22328	92.41212	1170960
805	648 025	28.37252	89.72179	1242236	855	731 025	29.24038	92.46621	1169591
806	649 636	28.39014	89.77750	1240695	856	732 736	29.25748	92.52027	1168224
807	651 249	28.40775	89.83318	1239157	857	734 449	29.27456	92.57429	1166861
808	652 864	28.42534	89.88882	1237624	858	736 164	29.29164	92.62829	1165501
809	654 481	28.44293	89.94443	1236094	859	737 881	29.30870	92.68225	1164144
810	656 100	28.46050	90.00000	1234568	860	739 600	29.32576	92.73618	1162791
811	657 721	28.47806	90.05554	1233046	861	741 321	29.34280	92.79009	1161440
812	659 344	28.49561	90.11104	1231527	862	743 044	29.35984	92.84396	1160093
813	660 969	28.51315	90.16651	1230012	863	744 769	29.37686	92.89779	1158749
814	662 596	28.53069	90.22195	1228501	864	746 496	29.39388	92.95160	1157407
815	664 225	28.54820	90.27735	1226994	865	748 225	29.41088	93.00538	1156069
816	665 856	28.56571	90.33272	1225490	866	749 965	29.42788	93.05912	1154734
817	667 489	28.58321	90.38805	1223990	867	751 689	29.44486	93.11283	1153403
818	669 124	28.60070	90.44335	1222494	868	753 424	29.46184	93.15562	1152074
819	670 761	28.61818	90.49862	1221001	869	755 161	29.47881	93.22017	1150748
820	672 400	28.63564	90.55385	1219512	870	756 900	29.49576	93.27379	1149425
821	674 041	28.65310	90.60905	1218027	871	758 641	29.51271	93.32738	1148106
822	675 684	28.67054	90.66422	1216545	872	760 384	29.52965	93.38094	1146789
823	677 329	28.68798	90.71935	1215067	873	762 129	29.54657	93.43447	1145475
824	678 976	28.70540	90.77445	1213592	874	763 876	29.56349	93.48797	1144165
825	680 625	28.72281	90.82951	1212121	875	765 625	29.58040	93.54143	1142857
826	682 276	28.74022	90.88454	1219654	876	767 376	29.59730	93.59487	1141553
827	683 929	28.75761	90.93954	1209190	877	769 129	29.61419	93.64828	1140251
828	685 584	28.77499	90.99451	1207729	878	770 884	29.63106	93.70165	1138952
829	687 241	28.79236	90.04944	1296273	879	772 641	29.64793	93.75500	1137656
830	688 900	28.80972	91.10434	1204819	880	774 400	29.66479	93.80832	1136364
831	690 561	28.82707	91.15920	1203369	881	776 161	29.68164	93.86160	1135074
832	692 224	28.84441	91.21403	1201923	882	777 924	29.69848	93.91486	1133787
833	693 889	28.86174	91.26883	1200480	883	779 689	29.71532	93.96808	1132503
834	695 556	28.87906	91.32360	1199041	884	781 456	29.83214	94.02127	1131222
835	697 225	28.89637	91.37833	1197605	885	783 225	29.74895	94.07444	1129944
836	698 896	28.91366	91.43304	1196172	886	784 996	29.76575	94.12757	1128668
837	700 569	28.93095	91.48770	1194743	887	786 769	29.78255	94.18068	1127396
838	702 244	28.94823	91.54234	1193317	888	788 544	29.79933	94.23375	1126126
839	703 921	28.96550	91.59694	1191895	889	790 321	29.81610	94.28680	1124859
840	705 600	28.98275	91.65151	1190476	890	792 100	29.83287	94.33981	1123596
841	707 281	29.00000	91.70605	1189061	891	793 881	29.84962	94.39280	1122334
842	708 964	29.01724	91.76056	1187648	892	795 664	29.86637	94.44575	1121076
843	710 649	29.03446	91.81503	1186240	893	797 449	29.88311	94.49868	1119821
844	712 336	29.05168	91.86947	1184834	894	799 236	29.89983	94.55157	1118568
845	714 025	29.06888	91.92388	1183432	895	801 025	29.91655	94.60444	1117318
846	715 716	29.08608	91.97826	1182033	896	802 816	29.93326	94.65728	1116071
847	717 409	29.10326	92.03260	1180638	897	804 609	29.94996	94.71008	1114827
848	719 104	29.12044	92.08692	1179245	898	806 404	29.96665	94.76286	1113586
849	720 801	29.13760	92.14120	1177856	899	808 201	29.98333	94.81561	1112347
850	722 500	29.15476	92.19544	1176471	900	810 000	30.00000	94.86833	1111111

(CONTINUED)

TABLE A.8 (CONTINUED)

N	N^2	$\sqrt{N}$	$\sqrt{10N}$	$1/N$ .00	N	N^2	$\sqrt{N}$	$\sqrt{10N}$	$1/N$ .00
900	810 000	30.00000	94.86833	1111111	950	902 500	30.82207	97.46794	1052632
901	811 801	30.01666	94.92102	1109878	951	904 401	30.83829	97.51923	1051525
902	813 604	30.03331	94.97368	1108647	952	906 304	30.85450	97.57049	1050420
903	815 409	30.04996	95.02631	1107420	953	908 209	30.87070	97.62172	1049318
904	817 216	30.06659	95.07891	1106195	954	910 116	30.88689	97.67292	1048218
905	819 025	30.08322	95.13149	1104972	955	912 025	30.90307	97.72410	1047120
906	820 836	30.09983	95.18403	1103753	956	913 936	30.91925	97.77525	1046025
907	822 649	30.11644	95.23655	1102536	957	915 849	30.93542	97.82638	1044932
908	824 464	30.13304	95.28903	1101322	958	917 764	30.95158	97.87747	1043841
909	826 281	30.14963	95.34149	1100110	959	919 681	30.96773	97.92855	1042753
910	828 100	30.16621	95.39392	1098901	960	921 600	30.98387	97.97959	1041667
911	829 921	30.18278	95.44632	1097695	961	923 521	31.00000	98.03061	1040583
912	831 744	30.19934	95.49869	1096491	962	925 444	31.01612	98.08160	1039501
913	833 569	30.21589	95.55103	1095290	963	927 369	31.03224	98.13256	1038422
914	835 396	30.23243	95.60335	1094092	964	929 296	31.04835	98.18350	1037344
915	837 225	30.24897	95.65563	1092896	965	931 225	31.06445	98.23441	1036269
916	839 056	30.26549	95.70789	1091703	966	933 156	31.08054	98.28530	1035197
917	840 889	30.28201	95.76012	1090513	967	935 089	31.09662	98.33616	1034126
918	842 724	30.29851	95.81232	1089325	968	937 024	31.11270	98.38699	1033058
919	844 561	30.31501	95.86449	1088139	969	938 961	31.12876	98.43780	1031992
920	846 400	30.33150	95.91663	1086957	970	940 900	31.14482	98.48858	1030928
921	848 241	30.34798	95.96874	1085776	971	942 841	31.16087	98.53933	1029866
922	850 084	30.36445	96.02083	1084599	972	944 784	31.17691	98.59006	1028807
923	851 929	30.38092	96.07289	1083424	973	946 729	31.19295	98.64076	1027749
924	853 776	30.39737	96.12492	1082251	974	948 676	31.20897	98.69144	1026694
925	855 625	30.41381	96.17692	1081081	975	950 625	31.22499	98.74209	1025641
926	857 476	30.43025	96.22889	1079914	976	952 576	31.24100	98.79271	1024590
927	859 329	30.44667	96.28084	1078749	977	954 529	31.25700	98.84331	1023541
928	861 184	30.46309	96.33276	1077586	978	956 484	31.27299	98.89388	1022495
929	863 041	30.47950	96.38465	1076426	979	958 441	31.28898	98.94443	1021450
930	864 900	30.49590	96.43651	1075269	980	960 400	31.30495	98.99495	1020408
931	866 761	30.51229	96.48834	1074114	981	962 361	31.32092	99.04544	1019368
932	868 624	30.52868	96.54015	1072961	982	964 324	31.33688	99.09591	1018330
933	870 489	30.54505	96.59193	1071811	983	966 289	31.35283	99.14636	1017294
934	872 356	30.56141	96.64368	1070664	984	968 256	31.36877	99.19677	1016260
935	874 225	30.57777	96.69540	1069519	985	970 225	31.38471	99.24717	1015228
936	876 096	30.59412	96.74709	1068376	986	972 196	31.40064	99.29753	1014199
937	877 969	30.61046	96.79876	1067236	987	974 169	31.41656	99.34787	1013171
938	879 844	30.62679	96.85040	1066098	988	976 144	31.43247	99.39819	1012146
939	881 721	30.64311	96.90201	1064963	989	978 121	31.44837	99.44848	1011122
940	883 600	30.65942	96.95360	1063830	990	980 100	31.46427	99.49874	1010101
941	885 481	30.67572	97.00515	1062699	991	982 081	31.48015	99.54898	1009082
942	887 364	30.69202	97.05668	1061571	992	984 064	31.49603	99.59920	1008065
943	889 249	30.70831	97.10819	1060445	993	986 049	31.51190	99.64939	1007049
944	891 136	30.72458	97.15966	1059322	994	988 036	31.52777	99.69955	1006036
945	893 025	30.74085	97.21111	1058201	995	990 025	31.54362	99.74969	1005025
946	894 916	30.75711	97.26253	1057082	996	992 016	31.55947	99.79980	1004016
947	896 809	30.77337	97.31393	1055966	997	994 009	31.57531	99.84989	1003009
948	898 704	30.78961	97.36529	1054852	998	996 004	31.59114	99.89995	1002004
949	900 601	30.80584	97.41663	1053741	999	998 001	31.60696	99.94999	1001001
950	902 500	30.82207	97.46794	1052632	1000	1 000 000	31.62278	100.00000	1000000

APPENDIX B

TABLE B.1

WHOLESALE PRICE INDEX, BY GROUP AND SUBGROUP OF COMMODITIES. [1967 = 100 UNLESS OTHERWISE SPECIFIED]*

Code	Commodity group	Annual average 1976	1976 June	July	Aug.	Sept.	Oct.	Nov.	Dec.	1977 Jan.	Feb.	Mar.	Apr.	May	June
	All commodities	183.0	183.2	184.4	183.8	184.8	185.3	185.6	187.1	188.0	190.0	191.9	194.3	195.2	194.4
	All commodities (1957–59 = 100)	194.2	194.4	195.6	195.0	196.1	196.6	196.9	198.5	199.5	201.6	203.6	206.2	207.1	206.3
	Farm products and processed foods and feeds	183.1	187.4	188.1	181.7	182.9	179.5	178.3	183.9	184.8	188.4	190.9	195.9	196.8	191.5
	Industrial commodities	182.4	181.5	182.7	183.8	184.8	186.3	187.1	187.4	188.4	189.9	191.6	193.2	194.2	194.6
	FARM PRODUCTS AND PROCESSED FOODS AND FEEDS														
01	Farm Products	191.1	196.5	196.9	189.7	191.9	186.7	183.6	191.6	193.5	199.0	202.4	208.1	204.3	192.7
01–1	Fresh and dried fruits and vegetables	178.4	160.7	164.6	159.3	180.2	192.4	166.5	174.5	198.4	212.6	219.1	205.6	201.8	176.2
01–2	Grains	205.9	225.1	224.3	207.6	205.5	186.7	175.4	180.6	184.9	185.8	183.4	184.4	171.2	157.7
01–3	Livestock	173.3	185.1	175.9	166.2	161.6	156.1	154.4	166.1	166.0	166.2	163.5	167.9	180.2	172.3
01–4	Live poultry	166.9	174.9	184.0	179.0	164.9	150.5	139.1	145.7	153.7	183.7	177.2	182.3	183.1	182.7
01–5	Plant and animal fibers	223.9	235.9	269.0	235.6	242.3	249.8	257.9	239.5	216.5	240.1	252.4	249.5	238.6	197.5
01–6	Fluid milk	201.2	190.2	193.1	200.6	203.5	206.7	204.4	202.8	200.2	198.4	195.2	197.7	198.3	199.3
01–7	Eggs	179.1	165.8	167.7	186.8	188.9	180.7	192.8	213.6	189.2	194.8	173.5	165.2	144.4	141.4
01–8	Hay, hayseeds, and oilseeds	210.4	219.5	238.4	224.4	231.7	220.1	231.9	242.3	244.9	254.5	278.0	322.4	289.8	270.2
01–9	Other farm products	223.4	224.4	218.6	227.5	252.7	256.9	271.3	290.6	295.2	313.1	367.3	385.0	357.5	341.1
02	Processed foods and feeds	178.0	181.8	182.6	176.7	177.2	174.9	174.8	179.0	179.3	181.9	183.9	188.5	192.0	190.1
02–1	Cereal and bakery products	172.1	173.7	173.6	170.2	169.7	169.9	168.7	168.6	168.4	169.9	171.5	171.6	172.0	171.3
02–2	Meats, poultry, and fish	181.6	190.0	185.3	174.7	176.1	168.5	168.4	176.9	176.6	177.4	174.2	174.9	183.8	183.4
02–3	Dairy products	168.5	167.2	170.2	173.9	170.5	169.8	168.1	167.3	166.8	166.9	168.1	173.6	174.2	174.3
02–4	Processed fruits and vegetables	170.2	168.8	169.7	171.1	172.1	174.4	175.7	175.6	175.4	182.9	184.2	185.6	185.8	187.8
02–5	Sugar and confectionery	190.9	197.4	201.3	184.9	167.3	176.4	171.4	170.5	171.9	177.6	180.2	186.4	184.4	176.3
02–6	Beverages and beverage materials	173.5	172.8	175.9	175.8	176.4	177.5	178.8	183.8	184.1	189.3	199.5	202.0	206.0	207.7
02–71	Animal fats and oils	210.2	189.0	215.8	216.7	243.4	221.4	231.5	230.9	240.9	253.0	253.0	305.8	307.7	279.9
02–72	Crude vegetable oils	162.5	148.6	186.7	163.8	186.9	170.4	186.0	178.4	171.8	190.0	222.9	253.7	248.6	229.6
02–73	Refined vegetable oils	187.5	170.3	200.3	179.8	199.3	186.4	190.3	185.2	185.7	204.9	219.9	229.1	228.9	219.2
02–74	Vegetable oil end products	174.2	170.6	173.1	172.7	180.4	177.6	180.6	178.3	177.9	182.7	187.8	206.3	214.1	216.3
02–8	Miscellaneous processed foods	174.7	172.7	173.0	173.7	175.4	177.8	178.1	182.8	182.1	183.8	183.8	184.8	192.4	192.9
02–9	Manufactured animal feeds	194.4	213.0	216.2	196.9	211.4	199.8	200.2	213.2	219.1	219.1	222.1	243.3	239.6	225.7
	INDUSTRIAL COMMODITIES														
03	Textile products and apparel	148.2	148.3	149.0	149.5	149.0	149.3	150.1	149.9	150.3	151.1	152.1	153.7	154.0	154.4
03–1	Synthetic fibers	102.4	102.5	102.6	103.3	103.1	101.9	101.7	101.6	102.6	103.4	103.4	106.6	107.0	109.5
03–2	Processed yarns and threads	99.6	99.7	101.2	99.6	98.9	98.0	97.5	97.2	96.6	97.2	98.7	101.1	102.3	103.4

Code	Commodity														
03–3	Grey fabrics	106.1	106.6	108.5	108.5	107.1	107.4	109.1	107.7	105.1	103.8	103.6	105.0	105.1	104.5
03–4	Finished fabrics	101.1	101.4	101.0	101.1	100.5	100.3	101.4	101.5	100.4	101.2	102.7	104.2	104.9	104.5
03–5	Apparel	139.9	139.8	140.2	141.5	141.3	142.2	142.8	142.9	144.8	145.6	146.0	146.4	146.6	147.2
03–6	Textile housefurnishings	159.3	158.9	159.6	160.7	161.1	162.4	163.2	162.7	165.5	167.1	169.6	170.4	169.7	169.7
04	Hides, skins, leather, and related products	167.8	168.1	170.3	171.6	173.6	170.9	169.8	171.5	174.5	176.7	177.6	180.1	181.9	179.7
04–1	Hides and skins	258.4	261.1	278.6	284.8	292.1	251.4	231.8	251.2	278.9	282.5	285.9	305.0	313.0	288.8
04–2	Leather	188.1	191.2	192.2	196.3	197.5	193.1	191.4	191.7	192.9	201.3	201.4	204.1	210.7	202.1
04–3	Footwear	158.9	158.7	160.7	161.2	162.5	162.6	162.9	163.8	164.5	165.9	166.7	167.7	168.2	168.6
04–4	Other leather and related products	152.9	152.2	152.3	152.7	155.2	155.8	155.9	156.2	159.5	159.8	161.0	163.1	163.7	163.7
05	Fuels and related products and power	265.6	260.5	265.3	269.2	271.2	277.1	281.6	279.0	278.7	289.0	293.4	298.6	302.3	304.0
05–1	Coal	368.7	366.6	367.7	367.8	368.0	368.4	369.1	374.0	376.3	377.5	378.9	380.0	386.9	390.6
05–2	Coke	346.8	346.0	346.0	346.0	346.0	349.7	349.7	363.4	367.3	367.3	367.3	372.9	375.1	386.1
05–3	Gas fuels	286.8	275.9	277.3	286.6	289.5	330.9	365.0	337.6	322.2	363.7	70.9	379.0	390.2	386.6
05–4	Electric power	207.6	206.3	210.1	213.6	214.5	213.2	214.0	211.5	214.0	219.8	223.5	229.6	230.7	234.4
05–61	Crude petroleum	253.6	248.1	254.3	254.3	254.3	264.4	264.4	264.4	262.9	274.2	270.0	270.0	271.0	271.8
05–7	Petroleum products, refined	276.6	270.6	276.8	280.7	283.7	285.0	285.8	287.6	289.2	295.1	301.3	306.4	310.1	311.6
06	Chemicals and allied products	187.2	187.3	187.1	188.0	188.6	188.6	188.6	188.2	188.9	189.8	191.1	192.6	193.8	193.9
06–1	Industrial chemicals	219.3	218.4	219.2	221.2	221.7	222.2	222.6	221.5	222.1	222.9	222.3	223.2	224.0	224.1
06–21	Prepared paint	174.4	173.9	173.9	175.7	176.2	176.9	177.3	177.3	177.3	178.9	180.6	181.7	181.7	182.3
06–22	Paint materials	189.8	190.5	194.7	195.7	198.9	198.3	200.3	200.7	200.5	202.3	206.1	209.4	210.1	209.3
06–3	Drugs and pharmaceuticals	134.0	134.4	134.7	135.2	135.4	135.4	135.9	136.4	137.5	138.4	139.0	139.5	139.7	140.8
06–4	Fats and oils, inedible	249.9	243.5	258.9	249.4	262.9	251.2	251.2	254.6	253.9	253.9	273.7	304.9	337.5	318.8
06–5	Agricultural chemicals and chemical products	188.3	188.1	184.5	186.2	186.9	186.8	184.1	183.4	182.2	183.5	186.9	188.1	187.7	189.0
06–6	Plastic resins and materials	194.0	195.2	197.6	197.1	195.9	195.6	195.4	194.9	193.4	192.9	194.4	195.9	196.6	197.6
06–7	Other chemicals and allied products	170.7	172.3	168.9	169.2	169.3	169.9	169.9	169.4	172.2	173.4	174.6	175.0	175.9	176.0
07	Rubber and plastic products	159.2	157.1	158.3	161.1	163.9	164.6	164.8	164.7	164.5	164.0	164.3	165.9	166.4	167.4
07–1	Rubber and rubber products	163.2	159.4	160.4	166.2	169.7	171.4	171.5	171.4	170.9	168.3	168.7	172.3	172.3	172.9
07–11	Crude rubber	161.0	163.7	161.4	162.6	162.2	164.0	164.9	163.2	163.5	164.8	166.5	168.2	169.7	171.4
07–12	Tires and tubes	161.5	155.5	157.3	167.4	170.9	172.9	172.1	172.3	170.0	163.6	163.6	170.1	167.8	167.8
07–13	Miscellaneous rubber products	163.9	160.7	161.7	164.0	168.8	170.0	170.8	170.9	172.5	173.0	173.3	174.1	176.3	177.3
07–21	Plastic construction products	127.2	126.1	128.9	129.3	129.2	129.2	129.4	129.4	129.5	129.6	129.8	129.0	130.5	134.1
07–22	Unsupported plastic film and sheeting	154.9	154.9	155.3	155.3	157.7	157.5	157.5	157.5	157.6	159.4	159.5	159.1	159.6	160.3
07–23	Laminated plastic sheets, high pressure	131.1	129.2	132.1	132.1	135.5	136.5	137.8	137.2	136.5	137.3	138.3	141.6	142.2	142.4
08	Lumber and wood products	205.6	199.9	203.7	207.5	212.8	213.6	214.3	220.0	222.7	224.2	228.7	229.6	229.3	228.7
08–1	Lumber	233.0	224.2	231.2	236.2	244.3	245.6	244.3	252.1	257.8	259.3	266.0	268.3	267.8	264.6
08–2	Millwork	176.9	178.1	177.4	178.2	180.2	181.5	183.0	183.1	183.4	185.6	188.0	190.7	191.5	192.4
08–3	Plywood	187.0	176.7	181.0	187.5	191.5	190.3	194.4	205.2	205.0	205.0	207.9	202.5	200.0	202.6
08–4	Other wood products	166.2	165.3	166.1	166.3	169.8	170.3	171.3	172.5	173.2	176.4	178.9	181.0	184.4	185.4

*Monthly Labor Review August 1977.

APPENDIX B.2

CONSUMER PRICE INDEX—U.S. CITY AVERAGE—GENERAL SUMMARY AND GROUPS, SUBGROUPS, AND SELECTED ITEMS (1967 = 100 UNLESS OTHERWISE SPECIFIED)*

General summary	Annual average 1976	1976							1977					
		June	July	Aug.	Sept.	Oct.	Nov.	Dec.	Jan.	Feb.	Mar.	Apr.	May	June
All items	170.5	170.1	171.1	171.9	172.6	173.3	173.8	174.3	175.3	177.1	178.2	179.6	180.6	181.8
All items (1957–59 = 100)	198.3	197.9	199.0	200.0	200.8	201.5	202.1	202.7	203.8	206.0	207.2	208.8	210.0	211.5
Food	180.8	180.9	182.1	182.4	181.6	181.6	181.1	181.7	183.4	87.7	188.6	190.9	191.7	193.6
Food at home	179.5	179.7	180.9	181.0	179.9	179.6	178.9	179.3	181.2	186.2	186.9	189.3	189.8	191.9
Food away from home	186.1	185.6	186.9	187.8	188.7	189.3	190.0	190.9	192.2	193.6	195.2	197.5	199.3	200.6
Housing	177.2	176.5	177.5	178.4	179.5	180.1	180.7	181.6	183.1	184.3	185.5	186.7	187.6	189.0
Rent	144.7	144.4	145.0	145.6	146.2	146.9	147.5	148.3	149.5	150.2	150.8	151.6	152.2	152.9
Homeownership	191.7	190.7	192.2	193.4	194.4	194.8	194.8	195.0	196.7	198.1	199.3	201.0	202.3	203.9
Apparel and upkeep	147.6	146.9	146.5	148.1	150.2	150.9	151.9	151.8	150.0	150.8	151.7	152.3	153.4	153.9
Transportation	165.5	165.9	167.6	168.5	169.5	170.9	171.4	171.4	172.1	173.3	174.8	176.8	178.2	179.2
Health and recreation	163.3	162.8	163.7	164.4	165.3	166.1	167.3	168.0	169.0	169.8	170.7	171.4	172.3	173.2
Medical care	184.7	183.7	185.5	186.8	187.9	188.9	191.3	192.3	194.1	195.8	197.6	199.1	200.5	201.8
Special groups														
All items less shelter	168.3	168.1	169.0	169.7	170.4	171.0	171.6	172.2	173.1	175.0	176.1	177.5	178.4	179.7
All items less food	167.5	167.0	167.9	168.9	170.0	170.8	171.6	172.2	172.9	174.0	175.1	176.3	177.3	178.4
All items less medical care	169.7	169.4	170.3	171.1	171.7	172.4	172.7	173.2	174.2	176.0	177.0	178.4	179.4	180.6
Appliances (including radio and TV)	123.3	123.3	123.5	123.6	124.2	124.4	124.8	124.7	124.8	125.0	125.0	125.2	125.6	125.7
Commodities	165.2	165.2	166.0	166.6	167.0	167.4	167.7	168.1	168.7	170.9	171.8	173.3	174.3	175.4
Nondurables	169.2	169.0	169.7	170.4	170.7	171.0	171.3	171.7	172.4	175.0	175.9	177.4	178.3	179.7
Durables	154.3	154.7	155.8	156.4	156.9	157.8	158.0	158.4	158.9	159.7	160.8	162.2	163.4	163.9
Services	180.4	179.5	180.7	181.8	183.2	184.1	185.1	185.8	187.5	188.7	190.0	191.3	192.3	193.7
Commodities less food	156.6	156.5	157.1	158.0	158.9	159.6	160.3	160.6	160.6	161.6	162.6	163.6	164.7	165.4
Nondurables less food	158.3	157.9	158.1	159.1	160.4	161.0	161.9	162.3	161.9	163.1	163.9	164.7	165.7	166.6
Apparel commodities	145.8	145.0	144.4	146.2	148.5	149.2	150.1	149.9	147.6	148.5	149.3	149.8	150.9	151.3
Apparel commodities less footwear	144.9	144.1	143.4	145.2	147.8	148.5	149.4	149.2	146.5	147.4	148.1	148.6	149.7	150.2
Nondurables less food and apparel	165.7	165.6	166.3	166.8	167.4	168.1	169.0	169.7	170.5	171.8	172.6	173.5	174.5	175.6
Household durables	146.0	146.1	146.5	146.3	146.7	147.2	147.8	148.2	148.4	148.8	149.7	150.7	151.2	151.6
Housefurnishings	150.7	150.9	150.9	150.8	151.7	152.2	152.9	153.2	153.0	153.7	154.7	155.5	155.9	156.6
Services less rent	186.8	185.8	187.2	188.4	189.8	190.8	191.8	192.6	194.4	195.6	197.1	198.4	199.5	201.1
Household services less rent	198.4	197.3	198.7	200.1	201.5	202.3	202.6	203.5	205.7	206.8	208.4	209.7	210.8	212.9
Transportation services	174.3	173.2	174.7	175.5	177.3	178.9	180.2	180.8	182.2	183.6	185.1	186.9	187.7	189.0
Medical care services	197.1	195.8	197.9	199.4	200.6	201.7	204.5	205.7	207.6	209.4	211.5	213.1	214.6	216.0
Other services	161.1	160.5	161.2	162.0	163.6	164.3	165.2	165.7	166.7	167.5	168.1	168.9	169.6	170.5

*Monthly Labor Review, August 1977.

APPENDIX B.3

Industrial production	*Seasonally adjusted, 1967 = 100* 1977 *Mar.*	*Apr.*	*May*[a]	*June*[b]	*Percent changes from—* *Month ago*	*Year ago*	*Q1 to Q2*
Total	135.2	136.2	137.6	138.6	.7	6.5	3.0
Products, total	134.9	136.2	137.2	138.3	.8	6.8	2.4
Final products	133.0	134.4	135.3	136.3	.7	6.8	2.6
Consumer goods	142.8	143.6	143.9	144.7	.6	5.0	2.0
Durable goods	152.4	152.0	152.6	155.2	1.7	7.6	3.7
Nondurable goods	139.0	140.3	140.4	140.7	.2	4.1	1.3
Business equipment	144.5	147.0	149.3	151.6	1.5	12.3	4.3
Intermediate products	141.9	142.8	144.6	145.8	.8	7.3	1.8
Construction supplies	136.4	137.4	139.6	141.1	1.1	7.1	2.4
Materials	135.5	136.4	137.9	138.9	.7	5.9	3.6

[a] Preliminary.
[b] Estimated.
Source: Federal Reserve Bulletin, August 1977.

INDEX

A

Absolute frequency distribution, 61–63
  see also Frequency, Frequency distribution
Acceptance region, defined, 314
Aggregative index number, defined, 147
Alpha (α)
  in hypothesis testing, 314
Alternate hypothesis, null hypothesis and, 289
Analysis
  correlation, *see* Correlation analysis
  preposterior, 516–517
  regression, *see* Regression analysis
Analysis of variance, 344–375
  computation with data in, 353
  defined, 344–345
  different situations in, 348
  error sum of squares in, 351
  least significant difference in, 359–361
  mean square in, 351
  notation for, 345–347
  one-factor, 354–359
  rationale for, 347–350
  sums of squares in, 350–354
  two-factor, 361–369
ANOVA, defined, 370
ANOVA table, 350–354
  for one-factor analysis of variance, 354, 359
  for two-factor analysis of variance, 365
Antilog, 26
Arithmetic mean, 80, 86–94
  defined, 100
  weighted, 86–87
Array
  defined, 54, 73
  frequency distribution and, 55–56
Association, rule of, 39–40
Attributes, defined, 9, 11
Average of relatives index, 137, 147
  see also Price index numbers

B

Base number, defined, 128
Base period, defined, 147
Bayes' criterion, in statistical decision theory, 494, 501–504, 510
  see also Decision theory
Bayes' risk of the decision rule, 518
Bayes' rule, 172–173, 513–514
Bernoulli population or process, 191
  defined, 209, 238, 266
Beta, 314
Bimodal distribution, 68
Binomial distribution, 191
  continuity correction for, 208–209
  cumulative, 193–194
  normal approximation to, 206–207

C

Categorical frequency distribution, 64
  see also Frequency distribution
Census, defined, 11
Central limit theorem, 229, 296
Central tendency, measures of, 80
Chance modes, defined, 508
Characteristic, of logarithm, 24–25
Chi-square distribution, 378–379
Chi-square statistic, 379
  in tests of independence, 389–394

Chi-square tests, 378–399
  contingency tables and, 389–394
  goodness-of-fit tests and, 381–389
Class boundaries, defined, 73
Classes
  in frequency distribution, 55–57
  open-ended, 61
Classification variable, defined, 389, 395
Class intervals
  defined, 73
  in frequency distribution, 59
  unequal, 61
Class limits
  for continuous data, 58
  in frequency distribution, 57
Class mark
  defined, 59–60, 73
  in frequency distribution, 60
  histogram and, 66
Clusters, defined, 227
Cluster sampling, 227
  defined, 240
Coefficient of correlation, 453
  defined, 432–433, 457
Coefficient of determination
  computation of, 447
  defined, 444
  in regression analysis, 443–446
Coefficient of multiple correlation, defined, 456–457
Coefficient of multiple determination, defined, 456
Coefficient of rank correlation, 416–418
  defined, 418
Coefficient of skewness, 118
Coefficient of variation, defined, 120
Combination(s)
  defined, 34, 37
  examples of, 37–39
Complement, defined, 28
Composite index number, 147
Computer, in statistics, 10
Conditioned probability, defined, 164–165
Confidence, defined, 277
Confidence interval, 254
  construction of, 263
  one- and two-sided, 256–260, 448
  for proportions, 266
  two-sided, 260, 448
Confidence interval estimate, 254
  defined, 277
Confidence intervals on a predicted *y*, 450–451
Confidence intervals on the predicted mean $\bar{y}_c$, 448–450
Confidence limit, defined, 277
Confidence multiplier, 254
  for one- and two-sided confidence intervals, 257
  sample size and, 273
Consistent estimator, 251
Constants
  addition of variable to, 18
  defined, 14
  sum of series of, 17–18
Consumer Price Index, 130
Contingency table, defined, 395
Continuous data, class limits for, 58
Continuity correction factor, in binomial distribution, 208–209
Continuous random variable, 183–184
  defined, 209
Continuous variable, defined, 9, 11
Correlation
  coefficient of, 432–436, 453
  defined, 429
  in regression analysis, 429–435
Correlation analysis, 457
  defined, 428
Correlation coefficient, 432–433
  inferences about, 435–436
  in multiple regression, 453
Counts, defined, 19
CPI, *see* Consumer Price Index
Critical region, defined, 314
Critical value, defined, 314
Crude mode, defined, 81, 100
Cumulative binomial distribution, 193
Cumulative frequency distribution, defined, 61, 73
  see also Frequency distribution
Cumulative relative frequency distribution, 63–64
  defined, 73
Cycle
  defined, 476, 488
  in time series, 476–480

D

Data, continuous, 58
Data organization, 52–76
  frequency distribution in, 53–65
  general purpose tables in, 71–72
  histograms in, 65–67
Decile, defined, 98, 100
Decision making
  risk in, 495
  states of nature and, 495–497
  terminal decision in, 509
Decision making under risk, defined, 519
Decision making under uncertainty, 495
  defined, 519
Decision rule
  defined, 314
  in hypothesis testing, 291, 304–305
  for one-tailed test, 308
  optimal, 494
Decision theory, 494–525
  decision trees in, 507–509
  expected value of perfect information in, 509–511

expected value of sample information, 517–519
likelihood in, 511
notation used in, 497
opportunity loss in, 504–507
posterior distribution in, 512
preposterior analysis in, 516–517
sampling and revising of prior distribution in, 511–516
Decision trees, 507–509
Decomposition, defined, 488
Deflation of time series, 145
Degrees of freedom, in *t* distribution, 236–238
Demand function, defined, 42
Dependent variable, 41
Descriptive measures, 80–122
Descriptive statistics, 3–4
defined, 11
Deseasonalizing
defined, 488
of time series data, 485
Deviation, standard, *see* Standard deviation
Discrete random variable, 183
defined, 209
Discrete variable, 9, 11
Distribution
bimodal, 68
binomial, 191
frequency, *see* Frequency distribution
hypergeometric, 195–199
normal, 201–203, 228
Poisson, 199–201
prior, 501–504
probability, *see* Probability distribution
standard normal, 202–205
Distribution-free statistics, defined, 402, 418
see also Statistics
Domain, defined, 39, 43
Domain set, 39
Dow-Jones average, 4

E

Elementary units, 5, 11
Elimination, rule of, 170–172
EMV, *see* Expected monetary value
EOL, *see* Expected opportunity loss
Error sum of squares, in analysis of variance, 351
Estimate
point and interval, 253–254
standard error of, *see* Standard error of the estimate
Estimation process, 248–283
Estimators
in analysis of variance, 348–349
consistency of, 251
defined, 250, 277
efficiency of, 251
properties of, 250–253
sufficiency of, 253
unbiasedness and, 250–251
Event, defined, 156, 173
Event set, defined, 173
Event space, defined, 156, 173
EVPI, *see* Expected value of perfect information
EVSI, *see* Expected value of sample information
Expected monetary value, 503
Expected net gain from sampling, defined, 519
Expected opportunity loss, 505
Expected risk of the decision rule, 518
Expected value
defined, 209
of perfect information, 509–511
of random variable, 187–189
of sample information in decision theory, 517–519
Experiment, defined, 173

F

Factor, in analysis of variance, 370
Factor level, 370
F distribution
and analysis of variance, 349
in hypothesis testing, 333–334
Finite population correction factor, 229
First decile, 98
Forecasting, time series and, 468–469, 487–488
Frequency
defined, 73
relative, 157
Frequency distribution
absolute, 61–63
array and, 55–56
categorical, 64–65
classes in, 56
class interval in, 59–60
class limit in, 57
class marks in, 60, 89
cumulative, 61–63, 73
cumulative relative, 63–64
defined, 53, 73
examples of, 53–61
histograms and, 66–69
mean of, 89, 91–92
open-ended classes in, 61
practical uses of, 67–69
relative, 73
special, 61–64
unequal class intervals in, 61
use of, 54–58
Frequency polygon, defined, 69, 73
Frequency table, defined, 53
Functions, 39–42

defined, 43
demand, 42

G

General purpose tables, 71–72
Geometric mean, 80, 96, 100
Goodness-of-fit tests, 381–389
defined, 395
degrees of freedom for, 382–389
Greek alphabet, 15

H

Histograms, 65–67
defined, 65
frequency distribution and, 66–69
practical uses of, 67–69
Hypergeometric distribution, 195–199
defined, 197
Hypothesis
defined, 286
null, *see* Null hypothesis
Hypothesis testing, 285–340
acceptance region in, 291
critical value in, 291
with critical values for z and t, 312–314
decision rule and, 291
defined, 286
of equality of variances, 334–335
errors in, 294–295
F distribution and, 333–334
general procedure for, 289–292
large vs. small samples in, 324–328
of mean, 309
null hypothesis in, 322, 349, 382, 385–386
one-sample tests in, 285–319
about population means, 296–302
power of test in, 310
about proportions, 302–306
rationale for, 287–289
rejection or critical region in, 291
significance level of, 288
and test of difference between propositions, 328–330
and tests using paired observations, 330–333
two sample tests in, 322–340
Type II error probability in, 306–311

I

Identical distributions, sign test for, 409
Independence tests
chi-square statistic and, 389–394
defined, 395
Independent trials, defined, 209
Independent variable, 41
Index numbers, 128–149
applications of, 145–147
base period in, 132
common examples of, 130–131
composite, 129–130
computational method for, 132–133
construction of, 131–133
defined, 128
item selection in, 131–132
in long-term lease or purchase contracts, 146–147
notation for, 133
quantity, 143–145
as relatives, 128
sample, 129–130, 148
weighting system for, 132
Index of Industrial Production, 131
Inference, statistical, 7, 248
Information
perfect, 509–511
sample, 517–519
Intersection symbol, 30
Interval estimates, 253–254
Inverse probability rule, 172
see also Bayes' rule

J

Joint probability, 165
Judgment sampling, defined, 222–223, 240

K

Kruskal–Wallace test, 413–416, 419, 423

L

Laspeyres index, 138
defined, 148
LCL, *see* Lower confidence limit
Least significant difference
in analysis of variance, 359–361
defined, 370
in two-factor analysis of variance, 366
Least squares trend line, 472–475
Likelihood, defined, 511
Location, measures of, 80
Logarithms, 24–27
antilog of, 26
calculations with, 26–27
characteristic of, 24
common, 24
examples of, 26–27
interpolation in, 24
mantissa and, 24
Long-term lease or purchase contracts, index numbers and, 146–147
Lower confidence limit, 254

M

MA, *see* Moving averages
Mann–Whitney U test, 409–410, 418
extension of, 413

Mantissa, of logarithm, 24–25
Marginal probability, defined, 165
Market research, statistical inference in, 4–6
Mean(s)
alternate values of, 309
arithmetic, *see* Arithmetic mean
calculation of, 88–89
characteristics of, 90–91
estimation of, 260–265
geometric, 80
for grouped data, 88–89
vs. median, 94–96
modified, 481–488
operating characteristic curve for test of, 309
power curve for test of, 310
rejection probabilities for alternate values of, 310
sample size for, 273–274
sampling distribution of, 229–233
test of difference between, 324–328
test statistic for tests of differences in, 323–324
weighted, 87–88
Mean square
in analysis of variance, 351
defined, 370
Means testing
for large samples, 296–299
for small samples, 299–302
Measurements
approximations and, 20
defined, 19
Measures of central tendency, 80
Median, 80
calculation of, 84
defined, 100
vs. mean, 94–95
Median class, 83
Modal class, 81
Mode
crude, 81, 100
defined, 80–82, 100
vs. mean and median, 94–96
usefulness of, 95
Modified mean
defined, 488
in time series, 481
Monthly Labor Review, 130
Moving averages, defined, 477, 488
Multiple correlation coefficient, defined, 457
Multiple regression, 452–457
and standard error of the estimate, 455–456
Multiple regression equation, by least squares, 454–455
Multiple-sample tests
by analysis of variance, 344–375
defined, 344
Multiplication rule
in permutations and combinations, 35
of probability, 169
Mutually exclusive events, 161

N

Node, of decision tree, 519
Nonparametric statistical methods, 402–425
defined, 402, 418
and Kruskal-Wallace rank sum test, 413–416
rank correlation in, 416–418
and rank sum test for two samples, 409–412
and runs test for randomness, 402–406
sign test in, 407–409
Normal distribution, 20, 203, 228
see also Distribution, Frequency distribution, Probability distribution
Null hypothesis, 289–290
acceptance or rejection of, 322, 358–360, 369
in analysis of variance, 349
chi-square tests and, 382, 385–386
in decision theory, 494
errors and, 294–295
in paired observations, 332
power of test and, 310
in test of difference between proportions, 328–330
for test of difference between two means, 324–326
test statistic and, 323
Types I and II errors and, 315
Number, logarithm of, 24–26
Numerical data, as statistics, 2–3
Numerical random variable, 182–183, 209
see also Random sample

O

OC curve, *see* Operating characteristic curve
Ogive, defined, 69–70, 73
One-factor analysis of variance, 354–359
see also Analysis of variance
One-tailed test
decision rule for, 308
defined, 292–293, 314
operating characteristic curve for, 307
Operating characteristic curve
defined, 314
for one-tailed test, 307
Operator, defined, 16
Opportunity loss
defined, 519

in decision theory, 504–507
Opportunity loss table, 504, 519
Optimal act, expected opportunity loss under, 510
Optimal decisions, 494
Outcome, defined, 156, 173

P

Paasche index, 138
defined, 148
Paired observations, in hypothesis testing, 330–332
Parameter
defined, 43
of probability distribution, 186–187, 209
Payoff, vs. opportunity loss, 504–505
Payoff table
defined, 519
expected monetary value and, 504
in statistical decision theory, 497–501
Percentile, defined, 98, 100
Perfect information, expected value of, 509–511
Permutations, 34–37
defined, 34
examples of, 35–37
number of, 36
Permutations and combinations, multiplication rule for, 35
Phi (ϕ), 15
Pi (π), as constant, 14
Point estimate, 253–254
defined, 277
Poisson distribution, 199–201, 228
Pooled estimate
defined, 336
of simple proportions, 329
Population
Bernoulli, 191, 209, 238, 266
defined, 11
differences among, 52
raw data in, 52
statistical, *see* Statistical population
Population means, *see* Mean(s)
Population proportions, hypothesis testing about, 302–306
Posterior distribution, 511–516
defined, 512, 519
Power, null hypothesis and, 314
Power curve, 310
null hypothesis and, 314
Predictions, confidence intervals for, 447–451
see also Forecasting, Regression analysis
Preposterior analysis
in decision theory, 516
defined, 519–520
Price index
see also Weighted average of relatives price index
unweighted aggregative, 134
unweighted average of relatives index and, 137–138
Price index numbers
base period in, 142–143
examples of, 133–143
Prior distributions, 501–504
defined, 511, 520
sampling and revising of, 511–516
Prior probability, decision trees and, 508
Probability, 155–174
basic rules of, 160–163
Bayes' rule in, 172–173
classical definition of, 156
conditional, 163
for continuous sample space, 158–159
defined, 154–157, 173
event and, 156
general additive rule of, 162
general multiplicative rule of, 167
joint, 165
marginal, 165
multiplicative rule of, 169
mutually exclusive events and, 161
nature of, 154–155
outcome and, 156
relative frequency definition of, 157
rule of elimination in, 170–172
special additive rule of, 161–162
statistical independence and, 168–170
uncertainty and, 154
Venn diagrams for, 166, 170
Probability distributions, 182–215
continuous, 201–209
defined, 182, 209
discrete, 191–201
parameters of, 186–187, 209
standard deviation and, 189
variances of, 189–190
Probability function
defined, 209
of random variable, 184–185
Probability sampling, 222–223
see also Random sampling
Proportions
confidence interval for, 266–267
hypothesis testing about, 302–306
sampling distribution of, 238–239
sample size for estimating of, 275
test difference between, 328–330
test statistic for tests of differences of, 323–324

Q

Qualitative characteristics, defined, 9
Quantity index numbers, 143–145
Quartile, defined, 98, 100

R

Randomness, runs tests for, 402–406
Random numbers, table of, 225
Random sample
  defined, 223
  in nonparametric statistical methods, 402–403
  selection of, 224
  simple, 223
Random sampling, 222–227
  defined, 240
  stratified, 227
Random variable(s)
  defined, 182–184, 209
  expected value of, 187–189
  numerical, 182–183, 209
  probability function of, 184–185
  statistics as, 287
Range
  defined, 43, 108, 120
  of function, 39
Range set, 39
Rank correlation, 416–418
Rank sum test
  see also Mann-Whitney *U* test
  for three or more samples, 413–416
  for two samples, 409–412
Raw data, defined, 8, 11
Reference tables, in data organization, 71–72
Regression
  multiple, 452–457
  simple linear, 436–447
Regression analysis, 428–464
  coefficient of determination in, 443–446
  and confidence intervals for predictions, 447–451
  correlation in, 429–435
  defined, 428, 457
  least squares criterion in, 439–440
  multiple regression in, 452–457
  primary purpose of, 447–448
  scatter diagrams in, 429–433
  simple linear regression in, 436–437
  and standard error of estimate, 441–443
Regression line, 437–438, 440
Rejection region, 314
Relative
  concept of, 128
  defined, 148
Relative frequency, probability as, 157, 173
Relative frequency distribution, 73
Relatives price index, 137–138
  weighted average of, 141–142
Relative variation, 116–117
Risk, decision making under, 495
Rounding, 19–24
  conventions in, 20
Rule of association, 39–40
Rule of Bayes, 172–173, 513–514
Rule of inverse probability, 172
Run, defined, 418
Runs test for randomness, in nonparametric statistical methods, 402–406

S

Sample
  defined, 5, 11
  large, 261
  representative, 218
  small, 264
Sample information, expected value of, 517–519
Sample mean, 219–220
Sample number
  defined, 238, 240
  successes and, 222
Sample proportions
  defined, 222, 240
  pooled estimate of, 329
Sample size
  for estimating proportions, 275
  for estimating standard deviations, 276
  estimation of, 272–277
  for estimating means, 273–274
Sample space, defined, 173
Sample statistics, 219–222
Sample survey, defined, 8, 11
Sample variance, 220–222
Sampling
  defined, 240
  expected net gain from, 519
  judgment, 222–223
  probability, *see* Probability sampling
  random, *see* Random sampling
  rationale for, 218–219
  systematic, 226–227
Sampling distribution of the mean, 229–233
Sampling distributions, 227–229
  probabilities and, 233–236
  of proportion, 238–239
Sampling error, defined, 233
Sampling fraction, defined, 209
Sampling frame, 224
Sampling interval, 226
Scatter diagrams, 429–433
  defined, 458
Seasonal index, in time series, 481–487
Seasonalizing, defined, 489
Seasonal variation, in time series, 469–470
Secular trend
  defined, 489
  in time series, 469, 472–476
Series, sum of, 17–19
Set(s)
  complement in, 28
  defined, 27, 43

elements vs. members of, 32
empty, 28
formulas for, 43
intersection of, 29–30
operations with, 28
rules pertaining to, 43
union of, 30
Significance level
defined, 314
in hypothesis testing, 288
Significant digits, 19–24
examples of, 20–22
Sign test, in nonparametric statistical methods, 407–409
Simple index number, defined, 148
Simple random sample, defined, 223
Simple random sampling, defined, 240
see also Random sampling
Skewness
coefficient of, 118–119
measure of, 118
Spearman's coefficient of rank correlation, 416–418
Standard deviation
calculation of, 113
defined, 189
formula for, 120
probability distribution and, 189
sample size for estimating of, 276
sample variance and, 220–222
Standard error, 228
Standard error of the estimate
defined, 458
in multiple regression, 455–456
in regression analysis, 441–443
Standard normal distribution, 202
States of nature
decision making and, 495–497
defined, 520
Statistic, defined, 240
see also Statistics
Statistical array, defined, 54
Statistical data
origins of, 8–9
raw data in, 8
types of, 9–10
Statistical decision theory, 494–525
see also Decision theory
and expected value of perfect information, 509–511
notation used in, 497
payoff tables and functions in, 497–501
prior distributions and Bayes' criterion in, 501–504
Statistical independence, 168–170
defined, 173
Statistical inference, 4–8
defined, 11, 248
uses of, 5
Statistical methods, nonparametric, *see* Nonparametric statistical methods
Statistical population, defined, 5, 11
see also Population
Statistical reports, primary and secondary sources in, 8–9
Statistical survey, defined, 11
see also Survey
Statistical universe, defined, 5, 11
Statistics
see also Statistical (*adj.*)
as branch of mathematics, 14
computer and, 10
defined, 2, 11, 240
descriptive, 3–4, 11
information obtainable from, 4–7, 52
language of, 14
prerequisites for, 14–49
as random variables, 287
reasons for studying, 2–3
terminology and notation in, 10–11, 14
Step deviation method, in calculating mean of frequency distribution, 91–94
Stratified random sampling, 227, 240
see also Random sampling
Student's t distribution, 236–238, 254, 264
Subscripts, defined, 16
Subsets
intersecting, 29
mutually exclusive, 30
Summation and subscripting operations, 15–19
Summation operator, rules governing, 17–18
Sums of squares
in analysis of variance, 350–354
defined, 370
Survey
defined, 8, 11
sample, 8
Survey of Current Business, 130–131
Symmetry
frequency tables and, 80
measures of, 53
Systematic sampling, 226
defined, 240–241

T

t distribution, degrees of freedom and, 236–238
see also Student's t distribution
Tenth percentile, 98
Test, power of, 310
see also Hypothesis testing, One-tailed test, Rank sum test, Two-tailed test
Test of independence, *see* Independence tests
Test statistic, 287
defined, 315
for tests of difference of means and proportions, 323–324

Time series, 468–491
  cycle in, 469–470, 476–480
  defined, 468, 489
  deflation in, 145
  forecasting and, 487–488
  modified mean in, 481
  moving averages in, 477–480
  random or irregular fluctuations in, 469–470
  seasonal index, 481–487
  seasonal variation in, 469–470
  secular trend in, 469
  trend and, 469, 472–476
Time series data, deseasonalizing of, 485–487
Time series model, 469–476
Total sum of squares, in analysis of variance, 351
  see also Sums of squares
Treatment(s), defined, 344–345, 370
Treatment combination, in analysis of variance, 361, 370
Treatment populations, 344
Trend
  defined, 489
  in time series, 469, 472–476
Two-factor analysis of variance, 361–369
  see also Analysis of variance
Two samples, rank sum test for, 409–412
Two-tailed test, defined, 292–293, 315
t values, in hypothesis testing, 312
  see also Student's *t* distribution
Type I error, 315
Type II error
  defined, 315
  probability of, 306–311

U

UCL, *see* Upper confidence limit
Unbiased estimator, defined, 277
Unbiasedness, 250–251
Uncertainty
  cost of, 510
  decision making under, 495
Union symbol, 30
Units, elementary, 5
Universal set, 28
Universe, defined, 11
  see also Statistical universe
Upper confidence limit, 254
U statistic, see Mann–Whitney *U* test

V

Variable(s)
  addition of constant to, 18
  continuous, 9
  defined, 9, 11, 14, 39
  dependent and independent, 41
  discrete, 11
Variables data, defined, 9
Variance
  analysis of, *see* Analysis of variance
  equality of, 334–335
  of probability distribution, 189–190
  standard deviation and, 109–116
Variance ratio, defined, 333, 336
Variation, 80
  defined, 120–121
  measures of, 108
  relative, 116–117
  sum of squares as measure of, 351
Venn diagram, 28–29
  examples of, 31, 34

W

Weighted aggregative price index, 138
Weighted arithmetic mean, defined, 100
Weighted average of relatives price index, 141–142
Wholesale price index, 130

Y

Yates correction factor, 392–394

Z

z values, in hypothesis testing, 312–314

A
B 8
C 9
D 0
E 1
F 2
G 3
H 4
I 5
J 6